Identification of Enterobacteriaceae

P. R. EDWARDS
W. H. EWING

Atlanta, Georgia

Third Edition

Authored by W. H. Ewing

Burgess Publishing Company
426 South Sixth Street ▪ Minneapolis, Minnesota 55415

W. H. Ewing is associated with the Center for Disease Control (formerly the Communicable Disease Center), U. S. Public Health Service, Atlanta, Georgia. From July 1, 1948 to June 30, 1962 he was microbiologist and Assistant Chief of the Enteric Bacteriology Unit, and from the latter date until August 30, 1969 he was Chief of that Unit. In order to devote more time to research and writing he requested that he be relieved of his position as Chief of the Unit effective August 30, 1969. His position now is Consulting and Research Microbiologist. He remains associated with the Enteric Bacteriology Unit.

This book was written by Dr. Ewing in his private capacity and no official support or endorsement by the Public Health Service is intended or should be implied.

Library of Congress Catalog Card Number 77-160829
SBN 8087-0516-4

2   3   4   5   6   7   8   9   0

Acknowledgments

The author takes this opportunity to acknowledge his indebtedness to many collaborators for aid, counsel, and encouragement through many years. Among these are Dr. E. S. Anderson, Professor R. E. Buchanan, Dr. K. P. Carpenter, Dr. W. W. Ferguson, Dr. D. C. Graham, Professor F. Kauffmann, Dr. L. J. Kunz, Professor L. Le Minor, Dr. E. Neter, Drs. F. and I. Ørskov, and Dr. J. Taylor. Many thanks also are due to several colleagues who, for many years, have been associated with the writer in the study of Enterobacteriaceae and allied bacteria: M. M. Ball, S. F. Bartes, B. R. Davis, M. A. Fife, O. Grados, G. J. Hermann, G. P. Huntley, W. J. Martin, A. C. McWhorter, J. V. Sikes, H. G. Wathen, and V. R. Wilson. The work of these associates has made possible much of the contents of this book.

The third edition of this book is dedicated to all those who found the earlier editions of value to them in their work.

Preface

This new edition was discussed with the late Dr. P. R. Edwards during 1965 and prior to his death on May 16, 1966. The decisions reached in those discussions regarding the general format, contents, nomenclature, and other important points have been followed in the preparation of this edition.

The intent is not to prescribe rules as to the media and methods that must be used by any worker. However, there are certain minimum standards below which work with Enterobacteriaceae should not fall, and there are a number of methods and media which have proved satisfactory in the hands of numerous competent and experienced workers. Some methods and media have become outmoded, while certain newer media and methods require evaluation. The purpose of this text is to provide recommended methods that may be employed in the isolation and study of Enterobacteriaceae, and to point out some pitfalls which the worker in even the smallest laboratory may avoid.

Although systematic bacteriology is neglected by most and complained about by many, it is essential for identification and communication. Labels of some sort must be available for the bacteria, and there must be a considerable amount of agreement regarding these labels and their use. Therefore, there must be rules of procedure for the selection and usage of such labels. Such rules should, however, be sufficiently flexible to allow for reasonable, substantiated, and justified change, for systematic bacteriology is an evolving discipline which should be permitted to continue its evolution.

The system of nomenclature and taxonomy employed in this book is a result of that evolution and is based on examination of the biochemical reactions of large numbers of cultures. It is felt that biochemical reactivities must form the primary basis for classification at the level of the tribe and the genus within the family Enterobacteriaceae. If we consider the community of antigens within the family and the intensive intergeneric antigenic relationships that are known, the need for the use of biochemical rather than serological reactions for classification at these levels is clear, and, with a few possible exceptions, this also holds true for classification at the species level.

The systematics of Enterobacteriaceae should continue to evolve, and it is reasonable to suppose that future generations of bacteriologists will continue to improve upon systems currently in use. Therefore, the best that present-day investigators can hope for is to provide a sound basis so that changes necessary in the future can be minimal and can be made gradually.

The discussions of the isolation of salmonellae from foods and feeds and of the bacteriophage typing of *Salmonella typhi* which appeared in the second edition have been omitted.

This edition contains a large amount of data on the biochemical reactions given by members of the genera of Enterobacteriaceae. These data are summarized in tabular form and the percentages of positive reactions, etc., are given for each substrate or test. Tests of particular value for differentiation of members of related genera and species are also tabulated. The chapters and their contents are arranged in such a way that the personnel of all laboratories of medical and public health bacteriology should be able to select tests and methods for presumptive identification, through intermediate stages, to complete biochemical characterization of an isolant. Partial or complete serological analyses may also be made in many instances.

Investigation of the antigenic properties of members of several of the genera within the family has been extended. Additional antigens and numerous serotypes have been characterized in genera previously studied intensively. In other genera where little or no work on antigenic properties had been done, knowledge has been expanded. While the antigenic schemata that have been established are not given in detail in every instance, references to advances are included and the work is described briefly. Until such time as it is demonstrated that serologic typing within a given genus or species can be used profitably in the study of infections, it does not seem worthwhile to catalogue long lists of antigens, type strains, and serotypes for all known schemata.

Discussion of several subjects is included in a new chapter (Chapter 4). Since the bases for serologic typing within the several genera of the family Enterobacteriaceae are similar, and since several of the variational phenomena that affect serotyping within the various genera are very much the same, discussion of these has been placed in the new chapter. Brief reviews of the literature on the immunochemistry of the antigens of Enterobacteriaceae and the literature that deals with the genetic changes that may affect the results of serotyping are included. Fluorescent antibody techniques have been applied to a variety of Enterobacteriaceae, and studies on the resistance (R) factors have been extended to many kinds of Enterobacteriaceae, as well as to other bacteria, hence these are mentioned in Chapter 4. Incorporation of these subjects in a single chapter avoids repetition in other chapters and provides additional basic information with which some readers may not be particularly familiar. In some instances this information is essential to the understanding and interpretation of results of serological analyses. The seemingly unrelated subject of the R factors is included because it is felt that all who work with Enterobacteriaceae should be at least casually acquainted with the subject.

The references cited should be sufficient to enable the reader to begin an independent search of the literature. Every care has been taken to eliminate errors, but both the author and the publisher shall be glad to be notified of any errors found by readers.

The importance of accurate identification of members of all genera of Enterobacteriaceae cannot be overstated. It is hoped that this text will prove to be a useful tool in this work.

Atlanta, Georgia   W. H. Ewing
March, 1972

Table of Contents

Preface   v

Chapter 1. Taxonomy and Nomenclature   1

Introduction, 1; Outline of Nomenclature, 1; Definitions, 3; References, 6.

Chapter 2. Isolation and Preliminary Identification   7

Collection and Processing Specimens, 7; Fecal Specimens, 7; Transport Solutions and Media, 7;
Enrichment Media, 8; Plating Media, 10; Isolation from Body Fluids, Exudates, and Tissues, 12;
Blood Cultures, 12; Urine Cultures, 13; Other Fluids, Exudates, and Tissue Specimens, 13;
Selection and Picking of Colonies, 14; Primary Differentiation, 14; Biochemical Reactions, 16;
Preliminary Serological Examination, 16; *Salmonella,* 16; *Shigella,* 18.

Chapter 3. Differentiation of Enterobacteriaceae by Biochemical Reactions   21

Chapter 4. The Antigens of Enterobacteriaceae   48

Introduction, 48; O Antigens, 48; H Antigens, 48; K Antigens, 48; Variational Phenomena
that Effect Serological Typing, 48; HO to O Variation, 48; Variations in H antigens, 49; S to R
Variations, 50; S-T-R Variation, 51; Form Variation, 51; KO to O Variations, 51; Nonfimbriate
to Fimbriate Variation, 53; Other Antigens, 54; X Antigen, 54; Common Antigen (CA), 54;
Immunochemical Investigations, 55; O, R, and Intermediate Antigens, 55; K Antigens, 57;
H Antigens, 59; Genetic Recombinations that may Effect Serological Typing, 59; The Resistance
(R) Factors, 62; Fluorescent Antibody Technics, 63; References, 63.

Chapter 5. The Genus *Escherichia*   67

Definition of Genus, 67; Biochemical Reactions of *Escherichia,* 67; Antigens of *Escherichia,* 67;
O Antigens, 73; Relationships of O Antigens, 73; Pooled K Antisera, 73; Absorption of O Antisera,
82; K Antigens, 87; Characteristics of K Antigens, 87; Relationships of K Antigens, 87; K Antigen
Determination, 90; Pooled K Antisera, 90; Absorption of K Antisera, 90; H Antigens, 91; H
Antigen Determination, 91; Pooled H Antisera, 94; Absorption of H Antisera, 94; Identification
of Serotypes of *E. coli* Associated with Diarrheal Disease, 95; General Considerations, Isolation and
Preliminary Examination, 95; Further Serological Examination, 97; The Alkalescens-Dispar (A-D)
Bioserotypes, 98; Production and Use of Antisera, 100; Serotypes of *E. coli* from Infections in
Animals, 101; Antiserum Production, 103; References, 105.

Chapter 6. The Genus *Shigella*   108

Classification and Nomenclature, 108; Definition of Genus, 108; Biochemical Reactions of *Shigella,*
113; Serological Identification of *Shigella,* 120; Production of Antisera, 120; Technic of Agglutinin
Absorption, 123; Use of Antisera, 123; Subgroup A, *S. dysenteriae,* 123; Subgroup B, *S. flexneri,*
126; Variation in Serotypes of *S. flexneri,* 128; Fimbrial Antigens of *S. flexneri,* 131; Subgroup C, *S.
boydii,* 131; Subgroup D, *S. sonnei,* 135; The K Antigens of *Shigella,* 136; Extrageneric Relationships
of Shigellae, 139; References, 139.

Chapter 7. The Genus *Edwardsiella* 143

Biochemical Reactions, 143; Serological Reactions, 145; References, 145.

Chapter 8. The Genus *Salmonella* 146

Definition of Genus, 146; Biochemical Identification of *Salmonella*, 146; Introduction, 146; Biochemical Reactions of *Salmonella*, 147; Serological Identification of *Salmonella*, 162; Variational Phenomena that Affect Serotypes, 162; HO-O Variation, 163; S-R Variation, 163; T Forms and T Antigens, 164; VW (Vi-O) Variation, 165; M-N Variation, 167; Nonfimbriate to Fimbriate Variation, 168; Antigenic Changes Induced by Bacteriophage, 168; Form Variation, 168; Phase Variation, 169; Irreversible Variation and Occurrence of Multiple Phases, 172; Antigenic Analysis, 174; Somatic Antigens of *Salmonella*, 174; Use of O Antisera, 176; Absorption of O Antisera, 178; Determination of Single O Antigen Factors, 183; Detection of Vi Antigen, 184; The H Antigens of *Salmonella*, 184; Use of H Antisera, 186; Absorption of H Antisera, 189; Isolation of Phases, Phase Reversal, 194; Stages in Serological Classification of *Salmonella*, 195; Production of Antisera, 198; O Antisera, 198; Vi Antiserum, 202; H Antisera, 202; Polyvalent Antisera, 203; References, 205.

Chapter 9 Antigenic Schema for *Salmonella* 208

Development of the Schema, 208; Alphabetical List of Serotypes, 209; Antigenic Schema, 209; References, 258.

Chapter 10. The Genus *Arizona* 259

Definition, 259; Biochemical Reactions of Cultures of *Arizona*, 259; Serological Characterization of Cultures of *Arizona*, 262; Determination of O Antigens, 262; Determination of H Antigens, 265; Antigenic Schema for the Genus *Arizona*, 265; Extrageneric Relationships, 265; Antiserum Production, 273; References, 274.

Chapter 11. The Genus *Citrobacter* 276

Definition, 276; Biochemical Reactions, 276; Serological Reactions, 282; Intergeneric Antigenic Relationships, 283; Production of Antisera, 287; References, 288.

Chapter 12. The Genus *Klebsiella* 290

Definition, 290; Biochemical Reactions, 290; Serological Reactions, 294; Capsular Antigens, 295; O Antigens, 295; Quellung Reactions, 296; Determination of Capsule Antigens, 296; Absorption Tests, 297; Irregularly Capsulated Cultures, 297; Production of Antisera, 299; References, 299.

Chapter 13. The Genus *Enterobacter* 301

Definition, 301; Biochemical Reactions, 302; Serological Investigations, 302; References, 307.

Chapter 14. The Genus *Serratia* 308

Definition, 308; Biochemical Reactions, 309; Serological Reactions, 311; References, 316.

Chapter 15. The Genus *Pectobacterium* 318

Definition, 318; Biochemical Reactions, 319; References, 319.

Chapter 16. The Genus *Proteus* 324

    Definition, 324; Biochemical Reactions, 324; Serological Reactions, 329; References, 329.

Chapter 17. The Genus *Providencia* 331

    Definition, 331; Biochemical Reactions, 331; Serological Reactions, 334; Intergeneric Antigenic Relationships, 334; References, 335.

Chapter 18. Media and Reagents 337

    Control of Media, 337; Preservative and Transport Media, 337; Enrichment Media, 339; Plating Media, 339; Differential Media, 342; Other Media and Solutions, 354; References, 354.

Index 357

Chapter 1

Taxonomy and Nomenclature

INTRODUCTION

The classification and nomenclatural system for the family ENTEROBACTERIACEAE used in this edition (v. inf.) is that proposed by Ewing (1963), as emended (1966, 1967, 1969, and 1970). This system resulted from comparative studies of the biochemical reactions given by relatively large numbers of cultures of each of the genera (e.g., Ewing, 1968).

From the beginning, the aim of the above-mentioned investigations was to compile numerical and percentage data that could be used to compare *all* of the reactions given by members of one genus with those of another, as well as to provide means for differentiation of genera and species. Therefore, the genera *Escherichia* and *Shigella* are placed in the same tribe because biochemically these two genera are more closely related to each other than either is related to any other genus in the family. The genera *Salmonella, Arizona,* and *Citrobacter* are placed in the same tribe for the same reason; and so on, throughout the family. It is the author's opinion that classification of the bacteria in this manner, i.e., on the basis of overall similarities, is much more logical and reasonable than classifications based on a single character, or on one or two criteria. As examples, *Shigella* and *Salmonella* are *not* placed in the same tribe because they share a negative character (failure to ferment lactose rapidly), as they are in the 7th edition of *Bergey's Manual* (Breed, Murray, and Smith, 1957). Further, a tribe SERRATEAE is not recognized, since pigment production on ordinary media is not a cardinal characteristic of *Serratia* (Davis et al., 1957 et seq.). The tribe ERWINEAE is not recognized as such, although part of it (the genus *Pectobacterium*) is incorporated into the tribe KLEBSIELLEAE for reasons given by Graham (1964) and Ewing (1967). The genus *Alginobacter* is not recognized because ENTEROBACTERIACEAE do not liquefy alginate and for other reasons (Davis and Ewing, 1964). At this late date it should not be necessary to add anything about the "genus" *Paracolobactrum* or the "paracolon group," since these have been dealt with by many investigators (e.g., Fields et al., 1967). However, for the record, the author will repeat the statement made in the second edition of this book: "The establishment of a single genus *(Paracolobactrum)* composed of diverse forms simply upon the basis of delayed fermentation of lactose is completely unjustified. For example, a culture classified as *Paracolobactrum coliforme* may be changed into a typical *Escherichia coli* culture simply by selection of components that rapidly ferment lactose."

The several classifications of ENTEROBACTERIACEAE proposed by Kauffmann (e.g., 1956, 1963, 1966) differ taxonomically, nomenclaturally, and in concept from that employed herein. However, it is not the author's intent to debate the merits of one system over another. The system employed in this edition is used not so much because it was proposed by the author as because it appears to be the best available at this time. Further, as has been said many times (Ewing, 1963, and since), acceptance of any system comes from usage. (no system can be legislated), and the classification used herein is gaining wide acceptance in this hemisphere, at least. To the author's knowledge the nomenclature employed is correct. With one exception, the emendations mentioned above, and made since 1963, involve changes in certain citations, and a specific epithet, necessitated by actions of the Judicial Commission of the International Committee on Nomenclature of Bacteria. The exception is the addition of the tribe EDWARDSIELLEAE (Ewing et al., 1965).

An outline of the nomenclature of the family ENTEROBACTERIACEAE, and definitions (Ewing, 1967; emended 1970) for the family, its tribes, and genera follow.

The Nomenclature of the Family ENTEROBACTERIACEAE

Family ENTEROBACTERIACEAE Rahn

Tribe I   ESCHERICHIEAE Bergey, Breed, and Murray

Genus   I *Escherichia* Castellani and Chalmers

        1. *Escherichia coli* (Migula) Castellani and Chalmers

Genus   II *Shigella* Castellani and Chalmers
    1. *Shigella dysenteriae* (Shiga) Castellani and Chalmers
    2. *Shigella flexneri* Castellani and Chalmers
    3. *Shigella boydii* Ewing
    4. *Shigella sonnei* (Levin) Weldin

Tribe II   EDWARDSIELLEAE Ewing and Mc-Whorter
    I *Edwardsiella* Ewing and McWhorter
        1. *Edwardsiella tarda* Ewing and McWhorter

Tribe III   SALMONELLEAE Bergey, Breed, and Murray
Genus   I *Salmonella* Lignières
    1. *Salmonella cholerae-suis* (Smith) Weldin
    2. *Salmonella typhi* (Schroeter) Warren and Scott
    3. *Salmonella enteritidis* (Gaertner) Castellani and Chalmers
Genus   II *Arizona* Ewing and Fife
    1. *Arizona hinshawii* (Ewing and Fife) Ewing
Genus   III *Citrobacter* Werkman and Gillen
    1. *Citrobacter freundii* (Braak) Werkman and Gillen

Tribe IV   KLEBSIELLEAE Trevisan
Genus   I *Klebsiella* Trevisan
    1. *Klebsiella pneumoniae* (Schroeter) Trevisan
    2. *Klebsiella ozaenae* (Abel) Bergey, Breed, and Murray
    3. *Klebsiella rhinoschleromatis* Trevisan
Genus   II *Enterobacter* Hormaeche and Edwards
    1. *Enterobacter cloacae* (Jordan) Hormaeche and Edwards
    2. *Enterobacter aerogenes* (Kruse) Hormaeche and Edwards
    3. *Enterobacter hafniae* (Moeller) Ewing
    4. *Enterobacter liquefaciens* (Grimes and Hennerty) Ewing
Genus   III *Pectobacterium* Waldee
    1. *Pectobacterium carotovorum* (Jones) Waldee
Genus   IV *Serratia* Bizio
    1. *Serratia marcescens* Bizio *(Serratia marcescens* subspecies *marcescens)*
        1a. *Serratia marcescens* (subspecies *kiliensis* Lehmann and Neumann) Ewing, Davis, and Johnson

Tribe V   PROTEEAE Castellani and Chalmers
Genus   I *Proteus* Hauser
    1. *Proteus vulgaris* Hauser
    2. *Proteus mirabilis* Hauser
    3. *Proteus morganii* (Winslow, Kligler, and Rothberg) Rauss
    4. *Proteus rettgeri* (Hadley et al.) Rustigian and Stuart
Genus   II *Providencia* Ewing
    1. *Providencia alcalifaciens* (DeSalles Gomes) Ewing
    2. *Providencia stuartii* (Buttiaux et al.) Ewing

Note: The first species listed in each genus is the type species.

The system given in the outline has been used by the author and colleagues for many years. Further, it has been adopted by the editors of certain manuals, by the author of at least one textbook, and it has appeared in a number of articles in well-known American journals. Therefore it may be recommended that laboratorians use it. Further, it is recommended that they inform physicians, nurses, epidemiologists, and others with whom they come in contact concerning the system so these persons can become familiar with it. Since there is very little that is new in the system, such familiarization should not take long or be difficult.

Only three species of *Salmonella* are recognized. The concept of limitation of the number of species of *Salmonella* to three apparently originated with Borman et al. (1944). This concept also was adopted by Kauffmann and Edwards (1952). Although alternatives were discussed, the three species concept was used by Ewing (1963). The late Dr. P. R. Edwards agreed, and stated (1962) that the proposal made with Professor Kauffmann in 1952 still was his choice as the most logical solution to this problem (see 1963 publication cited above). There is ample precedent for limiting the number of species in a genus. For example, only four species are recognized in the genus *Shigella* (see Chapter 6).

The three species recognized in the genus *Salmonella* are *Salmonella cholerae-suis* (the type species), *Salmonella typhi*, and *Salmonella enteritidis*. The name *S. enteritidis* was chosen for the third species because the epithet *enteritidis* was the oldest validly published (1888), legitimate epithet other than *cholerae-suis* and *typhi*. All salmonellae other than *S. cholerae-suis* and *S. typhi* are serotypes of *S. enteritidis*. It is emphasized that attention should be given to correct usage. Except where employed in a general way, as in an outline or summary of nomenclature or in a definition of the third species, the name *S. enteritidis* should not be used alone, since 1400 or more serotypes and bioserotypes are included in it. The designations given to the numerous serotypes of *S. enteritidis* are infrasubspecific and they have no standing in nomenclature (see International Code, 1966). It is of no consequence whether these infrasubspecific terms are italicized, since the use of italics, or other printing device, does not alter their status — they still are infrasubspecific designations, equivalent to, and used in lieu of, antigenic formulas.

Therefore the serotypes (ser) and bioserotypes (bioser) of *S. enteritidis* are recorded in the manner indicated in the examples that follow. The infrasubspecific designations are capitalized for reasons of clarity only. The names of the other two species *(S. cholerae-suis* and *S. typhi)* are unaffected and cultures of these are reported by these names as in the past.

Nomenclature used in this book	Old designation
S. enteritidis ser Typhimurium	*S. typhimurium*
S. enteritidis ser Heidelberg	*S. heidelberg*
S. enteritidis ser Enteritidis	*S. enteritidis*
S. enteritidis bioser Paratyphi-A	*S. paratyphi A*
S. enteritidis bioser Pullorum	*S. pullorum*

The term bioserotype is used in connection with certain serotypes of *S. enteritidis* that possess unique biochemical characteristics (see Chapter 8). Unnamed serotypes are recorded as in the following example: *S. enteritidis* ser 58:a:-.

It is recommended that in formal publications, authors give the complete name of each species (e.g., *Salmonella enteritidis*) the first time that each occurs. Thereafter, reference to serotypes of *S. enteritidis* may be made as follows: ser Enteritidis, ser Typhimurium, bioser Paratyphi-A, etc. Certain strains of serotypes of *Salmonella* used in research and for other purposes may be distinguished by numerals or letters following the serotypic designation (e.g., *S. enteritidis* ser Senftenberg 775W).

At first glance the above-mentioned system of nomenclature for the genus *Salmonella* may appear cumbersome. However, if it is used in the manner suggested it is simple and easy to use and has many advantages, as illustrated in Chapters 8 and 9. The author has adopted the three-species concept because it is practical, because no better system has been proposed, and because it is legitimate according to the rules of nomenclature. Further, the author is of the opinion that it is better to adopt and use a legitimate system than to continue indefinitely the use of an illegitimate one that appears to accord species status to all serotypes of salmonellae. Some workers may have initial difficulty in differentiating between the species *S. enteritidis* and the serotype Enteritidis (= 9, 12:g,m:-). The author believes that this initial difficulty will be overcome in a short time as workers become familiar with the system. However, it probably would be better to conserve the epithet *enterica* against the epithet *enteritidis* in the name of the third species of *Salmonella,* as suggested by Ewing (1963). If this

were done, the name of the third species would be *Salmonella enterica* and possible confusion of the sort mentioned would be obviated. Such conservation requires action by the Judicial Commission of the International Committee on Nomenclature of Bacteria.

Although the author has adopted the three-species concept of salmonellae for use herein, it should go without saying that any individual may use any system that is to his liking. The use of the above-mentioned system in this book should not detract in any way from whatever other value the book may have.

DEFINITIONS FOR THE FAMILY ENTEROBACTERIACEAE, ITS TRIBES, AND GENERA

The Family ENTEROBACTERIACEAE Rahn

The family ENTEROBACTERIACEAE consists of gram-negative, aerobic, facultatively anaerobic, asporogenous, rod-shaped bacteria that grow well on artificial media. Some species are atrichous, and nonmotile variants of motile species also may occur. Motile forms are peritrichously flagellated. Nitrates are reduced to nitrites, and glucose is utilized fermentatively with the formation of acid or of acid and gas. The indophenol oxidase test is negative and alginate is not liquefied. Pectate is liquefied by members of only one genus *(Pectobacterium).*

Tribe I ESCHERICHIEAE Bergey, Breed, and Murray

ESCHERICHIEAE are motile or nomotile bacteria that conform to the definition of the family ENTEROBACTERIACEAE. The methyl red reaction is positive and the Voges-Proskauer test is negative. Urease, phenylalanine deaminase, and hydrogen sulfide are not produced; sodium malonate is not utilized, gelatin is not liquefied, and growth does not occur on Simmons' citrate agar nor in medium containing potassium cyanide.

Genus I *Escherichia* Castellani and Chalmers

The genus *Escherichia* is composed of motile or nonmotile bacteria that conform to the definitions of the family ENTEROBACTERIACEAE and the tribe ESCHERICHIEAE. Both acid and gas are formed from a wide variety of fermentable carbohydrates but anaerogenic biotypes occur. Salicin is fermented by the majority of cultures but adonitol and inositol are utilized infrequently. Lactose usually is fermented rapidly but some strains utilize it slowly and some fail to ferment this substrate. Lysine, arginine, and ornithine are decarboxylated by the majority of cultures, acid is formed from sodium mucate, and sodium acetate is utilized as a sole source of carbon. The type species is *Escherichia coli* (Migula) Castellani and Chalmers.

Genus II *Shigella* Castellani and Chalmers

The genus *Shigella* is composed of nonmotile bacteria that conform to the definitions of the family ENTEROBACTERIACEAE and the tribe ESCHERICHIEAE. With the exception of certain biotypes of *Shigella flexneri* 6, visible gas is not formed from fermentable carbohydrates. Compared with *Escherichia, Shigella* are less active in their utilization of carbohydrates. Salicin, adonitol, and inositol are not fermented. Strains of *Shigella sonnei* ferment lactose upon extended incubation, but other species do not utilize this substrate in conventional medium. Lysine is not decarboxylated, the majority of strains do not possess a demonstrable arginine dihydrolase system, and ornithine is decarboxylated only by *S. sonnei* and *Shigella boydii* 13. The type species is *Shigella dysenteriae* (Shiga) Castellani and Chalmers.

Tribe II EDWARDSIELLEAE Ewing and McWhorter

EDWARDSIELLEAE are motile bacteria that conform to the definition of the family ENTEROBACTERIACEAE. Hydrogen sulfide is produced abundantly, indol is formed, the methyl red test is positive, the Voges-Proskauer reaction is negative, phenylalanine is not deaminated, and urea is not hydrolyzed. Gelatin is not liquefied, growth does not occur in Simmons' citrate medium, nor sodium acetate medium, nor in medium containing potassium cyanide. Lipase is not formed and arginine dihydrolase is not produced. Esculin is not hydrolyzed, and erithritol, and adonitol are not fermented.

Genus *Edwardsiella* Ewing and McWhorter

The genus *Edwardsiella* is composed of motile bacteria that conform to the definitions of the family ENTEROBACTERIACEAE and the tribe EDWARDSIELLEAE. Lysine and ornithine are decarboxylated, but neither malonate nor mucate is utilized. Glucose and maltose are fermented promptly, and with rare exceptions gas is formed from these substrates. Glycerol is utilized slowly by the majority of strains, but lactose, sucrose, mannitol, dulcitol, salicin, inositol, sorbitol, raffinose, rhamnose, xylose, cellobiose, and alpha methyl glucoside are not attacked. The type species is *Edwardsiella tarda* Ewing and McWhorter.

Tribe III SALMONELLEAE Bergey, Breed, and Murray

SALMONELLEAE are motile bacteria that conform to the definition of the family ENTEROBACTERIACEAE. The methyl red reaction is positive, the Voges-Proskauer test is negative, indol is not formed, and phenylalanine is not deaminated. Gelatin is not liquefied rapidly in nutrient medium. With few exceptions hydrogen sulfide is produced abundantly, growth occurs on Simmons' citrate and sodium acetate media, arginine dihydrolase is formed, and gas is formed from fermentable carbohydrates. Esculin is not hydrolyzed and erythritol and adonitol are not fermented.

Genus I *Salmonella* Lignières

The genus *Salmonella* is composed of motile bacteria that conform to the definitions of the family ENTEROBACTERIACEAE and the tribe SALMONELLEAE. Urease is not produced, sodium malonate is not utilized, gelatin is not liquefied, and growth does not occur in medium containing potassium cyanide. Lysine, arginine, and ornithine are decarboxylated. Acid is produced in Jordan's tartrate medium. Dulcitol is fermented and inositol is utilized by numerous strains. Sucrose, salicin, raffinose, and lactose are not fermented. The type species is *Salmonella cholerae-suis* (Smith) Weldin.

Genus II *Arizona* Ewing and Fife

The genus *Arizona* is composed of motile bacteria that conform to the definitions of the family ENTEROBACTERIACEAE and the tribe SALMONELLEAE. Urease is not produced and growth does not occur in medium containing potassium cyanide. Lysine, arginine, and ornithine are decarboxylated, sodium malonate is utilized, gelatin is liquefied slowly in nutrient medium, and lactose is fermented by the majority of cultures. With few exceptions acid is not produced in Jordan's tartrate medium. Dulcitol and inositol are not fermented and salicin is utilized infrequently. The type species is *Arizona hinshawii* (Ewing and Fife) Ewing.

Genus III *Citrobacter* Werkman and Gillen

The genus *Citrobacter* is composed of motile bacteria that conform to the definitions of the family ENTEROBACTERIACEAE and the tribe SALMONELLEAE. Lysine is not decarboxylated and less than 20 percent of strains possess ornithine decarboxylase. Urease is produced slowly by the majority of cultures, but the reactions are weak. Growth occurs in medium that contains potassium cyanide and acid is produced in Jordan's tartrate medium. Gelatin is not liquefied in nutrient medium. Dulcitol and cellobiose are fermented rapidly by the majority of cultures. Lactose is utilized but the reactions frequently are delayed. The type species is *Citrobacter freundii* (Braak) Werkman and Gillen.

Tribe IV KLEBSIELLEAE Trevisan

KLEBSIELLEAE are motile or nonmotile bacteria that conform to the definition of the family ENTERO-BACTERIACEAE. Hydrogen sulfide is not produced and urea is not hydrolyzed rapidly but delayed reactions may occur. With few exceptions indol is not produced, the methyl red test is negative, and the Voges-Proskauer reaction is positive. Growth occurs in Simmons' citrate medium and in medium containing potassium cyanide. Phenylalanine is not deaminated. Sodium alginate is utilized as a sole source of carbon by certain members of only one genus *(Klebsiella)*, and lipase is produced by only one species of *Enterobacter (E. liquefaciens)* and by members of the genus *Serratia.*

Genus I *Klebsiella* Trevisan

The genus *Klebsiella* is composed of nonmotile bacteria that conform to the definitions of the family ENTEROBACTERIACEAE and the tribe KLEB-SIELLEAE. The Voges-Proskauer test is positive, gelatin is not liquefied. Lysine decarboxylase is produced, but arginine dihydrolase and ornithine decarboxylase are not. The majority of cultures utilize sodium alginate as a sole source of carbon and esculin is hydrolyzed. Gas is formed from inositol and glycerol, and by the majority of strains from adonitol. Acid is produced from sorbitol, rhamnose, arabinose, and raffinose. The type species is *Klebsiella pneumoniae* (Schroeter) Trevisan.

Genus II *Enterobacter* Hormaeche and Edwards

The genus *Enterobacter* is composed of motile bacteria that conform to the definitions of the family ENTEROBACTERIACEAE and the tribe KLEB-SIELLEAE. The Voges-Proskauer reaction is positive, gelatin is liquefied slowly by the most commonly occurring forms *(Enterobacter cloacae).* Lysine decarboxylase is not produced by *E. cloacae,* but other species of the genus possess this enzyme system. Ornithine decarboxylase is produced. Sodium alginate is not utilized as a sole source of carbon. Gas is not formed from inositol and glycerol by cultures of *E. cloacae.* Acid is produced from sorbitol, rhamnose, arabinose, and raffinose by the majority of species. One species *(Enterobacter hafniae)* does not ferment sorbitol or raffinose. Only one species *(Enterobacter liquefaciens)* is lipolytic. The type species is *Enterobacter cloacae* (Jordan) Hormaeche and Edwards.

Genus III *Pectobacterium* Waldee

The genus *Pectobacterium* is composed of motile or nonmotile bacteria that conform to the definitions of the family ENTEROBACTERIACEAE and the tribe

KLEBSIELLEAE. Sodium pectate medium is liquefied. A minority of cultures produce indol, but the majority yield positive reactions in the methyl red test. Gelatin is liquefied although the reactions of a minority of strains may be somewhat delayed. Lysine, arginine, and ornithine are not decarboxylated. Sodium alginate is not utilized as a sole source of carbon. Gas is not formed from inositol or glycerol, and adonitol is not fermented. Sorbitol is fermented only very rarely but the majority of cultures produce acid from rhamnose, arabinose, and raffinose. The optimum growth temperature is about 25 C and cultures fail to grow or grow poorly at 37 C. The type species is *Pectobacterium caratovorum* (Jones) Waldee.

Genus IV *Serratia* Bizio

The genus *Serratia* is composed of motile bacteria that conform to the definitions of the family ENTEROBACTERIACEAE and the tribe KLEB-SIELLEAE. A positive Voges-Proskauer reaction is given by *Serratia marcescens* subsp. *marcescens,* but *S. marcescens* subsp. *kiliensis* gives negative results in this test. Lipase is produced, gelatin is liquefied rapidly, and lysine and ornithine are decarboxylated. Sodium alginate is not utilized as a sole source of carbon. When gas is formed from fermentable substrates the volumes are small (10 per cent or less). Acid is produced from sorbitol but rhamnose, arabinose, and raffinose are not fermented. The type species is *Serratia marcescens* Bizio.

Tribe V PROTEEAE Castellani and Chalmers

PROTEEAE are motile bacteria that conform to the definition of the family ENTEROBACTERIACEAE. The methyl red test is positive and with the exception of occasional strains of *Proteus mirabilis* the Voges-Proskauer reaction is negative. Phenylpyruvic acid is produced rapidly and abundantly from phenylalanine. Growth occurs in medium containing potassium cyanide, but sodium alginate is not utilized as a sole source of carbon. Gas volumes produced from fermentable substrates by aerogenic cultures are small (a bubble to about 15%). Urea is hydrolyzed rapidly and abundantly by members of one genus *(Proteus).*

Genus I *Proteus* Hauser

The genus *Proteus* is composed of motile bacteria that conform to the definitions of the family ENTEROBACTERIACEAE and the tribe PROTEEAE. Urea is hydrolyzed rapidly. Two species, *Proteus vulgaris* and *Proteus mirabilis,* produce hydrogen sulfide rapidly and abundantly, liquefy gelatin, and swarm on moist agar media. The majority of cultures of these two species are lipolytic. The other species, *Proteus morganii* and *Proteus rettgeri*, do not possess these particular characteristics. Ornithine is decarboxylated by two species *P. mirabilis* and *P. morganii*. Mannitol is fermented by the majority of strains of *P. rettgeri* but the remaining species fail to produce acid from this substrate. The type species is *Proteus vulgaris* Hauser.

Genus II *Providencia* Ewing

The genus *Providencia* is composed of motile bacteria that conform to the definitions of the family ENTEROBACTERIACEAE and the tribe PROTEEAE. Urea is not hydrolyzed and hydrogen sulfide is not produced. Indol is produced and growth occurs on Simmons' citrate medium. Gelatin is not liquefied, lysine, arginine, and ornithine are not decarboxylated, and lipase is not produced. With rare exceptions mannitol is not fermented and acid is not produced from alpha methylglucoside, erythritol, or esculin. The type species is *Providencia alcalifaciens* (De Salles Gomes) Ewing.

Additional references for citations, earlier reports, etc., may be found in the individual chapters, or in the publications cited.

REFERENCES

Bergey's Manual of determinative bacteriology, 7th ed., 1957. Edited by R. S. Breed, E. G. D. Murray, and N. F. Smith. Baltimore: Williams & Wilkins.

Borman, E. K., C. A. Stuart, and K. M. Wheeler. 1944. J. Bacteriol., **48**, 351.

Davis, B. R., W. H. Ewing, and R. W. Reavis. 1957. Int. Bull. Bacteriol. Nomen. Tax., **7**, 151.

Davis, B. R., and W. H. Ewing. 1964. J. Bacteriol., **88**, 16.

Ewing, W. H. 1963. Int. Bull. Bacteriol. Nomen. Tax., **13**, 95.

_____. 1966. Enterobacteriaceae: taxonomy and nomenclature. CDC Publ.*

_____. 1967. Revised definitions for the family Enterobacteriaceae, its tribes, and genera. CDC Publ.*

_____. 1968. Differentiation of Enterobacteriaceae by biochemical reactions. CDC Publ.* (Revised and emended, 1970.)

Ewing, W. H., A. C. McWhorter, M. R. Escobar, and A. H. Lubin. 1965. Int. Bull. Bact. Nomen. Tax., **15**, 33.

Fields, B. N., M. M. Uwaydah, L. J. Kunz, and M. N. Swartz. 1967. Amer. J. Med., **42**, 89.

Graham, D. C. 1964. Ann. Rev. Phytopathol., **2**, 13.

International Code of Nomenclature of Bacteria. 1966. Int. J. System. Bacteriol., **16**, 459.

Kauffmann, F. 1956. Zentralbl. f. Bakt. I. Orig., **165**, 344.

_____. 1963. Int. Bull. Bacteriol. Nomen. Tax., **13**, 187.

_____. 1966. The bacteriology of Enterobacteriaceae. Copenhagen: Munksgaard.

Kauffmann, F., and P. R. Edwards. 1952. Int. Bull. Bacteriol. Nomen. Tax., **2**, 2.

*CDC Publ., Publication from the Center for Disease Control (formerly the Communicable Disease Center), Atlanta, Ga. 30333.

Chapter 2

Isolation and Preliminary Identification

COLLECTION AND PROCESSING OF SPECIMENS

Isolation of members of the family Enterobacteriaceae from stools, from body fluids or tissues, and from environmental sources, and the differentiation of members of the various genera and species, pose problems encountered frequently by investigators in all bacteriological laboratories. If such work is undertaken it should be done in a systematic manner using acceptable methods; otherwise, not only is the purpose of the investigation defeated, but a false sense of security may be conferred. Failure to detect the typhoid bacillus or failure to diagnose correctly an active case of shigellosis, salmonellosis, or infantile diarrhea caused by a serotype of *E. coli* may have serious consequences. The purpose of this chapter is not to dictate the exact methods and materials each worker should use. On the contrary, the intent is to emphasize the principles involved and to outline acceptable practices that are known to have yielded satisfactory results in many laboratories.

From the standpoint of the effectiveness of the laboratory, nothing is more important than the adequacy and condition of the specimen received for examination. If specimens are not properly collected and handled or are not representative, the laboratory can contribute little or nothing to any investigation. This applies to specimens of all sorts.

FECAL SPECIMENS

Stools should be collected early in the course of enteric disease processes and before institution of treatment. Selected portions should be inoculated onto *adequate* plating and enrichment media as soon as possible after collection, since some etiological agents may decrease rather rapidly in numbers or may be overgrown by other bacteria. When present, pathologic constituents such as mucus and shreds of epithelium should be selected for cultural work. In bacillary dysentery, including the chronic form, immediate plating of swabs taken directly from lesions during proctoscopy is a most satisfactory method for isolation of *Shigella*.

Most competent investigators are agreed that the specimen of choice in enteric disease is a freshly passed stool. Rectal swabs may be used to collect specimens from persons who are acutely ill with diarrheal disease or dysentery, as in hospitals or in the examination of a sampling portion of those ill in a larger epidemic (unpublished data; Hardy et al., 1953). However, ordinary rectal swabs should not be relied upon to yield the maximum number of positive cultures (Shaughnessy et al., 1948; Thomas, 1954; McCall et al., 1966). Further, single rectal swabs are of little value in the examination of convalescent patients or in surveys for carriers. One hastily collected rectal swab, usually an anal swab, placed on a plate of selective medium can hardly be expected to yield numerous colonies of pathogens, especially of shigellae. Yet, this is precisely the procedure used as the basis for many reports recorded in the literature on the incidence of *Shigella* and other enteric bacteria in diarrheal disease.

Whenever possible, repeat specimens should be cultured, i.e., multiple specimens should be examined. Many investigators have demonstrated the value of this procedure. For example, it was demonstrated at least as early as 1916 by Ten Broeck and Norbury. Similar data later were reported by others (e.g., Floyd, 1954). In the examination of suspected carriers, purgation usually results in an increase in the number of positive cultures. This fact has been known for many years in connection with the isolation of *S. typhi* from carriers, for example, and more recently the advantages of purgation again have been demonstrated in the isolation of *Vibrio cholerae* (Gangarosa et al., 1966).

In the investigation of diarrheal diseases, the importance of obtaining stool specimens in the *acute stage* of the disease cannot be overemphasized. Usually the bacillary incitants of enteritis are present in large numbers at that time and often are the predominant organism in the stool. As the symptoms subside, the numbers of the causative agent rapidly decrease so that in cultures taken after the acute stage of the disease is past, the organisms responsible for the infection may be isolated only with difficulty or may not be found.

Transport Solutions And Media. Specimens that cannot be cultured very soon after collection should be placed in a transport solution or medium until they can be examined (or delivered to another laboratory). A number of transport media have been described; all of them are designed to hold the bacterial population in a specimen more or less stationary, and to prevent, as much as possible, overgrowth of a particular microorganism by others that may be present. One of the oldest known, and probably still the most widely used, is the buffered glycerol saline solution described by Teague and Clurman (1916) and modified by Sachs (1939). The final pH of this solution should be 7.4, and if it becomes acid it should be discarded. Approximately one gram of feces is added to 10 ml of solution and thoroughly emulsified in it.

Several additional methods for transport of stool specimens have been described. Stuart (1956, 1959) devised a semisolid agar menstrum, and advocated the use of charcoal-impregnated, buffer-treated swabs with it. Originally Stuart's medium was designed for transport of specimens from persons suspected of having gonococcal infection, but it has been used successfully in limited studies with stool specimens by Stuart (1956), Cooper (1957), and Ewing et al. (1966). Stuart's transport medium has been modified for various reasons by Cary and Blair (1964), Cary et al., (1965), and by Amies and Douglas (1965) and Amies (1967). These modifications have been used with success for the isolation of salmonellae and shigellae by the above-mentioned investigators and others. These semisolid transport menstra are inoculated with stool specimens collected on buffer-treated swabs, and the swabs are left in the tubes. The method for preparation of the buffer-treated swabs is given in Chapter 18.

Hajna (1955) devised a specimen preservative (SP) solution which in his hands doubled the number of isolations of shigellae and salmonellae when compared with buffered glycerol saline solution. Shipe and Fields (1956) investigated the use of a chelating agent, disodium ethylenediamine tetraacetate (EDTA), in the preservation of coliform bacteria in water samples. Later Shipe et al. (1960) used EDTA in a glycerolated preservative solution with good results. In a small investigation, Ewing et al. (1966) reported the results of comparative studies with 131 stool specimens from as many persons registered as carriers of S. typhi, using buffered glycerol saline solution, Stuart's method, and the EDTA glycerol saline solution of Shipe et al. The results of cultures of the fresh specimens, platings from each of the three transport media, and cultures from each transport medium following enrichment in selenite F broth (v. inf.) all were compared. All three transport methods yielded excellent results when cultured directly. About 3 per cent more positive cultures were obtained from specimens in EDTA solution under these conditions, but after specimens in EDTA medium were incubated in selenite F broth, the number of

positive cultures was reduced about 65 per cent. This effect of EDTA on selenite medium also was noted by Shipe et al. (1960).

Dold and Ketterer (1944), Lie (1950), and Bailey and Bynoe (1953), among others, reported that drying stool material on pieces of filter or blotting paper was a useful method for shipping specimens. This method may be of value when the use of other methods is impractical or impossible, as might be the case in obtaining specimens from very remote regions.

One other method of transporting specimens, of value only in investigations of outbreaks of salmonellosis, should be mentioned. Swabs (preferably buffer-treated) inoculated with stool specimen, or rectal swabs, are placed in tetrathionate brilliant green enrichment medium (v. inf.) in screw-capped tubes, packed carefully, and mailed. This method is satisfactory for most salmonelleae, but it is of little or no value when S. typhi, S. cholerae-suis, or S. enteritidis bioserotype Paratyphi-A is involved. McCall et al. (1966) employed this method with success during investigations of an outbreak of salmonellosis in which S. enteritidis serotype Derby was the etiological agent.

Methods for preparation and use of all of the transport solutions and media mentioned in the foregoing paragraphs are given in Chapter 18. It is evident that a number of such preservative menstra are available. However, a review of the subject of transport methods (Ewing, 1968) revealed a dearth of comparative data. Comparative studies of the preparations mentioned above by several groups of investigators working under various conditions are needed before an intelligent evaluation of their relative worth can be made. Some of the methods, e.g., Stuart's medium and some modifications thereof, show promise and should be evaluated further. A transport menstrum similar to that devised by Amies and Douglas (1965) and Amies (1967), but with something other than thioglycolate in it, might be a better choice. Such a medium might prove to be better not only for Enterobacteriaceae and allied bacteria, but for other microorganisms as well (e.g., Bacteriodes).

Enrichment Media. While it always is advisable to employ enrichment media in the examination of various kinds of specimens, their use is practically essential when dealing with fecal specimens from carriers or suspected carriers. Two such media have been employed extensively and may be recommended for general use. These are tetrathionate medium of Muller (1923) and selenite F broth devised by Leifson (1936). The combined enrichment medium of Kauffmann (1930) is a modification of Muller's medium which contains bile and brilliant green. The efficacy of this medium for the enrichment of *Salmonella* is attested to by the results of Kauffmann (1935), who found that when the medium was used in conjunction with brilliant green-phenol red agar of Kristensen, Lester, and Jurgens (1925) the isolations of *Salmonella enteritidis* serotype Paratyphi-B were increased 100 per cent and the isolations of *Salmonella* from acute gastroenteritis were increased 500 per cent over methods in which no enrichment was used. Galton and Quan (1944) attributed a 164 per cent increase in the isolation of *Salmonella* to the use of these media. (See also McCall et al., 1966.)

Tetrathionate base, which is Muller's medium with bile salts added, may be employed with all specimens, since salmonellae including *S. typhi* generally are greatly increased in numbers in this medium. Also some shigellae may be recovered from it. Tetrathionate medium with brilliant green added (1:100,000) is useful only for *Salmonella* other than *S. typhi*.

The selenite medium recommended by Leifson for use with feces (selenite F) was designed for enrichment of *Salmonella* including *S. typhi*, not for shigellae. However, the selenite media available from most commercial sources have been modified slightly and some shigellae may be recovered after plating from it. In fact, some investigators (Armstrong, 1954; Thomas, 1954) found selenite broth useful in the isolation of *Shigella sonnei*. Hobbs and Allison (1945) found selenite superior to tetrathionate in the isolation of *S. typhi* and equally as good as tetrathionate in the isolation of *S. enteritidis* serotype Paratyphi-B. These conclusions were confirmed by Cook et al. (1954). Smith (1952) emphasized the absolute necessity of enrichment media in the isolation of salmonellae other than *S. typhi* when the organisms were present in small numbers. He felt that selenite was slightly superior to tetrathionate. It should be noted, however, that Smith (1952, 1959) found it necessary to use brilliant green — MacConkey broth for the isolation of *S. cholerae-suis* since the organism failed to develop in selenite broth. This may explain the infrequent isolation of *S. cholerae-suis* from the intestinal tract and feces of man and swine. Both selenite and tetrathionate media reportedly are toxic for *S. cholerae-suis* and *S. enteritidis* ser Abortus-ovis (Smith, 1952), while tetrathionate broth inhibits the growth of *S. enteritidis* bioser Paratyphi-A (Banwart and Ayres, 1953). Leifson reported the high toxicity of selenite for *S. cholerae-suis* when he described the medium. The selenite F and tetrathionate media that are available in dehydrated form from several commercial outlets are quite efficient *when prepared in exact accordance with the directions that accompany them,* and these preparations can be recommended for use as outlined in the foregoing paragraphs.

Numerous modifications of each of these have been described (e.g., Knox et al., 1943; Rolfe, 1946; Nagel, 1950; Zschucke, 1951). The cystine-selenite medium of North and Bartram (1953), originally devised for the isolation of salmonellae from food products, apparently is of value in the isolation of salmonellae other than *S. typhi* from stool specimens. The author has no other knowledge of its value, or lack of it, in the isolation of *S. typhi*. Silliker et al. (1964) reported that the addition of an extract of feces to selenite medium enhanced the ability of this medium to select for salmonellae. These investigators also devised a medium for shigellae, to which autoclaved fecal extract was added. A measure of selectivity for shigellae apparently was effected. Silliker (personal communications, 1969, 1970) reported that an extract of Millorganite can be substituted for the above-mentioned extract of feces. Two hundred grams of Millorganite are added to 1000 ml of distilled water and sterilized at 121 C, 15 min. When cool this is filtered through paper (e.g., Whatman No. 2 or 3), and the filtrate is passed through a mixed bed ion exchange resin column. Following this, the extract of Millorganite is added to an equal volume of double strength selenite F medium, mixed, and dispensed in tubes in the usual manner. Millorganite contains quantities of inorganic salts, and relatively high concentrations of inorganic salts such as sodium chloride are known to affect the recovery rate of salmonellae that have been subjected to sublethal heat treatment (Clark and Ordal, 1969). Although salmonellae that have been injured in this way are more likely to occur in certain food and feed products than elsewhere, Silliker (personal communication, 1970) recommends preparation of the medium in the manner mentioned above regardless of the nature of the specimens under examination. This method is given in some detail since it is worthy of further investigation, especially in connection with isolation of salmonellae, and perhaps shigellae, from specimens in which these bacteria are present in small numbers or when the amount of specimens received for examination is small. Hajna (1955) devised a GN (gram-negative) broth, which was used in part as a preservative medium and in part as an enrichment medium for specimens preserved in SP medium (v. sup.). Increased numbers of salmonellae and shigellae were isolated by this method. Croft and Miller (1956) used GN broth successfully in the study of outbreaks of shigellosis, and Taylor and Schelhart (1969) apparently found it useful for shigellae. Rappaport et al. (1956) devised a magnesium chloride-malachite green enrichment broth which was inoculated with dilute (1-1000) suspensions of feces, and with which better results were obtained in the recovery of salmonellae other than *S. typhi* than with selenite or tetrathionate medium. They noted also that the choice of peptone influences the results, particularly the recovery of *S. enteritidis* bioserotype Paratyphi-A. The effect of various peptones on results obtained with selenite medium was reported earlier by North and Bartram (1953). Reports in the literature on the value of Rappaport's medium have differed. The literature was reviewed by Vassiliadis (1968) who pointed out that the inoculum recommended by Rappaport et al. (1956) was not used by all investigators. Taylor and Schelhart (1969) reported that Rappaport's enrichment medium was good for the purpose for which it was intended, i.e., the isolation of salmonellae other than *S. typhi*.

From the foregoing it is apparent that if a single enrichment must be selected for general purposes in clinical laboratories, selenite broth is the medium of choice. When possible, a combination of media, such as selenite and GN, should be used in such laboratories. In the experience of the author, ordinary nutrient broth is the best enrichment medium for shigellae, and, judging from the work of Thomson (1955), nutrient broth is useful for salmonellae as well. Enrichment media should be tubed in 8- to 10-ml amounts and inoculated with approximately 1 gram of formed stool or with 1 ml of liquid specimen. The inocula should be emulsified thoroughly in the media. After incubation at 35 to 37 C, (selenite 12 to 16 hr; tetrathionate 18 to 24 hr), cultures in enrichment media should be streaked on plates of the same sort used for primary inoculation of specimens (Figure 1). With noninhibitory and slightly inhibitory plating media, one loopful (3- to 4-mm loop) of inoculum may be used, but two or three loopsful may be streaked on moderately and highly selective media. In every instance plates must be streaked carefully in order to assure growth of well-isolated colonies.

For additional information regarding the numerous varieties and modifications of enrichment media readers are referred to the publications of Smith (1952). Banwart and Ayres (1953), Taylor and Schelhart (1969), the papers cited above, and earlier editions of this book.

Plating Media. To say that numerous media have been devised for isolation of pathogenic Enterobacteriaceae certainly is an understatement. Many, many such media have been described and numerous modifications have been devised for one purpose or another. It is beyond the scope and intent of this book to attempt to review, or even mention, all of these. On the contrary, only a brief review of important developments, and a categorization of several classes of isolation media, will be given.

In 1886 it was reported that *B. coli* fermented lactose, while the typhoid bacillus did not. Soon after, Wurtz devised a medium that contained lactose and litmus as an indicator. The practice of employing lactose in isolation media has continued to the present. The most notable exception to this was the introduction of bismuth sulfite agar by Wilson and Blair (1926, 1927, 1931). Isolation media that contained lactose for differentiation were the only ones available to earlier investigators. Such media were of value, and still are. However, the early use of lactose in this way undoubtedly led to the selection of "lactose negative pathogens," and strongly influenced earlier classifications. Had another carbohydrate been employed, a different set of pathogens might have been selected. Further, the early use of lactose in isolation media contributed to the development of the concept of paracolon bacteria and to the concept of coliform bacteria.

The media[1] in general use for the isolation of various Enterobacteriaceae may be divided into a number of categories according to their selectivity.

Noninhibitory media such as blood agar, plain infusion (IA) or nutrient agar (NA).

Noninhibitory differential media, e.g., bromthymol blue lactose (BTBL) and phenol red lactose (PRL) agars.

Differential media with little selectivity as regards Enterobacteriaceae. These include MacConkey's (MacC), eosin methylene blue (EMB), and desoxycholate (LD) agars.

Differential, moderately selective media such as Shigella-Salmonella (SS), and desoxycholate citrate (DC), Hektoen Enteric (HE), and Xylose Lysine Desoxycholate (XLD).

Highly selective media, which include bismuth sulfite (Wilson-Blair) agar (BS) and brilliant green agar (BG). The latter may be called a "one purpose" medium, particularly useful for the isolation of salmonellae. However, certain other bacteria (e.g., some pseudomonads and aeromonads) may grow on BG agar and form pink colonies.

The results that may be expected when the above-mentioned agar plating media are properly prepared and used for the isolation of *Shigella, S. typhi* and other *Salmonella* are as follows:

	Infusion agar (etc)	BTB (etc)	MacC (etc)	SS (etc)	BS	BG
Shigella	+	+	+	+	-	-
S. typhi	+	+	+	+	+	-
Other *Salmonella*	+	+	+	+	+	+

[1]Specific references to literature on plating media not cited here are given in Chapter 18.

Figure 1

ISOLATION AND PRELIMINARY IDENTIFICATION
OF *SALMONELLA* AND *SHIGELLA* CULTURES

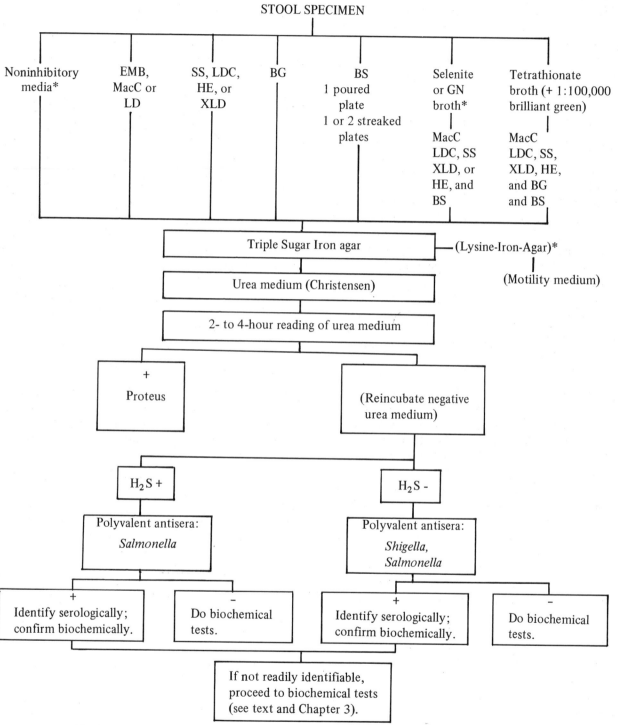

*See text.

EMB, Eosin methylene blue agar. MacC, MacConkey agar. LD, Leifson desoxycholate agar. SS, Shigella-Salmonella agar. LDC, Leifson desoxycholate citrate agar. HE, Hektoen enteric agar. XLD, xylose lysine desoxycholate agar. BG, brilliant green phenol red agar. BS, bismuth sulfite agar. GN, gram negative broth.

As might be expected, occasional strains of shigellae may not grow on MacConkey's, SS, or DC agar, or a particular *Salmonella* may not appear on brilliant green agar. Hence it is advisable to use a variety of media whenever possible. The procedure to be followed in any laboratory is dictated by facilities, personnel, and the amount of time available for work of this sort. In any search for salmonellae, including *S. typhi*, bismuth sulfite agar always should be used, since it is the most efficient medium yet devised for the isolation of *S. typhi*. For the maximum number of isolations many workers have found it advisable to use two plates. One plate should be streaked with feces or a heavy fecal suspension. The second plate is a poured plate in which has been placed 5 ml of a suspension of feces. In the stools of typhoid patients in which *S. typhi* is present in large numbers, typical colonies usually will be found on the streaked plate. In specimens from carriers, where the bacteria may be present in small numbers, the organism frequently may be isolated from subsurface colonies in the poured plate. Subsurface colonies obviously must be restreaked on a medium such as MacConkey or eosin methylene blue agar. The extreme variation in the numbers of *S. typhi* and *S. enteritidis* serotype Paratyphi-B excreted by different carriers is well illustrated by the work of Thomson (1955) who recommended the addition of a very lightly inoculated plate in the examination of carriers. He obtained good results by streaking 0.02 ml of a 1:1000 dilution of feces on desoxycholate citrate agar. *S. typhi* frequently does not produce typical blackening of bismuth sulfite agar when the colonies are very numerous and crowded. It is for this reason that the use of two plates or more is advisable. Strains of *Salmonella* usually grow well on bismuth sulfite agar and it is a good medium for their isolation. However, on this medium *Shigella* fail to develop except in very rare instances. Since bismuth sulfite agar is highly inhibitory, it is advisable to incubate plates 48 hr before they are discarded. Added reasons for the use of bismuth sulfite agar lie in the detection of members of the genus *Arizona* (Edwards and Fife, 1961), and the isolation of the very rare strain of *Salmonella* that ferments lactose rapidly. For the isolation of shigellae, most workers prefer to use one plate of a purely differential or only mildly selective medium and one plate of a more highly selective preparation. MacConkey's agar, desoxycholate agar, or eosin-methylene blue agar can be used for the first, and desoxycholate citrate or SS agar for the second. When possible it is advisable to employ a variety of media, and xylose lysine desoxycholate (XLD) or Hektoen enteric (HE) agar media may be included (Figure 1). Certain strains of *Shigella* grow very poorly on the more highly selective media and for that reason it is advisable to include one of the less inhibitive media. For example, Wheeler and Mickle (1945) demonstrated that *S. sonnei* form II does not grow well on selective media such as those just mentioned, and this form is prone to occur in convalescent patients and carriers (Branham et al., 1952). If XLD agar is used, it should be recalled that some shigellae ferment xylose.

In the experience of the author, the better way to isolate the maximum number of shigellae is to inoculate several plates of plain infusion agar, each with small inocula, and pick a number of each colonial type that appears on each plate.

Salmonellae usually can be isolated from any of the above-mentioned media, but it would not be advisable to rely on only one plate of a single medium. On the contrary, it is advisable to employ at least one plate each of the slightly selective, selective, and the highly selective media such as bismuth sulfite agar.

As mentioned, cultures in enrichment media should be inoculated onto the same kinds of plating media that are employed for fresh specimens. All isolation media should be inoculated in such a manner as to assure the maximum number of well-isolated colonies. All media employed for isolation should be freshly prepared in exact accordance with directions. One good plan is to prepare plating media one day and use it the next. With the exception of bismuth sulfite agar, the above-mentioned media may be kept in a refrigerator for about three days, provided that steps are taken to prevent evaporation. Bismuth sulfite agar should not be used after storage for more than 24 to 36 hours. When this medium becomes green it is too inhibitory, and one cannot expect good results.

Outlines of procedures for the isolation and preliminary identification of cultures of *Salmonella* and *Shigella* are given in Figure 1. *Members of other genera of Enterobacteriaceae also may be recovered and processed according to the procedures given in Figure 1.* However, an outline of procedures for isolation of enteropathogenic and other serotypes of *E. coli* is included in Chapter 5.

ISOLATION FROM BODY FLUIDS, EXUDATES, AND TISSUES

Blood Cultures. Samples of blood submitted to the laboratory for culture usually are derived from patients afflicted with febrile disease of unknown etiology. When possible, blood should be drawn during those periods when the patient's temperature is rising or falling, since the maximum number of positive cultures may not be obtained from specimens collected at the peak of a temperature curve. When possible multiple specimens should be collected.

Since the etiology usually is not known at this

point, media and methods that may be expected to support the growth of the more fastidious pathogenic bacteria such as brucellae or streptococci must be used. It is advisable to inoculate 100- to 150-ml amounts of an *adequate* broth medium with 10 to 15 ml of blood. The cultures should be examined after 24 hours of incubation at 35 to 37 C and, if negative, they should be reincubated for at least 3 weeks, with examination at frequent intervals. The author has found it advantageous to prepare pour plates with about two ml of patient's blood and an enriched infusion agar medium such as might be used for *Brucella.* When positive these plates yield valuable information at an early stage of investigation.

When blood is clotted, the clots should be macerated before the medium is inoculated. If it is suspected that blood may be contaminated, it is advisable to plate the blood culture on media designed for the isolation of enteric bacteria. Thomas et al. (1954) noted that larger numbers of *S. typhi* were recovered from sodium taurocholate broth when streptokinase (100 units per ml) was added to the medium. Further, the necessary incubation period was reduced. This effect was attributed to the lytic action of the enzyme on blood clots. Later it was found that the streptokinase neutralized the inhibitive action of blood serum and that 1.0% trypsin and other substances which inactivated complement had a like effect. Papaine also has been used for this purpose.

After incubation, blood cultures should be plated on infusion agar, MacConkey agar, or eosin methylene blue agar, and on bismuth sulfite agar, following which the procedures outlined in Figure 1 are used.

Of the Enterobacteriaceae, the bacterium that invades the bloodstream most often and produces enteric fever apparently is *S. typhi*. This is particularly true in adults. The microorganism often invades the blood before it appears in large numbers in stools. For this reason it is advisable to obtain blood cultures early in the course of disease, since the organism may disappear from the blood after a short period. If blood cultures taken during the 1st week of disease are negative, they should be repeated during the 2nd and 3rd weeks as necessary. This is especially desirable in severe cases during which the temperature remains elevated for a long period. Reviews of the literature (Shaw and MacKay, 1951) dealing with this probelm emphasize that several investigators have obtained the highest percentage of positive blood cultures during the 2nd week of the disease. Moreover, in those cases classed as severe, the highest percentage of positive cultures was obtained during the 3rd week after the appearance of symptoms (Shaw and MacKay, 1951).

Almost any other *Salmonella* occasionally may invade the blood and produce an enteric fever syndrome, and *S. cholerae-suis* has been reported frequently in children. Further, other Enterobacteriaceae may gain entrance to the blood, persist, and produce severe illness (e.g., *E. coli* 06:H1).

Urine Cultures. Midstream specimens collected after *thorough* cleansing of the genitalia with soap should be selected for cultural examination. The specimen or a representative portion of it should be centrifuged and a gram stain made of the sediment. Examination of gram stains gives information as to the predominant morphological types present and a general idea of their numbers. The former is of value in the selection of media and the latter aids in determining the point at which the plating of dilutions should be begun in quantitative work.

If gram-negative bacteria predominate, the sediment from a urine specimen should be inoculated directly onto the surface of MacConkey's agar or eosin methylene blue agar and onto bismuth sulfite agar. Blood agar also should be used when indicated. It is advisable also to inoculate 2 or 3 ml of urine into a tube of plain tetrathionate broth or of selenite medium.

In many instances quantification of the numbers of organisms present in a specimen of urine yields information of great value to the clinician. In such instances, viable counts may be made using the conventional dilution and plating technics employed in the examination of water or milk. When differential counts are required, dilutions may be made and 0.05-ml portions of each dilution spread over the surface of the plates of medium such as bromthymol blue lactose agar, phenol red lactose agar, or MacConkey agar.

When a search is being made for certain bacteria which may be present in small numbers, an entire urine specimen of 1 liter or more may be converted into a culture medium by the addition of 1.0% peptone and 0.1% glucose. The pH should be determined and adjusted to about 7.2 if necessary. Following incubation in a large flask, subcultures should be made to plating and enrichment media. Also, large amounts of urine may be examined by the use of membrane filter technics.

Other Body Fluids, Exudates, And Tissue Specimens. Duodenal drainage and bile should be examined according to the methods used for a liquid stool specimen. When gallbladders are submitted, a portion of the wall, as well as the bile, should be placed in enrichment broth. If gallstones are present, they may be ground aseptically and placed in a tube of enrichment medium. In the examination of bile or gallstones for *S. typhi,* it is preferable to use selenite broth for enrichment.

Purulent exudates from abscesses or joint cavities, sputa, cerobrospinal fluid, or other materials should be examined by the methods usually employed in the search of such materials for the presence of pathogenic bacteria. If bacteria cannot be detected in stained preparations of spinal fluid, the specimen itself should be incubated, reexamined, and cultured. In the examination of exudates or transudates derived from the pleural or peritoneal cavities, it is advisable to use a plate of MacConkey's agar and a plate of one of the more selective media, in addition to general-purpose media. Biopsy or necropsy materials such as spleen, liver, and lymph nodes should be ground aseptically and streaked on suitable plates, with a liberal portion placed in enrichment medium.

If specimens of the sort mentioned in the preceding paragraphs must be sent to another laboratory, the transport menstra used for stool specimens generally can be used. Exudates, sediment from spinal, pleural, or other fluids, macerated tissue, etc., may be collected on swabs, and these may be placed in transport medium. In addition, it is advisable to use the chopped meat medium commonly employed for anaerobes. A wide variety of Enterobacteriaceae and members of other families, as well as anaerobic bacteria, can be recovered from specimens transported in this medium (unpublished data, 1943-1945). Similar observations were made much earlier by Holman (1919).

The temperature of incubation employed for plating and enrichment media in most laboratories is 35 to 37 C, and a temperature within this range is recommended for general work. However, Harvey and Thomson (1953) and Georgala and Boothroyd (1965) reported more positive isolations of salmonellae from selenite broth incubated for 24 hr at 43 C than from the same medium incubated at 37 C. It should be noted that about 10 per cent of the selenite broth tubes inoculated with aliquots of fecal specimens by Harvey and Thompson (1953) failed to reveal the presence of *Salmonella* when this medium was incubated at either temperature. These specimens were proven to be positive by other methods. McCoy (1962) reported that tetrathionate medium was toxic for salmonellae when incubated at 43 C. Using modified selenite broth, tetrathionate, and brilliant green agar plating medium incubated at 41.5 C, Spino (1966) consistently recovered salmonellae from samples of surface waters. Salmonellae were not recovered from equal portions of the same samples incubated at 37 C. It seems clear that higher temperatures of incubation might yield higher percentages of positive results with salmonellae from certain materials at least, however incubation of equal portions of specimens at 35 to 37 degrees should not be omitted, certainly not in general work. Undoubtedly a number of factors are involved such as the nature of the material under examination, the modifications and kinds of media employed, and, as mentioned by Spino (1966), the combined effect of the media and the elevated temperature on competitive microorganisms. Further, the effects of incubation of various materials at elevated temperatures on the isolation of *Shigella* and other enteropathogenic bacteria requires additional investigation, including quantification of microorganisms recovered at the various temperatures.

SELECTION AND PICKING OF COLONIES

In the examination of plates inoculated with feces or other specimens, either directly or from enrichment media, a *careful* search should be made for significant organisms. Cursory examination will not suffice. For example, *Shigella* and *Salmonella* usually produce typical colonies on the various plating media but it must be remembered that their appearance may be altered by growth in close association with other organisms. At times these pathogens produce colonies of atypical appearance for reasons that are not entirely clear. Hence even persons of long experience in enteric bacteriology may be misled by colonial appearance. When searching for salmonellae or shigellae, it is advisable to pick at least two representatives of each type of colony other than those of frank lactose fermenters which appear upon the plates. Examination of the plates with a colony counter or magnifier is helpful in distinguishing the various colony types.

Many kinds of microorganisms do not develop on selective media and viable bacteria which have not grown, or which may have produced only microscopic colonies, are present on the surface of the agar. Therefore, the greatest care must be exercised in picking colonies if pure cultures are to be obtained. It is advisable to use a straight needle and to touch only the center of the selected colony. One should avoid touching the surface of the agar. Proper attention to this detail will result in fewer mixed cultures from plates inoculated with feces.

Primary Differentiation. Bacteria from selected colonies should be transferred to tubes of triple sugar iron (TSI) or Kliger's iron (KI) agar. The butt of the medium first is stabbed to the bottom of the tube, then the slant is streaked carefully over the entire surface. If good growth does not occur in the butt or on the slant of these media, atypical reactions may result. KI agar is recommended for use with specimens from extraintestinal sources, especially urine (v. inf.). It always is advisable to mark and number each colony picked and to place the same number on the corresponding TSI or KI agar tube. A record of the nature of the colony should be kept (i.e., typical, atypical, etc.).

If the reaction in TSI or KI agar medium following overnight (18 to 24 hr) incubation deviates from what might be expected from the colonial appearance, the culture may be mixed. In such instances replating for purity should be done, and a noninhibitory differential medium should be employed. Gram stains also are helpful for determination of purity. Some investigators advise the use of polyvalent antisera for testing the growth from such TSI or KI agar slants, but the culture still must be examined for purity, regardless of the outcome of such tests. The reactions given by various kinds of Enterobacteriaceae on TSI agar are recorded in Table 1.

Numerous complex differential media have been devised for the cultivation of colonies picked from isolation plates. A variety of biochemical and physiological properties may be determined in such media, some of which are tubed in two or even three layers, one superimposed upon another. Among these media are those of Colichon (1953), Rappaport et al. (1956), Gillies (1956) Velaudapillai (1957), Papadakis (1960),

and Monteverde et al. (1968). Some of these, e.g., dulcitol lactose iron agar (Taylor and Silliker, 1958), were designed for a specific purpose, in this instance for differentiation of salmonellae. While not discounting the value of such media, nor detracting from their performance, the author has preferred to use TSI (or KI) agar, in conjunction with lysine iron agar (v. inf.) for the cultivation of colonies on isolation plates, and to rely upon conventional methods for the determination of additional biochemical properties of isolants.

In the search for *Salmonella* or *Shigella* (Figure 1) tubes of TSI agar medium that exhibit an acid butt and an alkaline slant should be planted immediately on the urea agar of Christensen (1946). This medium is slanted so that the tubes have a deep butt and a short slant. The medium is inoculated heavily on the surface only. Cultures of *Proteus* produce marked alkalinity in Christensen's urea agar after very brief incubation. The cultures are examined after 2 to 4 hours and the negative tubes are reincubated. Cultures that belong to certain other genera produce varying degrees of alkalin-

TABLE 1
THE REACTIONS OF ENTEROBACTERIACEAE IN TRIPLE SUGAR
IRON (TSI) AGAR MEDIUM

Genera and Species	Slant	Butt	Gas	H_2S
Escherichia	A(K)	A	+(-)	-
Shigella	K	A	-	-
S. typhi	K	A	-	+(-)
Other *Salmonella*	K	A	+(-)	+++(-)
Arizona	K(A)	A	+	+++
Citrobacter	K(A)	A	+	+++
Edwardsiella	K	A	+	+++
Klebsiella	A	A	++	-
Enterobacter	A	A	++	-
E. hafniae	K	A	+	-
Serratia	K or A	A	-	-
Proteus vulgaris	A(K)	A	+	+++
P. mirabilis	K(A)	A	+	+++
P. morganii	K	A	-(+)	-
P. rettgeri	K	A	-	-
Providencia	K	A	+ or -	-

Adapted from Ewing (1966).

K, alkaline. A, acid

Symbols enclosed in parentheses indicate occasional reactions.

ity in Christensen's medium after 24 or more hours of incubation. If preferred, the rapid urease test of Stuart et al. (1945) may be substituted for Christensen's medium. The rapid urease test advocated by Stuart et al. utilizes the medium described by Rustigian and Stuart (1941, 1945), except that the buffer content is reduced to 1/100 (1%) of the amount contained in the original medium. A large inoculum from an agar slant is suspended in the medium and the tubes are incubated in a water bath at 37.5 C if necessary. Most cultures of the four species of *Proteus* are positive in 5 to 20 min and with very rare exceptions others are positive within an hour. Other urease test media have been described by Ferguson and Hook (1943), Elek (1948), and by others.

For a number of years, the author and co-workers (Ewing, 1966; Johnson et al., 1966) have used lysine iron agar (LIA, Edwards and Fife, 1961) in conjunction with TSI or KI agar as an aid in the examination of colonies picked from plating media. Others have been advised to try this method and have found it useful (e.g., Johnson et al., 1966). With a straight wire, inoculum is taken carefully from a selected colony and inoculated into TSI or KI agar in the usual way. Then, *without* going back to the colony, a tube of LIA is inoculated directly by stabbing the butt of the medium twice and streaking the slant. If the spot where the stab was made in the TSI agar medium is touched with the tip of the wire, a sufficient number of bacteria will be obtained for inoculation of the LIA medium. Inoculation of a tube of semisolid agar medium from LIA also is helpful. The combinations of reactions obtained through the use of this method yield a great deal of useful information early in the examination of a specimen. The use of a strip of filter paper previously impregnated with Kovacs' or Gillies' reagent and dried furnishes additional valuable information when the test is positive. The paper strips may be suspended over inoculated tubes of LIA medium, being held in place by the stopper. All negative tests for indol obtained by this method should be retested by conventional means. The reactions given by members of the family Enterobacteriaceae in LIA are listed in Table 2.

It should be emphasized that lysine iron agar (LIA) is not considered a substitute for the standard (Moeller) method for determination of lysine decarboxylation. While LIA is a valuable medium when used in the manner described in the preceding paragraph, its use should be limited to preliminary work. Some cultures may not yield evidence of lysine decarboxylation in the butt of LIA medium at 20 to 24 hr (Ewing, 1966).

Biochemical Reactions. Determination of their bio-

chemical reactions still is the only method available for recognition and differentiation of the various genera and species of Enterobacteriaceae. Any culture thought to be of significance, be it *Salmonella, Shigella,* or *Proteus,* for example, and irrespective of its source, should be identified, regardless of whether it is sent to a central laboratory.

Because of their importance, an entire chapter (Chapter 3) is devoted to differential reactions, and complete summaries of the biochemical reactions given by members of each of the genera are included in subsequent chapters. Methods for performance of tests are given in Chapter 18.

PRELIMINARY SEROLOGIC EXAMINATION OF CULTURES OF *SALMONELLA* AND *SHIGELLA*

Although the extent to which the personnel of a particular laboratory is able to pursue the serologic identification of cultures of *Salmonella, Shigella,* and other Enterobacteriaceae may be limited, the author is convinced that every laboratory of bacteriology, no matter at what level, should be prepared to identify completely certain species and serotypes. Beyond this, the personnel of each laboratory must decide how far they can go into serologic identification and how much time can profitably be devoted to it. Fortunately, there are a number of points at which more or less natural divisions can and should be made, and it is at one of these points that work with the majority of cultures may be stopped in a particular laboratory. If serological identification of strains is not completed, cultures may be forwarded to a central or reference laboratory, together with the minimal required information[2] regarding each isolate. At least minimal biochemical tests should be done before cultures are forwarded. Further, strains should be inoculated into tubes of stock culture medium for shipment. *Triple sugar iron (TSI) or Kligler's iron (KI) agar cultures should not be used for this purpose.*

SALMONELLA. As stated, it is the author's opinion that every laboratory of bacteriology should be equipped and prepared to identify completely certain important species and serotypes, as follows:

Salmonella typhi
Salmonella enteritidis, bioser Paratyphi-A
Salmonella enteritidis, ser Paratyphi-B
Salmonella cholerae-suis

---

[2]Identifying information: Name, age, and sex of individual(s) involved. Source of specimen: feces, urine, blood, food, other (specify). Clinical diagnosis: diarrheal disease, enteric fever, asymptomatic person, or other (specify). Results of biochemical and serological tests. Epidemiological relationships of cases, when known.

TABLE 2

THE REACTIONS OF ENTEROBACTERIACEAE IN LYSINE IRON
AGAR (LIA)

Genera and Species	Slant	Butt	Gas	H$_2$S
Escherichia	K	K or N	- or +	-
Shigella	K	A	-	-
Salmonella	K	K or N	-	+(-)
S. typhi	K	K	-	+ or -
Paratyphi-A	K	A	+ or -	- or +
Arizona	K	K or N	-	+(-)
Citrobacter	K	A	- or +	+ or -
Edwardsiella	K	K	- or +	+
Klebsiella	K or N	K or N	+ or -	-
Enterobacter				
cloacae	K or N	A	+ or -	-
aerogenes	K	K or N	+(-)	-
hafniae	K	K or N	- or +	-
Serratia	K or N	K or N	-	-
Proteus				
vulgaris	R	A	-	-(+)
mirabilis	R	A	-	-(+)
morganii	K or R	A	-	-
rettgeri	R	A	-	-
Providencia	R	A	-	-

Adapted from Ewing (1966)

An alkaline or neutral reaction in the *butt* of this medium indicated decarboxylation.

K = alkaline.   A = acid.
N = neutral.   R = red (oxidative deamination).

Salmonella enteritidis, ser Typhimurium

The first three microorganisms listed invade the bloodstream and usually cause enteric fever. The fourth may produce enteric fever, particularly in children (see report, Salmonella Committee, 1969), but also may cause only primary enteritis. These serotypes and species should be identified and a presumptive report issued as soon as possible so that appropriate treatment can be instituted. Hospital authorities, including the hospital epidemiologist, if one is available, also should be notified so that appropriate measures may be taken. Ser Typhimurium is mentioned because it occurs more frequently than any other serotype of *Salmonella;* hence, it should be identified. Only a few antisera are needed to perform complete antigenic analysis with cultures of each of the above-mentioned salmonellae. Methods for this are given detail in Chapter 8.

Cultures that yield reactions similar to those given by salmonellae on TSI and LIA agar media (Tables 1 and 2) should be tested in polyvalent antiserum for *Salmonella.* Such antisera may be produced in a number of ways (Chapter 8). However, one of the simplest contains agglutinins for the 0 antigens of serological groups A through E and for Vi antigen, similar to that described by Ewing and Bruner (1947). Approximately 95 per cent of the serotypes of *Salmonella* isolated from man and lower animals belong to these few serogroups (Kelterborn, 1967; Report, Committee on *Salmonella,* 1969).

Suspensions for agglutination tests with polyvalent antiserum may be prepared by suspending a loopful of growth from a TSI, or infusion, agar slant in a *small* amount of physiological saline solution or mercuric iodide saline solution. Such suspensions should be *dense.* A droplet of bacterial suspension is added to a droplet of properly diluted polyvalent antiserum on a slide, after which the slide is tilted back and forth several times. Positive tests are indicated by rapid, complete agglutination of the bacterial cells. If agglutination occurs, and the required antisera are available, serogrouping or complete serotyping may be done. If not, the biochemical reactions of the culture should be determined to make certain that it is a *Salmonella,* then it may be forwarded to a reference laboratory for complete antigenic analysis. However, if typical agglutination is obtained, and the results of preliminary biochemical tests are characteristic (24 to 48 hr readings), a report such as "*Salmonella,* serotype undetermined" may be issued. Strains that appear to be salmonellae with respect to their reactions on TSI and LIA media, but fail to react in polyvalent antiserum, may be encountered occasionally. Suspensions prepared from cultures of this sort should be heated in a beaker of boiling water for about 15 min, cooled, and retested in the antiserum. Some strains possess M antigens (Chapters 4 and 8), which inhibit agglutination of living (unheated) suspensions. Further, suspensions that do not react in the polyvalent antiserum being used may belong to a serogroup (beyond E, for example), agglutinins for which are not represented. The biochemical reactions of such a strain should be determined, and if the results indicate that the culture probably is a *Salmonella,* it may be sent elsewhere for further examination.

Isolates that yield reactions typical of *S. typhi* on TSI and LIA media should be examined by slide agglutination tests in the living (unheated) state in polyvalent, serogroup D, and Vi *(Citrobacter* 029*)* antisera, following which the suspensions should be heated in boiling water for about 15 min, cooled, and retested in the same antisera. If characteristic reversal of the reactions in O (serogroup D) and Vi antisera occurs (Chapter 8), a presumptive report of the isolation of *S. typhi* should be made as soon as possible. Confirmatory tests should follow, of course.

It is important to recall that strains of almost any serotype of *S. enteritidis,* as well as some cultures of *S. cholerae-suis* and *S. typhi,* may fail to form detectable amounts of hydrogen sulfide in TSI, KI, or LI agar media. Such cultures should not be discarded until it is certain that they are not salmonellae, or perhaps shigellae (see also Chapter 6).

SHIGELLA. The personnel of bacteriological laboratories should be prepared to determine the species to which cultures of *Shigella* belong. It would be most helpful if many more laboratories also were equipped to determine the serotype of strains that belong to each of the species.

The four species of *Shigella* are *S. dysenteriae, S. flexneri, S. boydii,* and *S. sonnei,* the first three of which are composed of a number of serotypes (Chapter 6). There is only one serotype of *S. sonnei,* although it exists in two forms (I and II).

An isolant that is anaerogenic, produces an alkaline slant and an acid reaction in the butt, fails to give evidence of hydrogen sulfide production in TSI or KI agar medium, yields a typical reaction in LIA (Table 2), and fails to hydrolyze urea rapidly should be considered as a possible *Shigella* and treated accordingly. Dense suspensions prepared in 0.5% sodium chloride solution or in mercuric iodide solution should be examined by slide tests for agglutination in absorbed polyvalent antisera (Chapter 6) for each of the four species of shigellae. Since certain anaerogenic, nonmotile biotypes of *E. coli,* i.e., members of the so-called Alkalescens-Dispar (A-D) group, particularly A-D 01, occur frequently and resemble shigellae on TSI agar, the use of a polyvalent antiserum for these microorganisms also is recommended. Cultures that are agglutinated rapidly and completely in one of the polyvalent antisera should be subjected to additional biochemical tests. If, after 20 to 24 hr of incubation, the results of these tests are compatible with those given by shigellae, a preliminary report may be made indicating the presumptive identification. If a suspension fails to react in one or another of the aforementioned polyvalent antisera, it should be heated in a beaker of boiling water for about 15 min, cooled, and retested in the same antisera. Many shigellae possess envelope or capsular antigens which inhibit agglutination of living (unheated) bacteria in O antisera. These antigens are inactivated by heat.

Certain biotypes of *S. flexneri* 6 (Newcastle and Manchester) produce small amounts of gas from fermentable substrates, and sometimes enough gas is evolved to become apparent in TSI and LI agar media.

A suspension that is not agglutinated in any of the polyvalent antisera for *Shigella* or in the A-D polyvalent antiserum should be tested in polyvalent and grouping antisera for salmonellae. In particular, such a suspension should be tested in group D and Vi antisera, since it may be a strain of *S. typhi* that fails to produce hydrogen sulfide.

Cultures that give the appearance of shigellae in TSI agar media, are urease negative, and fail to agglutinate in any of the antisera for shigellae mentioned in the preceding paragraphs should be subjected to additional biochemical tests. Such strains may be shigellae, agglutinins for which are not contained in the polyvalent antisera, or they may be members of another genus, e.g., *Providencia, Escherichia,* or *Salmonella.* The use of LIA in the manner outlined above should prove to be helpful in this connection.

Regardless of the outcome of serological tests, cultures suspected of being shigellae must be subjected to minimal biochemical tests at least (Chapter 3), since the O antigens of almost all of the serotypes of *Shigella* are identical with, or closely related to, those of various *E. coli* (see Chapter 6).

Cultures that are to be forwarded to a central or reference laboratory for additional study should be inoculated into stock culture medium, packed properly, and sent along with minimal information (see footnote, page 16).

REFERENCES

Amies, C. R. 1967. Can. J. Pub. Health, **58**, 296.

Amies, C. R., and J. I. Douglas. 1965. Ibid., **47**, 27.

Armstrong, E. C. 1954. Mo. Bul. Min. Health (Gt. Brit.). **13**, 70.

Bailey, W. R., and E. T. Bynoe. 1953. Can. J. Pub. Health, **44**, 468.

Banwart, G. J., and J. C. Ayres. 1953. Appl. Microbiol., **1**, 296.

Branham, S. E., S. A. Carlin, and D. B. Riggs, 1952. Amer. J. Pub. Health, **42**, 1409.

Cary, S. G., and E. B. Blair. 1964. J. Bacteriol., **88**, 96.

Cary, S. G., M. H. Fusillo, and C. Harkins. 1965. Amer. J. Clin. Pathol., **43**, 294.

Christensen, W. B. 1946. J. Bacteriol., **52**, 461.

Clark, C. W., and Z. J. Ordal. 1969. Appl. Microbiol., **18**, 332.

Colichon, H. 1953. Amer. J. Clin. Pathol., **23**, 506.

Cook, G. T., B. R. Frisbie, and W. H. H. Jebb. 1951. Mo. Bul. Min. Health (Gt. Brit.), **10**, 89.

Cooper, G. N. 1957. J. Clin. Pathol., **10**, 226.

Croft, C. C., and M. J. Miller. 1956. Amer. J. Clin. Pathol., **26**, 411.

Dold, H., and M. Ketterer. 1944. Z. f. Hyg., **125**, 444.

Edwards, P. R., and M. A. Fife. 1961. Appl. Microbiol., **9**, 478.

Elek, S. D. 1948. J. Pathol. and Bacteriol., **60**, 183.

Ewing, W. H. 1966. Differential reactions of Enterobacteriaceae. CDC Publ.*

_____. 1968. Transport methods for Enterobacteriaceae and allied bacteria. Formulas, summaries, and references. CDC Publ.*

Ewing, W. H., and D. W. Bruner. 1947. J. Bacteriol., **53**, 362.

Ewing, W. H., A. C. McWhorter, and T. S. Montague. 1966. Pub. Health Lab.,** **24**, 63.

Ferguson, W. W., and A. E. Hook. 1943. J. Lab. and Clin. Med., **28**, 1715.

Floyd, T. M. 1954. Amer. J. Trop. Med. Hyg., **3**, 294.

Galton, M. M., and M. S. Quan. 1944. Amer. J. Pub. Health, **34**, 1071.

Galton, M. M., J. E. Scatterday, and A. V. Hardy. 1952. J. Infect. Dis., **91**, 1.

Gangarosa, E. J., H. Saghari, J. Emile, and H. Siadat. 1966. Bull. Wld. Health Org., **34**, 363.

Georgala, D. L. and M. Boothroyd. 1965. Can. J. Res., **22**, 48.

Gillies, R. R. 1956. J. Clin. Pathol., **9**, 368.

Hajna, A. A. 1955a. Pub. Health Lab.,** **13**, 59.

_____. 1955b. Ibid., **13**, 83.

Hardy, A. V., D. C. Mackel, D. Frazier, and D. Hamerick. 1953. Armed Forces Med. J., **4**, 393.

Harvey, R. W. S., and S. Thomson. 1953. Mo. Bull. Min. Health (Gt. Brit.), **12**, 149.

Hobbs, B. C., and V. D. Allison. 1945. Mo. Bul. Min. Health (Gt. Brit.), **4**, 12, 40, 63.

Holman, W. L. 1919. J. Bacteriol., **4**, 149.

Johnson, J. G., L. J. Kunz, W. Baron, and W. H. Ewing. 1966. Appl. Microbiol., **14**, 212.

Kauffmann, F. 1930. Zentralbl. f. Bakt. I, Orig., **119**, 148.

_____. 1935. Z. f. Hyg., **117**, 26.

Kelterborn, E. 1967. Salmonella species. The Hague: Dr. W. Junk.

Knox, R., L. G. H. Gell, and M. R. Pollock. 1943. J. Hyg., **43**, 147.

Kristensen, M., V. Lester, and A. Jurgens. 1925. Brit. J. Exper. Pathol., **6**, 291.

Leifson, E. 1936. Amer. J. Hyg., **24**, 423.

Lie, Kian Joe. 1960. Nederl. Tijdschr. v. Geneesk., **94**, 1246. (Bull. Hyg. **25**, 1064.)

McCall, C. E., W. T. Martin, and J. R. Boring. 1966. J. Hyg., **64**, 261.

Monteverde, J. J., C. A. Hermida, N. Moran, and D. H. Simeone. 1968. Rev. Facult. Agron. Veterinar. Buenos Aires, **17**, 39.

Moeller, V. 1954. Acta Pathol. Microbiol. Scand., **34**, 115.

Muller, L. 1923. Compt. rend. Soc. de biol., **89**, 434.

Nagel, V. 1950. Ärztl. Wchnschr., 5, Heft. 51.

North, W. R., and M. T. Bartram. 1953. Appl. Microbiol., **1**, 130.

Papadakis, J. A. 1960. J. Hyg., **58**, 331.

*CDC Publ., Publication from the Center for Disease Control (formerly the National Communicable Disease Center), Atlanta, Ga. 30333.

**Public Health Laboratory: Journal of the Conference of Public Health Laboratory Directors.

Rappaport, F., N. Konforti, and B. Navon. 1956. J. Clin. Pathol., **9**, 261.

Rappaport, F., G. J. Stark, and N. Konforti. 1956. Appl. Microbiol., **4**, 157.

Rolfe, V. 1946. Mo. Bul. Min. Health (Gt. Brit.), **5**, 158.

Rustigian, R., and C. A. Stuart. 1941. Proc. Soc. Exper. Biol. and Med., **47**, 108.

————. 1945. J. Bacteriol., **49**, 419.

Sachs, A. 1939. J. Roy. Army Med. Corps, **73**, 235.

Salmonella, Committee on. 1969. An evaluation of the Salmonella problem, Publ. 1683, National Academy of Sciences–National Research Council. Washington, D. C.

Shaughnessy, H. J., F. Friewer, and A. Snyder. 1948. Amer. J. Pub. Health, **38**, 670.

Shipe, E. L., Jr., A. Fields, and J. R. Shea. 1960a. Pub. Health Lab.,** **18**, 95.

————. 1960b. Ibid. **18**, 104.

Silliker, J. H., and Taylor, W. I. 1958. Appl. Microbiol., **6**, 228.

Silliker, J. H., R. H. Deibel, and P. T. Fagan. 1964. Ibid., **12**, 110.

Smith, H. W. 1952. J. Hyg., **50**, 21.

————. 1959. Ibid., **57**, 266.

Spino, D. F. 1966. Appl. Microbiol., **14**, 591.

Stuart, C. A., E. Van Stratum, and R. Rustigian. 1945, J. Bacteriol., **49**, 437.

Stuart, R. D. 1956. Can. J. Pub. Health, **47**, 114.

————. 1959. Pub. Health Rpts., **74**, 431.

Taylor, W. I., J. H. Silliker, and H. P. Andrews. 1958. Appl. Microbiol., **6**, 190.

Taylor, W. I., and D. Schelhart. 1969. Ibid., **18**, 393.

Teague, O., and A. W. Clurman. 1916. J. Infect. Dis., **18**, 653.

TenBroeck, C., and F. G. Norbury. 1916. Boston Med. Surg. J., **174**, 785.

Thomas, J. C., K. C. Watson, and A. S. Hewstone. 1954. J. Clin. Pathol., **7**, 50.

Thomas, M. E. M. 1954. Brit. Med. J., ii, 394.

Thomson, S. 1954. J. Hyg., **52**, 67.

————. 1955. Ibid., **53**, 217.

Vassiliadis, P. 1968. J. Appl. Bacteriol., **31**, 367.

Velaudapillai, T. 1957. Proc. Ann. Meet. Soc. of Pathol. Gt. Brit.

Wheeler, K. M., and F. L. Mickel. 1945. J. Bacteriol., **51**, 257.

Wilson, W. J., and E. M. McV. Blair. 1926. J. Pathol. Bacteriol., **29**, 310.

————. 1927. J. Hyg., **26**, 374.

————. 1931. Ibid., **31**, 139.

Zschucke, J. 1951. Zentralbl. f. Bakt. I. Orig., **157**, 65.

Chapter 3
Differentiation of Enterobacteriaceae by Biochemical Reactions

In Chapter 2 it was mentioned that a study of their biochemical reactions is the only method available at present for recognition and differentiation of the genera and species to which cultures of Enterobacteriaceae belong. While it is convenient, even desirable in many instances, to perform serological examinations at an early stage, identification should not be made on the basis of serological tests alone.

Further, it was emphasized in Chapter 2 that isolation and preliminary identification of members of genera of Enterobacteriaceae other than *Salmonella* and *Shigella* also may be accomplished by the methods outlined, and that this applies to specimens from extraintestinal sources as well as stools. However, members of other families also may be recovered from one or another of the isolation media mentioned and listed in Figure 1 (Chapter 2). These bacteria not only must be differentiated from Enterobacteriaceae, but also should be identified when warranted by circumstances. Microorganisms such as pseudomonads (including *Pseudomonas alcaligenes*[1]), *Mima* and *Herellea*, aeromonads, and *Vibrio* may be mistaken for members of one or another of the genera of Enterobacteriaceae. Tests for reduction of nitrate, use of O-F medium (Hugh and Leifson, 1953), the oxidase test, decarboxylast tests, gram stains, and flagellar stains all are useful in the differentiation of the above-mentioned bacteria from Enterobacteriaceae, and from each other. Members of the genera *Aeromonas* and *Vibrio* utilize glucose fermentatively, but are oxidase positive as reported by Ewing and Johnson (1960), and members of these two genera yield different patterns of reactions in decarboxylase tests (Ewing et al., 1960). For additional information regarding the above-mentioned bacteria, readers are referred to the publications of Ewing et al. (1961, 1966), and to the chapters on *Pseudomonas* and *Vibrio* and on Miscellaneous Gram Negative Bacteria in Diagnostic Procedures for Bacterial, Mycotic, and Parasitic Infections, edited by Bodily, Updyke, and Mason (1970) or *Manual of Clinical Microbiology,* edited by Blair, Lennette, and Truant

[1] In the past, most bacteriologists have referred to this microorganism as *Alcaligenes faecalis.*

(1970).

In passing, strains of some bacteria that utilize glucose oxidatively, or are inactive toward it (e.g., certain pseudomonads, *Herellea*, etc.) may be confused with Enterobacteriaceae because of increased alkalinity in the area of the slant on TSI or KI agar. This may give the impression that acid has been produced in the butt of the medium. If cultures of this sort are compared with an uninoculated tube of the same medium, it is observed that there is no change in the butt of the medium.

Recommended or standard methods for preparation of media and use of various substrates and tests in the study of Enterobacteriaceae and certain allied bacteria are given in detail in Chapter 18. Where possible, alternate methods, which may be expected to yield comparable results, are outlined in that chapter. However, there are a few methods that should be emphasized at this point. Hydrogen sulfide production by Enterobacteriaceae should be judged by the action of the bacteria on TSI, KI, or peptone iron (PI) agar medium, since blackening of these media, or lack of it, corresponds closely to classical descriptions of members of the genera of the family. *Lead acetate papers should not be used for determination of hydrogen sulfide production by Enterobacteriaceae.* Methods that utilize lead acetate paper strips are overly sensitive, and frequently yield positive results with bacteria such as shigellae, for example. Tests for production of indol are made with Kovacs' reagent after the microorganisms have been cultivated for approximately 48 hr in 2% Bacto peptone water, or in a peptone water medium having a similar content of available tryptophane. The two-day incubation period is recommended for routine use, and this recommendation is based upon comparative studies made many years ago (unpublished data). However, the observation has been confirmed by Suassuna and Suassuna (1963). Under certain conditions tests may be made earlier or later, as mentioned in Chapters 2 and 18. *Motility of Enterobacteriaceae is determined by use of a semisolid agar medium.* This is a far more accurate method than direct microscopic examination. A medium that contains 0.4 per cent agar (Chapter 18) is recommended for general purposes. Cultures that appear nonmotile after overnight incubation at 35 to 37 C should be allowed to remain at room temperature (about 25 C) for two or three additional days, since some strains of certain genera may exhibit slight motility at the lower temperature. While the semisolid medium used for reversal of phases of cultures of *Salmonella* and *Arizona* may be used as a motility medium with strains that are actively motile at 35 to 37 C, it cannot be recommended for general use. This medium is a modification (Edwards and Bruner, 1942) of one described by Jordan et al. (1934). It contains gelatin, which becomes very viscid or may solidify when cultures in it are incubated at ambient temperatures in many parts of the world. This inhibits motility and defeats the purpose.

Tests for decarboxylation of lysine and ornithine, and for arginine dihydrolase, are of considerable value for differentiation in almost all areas of the family Enterobacteriaceae, and are helpful in characterization of certain genera within other families as well. Therefore, the use of all three of these tests, together with a control, is urged. The media devised by Moeller (1955) are regarded as standard. The reactions given by the genera and species of Enterobacteriaceae are listed in the tables that follow and in the individual chapters.

The tribe to which a member of the family Enterobacteriaceae belongs generally can be determined by employment of the substrates and tests listed in Table 3.

The majority of Enterobacteriaceae can be identified by means of the commonly employed tests and substrates listed in Table 4. The summary given in Table 4 is elaborated upon in Tables, 5, 6, 7, 8, and 9, in which percentage data are included. The tests and media recorded in Tables 5 to 9, inclusive, are regarded as essential for the family Enterobacteriaceae as a whole. However, in many instances it may not be necessary to perform all of the tests given in the above-mentioned tables for identification of a particular culture. That is, some tests obviously are of greater value in certain areas of the family than they are in others. In other cases it may be necessary to use additional tests for differentiation and identification. Biochemical tests that are of particular value in differentiation of members of the several genera of Enterobacteriaceae are listed in Tables 10 to 26, inclusive, together with percentage data. With the exception of Table 22, which has not been published previously in its present format, all of the aforementioned tables are adapted from publications by Ewing (1968, 1969). Readers are referred to the 1968 report for information regarding sources of the data in the tables. Table

22 is included because cultures are incubated at 35 to 37 C — in most medical and public health laboratories, in preliminary work at least — and because of the increased interest in the occurrence of *Klebsiella, Serratia,* etc. in hospital-acquired (nosocomial) infections.

The data presented in the tables are self-explanatory, but some additional comment about the signs, and the bases for their use, might be helpful. It is the author's opinion that unless percentages are included in tabular data, the resulting tables are of little value. As in the past, 90 per cent levels are employed in determining the signs to be applied. Some arbitrary level must be selected, and the author and colleagues believe that the 90 per cent level is the most reasonable and practical. Thus a + sign means that 90 to 100 per cent of strains tested on a particular substrate give positive results within one or two days of incubation. The addition of the actual percentage of such reactions (e.g., 91 per cent or 98 per cent) yields useful information. Conversely the same is true of negative results. Further, if percentages are included, the sign "d" (for different reactions) assumes meaning in many instances. For example, if 60 to 90 per cent of results obtained in a test are positive, then the "d" sign is meaningful since the majority of cultures yield positive results. Similarly, the "d" sign is useful if the majority of cultures give negative reactions on a particular substrate. When the percentages obtained are in the area between about 30 and 60 per cent the "d" sign usually indicates that tests that yield such figures are of little differential value, but this cannot be determined if the actual percentages of positive results are not listed. However, there are instances in which a "d" sign followed by about 40 per cent positive, for example, is of value as in the case of the differentiation of *S. cholerae-suis,* inositol, -:0%+, and *S. enteritidis,* inositol, d:39.1%+, or in the differentiation of commonly occurring salmonellae from members of the genus *Arizona*.

For many years the author and colleagues have categorized delayed reactions (+) according to whether they occurred within 3 to 7 days of incubation, 8 to 14 days, or 15 to 30 days. This has been done in all publications written during the past ten or more years (e.g. Ewing et al., 1965; Fife et al., 1965; Ewing, 1965; Martin et al., 1969). The reason for categorizing delayed reactions in the aforementioned manner was to determine the importance, or value, of incubating carbohydrate media, gelatin, etc., for periods longer than 14 days in most laboratories. All results obtained thus far indicate that with the exception of nutrient gelatin and perhaps lactose, incubation for periods longer than 14 days is unnecessary *in most labora-*

tories. In fact, essential results generally are available at 7 days. Obviously these statements do not apply to research or other laboratories in which complete characterization of individual strains or of collections is done.

The biochemical reactions of members of each genus are given in more detail in the individual chapters. In all instances, percentage data are included.

Numerous systems have been proposed by other investigators, e.g., Marymont et al. (1966), Carpenter et al. (1966), Douglas et al. (1966), Belliveau et al. (1968), Roundtable: How Far to Go with Enterobacteriaceae? (1969), and Domingue et al. (1969). All of these have merit and some workers, particularly those in hospital laboratories, may find some of the suggestions in the publications cited useful or helpful. Routine use of the ONPG test (see Chapter 18) is suggested in at least two of the papers cited above. The author sees no objection to this *provided* that the results obtained with the ONPG test are interpreted very carefully and as one part of the complete pattern of reactions given by a particular culture.

TABLE 3

Differentiation of the Tribes of *Enterobacteriaceae* By Biochemical Methods

	TRIBES				
Substrate or test	Escherichieae	Edwardsielleae	Salmonelleae	Klebsielleae	Proteeae
Hydrogen sulfide*	–	+	+	–	+ or –
Urease	–	–	–	– or (+)	+ or –
Indol	+ or –	+	–	–	+ or –
Methyl red	+	+	+	–	+
Voges-Proskauer	–	–	–	+	–
Citrate (Simmons')	–	–	+	+	d
KCN	–	–	– or +	+	+
Phenylalanine deaminase	–	–	–	–	+
Tartrate (Jordan's)	+ or –	–	d	+ or –	+ or –
Mucate	d	–	d	+ or –	–
Mannitol	+ or –	–	+	+	– or +

* TSI, KI, or PI agar (see Chapter 18)

NOTE: *Salmonella typhi, Salmonella enteritidis* bioserotype Paratyphi-A, and some rare biotypes fail to utilize citrate. Cultures of *S. enteritidis* bioser Paratyphi-A and some rare biotypes may fail to produce hydrogen sulfide. Some cultures of *P. mirabilis* may yield a positive Voges-Proskauer test.

   +   90% or more positive within 1 or 2 days.
   –   90% or more, no reaction.
   (+) delayed positive 3 or more days.
   d   different biochemical reactions: +, (+), –.
+ or – majority of strains +, some cultures negative.
– or + majority of cultures negative, some strains positive.

TABLE 4

Differentiation of Enterobacteriaceae by Biochemical Tests

TEST or SUBSTRATE	Escherichia	Shigella	Edwardsiella	Salmonella	Arizona	Citrobacter	Klebsiella	cloacae	aerogenes	hafniae 37C	hafniae 22C	liquefaciens 37C	liquefaciens 22C	Serratia	Pectobacterium 25C	vulgaris	mirabilis	morganii	rettgeri	alcalifaciens	stuartii
INDOL	+	- or +	+	-	-	-	- or +	-	-	-	-	-	-	-	- or +	+	-	+	+	+	+
METHYL RED	+	+	+	+	+	+	-	-	-	+ or -	-	+ or -	-	- or +	+ or -	+	+	+	+	+	+
VOGES – PROSKAUER	-	-	-	-	-	-	+	+	+	+ or -	+	- or +	+	+	- or +	-	- or +	-	-	-	-
SIMMONS'S CITRATE	-	-	-	d	+	+	+	+	+	+ or -	d	- or +	+	+	d	d	+ or (+)	-	+	+	+
HYDROGEN SULFIDE (TSI)	-	-	+	+	+	+ or -	-	-	-	-	-	-	-	-	-	+	+ or (+)	-	-	-	-
UREASE	-	-	-	-	-	d^w	+	+ or -	-	-	-	d	-	d^w	d^w	+	+	+	+	-	-
KCN	-	-	-	-	-	+	+	+	+	+	+	+	+	+	+ or -	+	+	+	+	+	+
MOTILITY	+ or -	-	+	+	+	+	-	+	+	+	+	d	+	+	+ or -	+	+	+	+	+	-
GELATIN (22 C)	-	-	-	-	(+)	-	+	(+) or -	- or (+)	-	-	+ or -	+	+	+ or (+)	+ or (+)	+	-	-	-	-
LYSINE DECARBOXYLASE	d	-	+	+	+	-	+	-	+	+	+	+ or -	-	+	- or +	-	-	-	-	-	-
ARGININE DIHYDROLASE	d	- or (+)	-	(+) or +	+ or (+)	d	-	+	- or (+)	-	-	-	-	-	- or +	-	-	-	-	-	-
ORNITHINE DECARBOXYLASE	d	d(1)	+	+	+	d	-	+	+	+	+	+	+	+	-	-	+	+	-	-	+
PHENYLALANINE DEAMINASE	-	-	-	-	-	-	-	-	-	-	-	-	-	-	-	+	+	+	+	+	+
MALONATE	-	-	-	-	+	d	+	+ or -	+ or-	+ or -	+ or -	-	+	-	- or +	-	-	-	-	-	+
GAS FROM GLUCOSE	+	-(1)	+	+	+	+	+	+	+	+	+	+	+	+ or -(3)	- or +	+ or -	+	d	- or +	+ or -	-
LACTOSE	+	-(1)	-	-	d	d	+	+ or (+)	+	× or (+)	- or (+)	d	(+)	- or (+)	- or +	-	-	-	- or +	-	-
SUCROSE	d	-	-	-	d	d	+	+	+	d	d	d	(+)	+	d	+	d	-	d	d	d
MANNITOL	+	+ or -	-	+	+	+	+	+	+	+	+	+	+	+	+	+	-	-	+	+	+
DULCITOL	d	d	-	d(2)	-	d	- or +	- or (+)	-	-	-	-	-	-	-	-	-	-	-	+ or -	d
SALICIN	d	-	-	-	d	d	+	+ or (+)	+	d	d	+	- or (+)	+	+	d	d	-	d	-	-
ADONITOL	-	-	-	-	-	-	+ or -	- or +	+	-	-	d	(+)	d	-	-	-	-	d	+	-
INOSITOL	-	-	-	d	-	-	+	d	+	-	-	+	+	+	-	-	-	-	+	d	d
SORBITOL	+	d	-	+	+	+	+	+	+	+	+	+	+	+	+	-	-	-	+	-	d
ARABINOSE	+	d	-	+(2)	+	+	+	+	+	+	+	+	+	-	+	-	-	-	-	-	-
RAFFINOSE	d	d	-	-	-	d	+	+	+	+	+	+	+	-	+ or (+)	-	-	-	-	-	-
RHAMNOSE	d	d	+	+	+	d	+	+	+	+	+	+	+	-	d	-	-	+ or -	+ or -	-	-

(1) Certain biotypes of S. flexneri produce gas; S. sonnei cultures ferment lactose and sucrose slowly and decarboxylate ornithine.
(2) S. typhi, S. cholerae-suis, S. enteritidis bioser. Paratyphi A and Pullorum, and a few others ordinarily do not ferment dulcitol promptly. S. cholerae-suis does not ferment arabinose.
(3) Gas volumes produced by cultures of Serratia, Proteus, and Providencia are small.

+, 90 percent or more positive in 1 or 2 days. —, 90 percent or more negative. d, different biochemical types [+, (+), –]. (+), delayed positive. + or –, majority of cultures positive.
– or +, majority negative. w, weakly positive reaction.

TABLE 5

Reactions of Members of the Tribes ESCHERICHIEAE and EDWARDSIELLEAE

Test or Substrate	ESCHERICHIEAE						EDWARDSIELLEAE		
	Escherichia			Shigella			Edwardsiella		
	Sign	%+	(%+)*	Sign	%+	(%+)*	Sign	%+	(%+)*
Indol	+	98.6		– or +	39.8		+	99.1	
Methyl red	+	99.9		+	100		+	100	
Voges-Proskauer	–	0		–	0		–	0	
Simmons' citrate	–	0.2	(0.4)	–	0		–	0	
Hydrogen sulfide (TSIA)	–	0		–	0		+	99.7	(0.3)
Urease	–	0		–	0		–	0	
KCN	–	2.4		–	0		–	0	
Motility	+ or –	69.1		–	0		+	98.2	
Gelatin (22C)	–	0		–	0		–	0	(0.3)
Lysine decarboxylase	d	88.7	(0.8)	–	0		+	100	
Arginine dihydrolase	d	17.6	(33.2)	– or (+)	9.5	(17.3)	–	0	(0.3)
Ornithine decarboxylase	d	64.2	(5.3)	d	20	(0.3)	+	100	
Phenylalanine deaminase	–	0		–	0		–	0	
Malonate	–	0		–	0		–	0	
Gas from glucose	+	91.1		–[1]	2.1		+	99.4	
Lactose	+	90.8	(5.1)	–[1]	0.3	(11.4)	–	0	
Sucrose	d	48.9	(5.6)	–[1]	0.9	(31.1)	–	0.3	
Mannitol	+	96.8		+ or –	80.5		–	0	
Dulcitol	d	49.5	(15)	d	5.4	(12.7)	–	0	
Salicin	d	40	(14)	–	0		–	0	(0.3w)
Adonitol	–	5.6	(0.7)	–	0		–	0	
Inositol	–	1.1	(0.2)	–	0		–	0	
Sorbitol	+	93.4	(1.1)	d	29.1	(21.9)	–	0.3	
Arabinose	+	99.4	(0.3)	d	67.8	(11.2)	–	9.3	(0.3)
Raffinose	d	50.9	(1)	d	20.7	(19.1)	–	0	
Rhamnose	d	82.3	(2.5)	d	16.6	(6.1)	–	0	

* Figures in parenthese indicate delayed reactions (3 days or more).

[1] Certain biotypes of S. flexneri 6 produce gas; S. sonnei cultures ferment lactose and sucrose slowly and decarboxylate ornithine.

+, 90% or more positive in 1 or 2 days. –, 90% or more negative. d, different biochemical types [+, (+), –] . (+), delayed positive, + or –, majority of cultures positive. – or +, majority negative. w, weakly positive reaction.

TABLE 6

Reactions of Members of the Tribe SALMONELLEAE

Test or Substrate	SALMONELLEAE								
	Salmonella			Arizona			Citrobacter		
	Sign	%+	(%+)*	Sign	%+	(%+)*	Sign	%+	(%+)*
Indol	-	1.1		-	2		-	6.7	
Methyl red	+	100		+	100		+	99.5	
Voges-Proskauer	-	0		-	0		-	0	
Simmons' citrate	d	80.1	(7)	+	98.7	(1.3)	+	90.4	(4.3)
Hydrogen sulfide (TSIA)	+	91.6	(0.8)	+	98.7		+ or -	81.6	(2.4)
Urease	-	0		-	0		dw	69.4	(6.9)
KCN	-	0.3	(0.3)	-	8.7		+	96.2	(0.9)
Motility	+	94.6		+	100		+	95.7	
Gelatin (22C)	-	0	(1.1)	(+)	4	(92)	-	0	(0.9)
Lysine decarboxylase	+	94.6		+	100		-	0	
Arginine dihydrolase	+ or (+)	58.5	(34)	(+) or +	12.7	(84.6)	d	43.6	(44.4)
Ornithine decarboxylase	+	92.7		+	100		d	17.2	(0.2)
Phenylalanine deaminase	-	0		-	0		-	0	
Malonate	-	0.5		+	92.6	(0.7)	d	21.8	(0.7)
Gas from glucose	+	91.9		+	99.3		+	90.9	
Lactose	-	0.8		d	61.3	(16.7)	(+) or +	39.4	(50.8)
Sucrose	-	0.5		-	4.7		d	15.3	(9.4)
Mannitol	+	99.7		+	100		+	99.8	(0.2)
Dulcitol	d¹	86.5	(2.7)	-	0		d	59.4	(0.7)
Salicin	-	0.8		-	4.7	(3.3)	d	4.1	(23)
Adonitol	-	0		-	0		-	0	
Inositol	d	34.5	(0.8)	-	0		-	3.3	(1.9)
Sorbitol	+	94.1	(4)	+	97	(2)	+	98	(1)
Arabinose	+ or (+)¹	89.2	(0.8)	+	98	(1)	+	100	
Raffinose	-	3	(0.3)	-	5	(1)	d	14.3	(0.8)
Rhamnose	+	90.3	(1.1)	+	93	(5)	+	99.4	

* Figures in parentheses indicate delayed reactions (3 days or more).

¹ S. typhi, S. cholerae-suis, S. enteritidis bioser Paratyphi-A and Pullorum, and a few others ordinarily do not ferment dulcitol promptly. S. cholerae-suis does not ferment arabinose.

+, 90% or more positive in 1 or 2 days. -, 90% or more negative. d, different biochemical types [+, (+), -]. (+), delayed positive, + or -, majority of cultures positive. - or +, majority negative. w, weakly positive reaction.

TABLE 7

Reactions of Members of the Tribe KLEBSIELLEAE

Test or Substrate	Klebsiella pneumoniae			Enterobacter cloacae			Enterobacter aerogenes			Enterobacter hafniae 37C			Enterobacter hafniae 22-25C		
	Sign	%+	(%+)*	Sign	%+	(%+)*	Sign	%+	(%+)*	Sign	%+	(%+)*	Sign	%+	(%+)*
Indol	−	6		−	0.5		−	0		−	0		−	0	
Methyl red	− or +	13.3		−	0.3		−	0		+ or −	54		−	1	
Voges-Proskauer	+	91.1		+	99.5		+	100		+ or −	65		+	99	
Simmons' citrate	+	97.7		+	99.5		+	93.7		(+) or −	0	(58)	d	3	(79)
Hydrogen sulfide (TSIA)	−	0		−	0		−	0		−	0		−	0	
Urease	+	94.5		+ or −	64.7		−	2.7		−	3w		−	5.4	
KCN	+	97.9		+	98		+	98.7		+	97		+	100	
Motility	−	0		+	94.5		+	97.3		+	93		+	98.6	
Gelatin (22C)	−	3.3		(+) or −	1	(96)	(+) or −	0	(77.3)				−	0	
Lysine decarboxylase	+	97.2		−	0.5		+	98.7		+	100		+	100	
Arginine dihydrolase	−	0.9		+	96.5		−	0		−	9		−	5	
Ornithine decarboxylase	−	0		+	96		+	98.7		+	100		+	100	
Phenylalanine deaminase	−	0		−	0		−	0		−	0		−	0	
Malonate	+	92.5		+ or −	80.6		+ or −	74.7		+ or −	74		d	70	(6)
Gas from glucose	+	96.5		+	100		+	100		+	100		+	100	
Lactose	+	98.2	(1.4)	+	93.5	(5.5)	+	92.1	(5.3)	− or (+)	0	(23)	−	0	(8.5)
Sucrose	+	98.9	(0.1)	+	96.5	(0.5)	+	100		d	12		− or (+)	0	(21)
Mannitol	+	100		+	100		+	100		+	100		+	100	
Dulcitol	− or +	31.5		− or +	12.9		−	4		−	1		−	0	(6.1)
Salicin	+	99.7	(0.3)	− or (+)	75.6	(18.4)	+	98.7	(1.3)	d	13	(8)	d	2	(12.6)
Adonitol	+ or −	87.7		− or +	28.4		+	98.7		−	0		−	0	
Inositol	+	97.9	(0.8)	d	21.9	(12.4)	+	100		−	0		−	0	
Sorbitol	+	99.4	(0.3)	+	94.5	(0.5)	+	100		−	0		−	0	
Arabinose	+	99.9		+	99.5		+	100		+	96		+	100	
Raffinose	+	99.7		+	97		+	96		−	0		−	0	
Rhamnose	+	99.3		+	92	(1.5)	+	98.7		+	93	(7)	+	91	

* Figures in parenthese indicate delayed reactions (3 days or more).

+, 90% or more positive in 1 or 2 days. −, 90% or more negative. d, different biochemical types [+, (3), −,] . (+), depayed positive. − or +, majority of cultures positive. − or +, majority negative. w, weakly positive reaction.

TABLE 8

Reactions of Members of the Tribe KLEBSIELLEAE

Test or Substrate	Enterobacter liquefaciens						Pectobacterium						Serratia		
	37C			22-25C			37C			22-25C					
	Sign	%+	(%+)*	Sign	%+	(%+)*	Sign	%+	(%+)*	Sign	%+	(%+)*	Sign	%+	(%+)*
Indol	-	0		-	0		- or +	20		- or +	21.4		-	0.2w	
Methyl red	+ or -	75		- or +	33.3		- or +	45.7		+ or -	75.7		- or +	17.7	
Voges-Proskauer	- or +	30.9		+ or -	79.4		- or +	22.9		- or +	47.1		+	100	
Simmons' citrate	+	91.2	(7.3)	+	96.3		d	45.7	(17.1)	+ or (+)	71.5	(20)	+	98.6	(0.5)
Hydrogen sulfide (TSIA)	-	0		-	0		-	0		-	0		-	0	
Urease	d	4.4	(19.1)				d	21.4	(7.1)	dw	30	(20)	dw	29.1	(26.8)
KCN	+	98.5					+ or -	56.1		+ or -	81.4		+	99.1	
Motility	+	97.1		+	100		+ or -	55.7		+ or -	87.2		+	98.6	
Gelatin (22C)	+ or -	82.4		+	98.5	(1.5)				+ or (+)	67.1	(32.9)	+ or (+)	86.9	(13.1)
Lysine decarboxylase	+	100		+	100		-	0		-	0		+	99.6	
Arginine dihydrolase	-	4.4		-	0		-	5.7		- or (+)	0	(12.9)	-	1.3w	
Ornithine decarboxylase	+	98.5		+	100		-	0		-	0		+	99.4	(0.1)
Phenylalanine deaminase	-	1.5					-	0		-	0		-	2.7w	
Malonate	-	7.4					- or +	21.4		- or +	22.9		-	1.7	
Gas from glucose	+	94.1	(1.5)				- or +	17.1		d	18.6	(25.7)	+ or -[1]	52.6	
Lactose	d	1.5	(29.4)	(+)	1.6	(91.6)	d	52.9	(18.5)	+ or (+)	75.7	(18.6)	-	2.2	(6.2)
Sucrose	+	100					+ or -	74.3		+	97.1	(2.9)	+	99.7	
Mannitol	+	100		+	100		+ or -	75.7	(11.4)	+	97.1	(2.9)	+	100	
Dulcitol	-	0					-	0		-	0		-	0	
Salicin	+	100					d	70	(1.4)	+	100		+	95.1	(1.7)
Adonitol	d	8.8	(2.9)				-	0		-	0		d	55.5	(16.4)
Inositol	+	97	(1.5)	+	97.9		d	4.3w	(11.4)	-	1.4	(4.3)	d	78.5	(8.2)
Sorbitol	+	97					-	2.8		-	1.4		+	98.3	(1.7)
Arabinose	+	92.6		+	94.6		d	70	(1.4)	+	95.7	(4.3)	-	0	
Raffinose	d	86.8	(2.9)	+	90.9	(7.2)	d	65.7	(2.9)	+ or (+)	84.3	(12.8)	-	1.7	(1.2)
Rhamnose	-	0					d	64.3	(5.7)	d	71.4	(14.3)	-	0	(0.3)

* Figures in parenthese indicate delayed reactions (3 days or more).

[1] Volumes of gas produced by Serratia are small (small bubble to 10%).

+, 90% or more positive in 1 or 2 days. -, 90% or more negative. d, different biochemical types [+, (+), -]. (+), delayed positive. + or -, majority of cultures positive.

- or +, majority negative. w, weakly positive reaction.

TABLE 9

Differentiation of Enterobacteriaceae by Biochemical Tests: Tribe PROTEEAE

Test or Substrate	Proteus								Providencia			
	vulgaris		mirabilis		morganii		rettgeri		alcalifaciens		stuartii	
	Sign	%+ (%+)*	Sign	%+ (%+)*	Sign	%+ (%+)*	Sign	%+ (%+)*	Sign	%+ (%+)*	Sign	%+ (%+)*
Indol	+	98.2	-	1.9	+	100	+	100	+	99.5	+	98.7
Methyl red	+	93	+	98.8	+	97.1	+	93.3	+	99.8	+	100
Voges-Proskauer	-	0	- or +	15.6	-	0	-	0	-	0	-	0
Simmons' citrate	d	10.5 (14.1)	+ or (+)	58.7 (37.1)	-	0	+	95.6 (3.3)	+	97.9 (1.3)	+	95.6 (3.1)
Hydrogen sulfide (TSIA)	+	94.7	+	94.2 (2.7)	-	0	-	0	-	0	-	0
Urease	+	94.7	+ or (+)	88.4 (1.9)	+	98.2 (0.9)	+	100	-	0	-	0
KCN	+	100	+	98.6	+	99	+	96.7	+	98.8	+	98.6
Motility	+	94.7	+	95.9	+ or -	87.7	+	94.4	+	96.5	+ or -	87
Gelatin (22C)	+	90.6 (9.4)	+	91.8 (5.7)	-	0	-	0 (2.3)	-	0 (1.4)	-	0 (6.8)
Lysine decarboxylase	-	0	-	0	-	0 (1w)	-	0	-	0 (0.9w)	-	0 (1)
Arginine dihydrolase	-	0	-	0	-	0	-	0	-	0 (0.5w)	-	0
Ornithine decarboxylase	-	0	+	99.2	+	97.1	-	0	-	1.4w	-	0
Phenylalanine deaminase	+	100	+	99.6	+	95.3	+	97.8	+	97.2	+	93.7
Malonate	-	0 (7.5)	-	1.5 (8.1)	-	4.8	-	1.2	-	0.7	-	1.2
Gas from glucose	+ or -	86	+	93.4 (0.4)	d	84.9 (0.9)	- or +	12.2	d	85.8 (0.6)	-	0
Lactose	-	0	-	1.5	-	0	-	8.9 (1.1)	-	0.3	-	3.8
Sucrose	+	94.7	d	18.9 (63.3)	-	1 (2.9)	d	13.3 (56.7)	d	13 (74.2)	(+) or +	2.6 (65.8)
Mannitol	-	0	-	0	-	0	+ or -	88.5	-	2 (0.2)	d	13.3 (1.3)
Dulcitol	-	0	-	0	-	0	-	0	-	0	-	0
Salicin	d	58.2 (10.9)	d	0.8 (29.8)	-	0	d	30 (6.6)	-	0.6 (0.3)	-	1.9
Adonitol	-	0	-	0	-	0	d	80.9 (5.6)	+	94.5 (0.2)	-	3.8
Inositol	-	0	-	0	-	0	+	93.3 (4.5)	-	0.6	+	97.5 (2.5)
Sorbitol	-	0	-	0	-	0	d	1.2 (9.6)	-	0 (1.2)	d	4.1 (40.2)
Arabinose	-	0	-	0	-	0	-	0	-	0.7 (0.7)	-	5 (3.3)
Raffinose	-	0	-	1	-	0	-	9.5	-	1 (1)	-	6.2 (1.6)
Rhamnose	-	9.4	-	1.5	-	0	+ or -	67.9	-	0 (1.3)	-	0

* Figures in parentheses indicate delayed reactions (3 days or more).

+ 90% or more positive in 1 or 2 days. -, 90% or more negative. d, different biochemical types [+, (+), -]. (+), delayed positive. - or +, majority negative. + or -, majority of cultures positive.

NOTE: Gas volumes produced by cultures of Proteus and Providencia are small (a bubble to 10% or 15%).

TABLE 10

Differentiation within the Tribe ESCHERICHIEAE

Substrate or test	Escherichia			Shigella		
	Sign	%+	(%+)*	Sign	%+	(%+)*
Gas from glucose	+	90.7		–	2.1[a]	
Lactose	+	90.8	(5.1)	–	0.3	(11.4)[b]
Sucrose	d	48.9	(5.6)	–	0.9	(31.1)[b]
Salicin	d	40	(14)	–	0	
Motility	+ or –	69.1		–	0	
Indol	+	99.2		– or +	37.8	
Lysine decarboxylase	d	87.9	(1.2)	–	0	
Arginine dihydrolase	d	17.2	(44.8)	d	9.5	(17.3)
Ornithine decarboxylase	d	63.4	(7.1)	–	20[b]	
Esculin	d	30.9	(19.7)	–	0	
Sodium acetate	+ or (+)	83.9	(9.7)	–	0	
Christensen's citrate	d	24.4	(21.2)	–	0	
Mucate	+	96.3		–	0	

* Figures in parentheses indicate percentages of delayed reactions (3 days or more).
[a] Certain biotypes of S. flexneri 6 form gas.
[b] S. sonnei strains usually ferment lactose and sucrose slowly and cultures of this species decarboxylate ornithine. Some strain S. sonnei utilize mucate.

NOTE: Obviously there is no difficulty in the differentiation of typical E. coli cultures and shigellae. However, the anaerogenic nonmotile varieties of E. coli, some of which are often referred to as Alkalescens-Dispar types, may require closer examination before they can be definitely classified as E. coli. In attempting to classify a particular strain as E. coli or as a member of the genus Shigella, the biochemical reactivites of the culture should be considered as a whole. Shigellae are much less reactive than E. coli strains and a culture that produces acid promptly (i.e., within 24 hrs.) from all, or most of a wide variety of carbohydra such as maltose, rhamnose, xylose, sorbitol, and dulcitol, undoubtedly is not a member of the genus Shigella.

+ 90% or more positive within 1 or 2 days. – 90% or more, no reaction. (+) Positive reaction 3 or more days. d Different bio-chemical reactions, +, (+), –. + or – Majority of strains +, some cultures negative. – or + Majority of cultures negative, some strains positive.

TABLE 11

Differentiation of *Escherichia* and *Edwardsiella*

Substrate or test	*Escherichia*			*Edwardsiella*		
	Sign	%+	(%+)*	Sign	%+	(%+)*
Hydrogen sulfide (TSI)	–	0		+	99.7	(0.3)
Mucate	+	91.6	(1.4)	–	0	(0.3)
Tartrate (Jordan's)	+	97.6	(1.9)	–	0	
Sodium acetate	+ or (+)	83.8	(9.7)	–	0	
Mannitol	+	97		–	0	
Sorbitol	+	93.4	(0.5)	–	0.3	
Rhamnose	d	81.8	(2.8)	–	0	
Xylose	d	82.4	(6.7)	–	0	
Trehalose	+	98.6	(1)	–	0.3	

* Figures in parentheses indicate percentages of delayed reactions (3 days or more).

+ 90% or more positive within 1 or 2 days.
– 90% or more, no reaction.
(+) Delayed positive 3 or more days.
d Different biochemical reactions, +, (+), –.
+ or – Majority of strains +, some cultures negative.
– or + Majority of cultures negative, some strains positive.

TABLE 12

Differentiation within the Tribe SALMONELLEAE

Test or substrate	Salmonella			Arizona			Citrobacter		
	Sign	%+	(%+)*	Sign	%+	(%+)*	Sign	%+	(%+)*
Urease	−	0		−	0		d	69.4	(6.9)
KCN	−	0.3	(0.3)	−	8.7		+	96.2	(0.9)
Gelatin (22 C)	−		(1.3)	(+)		(92)	−		(0.9)
Lysine decarboxylase	+	97.7		+	100		−	0	
Ornithine decarboxylase	+	100		+	100		d	17.2	(0.2)
Lactose	−	1		d	61.3	(16.7)	(+) or +	39.3	(50.9)
Sucrose	−	0.7		−	4.7		d	15.3	(9.4)
Dulcitol	+	98.3		−	0		d	59.4	(0.7)
Inositol	d	42.8	(1)	−	0		−	3.3	(1.9)
Cellobiose	(+) or +	5.4	(88.1)	d	1	(72)	+ or (+)	60.8	(38)
Malonate	−	0.7		+	92.6	(0.7)	d	21.8	(0.7)
Jordan's tartrate	+	92.5	(1.1)	−	5.3		+	100	
Beta galactosidase	−	1.5		+	92.8		+ or −	74.4	
Organic acids**									
citrate	+	96	(4)	+ or (+)	78.7	(19.3)	(+) or +	49.2	(49.5)
D-tartrate	+	91	(5.3)	(+) or −		(83.3)	(+)		(90.9)

* Figures in parentheses indicate percentages of delayed reactions (3 or more days).

** Organic acid media of Kauffmann and Petersen, 1956 (see also table 14).

+ Positive within one or two days' incubation (90% or more).
(+) Positive reaction after 3 or more days.
− No reaction (90% or more).
+ or − Majority of strains positive, some cultures negative.
− or + Majority of cultures negative, occasional strains positive.
(+) or + Majority of reactions delayed, some occur within 1 or 2 days.
d Different reactions: +, (+), −.

NOTE: The majority of salmonellae ferment dulcitol promptly, but *S. typhi, S. enteritidis* bioser Paratyphi-A and Pullorum, *S. cholerae-suis,* and a few others do not. Members of the genus *Arizona* are uniformly negative on this substrate. Bioser Paratyphi-A is lysine negative. *S. typhi* is ornithine negative. Final readings of Jordan's tartrate medium should be made at 48 hours.

TABLE 13

Differentiation of species of *Salmonella*

Test or substrate	S. cholerae-suis			S. typhi			S. enteritidis		
	Sign	%+	(%+)*	Sign	%+	(%+)*	Sign	%+	(%+)*
Hydrogen sulfide (TSI)	d	60	(10)	+^{W}	94.3		+	98	
Citrate (Simmons')	(+)		(90)	-	0		+	99.3	(0.7)
Ornithine decaroboxylase	+	100		-	0		+	100	
Gas from glucose	+	100		-	0		+	97.7	
Dulcitol	d	5	(15)	- or (+)	0	(31.3)	+	98.3	
Inositol	-	0		-	0		d	42.8	(1)
Trehalose	-	0		+	100		+	100	
Arabinose	-	0		-		(6.3)	+	99.3	
Rhamnose	+	100		-	0		+	95	
Cellobiose	-	0		d	37.5		(+)	5	(92.8)
Erythritol	(+^{W}) or -		(85)	-	0		-	0.6	
Sodium acetate	- or (+^{W})		(20)	-	0		+	92.4	(2.2)
Mucate	-	0		-	0		+ or (+)	88.3	(1.7)
Stern's glycerol fuchsin	-	0		-	0		+	98.2	(0.6)
Organic acids**									
citrate	-	10		-	10		+1 or 2da	96	(4)
D-tartrate	+1 or 2da	95		+1 or 2da	87.5	(6.3)	+	91	(5.3)
i -tartrate	(+) or -	0	(85)	-	0		d	4.7	(57.5)
l -tartrate	- or (+)	0	(35)	-	0		d	11.8	(75.2)

*Figures in parentheses indicate percentages of delayed reactions (3 days or more).
**Method of Kauffmann and Petersen, 1956. (Twenty hour readings except where indicated, see also table 14).

^{W} Weakly positive reactions.
+ Positive within one or two days' incubation.
(+) Positive reaction after 3 or more days.
– No reaction.
+ or – Majority of strains positive, occasional cultures negative.
– or + Majority of cultures negative, occasional strains positive.
(+) or + Majority of reactions delayed, some occur within 1 or 2 days.
d Different reactions: +, (+), –.

TABLE 14

Reactions of Members of the Tribe SALMONELLEAE in Organic Acid Media of Kauffmann and Petersen (1956)

Genus	No.	Citrate					D-tartrate				
		1*	2	5	14	—	1*	2	5	14	—
Salmonella	299	245 (81.9)	42 (14.1)	8 (2.7)	4 (1.3)	0	272 (91)	4 (1.3)	3 (1)	9 (3)	11 (3.7)
Arizona	150	7 (4.7)	111 (74)	23 (15.3)	6 (4)	3 (2)	0	0	68 (45.3)	57 (38)	25 (16.7)
Citrobacter	268	4 (1.5)	143 (53.3)	117 (43.7)	0	4 (1.5)	0	0	119 (44.4)	122 (45.5)	27 (10.1)

* Days of incubation

NOTE: Figures in parentheses indicate percentages.

TABLE 15

Biochemical reactions of *S. enteritidis* bioserotype Paratyphi-A

Substrate or test	Bioserotype Paratyphi-A			S. enteritidis		
	Sign	%+	(%+)*	Sign	%+	(%+)*
Hydrogen sulfide (TSI)	- or +^{w}	12.5		+	98	
Citrate (Simmons')	- or (+)	0	(25)	+	99.3	(0.7)
Lysine decarboxylase	-	0		+	99.7	
Inositol	-	0		d	42.8	(1)
Xylose	-	0		+	99	
Cellobiose	d	12.5	(6.2)	(+)	5	(92.8)
Glycerol	(+)	0	(100)	d	5.7	(7.2)
Stern's glycerol fuchsin	-	0		+	98.2	(0.6)
Jordan's tartrate	-	0		+	92.5	(1.1)
Sodium acetate	-		(6.2)	+	92.4	(2.2)
Mucate	-	0		+ or (+)	88.3	(1.7)
Organic Acid**						
citrate	-	0		+	96	(4)
D – tartrate	-	0		+	92.3	(4)
i – tartrate	-	0		d	4.7	(57.5)
l – tartrate	-	0		d	11.8	(75.2)

* Figures in parentheses indicate percentages of delayed reactions (3 or more days).
** Method of Kauffmann and Petersen, 1956.

+ Positive within 1 or 2 days' incubation (90% or more).
(+) Positive reaction after 3 or more days.
– No reaction (90% or more).
+ or – Majority of strains positive, occasional cultures negative.
– or + Majority of cultures negative, occasional strains positive.
(+) or + Majority of reactions delayed, some occur within 1 or 2 days.
d Different reactions: +, (+), –.
w Weakly positive reaction.

TABLE 16

Differentiation of *Salmonella enteritidis* bioserotypes Pullorum and Gallinarum

Substrate or test	Bioserotype Pullorum			Bioserotype Gallinarum		
	Sign	%+	(%+)*	Sign	%+	(%+)*
Jordan's tartrate	–	0		+	100	
Ornithine decarboxylase	+	100		–	0	
Mucate	–	0		+	90.3	
Gas from glucose	+	95.1		–	0	
Dulcitol	–	0		+	99	(1)
Maltose	– or (+)		(35)	+	98.1	(1.9)
Cellobiose	–	0		d	60	(30)
Glycerol	–	0		(+)		(90)
Cysteine-gelatin	–	0		+	98.1	(1.6)
Organic acids**						
citrate	–	0		– or (+)		(40)
D – tartrate	–	0		+	92.2	(3.9)
i – tartrate	–	0		(+)		(100)
l – tartrate	–	0		(+) or –		(80)

* Figures in parenthese indicate percentages of delayed reactions (3 days or more).
** Method of Kauffmann and Petersen, 1956.

N.B. This table is included primarily for the use of workers in veterinary bacteriology. However, others should remember that bioserotype Pullorum occasionally occurs in human infections.

+ Positive within 1 or 2 days' incubation (90% or more).
– No reaction (90% or more).
+ or – Majority of strains positive, occasional cultures negative.
– or + Majority of cultures negative, occasional strains positive.
(+) or – Majority of reactions delayed, some occur within 1 or 2 days.
d Different reactions: +, (+), –.

TABLE 17

Differentiation within the genus *Klebsiella*

Test or substrate	*K. pneumoniae*			*K. ozaenae*			*K. rhinoschleromatis*		
	Sign	%+	(%+)*	Sign	%+	(%+)*	Sign	%+	(%+)*
Urease	+	94.5		d	9.5	(10.3)	–	0	
Methyl red	– or +	13.3		+	99.1		+	100	
Voges-Proskauer	+	91.1		–	0		–	0	
Citrate (Simmons')	+	97.7		d	31.9	(31)	–	0	
Organic acids**									
citrate	+ or –	64.4		– or +	18		–	0	
D-tartrate	+ or –	67.1		– or +	36		–	0	
Malonate	+	92.5		–	4		+	95.5	
Mucate	+	92.8		– or +	24		–	0	
Lysine decarboxylase	+	97.2		– or +	48		–	0	
Gas from glucose	+	96.5		d	64	(2)	–	0	
Lactose	+	98.2	(1.4)	(+) or +	24.1	(70.7)	(+) or –		(72.8)
Dulcitol	– or +	31.5		–	0		–	0	

* Figures in parentheses indicate percentage of delayed reactions (3 or more days).
** Method of Kauffmann and Petersen (1956).

\+ Positive within 1 or 2 days' incubation (90% or more).
– No reaction (90% or more).
(+) Positive reaction after 3 or more days.
\+ or – Majority of strains positive, occasional cultures negative.
– or + Majority of cultures negative, occasional strains positive.
(+) or + Majority of reactions delayed, some occur within 1 or 2 days.
d Different reactions: + or (+), –.

TABLE 18

Differentiation of *Klebsiella pneumoniae* and *Enterobacter cloacae*

Test or substrate	K. pneumoniae			E. cloacae		
	Sign	%+	(%+)*	Sign	%+	(%+)*
Gas from:						
Inositol	+	91.9		–	4.5	
Glycerol	+	92.5		d	5.5	(15.9)
Adonitol	+ or –	83.7		– or +	28.4	
Esculin	+	98.9	(1.1)	– or +	29.3	
Lysine decarboxylase	+	97.2		–	0.5	
Arginine dihydrolase	–	0.9		+	96.5	
Ornithine decarboxylase	–	0		+	96	
Urease	+	94.5		+ or –	64.7	
Gelatin (22 C)	–	3.3		(+)		(96)
Motility	–	0		+	94.6	
Growth on synthetic aliginate medium	+ or (+)	88.5	(9.2)	–	0	

* Figures in parentheses indicate percentage of delayed reactions (3 or more days).

+ Positive within one or two days' incubation (90% or more).
(+) Positive reaction after 3 or more days.
– No reaction (90% or more).
+ or – Majority of strains positive, some cultures negative.
– or + Majority of cultures negative, some strains positive.
(+) or + Majority of reactions delayed, some occur within 1 or 2 days.
 d Different reactions: +, (+), –.

TABLE 19

Differentiation of *Enterobacter cloacae* and *Enterobacter aerogenes*

Test or substrate	E. cloacae			E. aerogenes		
	Sign	%+	(%+)*	Sign	%+	(%+)*
Urease	+ or –	64.7		–	2.7	
Lysine decarboxylase	–	0.5		+	98.7	
Arginine dihydrolase	+	96.5		–	0	
Jordan's tartrate	– or +	27.8		+ or –	89.3	
Adonitol acid	– or +	28.4		+	98.7	
gas	– or +	28.4		+	98.7	
Inositol acid	d	21.9	(12.4)	+	100	
gas	–	4.5		+	100	
Glycerol acid	d	43.3	(44.8)	+	100	
gas	d	5.5	(15.9)	+	98.7	(1.3)
Esculin	– or +	29.3		+	98	

* Figures in parentheses indicate percentage of delayed reactions (3 or more days).

\+ Positive within 1 or 2 days' incubation (90% or more).
(+) Positive reaction after 3 or more days.
– No reaction (90% or more).
\+ or – Majority of strains positive, occasional cultures negative.
– or + Majority of cultures negative, occasional strain positive.
(+) or + Majority of reactions delayed, some occur within 1 or 2 days.
d Different reactions: +, (+), –.

TABLE 20

Differentiation of *Enterobacter aerogenes* and *Enterobacter hafniae*

Substrate or test	E. aerogenes			E. hafniae		
	Sign	%+	(%+)*	Sign	%+	(%+)*
Adonitol						
acid	+	98.7		–	0	
gas	+	98.7		–	0	
Inositol						
acid	+	100		–	0	
gas	+	100		–	0	
Sorbitol	+	100		–	0	
Raffinose	+	96		–	0	
Salicin	+	98.7	(1.3)	d	13	(8)
Alpha methyl glucoside	+	96	(2)	–	0	
Esculin	+	98		–	6	(2)
Methyl red						
37 C	–	0		+ or –	54	
22 C				–	1	
Voges-Proskauer						
37 C	+	100		+ or –	65	
22 C				+	99	
Citrate (Simmons')						
37 C	+	93.7		(+) or –		(58)
22 C				d	3	(79)
Gelatin (22 C)	(+) or –		(77.3)	–	0	
Mucate	+	94.7		–	0	

* Figures in parentheses indicate percentage of delayed reactions (3 or more days).

+ Positive within 1 or 2 days' incubation (90% or more).

– No reaction (90% or more).

(+) Positive reaction after 3 or more days.

+ or – Majority of strains positive, occasional cultures negative.

– or + Majority of cultures negative, occasional strains positive.

(+) or + Majority of reactions delayed, some occur within 1 or 2 days.

d Different reactions: +, (+), –.

TABLE 21

Differentiation of *Enterobacter liquefaciens* and *Serratia marcescens* subspecies *marcescens*

Substrate or test	E. liquefaciens			S. marcescens, marcescens		
	Sign	%+	(%+)*	Sign	%+	(%+)*
Glucose						
Acid	+	100		+	100	
Gas	+	94.1		+[b] or –	52.6	
Inositol						
Acid	+	97	(1.5)	d	78.5	(8.2)
Gas	d	1.5	(22)	–	0	
Glycerol						
Acid	+	98.5	(1.5)	+	97	(2.6)
Gas	d	45.6	(38.2)	–	0	
Cellobiose						
Acid	d	26.5	(44.1)	d	20.8	(33.4)
Gas	d	5.9	(33.8)	–	0	
Esculin						
Acid	d	75	(1.6)	+	90.8	
Gas	– or +[a]	37.5		–	0	
Raffinose						
Acid	d	86.8	(2.9)	–	1.7	(1.2)
Gas	d	17.6	(60.3)	–	0	
Arabinose						
Acid	+	92.6		–	0	
Gas	d	23.5	(14.7)	–	0	
Xylose						
Acid	+	92.6	(1.5)	d	8	(18.3)
Gas	d	30.9	(23.5)	–	0	
Erythritol						
Acid	–	0		d	1.7	(22.8)
Alpha methyl glucoside						
Acid	– or +	21.7		–	0.9	(0.6)
Methyl red						
(37 C)	+ or –	75		– or +	17.7	
(22 C)	– or +	33.3		– or +	8.8	
Voges-Proskauer						
(37 C)	– or +	30.9		+	100	
(22 C)	– or +	79.4		+	100	

[a] Gas volumes: bubble to 10%.

[b] Gas volumes: 10% or less.

* Figures in parentheses indicate percentage of delayed reactions (3 or more days).

+ Positive within 1 or 2 days' incubation (90% or more).

– 90% or more, no reaction.

(+) Positive reaction 3 or more days.

d Different biochemical reactions: +, (+), –.

+ or – Majority of strains +, some cultures negative.

– or + Majority of cultures negative , some strains positive.

N.B. The only important difference between *S. marcescens* subspecies *marcescens* and *S. marcescens* subspecies *kiliensis* is their reactions in the Voges-Proskauer test. Cultures of the latter are V-P negative.

TABLE 22

Differentiation of members of the tribe KLEBSIELLEAE

Test or substrate	Klebsiella pneumoniae			Enterobacter cloacae			aerogenes			hafniae			liquefaciens			Pectobacterium			Serratia marcescens, marcescens		
	Sign	%+	(%+)*	Sign	%+	(%+)*	Sign	%+	(%+)*	Sign	%+	(%+)*	Sign	%+	(%+)*	Sign	%+	(%+)*	Sign	%+	(%+)*
Gas from:																					
glucose	+	100		+	100		+	100		+	100		+	94.1	(1.5)	- or +	17.1		+w or -	52.6	
adonitol	d	83.4	(0.3)	- or +	28.4		+	98.7		-	0		-	1.5	(1.5)	-	0		-	0	
inositol	+	91.9	(2)	-	4.5		+	100		-	0		d	1.5	(22)	-	0		-	0	
glycerol	+	92.5	(3.7)	d	5.5	(15.9)	+	98.7	(1.3)	+	95	(5)	d	45.6	(38.2)	-	0		-	0	
cellobiose	+	95.7	(0.3)	+	100		+	98.7		d	67	(22)	d	5.9	(33.8)	d	5.7	(14.3)	-	0	
Sorbitol	+	99.4	(0.3)	+	94.5	(0.5)	+	100		-	0		+	97		-	2.8		+	98.3	
Raffinose	+	99.7		+	97		+	96		-	0		d	86.8	(2.9)	d	65.7	(2.9)	-	1.7	(1.2)
Rhamnose	+	99.3	(0.4)	+	92	(1.5)	+	98.7		+	93	(7)	-	0		d	64.3	(5.7)	-	0	(0.3)
Arabinose	+	99.9		+	99.5		+	100		+	96		+	92.6		d	70	(1.4)	-	0	
Methyl red, 37C	- or +	13.3		-	3		-	0		+ or -	54		+ or -	75		- or +	45.7		- or +	17.7	
22C										-	1		- or +	33.3		+ or -	75.7				
Voges-Proskauer, 37C	+	91.1		+	99.5		+	100		+ or -	65		- or +	30.9		- or +	22.9		+	100	
22C				+	99.5		+	100		+	99		+ or -	79.4		- or +	47.1				
Lysine	+	97.2	(2.8)	-	0.5		+	98.7		+	100		+ or -	82.4		-	0		+	99.6	
Arginine	-	0.9		+	96.5		-	0		-	9		-	4.4		-	0	(5.7)	-	1.3w	
Ornithine	-	0		+	96		+	98.7		+	100		+	98.5		-	0		+	99.4	(0.1)
Malonate	+	92.5		+ or -	80.6		+ or -	74.7		+ or -	74		-	7.4		- or +	21.4		-	1.7	
Mucate	+	92.8		+ or -	75.6		+	94.7		-	0		-	0		d	45.7	(5.7)	-	0	
Urease	+	94.5		+ or -	64.7		-	2.7		-	3		d	4.4	(19.1)	d	21.4	(7.1)	d	29.1	(26.8)
Gelatin, 22C	-	3.3		(+)	1	(96)	(+) or -	0	(77.3)	-	0		+	98.5	(1.5)	+ or (+)	67.1	(32.9)	+ or (+)	86.9	(13.1)
Motility	-	0		+	94.5		+	97.3		+	93		d	79.4	(17.7)	d	55.7	(27.1)	+	98.6	
Alginate	+ or (+)	88.5	(9.2)	-	0		-	0		-	0		-	0		-	0		-	0	(3.7)
DNase	-	0		-	0		-	0		-	0		+ or -	69.4		- or +			+	96.7	
Pectate, 37C	-	0		-	0		-	0		-	0		-	0		d	41.5	(17.1)	-	0	
22C																+ or (+)	75.7	(24.3)			

TABLE 23

Differentiation of *Proteus vulgaris* and *Proteus mirabilis* from *Proteus morganii* and *Proteus rettgeri*

ostrate or test	*P. vulgaris* and *P. mirabilis*			*P. morganii* and *P. rettgeri*		
	Sign	%+	(%+)*	Sign	%+	(%+)*
drogen sulfide (TSI)	+	94.5	(2.6)	–	0	
latin (22 C)	+	91.6	(6.4)	–	0	
ase (corn oil)	+	89.6	(5.2)	–	0	
arm (2% agar)	+	94	(1)	–	0	

*Figures in parentheses indicate percentage of delayed reactions (3 or more days).

+ Positive within 1 or 2 days' incubation (90% or more).
(+) Positive reaction after 3 or more days.
 – No reaction (90% or more).
+ or – Majority of strains positive, occasional cultures negative.
 – or + Majority of cultures negative, occasional strains positive.
(+) or + Majority of reactions delayed, some occur within 1 or 2 days.
 d Different reactions: +, (+), –.

TABLE 24

Differentiation of *Proteus vulgaris* and *Proteus mirabilis*

Test or substrate	P. vulgaris			P. mirabilis		
	Sign	%+	(%+)*	Sign	%+	(%+)*
Indol	+	98.2		–	1.9	
Voges-Proskauer						
37 C	–	0		– or +	15.6	
22 C	– or +	11.3		+ or –	51.6	
Citrate (Simmons')	d	10.5	(14.1)	+ or (+)	58.7	(37.1)
Ornithine decarboxylase	–	0		+	99.2	
Sucrose	+	94.7		d	18.9	(63.3)
Maltose	+	96.2	(1.9)	–	0.9	(0.4)
Salicin	d	58.2	(10.9)	d	0.8	(29.8)
Alpha methyl glucoside	d	79.5	(5.1)	–	0	
Esculin	d	59	(2.6)	–		(0.9)
DNase	+ or –	60		–	0	

*Figures in parentheses indicate percentage of delayed reactions (3 or more days).

+ Positive within 1 or 2 days' incubation (90% or more).
(+) Positive reaction after 3 or more days.
– No reaction (90% or more).
+ or – Majority of strains positive, occasional cultures negative.
– or + Majority of cultures negative, occasional strains positive.
(+) or + Majority of reactions delayed, some occur within 1 or 2 days.
d Different reactions: +, (+), –.

TABLE 25

Differentiation of *Proteus morganii* and *Proteus rettgeri*

Test or substrate	P. morganii			P. rettgeri		
	Sign	%+	(%+)*	Sign	%+	(%+)*
Citrate (Simmons')	–	0		+	95.6	(3.3)
Ornithine decarboxylase	+	97.1		–	0	
Gas from glucose	d	84.9	(0.9)	– or +	12.2	
Sucrose	–	1	(2.9)	d	13.3	(56.7)
Mannitol	–	0		+ or –	88.5	
Adonitol	–	0		d	80.9	(5.6)
Inositol	–	0		+	93.3	(4.5)
Salicin	–	0		d	30	(6.6)
Erythritol	–	0		d	78.3	(6.5)
Esculin	–	0		d	30.4	(8.7)
Xylose	–	0		– or +	15.1	
Cellobiose	–	0	(1.9)	d	3.7	(30.4)

*Figures in parentheses indicate percentage of delayed reactions (3 or more days).

+ Positive within 1 or 2 days' incubation (90% or more).
(+) Positive reaction after 3 or more days.
– No reaction (90% or more).
+ or – Majority of strains positive, occasional cultures negative.
– or + Majority of cultures negative, occasional strains positive.
(+) or + Majority of reactions delayed, some occur within 1 or 2 days.
d Different reactions: +, (+), –.

TABLE 26

Differentiation of *Proteus morganii* and *Proteus rettgeri* from *Providencia alcalifaciens* and *Providencia stuartii*

Substrate or test	P. morganii		P. rettgeri		P. alcalifaciens		P. stuartii	
	Sign	%+ (%+)*	Sign	%+ (%+)*	Sign	%+ (%+)*	Sign	%+ (%+)*
Urease	+	98.2 (0.9)	+	100	-	0	-	0
Ornithine decarboxylase	+	97.1	-	0	-	0	-	0
Gas from glucose	d	84.9 (0.9)	- or +	12.2	d	85.8 (0.6)	d	13.3 (1.3)
Mannitol	-	0	+ or -	88.5	-	2 (0.2)	-	3.8
Adonitol	-	0	d	80.9 (5.6)	+	94.5 (0.2)		
Inositol	-	0	+	93.3 (4.5)	-	0.6	+	97.5 (2.5)
Erythritol	-	0	d	78.3 (6.5)	-	0	-	0
Esculin	-	0	d	30.4 (8.7)	-	0	-	0
Cellobiose	-	0 (0.9)	d	3.7 (30.4)	-	1.5 (3)	d	12.5 (68.7)

* Figures in parenthese indicate percentage of delayed reactions (3 or more days).

+ Positive reaction within 1 or 2 days' incubation (90% or more).

- No reaction (90% or more).

+ or - Majority of strains positive, occasional cultures negative.

- or + Majority of cultures negative, occasional strains positive.

(+) or + Majority of reactions delayed, some occur within 1 or 2 days.

d Different reactions: +, (+), -.

REFERENCES

Belliveau, R. R., J. W. Grayson, and T. J. Butler. 1968. Amer. J. Clin. Pathol., **50**, 126.

Blair, J. E., E. H. Lennette, and J. P. Truant. (eds.) 1970. Manual of clinical microbiology. American Society for Microbiology. Baltimore: Williams & Wilkins.

Bodily, H. L., E. L. Updyke, and J. O. Mason (eds.). 1970. Diagnostic procedures for bacterial, mycotic, and parasitic infections. New York: American Public Health Association.

Carpenter, K. P., S. P. Lapage, and K. L. Steel. 1966. *In* Identification methods for microbiologists. London: Academic Press.

Domingue, G. J., F. Dean, and J. R. Miller. 1969. Amer. J. Clin. Pathol., **51**, 62.

Douglas, G. W., C. O'Connor, and V. M. Young. 1966. Ibid., **45**, 497.

Edwards, P. R., and D. W. Bruner. 1942. Ky. Agric. Exper. Sta. Cir., 54.

Ewing, W. H. 1965. Differentiation of members of the genera *Salmonella, Arizona,* and *Citrobacter* by biochemical methods. CDC Publ.*

————. 1968. Differentiation of Enterobacteriaceae by biochemical reactions. CDC Publ.* (revised 1970).

————. 1969. Biochemical reactions given by Enterobacteriaceae in commonly used tests. CDC Publ.*

Ewing, W. H., and Johnson, J. G. 1960. Int. Bull. Bacteriol. Nomen. Taxon., **10**, 223.

Ewing, W. H., B. R. Davis, and P. R. Edwards. 1960. Pub. Health. Lab.,** **18**, 77.

Ewing, W. H., R. Hugh, and J. G. Johnson. 1961. Studies on the *Aeromonas* group. CDC Publ.

Ewing, W. H., M. A. Fife, and B. R. Davis. 1965. The biochemical reactions of *Arizona arizonae*. CDC Publ.*

Ewing, W. H., B. R. Davis, and W. J. Martin. 1966. Outline of methods for the isolation and identification of *Vibrio cholerae*. CDC Publ.*

Fife, M. A., W. H. Ewing, and B. R. Davis. 1965. The biochemical reactions of the tribe *Klebsielleae*. CDC Publ.*

Hugh, R., and E. Leifson. 1953. J. Bacteriol., **66**, 24.

Jordan, E. O., M. E. Caldwell, and D. Reiter. 1934. Ibid, **27**, 165.

Kauffmann, F., and A. Petersen. 1956. Acta Pathol. Microbiol. Scand., **38**, 481.

Martin, W. J., W. H. Ewing, A. C. McWhorter, and M. M. Ball. 1969. Pub. Health Lab.,** **27**, 61.

Marymont, J. H., U. Amanna, and B. H. Lloyd. 1966. Amer. J. Clin. Pathol., **46**, 702.

Moeller, V. 1955. Acta Pathol. Microbiol. Scand., **36**, 158.

Roundtable: How far to go with Enterobacteriaceae? 1969. J. Infect. Dis., **119**, 197.

Suassuna, I., and Suassuna, I. R. 1963. An de Microbiol., **11**, 105.

*CDC Publ., Publication from the Center for Disease Control (formerly the Communicable Disease Center), Atlanta, Ga. 30333.
**Public Health Laboratory: Journal of the Conference of Public Health Laboratory Directors.

Chapter 4
The Antigens of Enterobacteriaceae

INTRODUCTION

The bases for serologic typing within the various genera of Enterobacteriaceae are similar, although different sets of antisera are employed and in some instances slightly different procedures are used. Complete serotyping depends upon the determination of the O antigens, O antigen factors, flagellar (H) antigens, H antigen factors, and K (capsular or envelope) antigens.

O Antigens. These are the heat-stable somatic antigens which are composed of phospholipid-polysaccharide complexes. Analyses of the O antigens generally reveal polysaccharide (c. 60%), lipid (20% to 30%), and hexosamine (3.5% to 4.5%). It is the nature of the terminal groups and the order in which they occur in the repeating units of the polysaccharide chain that render specificity to the numerous kinds of O antigens. These antigens are resistant to alcohol and dilute acid. Agglutination reactions, as in tube tests, occur relatively slowly and the aggregates are granular (Polagglutination). O antigens are subject to smooth (S) to rough (R) and form variation. Some are subject to lysogenic conversion.

H Antigens. These are heat-labile antigens that occur in the flagella of the bacteria. They are protein in nature and are composed of flagellins. The amino acid content and the order in which these acids occur in the flagellins determines the specificity of the many different H antigens of Enterobacteriaceae. Flagella are inactivated slowly by alcohol. Flagellar agglutination occurs very rapidly and the aggregates formed are loosely knit and floccular. Phase variation of Andrewes occurs in the genera *Salmonella* and *Arizona*. In certain other genera (e.g., *Citrobacter*) another sort of variation occurs in the flagellar antigens. This is a segregation or subdivision of antigens which gives rise to two or more variants, but which is not known to be reversible. Loss variation also may occur in any motile bacteria (HO to O).

K Antigens. These are the somatic antigens of the bacteria that occur as capsules or as envelopes. When present in sufficient amounts these antigens inhibit the agglutination of living or unheated bacterial suspensions in O antisera. The K (from Kapsule) antigens form a class which may be subdivided according to their physical and chemical characteristics. Contrary to statements frequently seen in the literature, most of the varieties of K antigens are not destroyed by heat. However, some of their characteristics are altered by heat at various temperatures for specified periods of time. The K antigens also are polysaccharides (v.inf.). Agglutination reactions in tube tests occur slowly, and the titers of K antisera are relatively low. Complete agglutination of K forms yields an aggregate that resembles a disc or membrane. This class of antigens includes the L, A, and B antigens of *E. coli*, the Vi antigen of *S. typhi* and certain serotypes of *Citrobacter*, the B antigens of shigellae, and the M antigens of some strains of *Salmonella* and *Arizona*, to list a few. Loss variation (partial or complete) occurs in some instances, i.e., KO to O forms.

The general characteristics of O, K, and H antigens are of cardinal importance in the production and use of antisera prepared with members of the genera of Enterobacteriaceae. Details for both production and use of antisera are given in the individual chapters.

Where applicable and necessary the kinds of variation mentioned in the foregoing paragraphs, as well as a few others, are mentioned again in the individual chapters that follow. However, since several of these variational phenomena occur in members of all of the genera, the author elected to discuss them in a single chapter in order to avoid repetition as much as possible. For the same reason, certain other antigens (not mentioned above) and several other topics of interest are reviewed or mentioned in this chapter. These discussions are necessarily brief, but the literature cited is such that readers who are interested in pursuing the subjects further should have no difficulty in doing so.

VARIATIONAL PHENOMENA THAT AFFECT SEROLOGIC TYPING OF ENTEROBACTERIACEAE

HO to O Variation. It has been known for many years that the type of agglutination reaction given by motile and nonmotile variants of the same bacterium is different. Smith and Reagh (1903) and Beyer and Reagh

(1904) reported that motile cultures of *S. cholerae-suis* gave loose, floccular agglutination and that nonmotile variants yielded granular aggregates. It was established that the antigen that occurred only in the flagellated forms was heat-labile and was damaged by heat at 60 to 70 C, whereas the somatic material present in both motile and nonmotile variants was thermostable. However, antibodies for the flagellar material were more resistant to heat than agglutinin for the somatic substance. Joos (1903) made similar observations with cultures of *S. typhi*.

The importance of these findings was not fully realized and they attracted little attention until Weil and Felix (1917) published their observations on the differences in the form of growth and agglutination of motile and nonmotile strains of *Proteus* X19. These workers found that both motile and nonmotile cultures of these bacteria were agglutinated in a granular manner in sera from patients with typhus fever. However, in the antiserum from rabbits vaccinated with motile forms, the homologous strain was agglutinated in large fluffy clumps, whereas nonmotile forms behaved as they did in patients' sera, i.e., they were agglutinated in small granular clumps. The motile variant, which they called H (*Hauch,* cloud or film), grew as a spreading film on agar while the nonmotile form, called O (*ohne Hauch,* without cloud or film), grew in discrete colonies. The O antigens resisted heating at 100 C while the H antigens were destroyed at that temperature. In an extensive series of experiments these workers confirmed all the observations of earlier investigators and established the nature of somatic and flagellar antigens and agglutinins. Later Weil and Felix (1920) extended these observations to *Salmonella.* As a result of this work the essential differences in somatic and flagellar antigens generally have been accepted and the terms applied by Weil and Felix have come into common use. The designation O is applied to the heat-stable somatic (body) antigens and their corresponding agglutinins, while H is used to denote antigens associated with the flagella and the antibodies induced by them.

Motile microorganisms possess at least two types of antigens which differ in their physical characteristics. Each of these stimulates, in vivo, its own specific antibody with which it reacts. One of these, the H antigen, is associated with the flagella and is not found in nonflagellated (atrichous) microorganisms. It is heat-labile and is inactivated progressively at temperatures above 60 C. After exposure to 100 C for 1 hr its agglutinogenic properties usually are destroyed. Heat treatment at 100 C for 2½ hr definitely destroys the agglutinogenic propensities of the H antigens. H antigens are inactivated by acids and by alcohol. When

placed in contact with corresponding (homologous) agglutinins, H antigens are flocculated rapidly in fluffy clumps which are dispersed easily. In 1924 Orcutt (quoted by Harvey, 1929) reported that injection of flagella removed from bacteria resulted in an antiserum that reacted with motile forms only and yielded the aforementioned loosely knit, floccular type of agglutination. Other investigators since have produced H antisera by injection of washed flagella removed from the cells. The O antigens are present in both motile and nonmotile microorganisms. They are resistant to prolonged treatment with heat at 100 C and to treatment with alcohol and dilute acids. O agglutinins react much more slowly with their respective antigens than do H agglutinins, irrespective of the motility or nonmotility of the bacteria. In contrast to the loose, floccular character of H agglutination, O agglutinins produce finely granular clumps which are dispersed only with difficulty. *It should be emphasized that the term O antigen does not include all antigens which may be present in nonmotile strains but only those heat stable antigens present in the body of smooth organisms.* Antigens present in rough cultures, and heat labile antigens present in envelopes or capsules must be considered separately.

If dilute O antiserum and a dilute O antigen suspension of the homologous bacterium are mixed on a slide, and the preparation is examined microscopically, it may be observed that agglutination of the bacteria first takes place at the poles of the cells (Polagglutination). This results in the granular aggregates mentioned above. When a similar mixture of dilute H antiserum and homologous motile bacteria is examined, the flagella of one become entangled with those of other cells, the bacteria become immobilized, and the cells actually are held apart by entangled, agglutinated flagella. Hence, the fluffy, floccular type of agglutination results, which is characteristic of flagellar antigens. As stated, these loosely knit aggregates are easily dispersed. However, they reform quickly upon reincubation.

From the above-mentioned facts it is clear that HO to O variation is a loss variation that results in the development of nonmotile forms or mutants from a motile parent culture. It is apparent that the genetic material responsible for development of the enzyme systems that synthesize the flagellin is lost or altered in some way. This results in the production of nonmotile forms.

Variations in H Antigens. In Enterobacteriaceae the term "phase variation" applies to the reversible variation of H antigens described by Andrewes (1922). This is known to occur only in members of the genera

Salmonella and *Arizona* and its application to serological typing of these microorganisms is discussed in detail in Chapters 8 and 10. Suffice it to say here that the observations made by Andrewes were essential to the development of the antigenic schemata for these bacteria. As mentioned at the beginning of this chapter, another kind of variation occurs in the flagellar antigens of certain other Enterobacteriaceae. This was first described as zeta (term no longer used) variation in bacteria that are now known to be members of the genus *Citrobacter* (see Chapter 11). This variation amounts to a subdivision or segregation of the H antigens of the parent culture into variants which apparently are stable (Edwards, 1946). This kind of variation may occur in members of other genera (unpublished observations), but studies have not progressed far enough to be certain. Genetic studies apparently have not been made on this type of variation.

S to R Variations. Smooth (S) to rough (R) variations occur in members of all of the genera of Enterobacteriaceae, although much of the early work was done with salmonellae and shigellae. At the outset it should be emphasized that S to R variations ordinarily are not abrupt. On the contrary, they occur gradually and intermediate degrees of roughness exist in the transition from a smooth to a rough form.

Baerthlein (1912) observed differences in colonial form which were associated with changes in the cellular morphology. This investigator described several colony forms in cultures held without transfer as well as in freshly isolated strains. Von Lingelsheim (1913) described S and R forms although the latter were designated Q from the strain from which they were derived. This worker noted differences in the two forms in colony type, morphology, and growth in broth. Both Baerthlein and von Lingelsheim claimed that reversions from R to S took place. This now seems improbable, unless the cultures were not completely rough. Gildemeister (1916a, 1916b, 1917) observed a number of colony forms, which probably corresponded to S, R, rho, and dwarf, in freshly isolated cultures from patients and carriers. Gildemeister directed attention to spontaneous agglutination, variability of agglutination by the several forms, and the inability of von Lingelsheim's Q (SR or R) forms to absorb all agglutinin from antisera prepared with the normal (S) form. He also noted that several forms (S, R, etc.) differed in their susceptibility to agglutination by acids (Michaelis test).

Arkwright (1920, 1921) studied certain shigellae, *E. coli*, and salmonellae and reported changes in colonial morphology which were associated with alterations in the characteristics of growth in broth and stability of saline suspensions. These changes were regarded as a degenerative process and were referred to as roughness since colonies of the changed races possessed roughened surfaces and irregular borders. The original condition of the cultures was referred to as the smooth or S state and the altered cultures as the rough or R state. R cultures had a tendency to produce granular growth in broth, and saline suspensions prepared from R cultures were much less stable than those prepared from S forms. Arkwright recognized that the serological properties of antigens of R forms differed from those of S forms and demonstrated this by agglutinin absorption. This investigator was able to do agglutination tests with R strains by using reduced concentrations of sodium chloride (e.g., 0.2%) rather than an isotonic solution. Schuetze (1921) also found that R cultures had a tendency to cross-agglutinate, regardless of whether their S parents were related. This relationship was expressed by Schuetze as "the serological cosmopolitanism of rough variants." However, it is emphasized that the change from S to R is not a dramatic transformation. Usually it is a gradual process, and there are different degrees of roughness. Further, the R antigens of all salmonellae, for example, are not identical serologically. Although changed colony formation, stability in broth, and altered antigenic characteristics generally are correlated, many cultures produce colonies that are quite rough in appearance yet their broth cultures are stable and they retain most of their normal O antigen complex. The O antigens of cultures which have a tendency to granular growth in broth often are still recognizable though somewhat changed. Conversely, the antigens of other cultures which produce colonies that do not appear to be abnormal, and which have little tendency to produce granular growth in broth, may be so changed that they cannot be recognized (e.g., see T antigens). Therefore colonial morphology and stability in broth are not always accurate indications of antigenic composition.

In several of his publications White (1926, 1927, 1928, 1929a, 1929b, 1931, 1932, 1933) described investigations into the characteristics of S, R, and other forms in a variety of bacteria. Among these was the ρ (rho) form which occurred in old strains from which neither S nor R forms could be isolated. By extraction of S, R, or ρ with acidified alcohol a protein, labeled Q, was identified. This protein was antigenic and produced antibody that reacted well with ρ and poorly with R, but did not agglutinate S forms (White, 1932). Still another antigenic protein (T) was extracted with 75% alcohol. The T substance gave rise to antibodies when injected and these agglutinated ρ and R, but not S forms (White, 1933). White (1928) demonstrated another substance, apparently a phospholipid, in both

S and R forms. This substance was extractable with alcohol and chloroform. It possessed hydrophobic properties which contributed to the instability of certain cell suspensions. White's use of alcohol treatment in the preparation of suspensions of R and partially R cultures is the basis for the use of alcohol-treated suspensions mentioned in Chapter 8. The flagella of the bacteria also are inactivated partially by the combined heat and alcohol treatment (Chapter 8) as shown by White and by others.

Kauffmann (1941) stated that serological methods are the most reliable index of roughness, and this point should be emphasized (see also Chapter 8). As mentioned, White and others reported upon the serological differences between S and R cultures. Many other investigators also have reported the results of analogous or similar studies with a number of members of the family Enterobacteriaceae. According to Kauffmann (1954, p. 168), Moeller (1948) demonstrated two kinds of R antigens in cultures of *E. coli* O antigen groups 8 and 9. Each of these R antigens was found in members of each of the two O groups. The investigations of Kroeger (1953) and Rische et al. (1964) also should be cited.

The R antigens differ from S in many ways other than those mentioned above. For references to the biological characteristic of R forms the reader is referred to the publications of Wilkinson (1958), Rowley (1968), and Luederitz et al. (1966, 1968).

S-T-R Variation. Kauffmann (1956, 1957) described somatic antigens which he labeled T (from transient). Two such antigens, T_1 and T_2, were characterized in certain salmonellae. T variation as described by Kauffmann occurs only in salmonellae as far as is known (although variations of an analogous sort have been known in shigellae for many years). For this reason T variation is discussed in more detail in Chapter 8. Although chemical analyses (Luederitz et al. 1966) indicate that the T antigens of Kauffmann and R forms derived from the same serotypes belong to the same chemotype (i.e., their polysaccharides are of similar carbohydrate composition), immunological specificity exists in both T and R forms.

The designation SR also has been used for many years for forms that are intermediate between S and R. These (SR) forms probably represent various stages in the progression from the smooth to the rough state. Naide et al. (1965) have used the designation SR for certain kinds of semirough mutants of salmonellae (v. inf.).

Form Variation. This is a quantitative variation in the *amount* of O antigen present in the progeny of a strain, e.g., O antigen 6_1 in salmonellae of serogroup C_1. This variation is known to occur only in members of the genus *Salmonella* and for that reason it is discussed in Chapter 8. However a variation that is somewhat analogous to form variation is known to occur in certain shigellae (Chapter 6), and it seems reasonable to suppose that this sort of variation, or something like it, might occur in members of other genera of Enterobacteriaceae.

KO To O Variations. Variation of this kind involves the loss, completely or partially, of the capsule or envelope (microcapsules of Wilkinson, 1958) of the bacteria with the result that the bacteria become agglutinable in their respective O antisera. That is, suspensions become O-agglutinable without being subjected to treatment with heat. In some instances, KO to O variation results in loss of the K antigen, on which occasions a smooth O form results. In other cases the K antigen is not lost completely and although suspensions prepared from the progeny of variant colonies become O-agglutinable, the presence of K antigen in the variant can be demonstrated by other methods, e.g., in agglutinin absorption tests or by immunogel-precipitation (Grados and Ewing, 1969 and unpublished data). In still other instances the variation proceeds from K to R, and smooth forms cannot be demonstrated by conventional methods.

The occurrence of mucoid (M) variants of certain Enterobacteriaceae has been known for many years. Most of the early work on these variants dealt with salmonellae (see White, 1929a, 1929b). Kauffmann (1936) used the terminology M-N to denote variation between mucoid (M) and normal (N) forms of salmonellae. The M form of growth is most likely to occur in freshly isolated cultures and has been observed more often in *S. enteritidis* ser Paratyphi-B than in other serotypes. Fresh isolants of ser Paratyphi-B, when incubated overnight at 37 C and then left for 1 or 2 days at 22 C, develop a raised, moist rim around the circumference of the colonies, the so-called slime-wall mentioned in the older German literature. Kauffmann (1935, 1936) demonstrated a special antigen (M antigen) in mucoid cultures. Mucoid forms are not agglutinated by O antisera but if living mucoid cultures are injected into rabbits, agglutinins for the M form are produced. M agglutinins usually are low in titer and produce an agglutination which is firm and disclike in character. It is possible also to produce M agglutinins by injection of formalinized cultures and cultures which have been heated at 60 C for one hour. However, Kauffmann found that mucoid cultures which had been heated at 100 C for 2 hr, or treated with 96% alcohol or N/1 HCl for 20 hr at 37 C did not produce M agglutinins. By absorption of an anti-M serum with a nonmucoid form of the serotype with which the serum was prepared, it is possible to produce a pure anti-M serum which agglutinates mucoid forms but not normal, nonmucoid cultures.

Definite capsules can be demonstrated in some, but not all, mucoid strains of *Salmonella*. When present, they are demonstrable in moist India ink preparations. Capsulated forms give definite quellung (capsular) reactions with anti-M serum or with antiserum for type 13 *Klebsiella* (Perch, 1950). Birch-Hirschfeld (1935) examined the M antigen of ser Paratyphi-B by chemical methods and reported that it was a nitrogen-free polysaccharide, which upon hydrolysis yielded 40 per cent glucose.

Mucoid forms have been reported in cultures of *E. coli,* certain serotypes of *Shigella, Arizona,* and in members of certain other genera in addition to klebsiellae. These are mentioned in the individual chapters.

Anderson (1961) employed nutrient agar, to which Sorensen buffer (pH 7.0, in concentration of 5/M) was added, for study of M forms. This investigator reported that production of M substance (slime) was greatest when the reaction of the medium was between pH 7 and 8. Using the phosphate agar at pH 7.0, and incubation at 37 C overnight, followed by incubation for another 24 hr at 21 C, Anderson was able to demonstrate the occurrence of M substance in all of the serotypes of *Salmonella* examined, and in members of several other genera as well.

From what is known of them, it is apparent that the M antigens belong to the general class of bacterial substances known as K antigens.

The Vi antigen, which occurs in *S. typhi* and certain serotypes of *Citrobacter,* also is a K antigen. A related antigen occurs in *S. enteritidis* ser Paratyphi-C, and the occurrence of Vi in certain strains of *S. enteritidis* ser Dublin has been reported (LeMinor and Nicolle, 1964). Vi forms are inagglutinable in O antiserum, a fact that apparently was first noted by Bensted (1929).

Although the Vi and Vi-like antigens of the above-mentioned bacteria are related, they are not identical either immunologically or biologically. Seven varieties of Vi antigen are known in strains of *S. typhi,* and only one of these (II) is employed in bacteriophage typing of these bacteria. The Vi antigen of *S. typhi* is a highly polymerized, acidic polysaccharide believed to be composed of units of *N*-actyl-aminohexuronic acid (v. inf.). Loss of Vi antigen usually gives rise to O forms and this kind of variation has been designated V-W by Kauffmann (1935), but it also is known as Vi-O variation. Since the Vi antigen is so important in identification and bacteriophage typing of strains of *S. typhi,* it is dealt with in more detail in Chapter 8. Contrary to statements often seen in the literature, Vi antigen is not extremely labile nor is it identical with the somatic antigen labeled 5 in the Salmonella Schema.

Other sorts of KO-O variation occur in cultures of *E. coli.* Members of this species possess at least three kinds of K antigens, labeled L, A, and B. These are discussed ·in some detail in Chapter 5. It is sufficient to note here that the term K antigen first was used in connection with *E. coli* (Kauffmann, 1943), and that much is being learned about the K (and other) antigens of *E. coli* and other Enterobacteriaceae from analyses made by immunogel precipitation and immunoelectrophoresis (for references see Grados and Ewing, 1969, 1970).

Namioka and Sakazaki (1959) characterized another variety of K antigen which they labeled C antigen. This antigen was reported to occur in certain strains of *Proteus.*

The alpha antigen (Stamp and Stone, 1944) possesses the general characteristics of K antigens but differs from those mentioned above in at least one important respect. The titers for alpha antigen in sera that contain alpha agglutinins usually are high (1:25,000 or more). The aggregates are formed rapidly and resemble those obtained with K forms of *E. coli.* Alpha antigen is inactivated by heat treatment for 15 minutes at 100 C and by treatment with alcohol (Stamp and Stone, 1944). Alpha agglutinins have been detected in sera from normal rabbits and in some antisera produced with a variety of Enterobacteriaceae. The occurrence of alpha antigens is known to be variable in certain members of the genera *Citrobacter* and *Providencia* (Chapters 11 and 17 respectively). That is, α O to O variation occurs in strains that possess alpha antigen. However, it is clear that all alpha antigens are not antigenically identical.

Since alpha agglutinins may occur in rabbits that have not been vaccinated, it is advisable to test all newly prepared antisera with a culture that contains alpha antigen and one of the same O antigen group that is free of alpha antigen. If alpha agglutinins are present in an antiserum, they may be removed easily by absorption. In the experience of the author, agglutinin for alpha antigen does not occur frequently enough in normal rabbits to warrant testing of all rabbits prior to vaccination.

The beta (β) antigen described by Mushin (1949, 1955) also appears to belong to the K class of antigens. This antigen occurs in a number of otherwise unrelated bacteria. The beta antigen resembles alpha in most of its characteristics. It is thermolabile (completely or partially inactivated by heat treatment for 1 hr at 100 C), it is inactivated by alcohol, and agglutinin titers of antisera usually are very high. Agglutination reactions occur rapidly, but the aggregates formed resemble those obtained with H antigens, i.e., the aggregates are loosely knit and floccular. Variation from βO to O occurs, but, as pointed out by Mushin (1955), this variation is by no means as predictable as the other K variations mentioned above. This investigator also mentioned (1955) that in the cultures of *E. coli* 0111:B4:H2 studied, beta antigen could not be demonstrated directly by agglutination, but its presence in the strains was detected by their ability to absorb beta antibodies from antisera.

All of the above-mentioned variational phenomena, and the different categories of antigens, are of the greatest importance in the preparation and absorption of antisera and in serological typing of Enterobacteriaceae. It is clear that most, if not all, of these antigens are synthesized within the cell and that the enzyme systems responsible for their synthesis are genetically controlled.

Nonfimbriate — Fimbriate Variation. In addition to flagella, bacteria may produce other filamentous appendages of an entirely different sort. These can be seen only in electron photomicrographs. These appendages were observed first by Anderson (1949), Houwink (1949), and Houwink and van Iterson (1950). Duguid and his associates (Duguid et al., 1955; Duguid and Gillies, 1956, 1957, 1958; Gillies and Duguid, 1958; Duguid and Wright, 1959) studied these appendages systematically and suggested the name fimbriae for them. In dried preparations, fimbriae usually are approximately 0.01 micron in width and 0.3 to 1.0 micron in length, rarely reaching a length of 4.0 microns. They are much smaller and shorter than flagella and lack the typical wavy spirals seen in stained preparations of flagella. They often are very numerous (100 to 250 per cell) and may be seen on the surface of the bacterium and its cell wall and extending from them. Fimbriae have been seen in members of other families of bacteria, and in the Enterobacteriaceae they have been observed in members of the genera *Escherichia, Shigella, Salmonella, Klebsiella, Serratia,* and *Proteus.* Duguid et al. (1955) and Duguid and Gillies (1957) clearly showed that fimbriate cultures underwent reversible variation between fimbriate and nonfimbriate forms of growth and that this variation generally could be controlled by the conditions under which the bacteria were cultivated. This variation was studied carefully by Duguid and Gillies (1957) in shigellae. It was found that the fimbriate form became dominant when cultures were transferred at 48 hour intervals (a) aerobically in unagitated tubes of broth, (b) aerobically in unagitated shallow layers of broth, (c) aerobically on agar slants with excess condensation water, and (d) anaerobically on agar slants. The non-

fimbriate form became dominant on serial transfer (a) aerobically on agar plates or dry slants, (b) aerobically in shallow broth layers rotated continuously, (c) anaerobically in tubes of broth, and (d) aerobically in tubes of glucose broth. Conversion from one form to the other usually required several subcultures under the determining conditions. Fimbriation often was associated with the formation of a thin surface pellicle.

One of the most prominent characteristics of fimbriate cells is their power to agglutinate red blood cells of the guinea pig and to a lesser degree, red cells of most other animals. Duguid and his associates found this hemagglutinating or adhesive property very closely associated with fimbriation, fimbriate cells producing hemagglutination, whereas nonfimbriate cells did not. There was no correlation between fimbriation and the production of capsules, flagella, or hemolysin. The hemagglutinins or adhesins of fimbriate cells were of two kinds: a mannose-sensitive (MS) adhesin which was inhibited by the presence of a small amount of D-mannose and which was demonstrable with untreated guinea pig cells, and a mannose resistant (MR) adhesin which was not inhibited by D-mannose and failed to agglutinate untreated red cells, but which adhered to and agglutinated tannic acid treated red cells of the ox and other animals or red cells heated to 70 C. The hemagglutinating activity of fimbriate cells probably is the most practical method of determining the presence of fimbriae.

Duguid and Gillies (1956) and Gillies and Duguid (1958) observed that fimbriation interfered with O agglutination. Agglutination of richly fimbriated cells in antisera produced from nonfimbriate cells was low in titer and consisted of the formation of very fine granules adhering to the walls of the tube. The same workers demonstrated that fimbriae were antigenic and gave rise to antisera having unusually high titers, some having titers in excess of 1:100,000. Antifimbrial sera were produced by the injection of living suspensions of fimbriate cells. Pure antifimbrial antisera were prepared by absorption of crude antisera with afimbriate suspensions of cells of the homologous cultures. Heating cultures at 60 C did not alter their fimbriate state and cells so treated still were agglutinable in antifimbrial sera, absorbed antifimbrial sera, and produced hemagglutination. Practically all cells of cultures heated at 100 C for 1 hr or 120 C for 30 min were freed of fimbriae and such cells centrifuged from suspension and resuspended in fresh saline no longer agglutinated in antifimbrial sera nor absorbed antifimbrial agglutinins. However, before centrifugation suspensions heated at 100 C for 1 hr contained many detached, apparently unaltered fimbriae. Although uncentrifuged suspensions were only slightly agglutinable in antifimbrial sera they were able to absorb the homologous agglutinins. The same observations were made with suspensions heated at 120 C. While hemagglutinating properties were lost by heating at 90 C or 120 C, the uncentrifuged suspensions retained the power to absorb the hemagglutination inhibiting factors found in antifimbrial sera.

Fimbrial agglutination occurs with relative rapidity and is loosely floccular in appearance. Therefore, it may be mistaken for flagellar agglutination. However, fimbrial agglutinability of cells is not annulled by treatment with 0.005 N HCl or 50% ethyl alcohol, as is flagellar agglutinability. Treatment with 1.0 N HCl for 20 hr at 37 C renders the cells inagglutinable in antifimbrial sera and inactivates their hemagglutinating properties.

The above-mentioned work with fimbriae has been extended and these appendages have been found in members of the genus *Arizona* in addition to the genera mentioned (Duguid et al., 1966, and Olds et al., 1968). Further, Duguid et al. (1966) have characterized six types of fimbriae in various Enterobacteriaceae. These six types differ in certain characteristics, e.g., some do not cause hemagglutination.

Duguid et al. (1966) mention a seventh kind of fimbria and discuss the F fimbria (or F pilus). Here F is the symbol used by bacterial geneticists for the fertility or sex factor (e.g., see Demerec et al., 1968; also v. inf. Resistance Factors). The term pili (pilus) was introduced by Brinton in 1959 and Duguid et al. (1966) state that Crawford and Gesteland (1964) and Brinton (1964) described the F+ fimbria (or F+ pilus) together with the type 1 fimbriae in F+ and Hfr (v. inf.) strains of *E. coli*. According to Duguid et al., this F+ fimbria (or pilus) "is morphologically identical with the type 1-fimbriae, but only 1-4 filaments are present on a bacillus, and unlike the type 1-fimbria, it carries the specific adsorption sites for male-specific bacteriophages." Photomicrographs that illustrate this phenomenon are included in the publication by Datta et al. (1966).

Some authors have used the terms fimbriae and pili as synonyms, and it appears that they are. With a few exceptions, bacterial geneticists and those concerned with the development of resistance to antimicrobial agents have employed the term pilus (pili). That the sexual fimbriae (or pili), of which two kinds are known, are important in transmission of resistance factors appears to have been demonstrated quite clearly (v. inf.). The sexual fimbriae are limited in number (usually one to four per cell) and differ morphologically from ordinary fimbriae which are very numerous (e.g., see Datta et al., 1966).

Since variation from the nonfimbriate to fimbriate form largely depends upon the conditions of growth, fimbriae are important to those who produce antisera for use in serological typing of Enterobacteriaceae. Fimbrial agglutinins are unlikely to be stimulated by injection of the heated suspensions employed in production of O antisera. However, fimbrial agglutinins may be formed when unheated, formalinized broth cultures are used in production of K or H antisera, particularly if steps are not taken to minimize this possibility. Taylor (personal communication, 1960) observed very high titer agglutinins (1:256,000) in certain H antisera. These agglutinins were thought to be fimbrial. (Thus there is an additional reason for titration of all antisera against homologous and heterologous antigens before they are put to use in serotyping.) When unheated suspensions are employed for antiserum production, such suspensions should be prepared from young agar slant cultures or young (4 to 6 hr in a waterbath) broth cultures. In either instance serial transfer should be avoided.

OTHER ANTIGENS

The X Antigen. In 1924 Topley and Ayrton described a thermostable antigen which developed in certain cultures of salmonellae. This antigen, which these investigators labeled X, caused cross-agglutination of unrelated serotypes. They reported that the X antigen was formed in cultures incubated at 37 C, and that extended incubation resulted in increased production of the antigen. Formation of X antigen could be avoided by incubating cultures at 22 C. Topley and Ayrton also reported that antibodies for the X antigen could be removed from antisera by absorption with appropriately selected cultures. Cruickshank (1939) studied the X antigen and confirmed the fact that it was thermostable (not inactivated by steaming cultures for 30 min), and reported that X antigen occurred in both S and R forms of certain salmonellae, that it produced a granular type of agglutination, and that it might interfere in serological tests with patients' sera. To the author's knowledge the X antigen cannot be equated with any of the antigens mentioned in the preceeding pages.

The Common Hapten or Common Antigen (CA). This antigen first was described in cultures of *E. coli* O antigen group 14 by Kunin et al. (1962) and Kunin (1963). It was referred to as the common antigen by Kunin et al., and as heterogenic enterobacterial antigen (Kunin) by Whang and Neter (1962). The CA is common to most gram negative bacteria but not to all.

Further, Whang and Neter (1963) reported that antibodies for the CA were demonstrable in the sera of healthy blood donors and in commercially available gamma globulin preparations, although the titers usually were low. As mentioned by Domingue and Neter (1966), the CA probably was not detected earlier because it cannot be demonstrated by agglutination or precipitation technics. However, CA can be absorbed onto the surfaces of erythrocytes and such cells then are agglutinated by antibodies in antiserum for *E. coli* O group 14. Intravenous injection of enteric bacteria other than *E. coli* 014 does not elicit antibodies for the CA. However, antibodies for CA can be demonstrated in antisera produced with other gram-negative bacteria when crude extracts of CA are injected into the footpads of rabbits together with Freund's adjuvant, or by similar injection of cell-attached rather than soluble CA preparations (Domingue and Neter, 1966). The CA can be separated, in part, from the O antigen by extraction with 85% ethyl alcohol (Suzuki et al., 1964). The alcohol-soluble CA proved to be highly immunogenic when injected intravenously into rabbits (Domingue and Neter, 1966). Whang and Neter (1964) reported that a strain of *Pseudomonas,* which they isolated fortuitously, produced a factor (possibly an enzyme) that destroyed the CA but left O antigen intact. That CA is a cell wall antigen that can be detected by a fluorescent antibody technic was demonstrated by Aoki et al. (1966).

The CA has been studied extensively by Neter and co-workers and these investigators have characterized a number of its biological properties (for example, see Domingue and Neter, 1966; Kessel et al., 1966; and Gorzynski and Neter, 1970).

IMMUNOCHEMICAL INVESTIGATIONS

Extensive immunochemical analyses have been made with the O, K, H, and R antigens of members of several genera of Enterobacteriaceae.

The O, R, and Intermediate Antigens. The cell walls of both gram-negative and gram-positive bacteria are known to contain units of *N*-acetylglucosamine and a muramic acid peptide in the form of a rigid polymer (see Perkins, 1963 or Salton, 1964). Similarity between these two major groups of bacteria ends at this point, however. Unlike the gram-positive bacteria, which posses teichoic acid in their cell walls, the gram-negative species contain complex lipopolysaccharides, phospholipids, and protein. All of these chemical components apparently lie adjacent to the aforementioned rigid polymer of mucopeptide. Much time and effort has been devoted to investigation of the lipopolysaccharide complexes of gram-negative bacteria. According to the pioneer work of Boivin et al. (1933), the chemical characteristics of endotoxin largely are those of the protein-lipid-lipopolysaccharide complex,

which also has been equated with the O antigen. This same complex apparently is responsible for a wide variety of physiological effects usually incurred by injection of heat-killed bacteria. Among these effects are fever, leukopenia, nonspecific enhancement of resistance, shock, internal hemorrhage etc. This complex is highly antigenic and, as mentioned, comprises the specific O antigens of many, if not all, gram-negative bacteria, including Enterobacteriaceae.

Several procedures for the isolation of O antigen have been described since the report of Boivin et al. (1933) appeared.[1] Their trichloroacetic acid extraction method enabled them to isolate toxic and immunogenic fractions from many gram-negative bacteria. These investigations, as well as the early work of Raistrick and Topley (1934) and Morgan (1937), indicated that both the antigenic and toxic properties of endotoxin were associated with the lipopolysaccharide complex. In addition, the O antigen specificity of the microorganism was shown to reside in the lipopolysaccharide fraction, but this fraction was nontoxic. According to the work of Westphal and Luederitz (1954), the lipid fraction (lipid A) of the lipopolysaccharide molecule appears to be a phosphorylated lipid complex containing glucosamine and β-hydroxymyristic acid. As to toxicity, it still is not clear whether this lipid is fully responsible for all of the biological manifestations usually accorded to endotoxin. Westphal (1960) and Kauffmann (1960) reported that varying degrees of toxicity could be attributed to their lipid A preparations, whereas Ribi et al. (1961) maintain that the polysaccharide moiety rather than the lipid is responsible for toxicity. The work of Herzberg and Green (1964) tends to support the view of Ribi et al.

The reactive sites of the O antigen of a particular microorganism appear to be located in a single polysaccharide molecule. Lipopolysaccharides of salmonellae, as well as several other Enterobacteriaceae, can be obtained from intact bacterial cells, or from purified cell wall preparations, by means of the phenol-water extraction method described by Westphal et al. (1952). This procedure separates protein from the lipopolysaccharide (LPS). The polysaccharide portion of this complex has been found (by mild acid hydrolysis) to be linked covalently to the glucosamine-lipid, lipid A (Westphal and Luederitz, 1954). In analyzing similar soluble polysaccharides, Kauffmann et al. (1960a,

---

[1]Pyrolysis-gas-liquid chromatography (PLGC) is another method of analysis which seems to be promising (e.g., see Reiner and Ewing, 1968). Although the method does not appear to have been used extensively with Enterobacteriaceae its potentialities appear to be great, especially when used in conjunction with other methods of analysis.

1960b) delineated from five to eight different carbohydrates (sugars). All of the salmonellae examined contained glucose, glucosamine, galactose, an aldoheptose, and moderate amounts of organic phosphate. Notably, almost all of the *E. coli* O antigen groups studied contain the same sugars in their polysaccharides. Heath and Ghalanbor (1963) identified another sugar, 2-keto-3-deoxyoctonate (KDO), as a basic component of the LPS of *E. coli* O group 111. This unusual sugar also is a constituent of the polysaccharides in all members of the genera *Salmonella* and *Escherichia* that have been studied. It is apparent that KDO is widely distributed in the cell walls of gram negative bacteria, presumably as one of the components of the cell wall LPS (Ellwood, 1966, 1970; Luederitz et al., 1966, 1968).

The basal components (glucose, glucosamine, galactose, aldoheptose, KDO, and organic phosphate) frequently are referred to as core polysaccharide. In addition to this core structure, smooth, naturally occurring bacteria of the genera of *Salmonella, Escherichia* and certain others that have been studied contain additional sugars in numerous combinations and sequences. Among these sugars are galactosamine, fucose, rhamnose, and one of the 3, 6-dideoxyhexoses such as abequose, paratose, colitose, or tyvelose. These additional sugars compose the O antigen side chain which is linked to the basal core of the cell wall. The O side chain may have as many as 30 repeating units, each of which is composed of several sugars arranged in a definite combination and sequence. One of the sugars in the unit usually is immunodominant. Thus the composition of the basal core and in particular, of the side chains determines the specificity of each of the numerous O antigens.

The composition of extracted lipopolysaccharides from smooth (S) and rough (R) forms of gram-negative bacteria, and the immunological specificities related to the sugar composition of the lipopolysaccharides, have been investigated intensively by a number of investigators (e.g., see Luederitz et al., 1966, 1968). Other workers have been concerned with the endotoxic and antigenic portions of washed cell wall preparations, as opposed to extracted LPS (e.g., Ribi et al., 1959; Herzberg and Green, 1964).

The greatest volume of immunochemical work with O and R antigens of Enterobacteriaceae has been done with members of the genera *Salmonella* and *Escherichia* (see Staub and Forrest, 1963; Luederitz et al., 1966; Raff and Wheat, 1967; Wheat et al., 1968; Roantree, 1967; and Fuller et al., 1968 for references), but investigations also have been made with members of the genus *Arizona* (Kauffmann et al., 1962; Luederitz et al., 1966), the genus *Shigella* (Iseki et al.,

1961; Itikawa, 1964; Simmons, 1966; Seltmann and Hoffmann, 1966; Kontrohr and Westphal, 1967; Romanowska and Mulczyk, 1967; Slopek et al., 1967; Johnson et al., 1967; Simmons, 1969), some serotypes of the genus *Citrobacter* (Luederitz et al., 1966; Raff and Wheat, 1967, 1968), the genus *Serratia* (Nowotny, 1963; Tripodi and Nowotny, 1966, and members of some other genera (see Luederitz et al., 1966).

According to these investigators, more than 15 different monosaccharides are known to occur as constituents of the O antigens of *Salmonella* (see also Kauffmann et al., 1960, 1962). The O antigens of serogroups of *Salmonella* have been divided into 17 chemotypes (see Luederitz et al., 1966). Each chemotype is composed of antigens that have the same qualitative sugar composition, from chemotype I, which is the simplest, to chemotypes XIV to XVI, which contain eight different sugars. For example, members of serogroups 17, 44, 47, and 48 of *Salmonella* belong to chemotype I. The polysaccharides of this chemotype possess only the five qualitative sugars: glucose, glucosamine, galactose, aldoheptose, and KDO. The polysaccharides of members of serogroup D contain D-mannose, L-arabinose, and tyvelose in addition to the five basal sugars. Serotypes of *Salmonella* that belong to serogroups B, C_2, and (8), 20 contain the five basal sugars, D-mannose, L-rhamnose, and abequose. Therefore, immunochemical analyses indicate that qualitatively the sugar content of the O antigens of serotypes of *Salmonella* that belong to the same serogroup is identical. However, the O antigens of serotypes that are members of two different serogroups may belong to the same chemotype. It is assumed that in these instances the same sugars are linked at least partially in different ways (Luederitz et al., 1966). This would explain the known differences in specificity.

Similar studies with more than 100 serogroups of *E. coli* indicated that the O antigens of these microorganisms could be classified into 20 chemotypes (Luederitz et al., 1966). With the exception of chemotypes XVII and XIX, which lacked galactose, these chemotypes possessed the same five basal qualitative sugars as did salmonellae. Further, twelve of the chemotypes of *Salmonella* also were identified among the *E. coli* cultures. Some chemotypes were found in *E. coli* O antigen groups which did not occur among salmonellae and vice versa.

As might be expected, intergeneric relationships often may be correlated with similar or identical sugar composition of the respective antigens. This was reported in a number of instances in which the O antigens of certain salmonellae are known to be related to those of serotypes of *E. coli, Arizona, Citrobacter* etc. (see Luederitz et al., 1966).

Insofar as they have been studied the results of immunochemical analyses of *Shigella, Citrobacter* etc. (v. sup. for references) are analogous to those obtained from work with salmonellae and *Escherichia.* The lipopolysaccharides of members of some of the other genera contain other compounds not present in the chemotypes of *Salmonella* and *E. coli* (see also Luederitz et al., 1968).

Immunochemical analyses also have been made with the lipopolysaccharides extracted from rough (R) forms derived from smooth (S) cultures of salmonellae and *E. coli* (Kauffmann et al., 1960a, 1960b; Luederitz et al., 1966). The R polysaccharides may be extracted by means of the phenolwater procedure in the same manner that S polysaccharides are extracted. However, certain other extraction procedures are not suitable for work with R forms (Luederitz et al., 1966). The amount of polysaccharide extracted from R forms is smaller (1% of dry weight) than that extracted from the corresponding S forms. The R polysaccharides are less soluble in water than S and may contain 50 per cent or more of a material with the characteristics of lipid A.

Kauffmann et al. (1960a, 1960b) recovered R forms from 25 serotypes of *Salmonella* that belonged to numerous serogroups and chemotypes. The specific sugars present in the corresponding S lipopolysaccharides were absent from the R lipopolysaccharides and all R forms of the aforementioned 25 salmonellae belonged to chemotype I (i.e., contained only the five basal sugars) regardless of the chemotype of the parent S forms. However, analyses of the R polysaccharides derived from R lipopolysaccharides showed differences in their sugar constituents. The first groups of R lipopolysaccharides (chemotype Ra) contained glucosamine as a constituent, while in a second group (chemotype, Rb) this substance was present only in the lipid fraction (see Luederitz et al., 1966). Since S forms of certain serogroups of salmonellae belong to chemotype I, and the R forms of many salmonellae (chemotype Ra) contain the same qualitative basal sugars identified in S forms of chemotype I, it would seem that there must be quantitative, and perhaps other, differences in the sugars present in the lipopolysaccharides of these particular S and R forms (S forms of chemotype I and corresponding R forms of chemotype Ra). Kauffmann et al. (1961) found that it was very difficult to obtain R forms from S forms that belonged to chemotype I. The R forms derived from salmonellae of chemotypes other than I are the result of loss of the specific side chains characteristic of each of the several S chemotypes (Kauffmann et al., 1961; Luederitz et al., 1966).

Immunochemical analyses of the R forms of *Salmo-nella* have shown their R lipopolysaccharides can be classified into five groups, each of which is a distinct chemotype. As mentioned, the first group of R antigens is composed of the five basal qualitative sugars (chemotype Ra). The R antigens of the second, third, fourth, and fifth groups of R forms contain four, three, two or one different sugars (respectively) as constituents of their R polysaccharides, with glucosamine being present only in lipid A. These R antigens represent chemotypes that are simpler than chemotype Ra (which is equivalent, qualitatively, to chemotype I). These additional chemotypes have been labeled Rb, Rc, Rd, and Re, respectively (Luederitz et al., 1966). Rough forms of chemotype Rb have been found in salmonellae (e.g., in serotypes Minnesota and Typhimurium), of chemotype Rc in *E. coli* and in serotypes Enteriditis and Typhimurium, of Rd in *E. coli* strains and certain serotypes of salmonellae (e.g., ser Minnesota), and chemotype Re in *S. sonnei* and some strains of *E. coli* (Luederitz et al., 1966). Further, Luederitz and Westphal (1966, see Luederitz et al., 1966) obtained mutants from a strain of ser Minnesota that belong to each of the chemotypes, Ra to Re, inclusive.

The antigens designated RI and RII by Beckmann et al. (1964) have been shown to belong to chemotypes Rb and Ra, respectively (Luederitz et al., 1966).

According to Luederitz et al. (1966), the biosynthesis of O antigen specific polysaccharide occurs in two principal steps. The monosaccharide constituents of the antigens are synthesized and, with the aid of specific transferases, these constituents are transferred to a growing polysaccharide acceptor in a specific, genetically determined sequence. Rough forms arise when there is a block in the activity of a transferase or one of the enzymes (synthetases) that activate the synthesis of a sugar. Incomplete synthesis or partial transfer of the sugar constituents of the polysaccharide gives rise to such forms as the T forms of Kauffmann (1956, 1957; Wheat et al., 1967) and the semirough forms described by Naide et al. (1965).

K Antigens. Several members of the Enterobacteriaceae are known to produce more than one kind of specific polysaccharide. These additional specific substances comprise the K antigens (L, A, and B) of *E. coli,* the Vi antigen of *S. typhi,* and the M antigens of *Salmonella, Arizona, E. coli* etc., and the capsular antigens of *Klebsiella.* These antigens occur in the capsules or envelopes of the bacteria and have been referred to as extracellular polysaccharides by Wilkinson (1958).

The immunochemistry of the K antigens of a number of serotypes of *E. coli* have been studied by Ørskov and co-workers. Ørskov et al. (1963) examined the K antigens of *E. coli* serotypes 0141:K85(B):H4 and 08:K42(A):NM (nonmotile). Acidic polysaccharides were derived from each of these. The first, from K antigen 85a, 85b consisted primarily of mannose (33%), rhamnose (18%), glucuronic acid (19%), and glucosamine (17%). In addition, the radicals *O*-acetyl (4%), *N*-acetyl (4%), and *N*-acyl (6%), calculated as β-hydroxymyristic acid) were identified. The acidic capsular polysaccharide derived from K antigen 85a, 85c was of the same qualitative composition as that from K85a, 85b. After purification, the acidic capsular polysaccharide extracted from K antigen 42(A) still contained 3 to 5 per cent protein, which was removed by appropriate procedures. The resultant polysaccharide then contained galactose (37%), fucose (18%), galacturonic acid (33%), glucosamine (2.5%), and the following radicals: *O*-acetyl (6.4%), *N*-acetyl (0.8%), calculated as β-hydroxymyristic acid. Thus it was demonstrated that the acidic polysaccharides can be differentiated from the O lipopolysaccharides by the absence of constituents such as heptose, glucose, mannose etc. Further, it was shown that it was possible to differentiate between the acidic capsular polysaccharide and the polysaccharide of the M substance by differential precipitation and identification of the constituent of each fraction. Three separate polysaccharides were characterized in *E. coli* 059:K?:H19. These were the polysaccharides of the O antigen and those of the K and of the M antigens. These studies were extended by Jann et al. (1965), who also reviewed the earlier work of T. Smith et al. (1927), D. E. Smith (1927), and Wiley and Scherp (1958).

Hungerer et al. (1967) analyzed an acidic polysaccharide derived from *E. coli* serotype 09:K30(A):H12 and reported that it consisted of equimolar amounts of glucuronic acid, galactose, and mannose, and *O*-acetyl groups (2.5%). Also Jann et al. (1968) studied the acidic capsular polysaccharide extracted from *E. coli* serotype 08:K27(A):NM and found that it was composed of equimolar amounts of D- glucuronic acid, D-glucose, D-galactose, L-fucose, and *O*-acetyl groups (5.2%).

Stirm et al. (1967) reported that K antigen 88 (L), extracted from a nonmotile strain of *E. coli* O antigen group 8, was a protein. This was the first published report of the protein nature of a K antigen of *E. coli*.

The capsular antigens of strains of *Klebsiella* are K antigens similar to the A antigens of *E. coli*. Analysis of the capsular antigens of cultures of *Klebsiella* of types 8, 29, 36, 54, and 57 (Dudman and Wilkinson, 1956) revealed three component sugars in each. One of these invariably was a uronic acid. Barker et al. (1963) isolated a similar acidic polysaccharide from a strain of capsule type 1 of *Klebsiella*. Analysis showed the presence of glucuronic acid, glucose, and fucose.

As noted in the section on Vi to O (V-W) variation, Vi antigen occurs in strains of *S. typhi* and in certain serotypes of *Citrobacter*. It also has been reported in occasional cultures of ser Dublin, and is said to occur in strains of ser Paratyphi-C. The earlier work on the chemical nature of Vi antigen was reviewed by Webster et al. (1952), by Jarvis et al. (1967) and Martin et al. (1967). From the numerous studies reviewed it appears that Vi antigen is a polymer of *N*-acetyl-D-galacturonic acid. The native molecule contains *O*-acetyl and is highly polymerized (Martin et al., 1967). According to these investigators and Jarvis et al. (1967), the molecular weight of intact Vi antigen from a strain of *Citrobacter* was between 1×10^6 and 3.9×10^4. Martin et al. (1967) reported that sonically treated Vi antigen from this strain of *Citrobacter* was only about 1 per cent as effective as the original (untreated) antigen in stimulation of protection in mice against challenge with *S. typhi*. Antigen treated sonically elicited lower antibody titers in mice and rabbits but no loss of ability to precipitate antibody, or to sensitize erythrocytes, was noted.

It would seem that more immunochemical and physical chemical work has been done with Vi antigen derived from the other sources mentioned in the preceding paragraph than from *S. typhi*. However, the author can find no definite evidence in the literature that the Vi antigens from the several sources are identical either chemically or immunologically. On the contrary, Whiteside and Baker (1961) demonstrated physical and chemical differences in Vi antigens derived from a serotype of *Citrobacter* and from *S. typhi*. These differences did not appear to be related to the serological specificity of the derivatives. However, it is known from immunoelectrophoretic analyses that the Vi antigen of *S. typhi* consists of at least six fractions[2] which migrate cathodically, and the Vi antigen of serotypes of *Citrobacter* shares both thermolabile fractions and two of the thermohaptens (Grados and Ewing, unpublished data). The so-called Vi antigen of ser Paratyphi-C actually is a heat-labile L-like antigen, proteinlike in behavior, and consisting of two fractions which are related to two of the fractions of the Vi antigen of *S. typhi* and the serotypes of *Citrobacter*.

Somatic antigen 5 of salmonellae also may be regarded as a K antigen. Kauffmann demonstrated that factor 5 was more labile than O antigen 4, for example (see Kauffmann, 1966, pp. 70 and 74). The agglutinability, antibody binding power, and agglutinogenic properties of antigen 5 were inactivated by treatment with normal HCl (20 hr at 37 C) as were the M, Vi, and H antigens. The 5 factor was not inactivated by heat treatment at 100 C, but was inactivated by heat at

---

[2]Two thermolabile fractions and four thermohaptens.

120 C. Kotelko et al. (1961, cited by Luederitz et al., 1966) reported that the polysaccharide derived from a serotype that possessed antigen 5 contained *O*-acetyl groups, whereas the polysaccharide of strains of the same serotype which lacked antigen 5 did not contain these groups.

The biological properties of K antigens were reviewed by Wilkinson (1958) and by Luederitz et al. (1966).

H Antigens. The flagella of Enterobacteriaceae originate within the cell where each is attached to a basal granule. Flagella are composed of a protein called flagellin, which belongs to a class of proteins known as the keratin-myosin-epiderm-fibrinogen group (see Stodola, 1958). The molecular weight of flagellin from certain salmonellae is about 40,000 (McDonough, 1965). Since flagella are only about 100 A wide, they are among the smallest known organelles of motility.

Immunochemical analyses have been made with flagella from *Proteus vulgaris* (Stodola, 1958), serotypes of *Salmonella* (McDonough, 1965), and other bacteria. The work of McDonough (1965) is of particular interest. The investigator examined the flagellins of several salmonellae (H antigens 1, 2; i; a; b; r; e, h; e, n, x; g, m; g, p; and g, s, t) and those of a serotype of *Arizona*. Most of these flagellins contained high proportions of aspartic acids and glutamic acids or amides, threonine and alanine, and smaller amounts of tyrosine, phenylalanine, methionine, proline, and histidine. Cysteic acid or cysteine and tryptophan were not detected in any of the flagellins examined. Histidine was not present in antigens of the g . . . complex (g, p; g, m; g, m, s). Similar analyses were obtained in repeat examinations of antigens from the same serotype, and analyses of particular antigens (e.g., 1,2) from different serotypes also yielded similar amino acid residues. In other words, McDonough (1965) found that the amino acid composition of a particular antigenic type was constant while the amino acid content of antigenically distinct flagellins differed in varying degrees. From this it was concluded that differences in antigenic character were reflections of differences in the amino acid composition of the flagellins.

Although the analyses of the 1,2 antigens from ser Typhimurium and ser Paratyphi-B yielded very similar amino acid residues, it is known that immunologically these antigens from these two sources are not identical. The flagellar antigens of the second phase of ser Typhimurium are (factors) 1,2,3, although factor 3 is not expressed in the schema, and in ser Paratyphi-B only the 1,2 factors are present (see Chapter 8).

It seems clear that it is the amino acid content and the sequence in which these acids appear that determines the specificity of each of the numerous flagellar antigens. Spacial arrangement of these acids in a molecule of flagellin also may be of importance in specificity.

GENETIC RECOMBINATIONS THAT MAY AFFECT SEROLOGICAL TYPING

Genetic changes of one kind or another have been reported in members of all of the genera of Enterobacteriaceae that have been investigated in this regard, and such changes are known to occur in many other families of bacteria, fungi, and protozoa (Lederberg, 1951; Jacob and Wollmann, 1961; Peters, 1959; Hayes, 1964; Iino and Lederberg, 1964; and Ravin, 1965).

Much of the progress made in bacterial genetics during the past twenty years has involved the use of members of the various genera of Enterobacteriaceae. Since considerable information already was available that dealt with the natural variational phenomena of the O (somatic) and H (flagellar) antigens of salmonellae, and since an antigenic schema in which the serotypes could be oriented was available, the potentialities of salmonellae in genetic investigations quickly were recognized. Studies made in 1952 and 1953 dealt with transduction, and marked the beginning of an era of investigation of the genetics of salmonellae (Zinder and Lederberg, 1952; Lederberg and Edwards, 1953). Since then serotypes of *Salmonella* have been employed in investigations of the genetics of O, H, and R (rough) variations, as well as resistance to antimicrobial agents, and hundreds of publications have resulted.

Variations mediated by bacteriophage or other mechanisms. There are at least three kinds of genetic mechanisms that affect salmonellae, but not salmonellae alone (Marmur et al., 1963). These are transduction, lysogenic (or phage) conversion, and conjugation.

1. Zinder and Lederberg (1952) proved that a temperate phage called PLT 22, affecting members of serogroups A, B, and D of salmonellae, could introduce genetic material into susceptible cells and hence effect transfer (transduction) of H antigens or of physiological characters from donor to recipient cells. Lederberg and Edwards (1953) produced a number of recombinants using PLT 22, and Kauffmann (1953) effected similar changes in members of serogroups A and B with different temperate phages. Edwards et al. (1955) characterized temperate phages that were capable of transducing H antigens in a number of serogroups other than those mentioned above. Several investigators were able to accomplish transductions with virulent phages by means of specialized techniques (Zinder, 1955, 1957; Uetake et al., 1955; Stocker, 1958).

Transduction has been put to practical use in the

production of mutants that are useful in production and absorption of antisera (Edwards et al., 1955) and in determination of the H antigens of nonmotile strains (Bailey, 1956). See also Chapter 8.

2. In lysogenic conversion (Lederberg, 1955; Barksdale, 1959) the phage genes as such are thought to function as part of the bacterial cells. The state of lysogenicity per se, caused by infection with a phage that has converting properties, brings about changes in the O antigens. As long as the organism remains lysogenic (i.e., as long as the phage is present) these changes persist, and the particular O antigen can be demonstrated. Conversely, when the state of lysogenicity is lost (i.e., when the converting phage is lost) the O antigen involved then cannot be detected. It has long been known that O antigen, 1, for example, is present in certain bacteria within serogroup A of *Salmonella* and absent in variant forms of the same bacteria. O antigen 1 also may occur in serotypes of other serogroups or may be absent in variant forms of the same serotypes. This kind of variation now is known to be dependent upon whether the particular culture has been lysogenized. However, this should not be confused with form variation (Chapter 8). In transduction experiments in which H antigens were transduced to strains of serogroup A, Kauffmann (1953) noted that O antigen 1 also was transferred. The majority of the O antigens of *Salmonella* that are known to be subject to lysogenic conversion are recorded in table 12 (p. 214) of the review by Leuderitz et al. (1966) together with references. (The phages listed as xi in table 12 of that review should be epsilon, however.) The O antigens listed there are 1, 6, 12, 14, 15, 20, 27, 34, 37, and 42_2, and the serogroups involved are A, B, C_1, C_2, D, E, G, 18, 40, and 42. Since the above-mentioned review was written, additional papers have appeared that deal with lysogenic conversion. For example, Le Minor (1966) reported that cultures of O group 51 that possessed O antigen 1 were lysogenic, and that the phage released by these cultures lysogenized strains of 051 which then acquired factor 1. However, the 1 factor of 1, 51 strains was not identical with that of other serogroups in which factor 1 occurred. Extensive investigations of members of serogroups C_1, C_2 and other serogroups have been described (for references see: Zinder, 1957; Iseki and Kashiwaga, 1957; Le Minor, 1968a, 1968b; Mäkelä, 1965, 1966a, 1966b; Ball and Ewing, 1966; Johnson, 1967; Escobar and Edwards, 1968).

3. Recombination of characteristics also occurs by means of conjugation of *E. coli* and *Salmonella* as shown by Baron et al (1959a) and Zinder (1960), and by conjugation between different salmonellae (Baron et al., 1959b). Hybrids produced by conjugation usually have been selected and recognized by means of biochemical characters. However, transfer of antigenic characters through sexual processes, i.e., the transfer of the H antigens of both phase 1 and 2 of *S. enteritidis* serotype Abony (4, 5, 12:b:e, n, x) to *E. coli* has been demonstrated (Mäkelä, 1964). Also changes in the O antigens have been effected by conjugation (Mäkelä, 1965, 1966; Mäkelä and Mäkelä, 1966), and conjugation has been utilized in studies on S, R, S-R, and T antigens of salmonellae (Naide et al., 1965; Sarvas and Mäkelä, 1965).

In conjugation, genetic material flows from one conjugant to the other, but only in one direction. Progeny bearing recombined characters of the two parents are found in only one of two clones. The donor bacterium possesses an agent called the fertility factor (F+), and the F+ factor is responsible for changes in the protein constituents of the cell membrane of the donor. The protein differences in the membranes of donor and recipient bacteria are believed to account for the coupling phenomenon that precedes transfer of genetic material. Most donors transmit nothing more than F+ factors to the conjugating F- recipient. Rare mutant donors transmit large groups of genes, and at the same time transmit F+ factors at a greatly reduced rate. These mutant donors are called "Hfr" because, when isolated, they can transfer genetic material at high frequency. In the usual F+ population the frequency of recombinants bearing a given donor gene is in the order of 10^{-5} whereas the frequency of such recombinants obtained when an Hfr donor is used may be as high as 10^{-1} couplings (see Marmur, 1963; Ravin, 1965). Fertility factors sometimes carry genes (or genetic material), and when they do, they are capable of transferring by infection the specific genetic material that they carry. Such transfer through fertility factors is called sex-duction. Fertility factors and phages form a class of genetic agents called episomes.

Sex-duction, or F-duction, (Ravin, 1965; Marmur et al., 1963) may differ from conjugation only in the amount of the chromosomal segment transferred. Certain genetic elements, which may be transferred during conjugation, have been shown to be independent of chromosomal transfer. Such elements (episomes) may alternate between a state of fixed attachment to the chromosome or autonomy.

Transmission of colicinogeny between certain strains of ser Typhimurium also has been demonstrated (Ozeki et al., 1962). The ability to produce certain colicines was transferred to ser Typhimurium (strain LT 2) by growth in broth with colicinogenic cultures of *E. coli* and *Shigella sonnei*. Some colicine factors were readily transmissible, some were transmitted at a very low rate, and some were not transferred to a detectable extent. All of the 20 wild or naturally occurring cultures of ser Typhimurium (made colicinogenic and found to produce colicine I) readily transferred colicinogeny to ser Typhimurium strain LT 2 *cys* D- 36 *str-r,* but none of the 12 wild strains that were found to produce colicine E2 transmitted colicinogeny to a detectable extent. "Only a fraction (10^{-3} to 10^{-4}) of the bacteria in a broth culture of an LT 2 strain carrying *col I* actually transmits the factor and it appears that during long incubation of a mixed culture, *col I* spreads 'epidemically' in the acceptor population." (Ozeki et al., 1962.)

It is evident that the effects of genetic recombination cannot be ignored, especially in serotyping. The roles of the phenomena mentioned in the origin of serotypes of *Salmonella* have been the subject of much speculation.

Of the mechanisms mentioned, lysogenic conversion probably has the most bearing on serotypic identity and epidemiology, since each lysogenized organism undergoes antigenic change. A classic example was the presence of both *S. enteritidis* ser Anatum and *S. enteritidis* ser Newington in the outbreak of disease in which both types first were isolated (Rettger and Scoville, 1920; Edwards, 1937). The two serotypes often have been found in association and now it is apparent that the factor determining their identity is infection of ser Newington by phage epsilon 15. Likewise, the presence or absence of O antigens 1 and 27 in certain serotypes is not necessarily of epidemiologic significance. The list of antigens in this category is increasing and other instances involving lysogenic conversion are known (v. sup. for references). A few examples follow:

6, 7	6, 7, 14	H antigens
ser Lille	ser Bornum	z_{38}
ser Livingstone	ser Eimsbuettel	d:1, w
ser Ohio	ser Nienstedten	b:1, w
ser Oranienburg	ser Thielallee	m, t
ser Amersfoort	ser Omderman	d:e, n, x

Lysogenic conversion also has been described in serotypes of *Shigella flexneri* (Itakawa, 1964; Mulczyk, 1967). The work of Weil and Binder (1947) is notable in this regard. These investigators inoculated separate filtrates of cultures of each of the several serotypes of *S. flexneri* with individual strains of these serotypes. After incubation, the mixtures were inoculated onto plates of infusion agar, and isolated colonies were examined by serological and biochemical methods. Evidence of antigenic change was obtained in three of 225 attempts:

filtrate of ser 2b inoculated
with ser 1a ——————→ ser 1a, 2b

filtrate of ser 2a inoculated
with ser 5 ——————→ ser 5, 2a

filtrate of ser 2a inoculated
with X ——————→ ser 2a, X.

The mutants obtained were stable. It is impossible to be certain, of course, but everything considered it would appear that the first two mutants, at least, probably were the result of lysogenic conversion.

It seems likely that lysogenic conversion occurs in most, if not all Enterobacteriaceae.

The extent to which transduction and conjugation enter into serotypic changes that have epidemiological implications is debatable. Within recent years, unusually complex serologic types have been encountered and such instances certainly suggest that recombination occurred. The first of these to be described was *S. enteritidis* ser Salinatis (4, 12:d, e, h:d, e, n, z_{15}) reported by Edwards and Bruner (1942) which possessed the major H antigen d in both phases. More recently a number of similar types have been found (Douglas and Edwards, 1962; McWhorter and Edwards, 1963; McWhorter et al., 1964). These complex serotypes spontaneously may lose a major antigen and then be indistinguishable from well-known normal serotypes of *Salmonella.* The most frequently encountered example of this is ser Senftenberg (1, 3, 19:g, s, t:-) which has been found repeatedly with one or another of several different H antigens (z_{27}, z_{37}, z_{43}, z_{45}, or z_{46}). While these antigens completely mask agglutination in g, s, t antiserum (Taylor et al., 1960), they may be lost spontaneously and the cultures then behave as typical strains of ser Senftenberg.

Occasionally, cultures possessing three or even four reversible H phases have been found in nature (Edwards et al., 1962). These have been observed to occur during experiments in transduction (Spicer and Datta, 1959) and in conjugation (Hirokawa and Iino, 1961). Triphasic or quadriphasic cultures spontaneously may lose the ability to produce one or two phases and return to a stable diphasic state. Although such strains rarely are found they may cause confusion in serotyping, and unless the possibility of such occurrences is realized, possible epidemiologic connections between certain strains may not be recognized. The phenomena mentioned above may be related to the duplication of phases discussed by Lederberg (1961).

Velaudapillai (1960) reported that phage Pl 22, cultivated on *S. enteritidis* ser Chester (4, 5, 12:e, h:e, n, x), mediated transduction of somatic antigen 5 to a nonmotile strain of ser Paratyphi-C (4, 12:-:-) in embryonated and nonembryonated eggs, but not *in vitro.* Transduction of somatic antigen 5 in this manner reportedly was effected in 8 of 57 attempts. Transfer of H antigens, including those normally found in ser Paratyphi-B, also was reported.

The above-mentioned effects of genetic change on

Salmonella do not minimize the value of serological analysis to epidemiology in any way. Exceptional cultures such as those mentioned occur rarely, and the salmonellae seen in daily practice ordinarily are typical in their serological characteristics. Even such organisms as serotypes Anatum and Newington, which are lysogenic counterparts, occur separately much more frequently than in association.

Similarly, the occurrence of cultures of *Salmonella*[3] that are atypical biochemically is rare (Ewing and Ball, 1966). Attention is directed, perhaps unduly, to atypical forms because papers are written about them, whereas little is said about the thousands and thousands of typical strains. It should be emphasized that a single atypical strain may become epidemic and may be spread through a hospital or an urban community, but it still is a single strain giving rise to multiple isolations. Further, a strain of serotype Newington that ferments lactose, for example, produces symptoms in the same manner as one that does not (Report, No. 57, 1967).

[3]Part of the explanation for the occurrence of atypical strains of *Salmonella, Proteus,* and members of certain other genera of Enterobacteriaceae which ferment lactose lies in the fact that under certain conditions a fragment of genetic material (an episome) containing regulators for lactose fermentation is transferred from a bacterium that normally ferments lactose to one that normally does not (Wohlhieter et al., 1964; Falkow et al., 1964; and Easterling et al., 1969). The latter microorganism then may transfer the episome to other cells and the end result is an atypical culture of *Proteus,* for example, that ferments lactose. From experience it is known that the frequency of occurrence of the original recombinant mentioned above must be very low since they are rarely seen in nature. Further production of the recombinants in the laboratory also proceeds at a low rate and normal forms (lactose -) occur in the population along with the recombinants (lactose +). Characters acquired in this manner may be lost, in which instances the episome is lost, restricted, or destroyed, and the bacterium becomes normal again.

Another example that may be cited is the occasional occurrence of a culture of *E. coli* that produces hydrogen sulfide in media such as triple sugar iron agar. It has been known for many years that occasional strains of certain biotypes of *E. coli* produce hydrogen sulfide in such media (Galton and Hess, 1946; author's unpublished data, 1949-1950). However, hydrogen sulfide + variants of these particular biotypes have not been seen in recent years. Since 1962 the author and colleagues occasionally have received cultures that appear to be *E. coli,* but which produce hydrogen sulfide abundantly. Such strains were not seen for many years either by the author or by other investigators. It now is known (personal communication, Dr. I. Ørskov, 1970) that hydrogen sulfide production in such strains of *E. coli* is mediated by an episome. In one of 25 attempts Dr. Ørskov was able to demonstrate transfer of this character from an atypical (H_2S +) culture of *E. coli* to a normal one.

Thus the appearance of an atypical character in a strain that otherwise is typical may be explained, in many instances at least, by recombinations in which an episome carrying genetic material for that character is involved. In addition to the examples cited above this mechanism may be responsible for the appearance of other aberrancies that have been reported, e.g., the cultures of *S. sonnei* reported in Europe which ferment salicin (see Chapter 6).

Similarly, a culture of *S. typhi* into which ability to ferment lactose had been induced produced typical typhoid fever (Kunz and Ewing, 1965).

THE RESISTANCE (R) FACTORS

Susceptibility or resistance of serotypes of *Salmonella* or other Enterobacteriaceae to various antimicrobial agents and their ability, or lack of it, to produce bacteriocinlike substances have been used extensively as epidemiological markers. In some instances this led to confusion because of the appearance of multiple resistance in strains isolated from animals or humans that were not receiving all the agents for which resistance appeared (Anderson and Datta, 1965; Anderson, 1968). No doubt the apparent epidemiologic discrepancies were caused in many instances by genetic transfer by conjugation or by transduction of episomes that regulate the factors in question (Smith and Stocker, 1962; Dubnau and Stocker, 1964; Watanabe, 1963; Watanabe et al., 1968). In ser Typhimurium resistant to several antibiotics the transfer occurs either for the resistance to individual antibiotics or, more frequently, for combinations of resistance determinants. R factors determine resistance to tetracycline, chloramphenicol, ampicillin, streptomycin, kanamycin, and sulfonamides. Further, genetic transfer of resistance from ser Typhimurium to *E. coli* and vice versa has been demonstrated (Anderson and Lewis, 1965).

Resistance to tetracyclines has increased rapidly in cultures from both man and animals within the past few years (McWhorter et al., 1963). Multiple resistant salmonellae, many of which are capable of transferring resistance to other microorganisms such as *E. coli,* have been reported in the United States (Smith, 1966; Gill and Hook, 1966).

Multiple resistance of shigellae to various antibiotics was noted in Japan in 1955 (see Mitsuhashi et al., 1967). Since then it has been reported in most Enterobacteriaceae and in certain other gram-negative bacteria as well (see Watanabe, 1963; Williams and Ewing, 1964; Vivona et al., 1966; Mitsuhashi et al., 1967). Current knowledge indicates that R factors occur only in gram-negative bacteria. These episomal factors are responsible for development of multiple resistance of microorganisms to antimicrobial agents. Transfer of R factors takes place primarily through conjugation (see F-duction and sex-duction, above), but transduction is involved in some instances at least (Harada et al., 1964; Watanabe et al., 1964; Watanabe et al., 1968). The transfer of resistance occurs from donor cells to recipient cells. The donor cells are of two types Fi+ and Fi-. The transfer occurs usually at very low frequency, 10^{-6} − 10^{-5} of the donor cells, except several cell generations after the infection of the acceptor cells, which become high frequency donors, thus contributing to the infectious spread of the R factors. Physical transfer of the R factor material from cell to cell apparently takes place by means of sexual fimbriae, the presence of which also is genetically controlled. The R factors consist of deoxyribonucleic acid and resemble other bacterial episomes and plasmids. They are of two kinds or are composed of two parts: the resistance transfer factor (RTF) and the genetic determinants for drug resistance. The RTF contains the genetic material that confers the ability to conjugate and to transfer the factor to a new host. In other words, transfer of the genetic material responsible for resistance is mediated by the RTF. Transfer factors may acquire drug resistance determinants by recombination with nontransferable determinants. (See Watanabe et al., 1964; Ginoza and Painter, 1964; Falkow et al., 1966; Harold and Baldwin, 1967; Editorial, 1967; Watanabe et al., 1968; and Anderson, 1968). Transfer of R factors in nature is limited by the rate of conjugation and by the rare chances of contact between cells that possess the transfer factor and cells that possess a resistance factor. However, when a drug is introduced into a heterogeneous population of gram-negative bacteria, selective pressure (called antibiotic pressure) is exerted upon the population. This pressure facilitates the transfer and development of multiple resistant strains of bacteria, which then may become predominant. This occurs in hospitals and other institutional situations. It also occurs in animals under analogous circumstances, as shown by Anderson (1968). Other transfer factors mediate transfer of different characters.

This brief description of R factors undoubtedly is an oversimplification. Much additional information is available in the literature and interested readers are referred to the publications of Datta, 1965; Adelberg and Pittard, 1965; Lark, 1966; Pearce and Meynell, 1968; and Datta, 1968, in addition to those cited in the foregoing paragraphs.

The importance of the R factors and of the production of multiple resistant strains of various kinds of gram-negative bacteria in hospitals should be obvious. Their importance to veterinary medicine also is clear (see Symposium, 1969). Among others, Smith (1966) has said that the widespread use of antibiotics in the United States, both clinically and in animal husbandry, leads to the prediction that transferable R factors are widespread in gram-negative bacteria in this country and constitute a major public health problem.

APPLICATION OF FLUORESCENT ANTIBODY (FA) TECHNICS

The use of antibody globulins labeled with fluorescent compounds has been advocated for identification of the etiological agents in a wide variety of infectious diseases and for the detection of certain microorganisms in foods, food products, etc. Since FA methods are specialized, and since these methods and the applications and interpretations of FA techniques are given in detail in other publications, the author elected to omit them rather than include only a cursory outline or discussion. Readers are referred to the publications of Beutner (1961), Hall and Hansen (1962), Davis and Ewing (1963), Trabulsi and Camargo (1965), Cherry and Moody (1965), Ajello et al. (1966), and Cherry and Thomason (1969). These publications contain extensive bibliographies.

REFERENCES

Adelberg, E. A., and J. Pittard. 1965. Bacteriol. Revs., **29**, 161.

Ajello, G., J. Hill, and S. S. Ambrose. 1966. Investigative Urology, **3**, 486.

Anderson, E. S. 1961. Nature, **190**, 284.

————. 1968. Ann. Rev. Microbiol., **22**, 131.

Anderson, E. S., and N. Datta. 1965. Lancet, **1**, 407.

Anderson, E. S., and M. J. Lewis. 1965. Nature, **206**, 579.

Anderson, T. F. 1949. *In* The nature of the bacterial surface, ed. A. A. Miles and N. W. Pirie, p. 76. New York: Oxford University Press.

Andrewes, F. W. 1922. J. Pathol. Bacteriol., **25**, 505.

Aoki, S., M. Merkel, and W. R. McCabe. 1966. Proc. Exp. Biol. Med., **121**, 230.

Arkwright, J. A. 1920. J. Pathol. Bacteriol., **23**, 358.

————. 1921. Ibid., **24**, 36.

Baerthlein, K. 1912. Arbeit. aus dem Kaiserlichen Gesundheitsamte, **40**, 435.

Bailey, W. R. 1956. Can. J. Microbiol., **2**, 549.

Ball, M. M., and W. H. Ewing. 1966. Lysogenic conversion of somatic antigens in *Salmonella* group C. CDC Publ.*

Barker, S. A., J. S. Brimscombe, J. L. Ericksen, and M. Stacey. 1963. Nature, **197**, 899.

Barksdale, L. 1959. Bacteriol. Revs., **23**, 202.

Baron, L. S., W. F. Carey, and W. M. Spillman. 1959a. Proc. Nat. Acad. Sci., **45**, 1752.

————. 1959b. Science, **130**, 566.

Bayer, H. G., and A. L. Raegh. 1904. J. Med. Res., **12**, 313.

Beckmann, I., T. V. Subbaiah, and B. A. D. Stocker. 1964. Nature, **201**, 1299.

Bensted, H. J. 1929. *Bacillus typhosa*. A system of bacteriology, Vol. 4, Chapter I, p. 74. Privy Council.

*CDC Publ., Publication from the Center for Disease Control (formerly the Communicable Disease Center), Atlanta, Ga. 30333.

Med. Res. Council. London: H. M. Stationery Office.

Beutner, E. H. 1961. Bacteriol. Revs., **25**, 49.

Birch-Hirschfeld, L. 1935. Z. f. Hyg., **117**, 626.

Boivin, A. L., J. Mesrobeanu, and L. Mesrobeanu. 1933. Compt. Rend. Soc. Biol., **113**, 490.

Brinton, C. C. 1959. Nature, **183**, 782.

Cherry, W. B., and M. D. Moody. 1965. Bacteriol. Revs., **29**, 222.

Cherry, W. B., and B. M. Thomason. 1969. Pub. Health Rep., **84**, 887.

Cruickshank, J. C. 1939. J. Hyg., **39**, 224.

Datta, N. 1965. Brit. Med. Bull, **21**, 254.

———. 1969. Brit. Med. J., **2**, 407.

Datta, J., A. M. Lawn, and E. Meynell. 1966. J. Gen. Microbiol., **45**, 365.

Davis, B. R., and W. H. Ewing. 1963. J. Clin. Pathol., **39**, 198.

Domingue, G. J., and E. Neter. 1966. J. Bacteriol., **91**, 129.

Douglas, G. W., and P. R. Edwards. 1962. J. Gen. Microbiol., **29**, 367.

Dubnau, E., and B. A. D. Stocker. 1964. Nature, **204**, 1112.

Dudman, W. F., and J. F. Wilkinson. 1956. Biochem. J., **62**, 289.

Duguid, J. P. 1959. J. Gen. Microbiol., **21**, 271.

Duguid, J. P., I. W. Smith, G. Dempster, and P. N. Edmunds. 1955. J. Pathol. Bacteriol., **70**, 335.

Duguid, J. P., and R. R. Gillies. 1956. J. Gen. Microbiol., **15**, 6.

———. 1957. J. Pathol. Bacteriol., **74**, 397.

———. 1958. Ibid., **75**, 519.

Duguid, J. P., and H. A. Wright. 1959. Ibid., **77**, 669.

Duguid, J. P., E. S. Anderson, and I. Campbell. 1966. Ibid., **92**, 107.

Easterling, S. B., E. M. Johnson, J. A. Wohlhieter, and L. S. Baron. 1969. J. Bacteriol., **100**, 35.

Editorial. 1967. J. Amer. Med. Ass., **200**, 42.

Edwards, P. R. 1937. J. Hyg., **37**, 384.

———. 1946. J. Bacteriol., **51**, 523.

Edwards, P. R., and D. W. Bruner. 1942. Ibid., **44**, 289.

Edwards, P. R., B. R. Davis, and W. B. Cherry. 1955. Ibid., **70**, 279.

Edwards, P. R., R. Sakazaki, and I. Kato. 1962. Ibid., **84**, 99.

Ellwood, D. C. 1966. Biochem. J., **99**, 55.

———. 1970. J. Gen. Microbiol., **60**, 373.

Escobar, M., and P. R. Edwards. 1968. Can. J. Microbiol., **14**, 453.

Ewing, W. H., and M. M. Ball. 1966. The biochemical reactions of members of *the genus Salmonella*. CDC Publ.*

Falkow, S., J. A. Wohlhieter, R. V. Citarella, and L. S. Baron. 1964a. J. Bacteriol., **87**, 209.

———. 1964b. Ibid., **88**, 1598.

Falkow, S., R. V. Citarella, J. W. Wohlhieter, and T. Watanabe. 1966. J. Mol. Biol., **17**, 102.

Fuller, N. A., V. Etievant, and A. M. Staub. 1968. European J. Biochem., **6**, 525.

Galton, N. M., and M. E. Hess. 1946. J. Bacteriol., **52**, 143.

Gildemeister, E. 1916a. Zentralbl. f. Bakt. I. Orig., **78**, 129.

———. 1916b. Ibid., **78**, 209.

———. 1917. Ibid., **79**, 49.

Gill, F. A., and E. W. Hook. 1966. J. Amer. Med. Assoc., **198**, 1267.

Gilles, R. R., and J. P. Duguid. 1958. J. Hyg., **56**, 303.

Ginoza, H. S., and R. B. Painter. 1964. J. Bacteriol., **87**, 1339.

Gorzynski, E. A., and E. Neter. 1970. Bacteriol. Proc., p. 112.

Grados, O., and W. H. Ewing. 1969. Technics for characterization of soluble antigens of Enterobacteriaceae. CDC Publ.*

———. 1970. J. Infect. Dis., **122**, 100.

Hall, C. T., and P. A. Hansen. 1962. Zentralbl. f. Bakt. I. Orig., **184**, 548.

Harada, K., M. Kameda, M. Suzuki, and S. Mitsuhashi. 1964. J. Bacteriol., **88**, 1257.

Harold, L. C., and R. A. Baldwin. 1967. FDA papers, **1**, 20.

Harvey, D. 1929. *Bacillus typhosa*. A system of bacteriology, Vol. 4, Chapt. I., p. 28. Privy Council, Med. Res. Council. London: H. M. Stationery Office.

Hayes, W. 1964. The genetics of bacteria and their viruses. New York: John Wiley & Sons.

Heath, E. C., and M. A. Ghalonbor. 1963. Biochem. Biophys. Res. Commun., **10**, 340.

Herzberg, M., and J. H. Green. 1964. J. Gen. Microbiol., **35**, 421.

Hirokawa, H., and T. Iino. 1961. Ann. Rept. Inst. Genetics (Japan), **12**, 81.

Houwink, A. L. 1949. *In* The nature of the bacterial surface, ed. A. A. Miles and N. W. Pirie., p. 92. Oxford University Press.

Houwink, A. L., and W. van Iterson. 1950. Biochem. Biophys. Acta, **5**, 10.

Hungerer, D., K. Jann, B. Jann, F. Ørskov, and I. Ørskov. 1967. European J. Biochem., **2**, 115.

Iino, T., and J. Lederberg. 1964. Genetics of *Salmonella. In* World problem of salmonellosis, ed. E. van Oye. The Hague: Dr. W. Junk.

Iseki, S., and K. Kashiwagi. 1957. Proc. Japan Academy, **33**, 481.

Iseki, S., H. Ichikawa, and K. Fujisawa. 1961. Ibid., **37**, 651.

Itikawa, H. 1964. Jap. J. Genetics, **38**, 317.

Jacob, F., and E. L. Wollman. 1961. Sexuality and the genetics of bacteria. New York : Academic Press.

Jann, K., B. Jann, F, Ørskov, I. Ørskov, and O. Westphal. 1965. Biochem. Z., **342**, 1.

Jann, K., B. Jann, K. F. Schneider, F. Ørskov, and I. Ørskov. 1968. European J. Biochem., **5**, 456.

Jarvis, F. G., M. T. Mesenko, D. G. Martin, and T. D. Perrine. 1967. J. Bacteriol., **94**, 1406.

Johnson, E. M. 1967. Ibid., **94**, 2018.

Johnson, J. H., R. J. Johnson, and D. A. R. Simmons. 1967. Biochem. J., **105**, 79.

Joos, A. 1903. Zentralbl. f. Hyg., **33**, 617.

Kauffmann, F. 1935. Z. f. Hyg., **116**, 617.

———. 1936. Ibid., **118**, 318.

———. 1941. Die Batkeriologie der Salmonella-Gruppe. Ann Arbor, Michigan: (reprinted by) Edwards Bros.

———. 1943. Acta Pathol. Microbiol. Scand., **20**, 21.

———. 1953. Ibid., **33**, 409.

———. 1954. Enterobacteriaceae, 2nd ed. Copenhagen: Munksgaard.

———. 1956. Acta Pathol. Microbiol. Scand., **39**, 299.

———. 1957. Ibid., **40**, 343.

———. 1966. The bacteriology of Enterobacteriaceae. Copenhagen: Munksgaard.

Kauffmann, F., O. Luederitz, H. Stierlin, and O. Westphal. 1960a. Zentralbl. f. Bakt. I. Orig., **178**, 442.

Kauffmann, F., O. H. Braun, O. Leuderitz, H. Stierlin, and O. Westphal. 1960b. Ibid., **180**, 180.

Kauffmann, F., L. Krueger, O. Luederitz, and O. Westphal. 1961. Ibid., **182**, 57.

Kauffmann, F., B. Jann, L. Krueger, O. Luederitz, and O. Westphal. 1962. Ibid., **186**, 509.

Kessell, R. W. I., E. Neter, and W. Braun. 1966. J. Bacteriol., **91**, 465.

Kontrohr, T., and O. Westphal. 1967. Acta Microbiol. Acad. Sci. Hung., **14**, 205.

Kroeger, E. 1953. Z. f. Immunitätsf., **110**, 414.

Kunin, C. M. 1963. J. Exptl. Med., **118**, 565.

Kunin, C. M., M. V. Beard, and N. E. Halmagyi. 1962. Proc. Soc. Exptl. Biol. Med., **111**, 160.

Kunz, L. W., and W. H. Ewing. 1965. J. Bacteriol., **89**, 1629.

Lark, K. G. 1966. Bacteriol. Revs., **30**, 3.

J. Lederberg. 1951. Papers in microbial genetics. Madison, Wisconsin: University of Wisconsin Press.

———. 1955. J. Cell Comp. Physiol., **45** (Suppl. 2), 75.

———. 1961. Genetics, **46**, 1475.

Le Minor, L. 1966. Ann. Inst. Pasteur, **110**, 562.

———. 1968a. Ibid., **114**, 49.

———. 1968b. Ibid., **115**, 62.

Le Minor, L., and P. Nicolle. 1964. Ibid., **107**, 550.

Luederitz, O., A. M. Staub, and O. Westphal. 1960.

Zentralbl. f. Bakt. I. Orig., **179**, 180.

———. 1966. Bacteriol. Revs., **30**, 192.

Luederitz, O., K. Vann, and R. Wheat. 1968. *In* Comprehensive biochemistry, ed. M. Florkin and E. H. Stotz. Vol. 26A. New York: Elsevier Publishing Co.

McDonough, M. W. 1965. J. Mol. Biol., **12**, 342.

McWhorter, A. C., M. C. Murrell, and P. R. Edwards. 1963. Appl. Microbiol., **11**, 368.

McWhorter, A. C., and Edwards. P. R. 1963. J. Bacteriol., **85**, 1440.

McWhorter, A. C., M. M. Ball, and B. O. Freeman. 1964. Ibid., **87**, 967.

Mäkelä, P. H. 1964. J. Gen. Microbiol., **35**, 503.

———. 1965. Ibid., **41**, 57.

———. 1966. J. Bacteriol., **91**, 1115.

Mäkelä, P. H., and Mäkelä, O. 1966. Ann. Med. Exp. Fenn., **44**, 310.

Marmur, J., S. Falkow, and M. Mandel. 1963. Ann. Rev. Microbiol., **17**, 329.

Martin, D. G., F. G. Jarvis, and K. C. Milner. 1967. J. Bacteriol., **94**, 1411.

Mitsuhashi, S., H. Hashimoto, R. Egawa, T. Tanaka, and R. Nagai. 1967. Ibid., **93**, 1242.

Morgan, W. T. J. 1937. Biochem. J., **31**, 2003.

Mulczyk, M. 1967. Arch. Immunol. Ther. Exp., **15**, 636.

Mushin, R. 1949. J. Hyg., **47**, 227.

———. 1955. Ibid., **53**, 297.

Naide, Y., H. Nikaido, P. H. Makela, R. G. Wilkinson, and B. A. D. Stocker. 1965. Proc. Nat. Acad. Sci., **53**, 147.

Namioka, S., and R. Sakazaki. 1959. J. Bacteriol., **78**, 301.

Nowotny, A. 1963. Ibid., **85**, 427.

Ørskov, I., F. Ørskov, B. Jann, and K. Jann. 1963. Nature, **200**, 144.

Old, D. C., I. Corneil, L. F. Gibson, A. D. Thomson, and J. R. Duguid. 1966. J. Gen. Microbiol., **51**, 1.

Ozeki, H., B. A. D. Stocker, and S. M. Smith. 1962. Ibid., **28**, 671.

Pearce, L. E., and E. Meynell. 1968. Ibid., **50**, 159.

Perch, B. 1950. Acta Pathol. Microbiol. Scand., **27**, 565.

Perkins, H. R. 1963. Bacteriol. Revs., **27**, 18.

Peters, J. A. 1959. Classic papers in genetics. Englewood Cliffs, N. J. : Prentice-Hall.

Raff, R. A., and R. W. Wheat. 1966. Biochim. Biophys. Acta, **127**, 271.

———. 1968. J. Bacteriol., **95**, 2035.

Raistrick, H., and W. W. C. Topley. 1934. Brit. J. Exptl. Pathol., **15**, 113.

Ravin, A. W. 1965. The evolution of genetics. New York : Academic Press.

Reiner, E., and W. H. Ewing. 1968. Nature, **217**, 191.

Report No. 57, Salmonella Surveillance. 1967. CDC Publ.*

Rettger, L. F., and M. M. Scoville. 1920. J. Infect. Dis., **26**, 217.

Ribi, E., K. C. Milner, and T. D. Perrine. 1959. J. Immunol., **82**, 75.

Rische, H., E. Tal, and G. Seltmann. 1964. Zentralbl. f. Bakt. I. Orig., **195**, 206.

Roantree, R. J. 1967. Ann. Rev. Microbiol., **21**, 443.

Romanowskaya, E., and M. Mulczyk. 1967. Biochim. Biophys. Acta, **136**, 312.

Rowley, D. 1968. J. Bacteriol., **95**, 1647.

Salton, M. R. J. 1964. The bacterial cell wall. Amsterdam: Elsevier.

Sarvas, M., and P. H. Mäkelä. 1965. Acta. Pathol. Microbiol. Scand., **65**, 654.

Schuetze, H. 1921. J. Hyg., **20**, 330.

Seltmann, G., and S. Hoffmann. 1966. Zentralbl. f. Bakt. I. Orig., **199**, 497.

Simmons, D. A. R. 1966. Biochem. J., **98**, 903.

_____. 1969. European J. Biochem., **11**, 554.

Slopek, S., H. Godinska, E. Romanowska, and M. Mulczyk. 1967. Arch. Immunol. Ther. Exper., **15**, 612.

Smith, D. H. 1966. N. Eng. J. Med., **275**, 626.

Smith, S. M., and B. A. D. Stocker. 1962. Brit. Med. Bull., **18**, 46.

Smith, T., and A. L. Reagh. 1903. J. Med. Res., **10** (N.S.5), 89.

Spicer, C. C., and N. Datta. 1959. J. Gen. Microbiol., **20**, 136.

Stamp, L., and D. M. Stone. 1944. J. Hyg., **43**, 266.

Staub, A. M., and N. Forrest. 1963. Ann. Inst. Pasteur, **104**, 371.

Stirm, S., F. Ørskov, I. Ørskov, and B. Mansa. 1967. J. Bacteriol., **93**, 731.

Stocker, B. A. D. 1955. Proc. Soc. Gen. Microbiol., J. Gen. Microbiol., **18**, IX.

Stodola, F. H. 1958. Chemical transformations by microorganisms. New York: John Wiley & Sons.

Suzuki, T., B. A. Gorzynski, and B. Neter. J. Bacteriol., **88**, 1240.

Symposium. 1969. The use of drugs in animal feeds. Nat. Acad. Sci., Nat. Res. Council, Publ. No. 1679. Washington, D. C.

Taylor, J., M. M. Lee, P. R. Edwards, and C. H. Ramsey. 1960. J. Gen. Microbiol., **23**, 583.

Topley, W. W. C., and J. Ayrton. 1924. J. Hyg., **23**, 198.

Trabulsi, L. R., and M. E. Camargo. 1965. Rev. Inst. Med. Trop., Sao Paulo, **7**, 65.

Tripodi, D., and A. Nowotny. 1966. Ann. N. Y. Acad. Sci., **133**, 604.

Uetake, H., T. Nakagawa, and T. Akiba. 1955. J. Bacteriol., **69**, 571.

Velaudapillai, T. 1960. Z. f. Hyg., **146**, 548.

Vivona, S., T. T. Minh Ha, F. L. Gibson, and D. C. Cavenaugh. 1966. Mil. Med., **131**, 68.

Von Lingelsheim. 1913. Zentralbl. f. Bakt. I. Orig., **68**, 577.

Watanabe, T. 1963. Bacteriol. Revs., **27**, 87.

Watanabe, T., H. Nishida, C. Ogata, T. Arai, and S. Sato. 1964. J. Bacteriol., **88**, 716.

Watanabe, T., C. Furuse, and S. Sakaizumi. 1968. Ibid., **96**, 1791.

Webster, M. E., M. Landy, and M. E. Freeman. 1952. J. Immunol., **69**, 135.

Weil, E., and A. Felix. 1917. Wien. Klin. Wchnschr., **30**, 393.

_____. 1920. Z. f. Immunitätsf., **29**, 24.

Weil, A. J., and Binder, M. 1947. Proc. Soc. Exptl. Biol. Med., **66**, 349.

Westphal, O. 1960. Ann. Inst. Pasteur, **98**, 789.

Westphal, O., O. Luederitz, and F. Bister. 1952. Z. Naturforsch., **7b**, 148.

Westphal, O., and O. Luederitz. 1954. Angew. Chem., **66**, 407.

Whang, H. Y., and E. Neter. 1962. J. Bacteriol., **84**, 1245.

_____. 1963. J. Pediatrics, **63**, 412.

_____. 1964. J. Bacteriol., **88**, 1244.

Wheat, R. W., M. Berst, E. Ruschmann, O. Luederitz, and O. Westphal. 1967. Ibid., **94**, 1366.

White, P. B. 1926. Med. Res. Council Gt. Brit., Spec. Rept. Ser. No. 103.

_____. 1927. J. Pathol. & Bacteriol., **30**, 113.

_____. 1928. Ibid., **31**, 423.

_____. 1929a. Ibid., **32**, 85.

_____. 1929b. The Salmonella group. A system of bacteriology, Vol. 4, Chapt. II. Privy Council, Med. Res. Council. London: H. M. Stationery Office.

_____. 1931. J. Pathol. Bacteriol., **34**, 325.

_____. 1932a. Ibid., **35**, 77.

_____. 1932b. Ibid., **36**, 65.

Whiteside, R. E., and E. E. Baker. 1961. J. Bacteriol., **86**, 538.

Wilkinson, J. F. 1958. Bacteriol. Revs., **22**, 46.

Williams, R. B., and W. H. Ewing. 1964. The susceptibility of *Shigella* and *Escherichia* to antimicrobial agents. CDC Publ.*

Wohlhieter, J. A., S. Falkow, R. V. Citarella, and L. S. Baron. 1964. J. Mol. Biol., **9**, 575.

Zinder, N. D. 1955. J. Cell. Comp. Physiol., **45** (Suppl. 2), 23.

_____. 1957. Science, **126**, 1237.

_____. 1960. Ibid., **131**, 813.

Zinder, N. D., and J. Lederberg. 1952. J. Bacteriol., **64**, 679.

Chapter 5
The Genus *Escherichia*

The genus *Escherichia* is composed of bacteria that give similar biochemical reactions but which may be separated into many serotypes and bioserotypes. Although many definitions of this genus may be found in the literature (e.g., see Reports, 1958, International Subcommittee on Enterobacteriaceae), the author elected to employ that given by Ewing (1967) which follows:

The genus *Escherichia* is composed of motile or nommotile bacteria that conform to the definitions of the family ENTEROBACTERIACEAE and the tribe ESCHERICHIEAE. Both acid and gas are formed from a wide variety of fermentable carbohydrates but anaerogenic biotypes occur. Salicin is fermented by the majority of cultures but adonitol and inositol are utilized infrequently. Lactose usually is fermented rapidly but some strains utilize it slowly and some fail to ferment this substrate. Lysine, arginine, and ornithine are decarboxylated by the majority of cultures, acid is formed from sodium mucate, and sodium acetate is utilized as a sole source of carbon. The type species is *Escherichia coli* (Migula) Castellani and Chalmers.

Definitions for the family and the tribe may be found in Chapter 1. The generic definition should be used in conjunction with that of the tribe. (See also episomal transmission of aberrant characters, Chapter 4.)

BIOCHEMICAL REACTIONS OF ESCHERICHIA

The biochemical reactions given by cultures of *E. coli* in commonly used tests are given in Tables 4 and 5 (Chapter 3) and certain tests and substrates that are of particular value in the differentiation of *E. coli* are presented in Tables 10 and 11 (Chapter 3) and Tables 60a and 60b (Chapter 6). Further, the reactions given by 1021 strains taken at random and by 210 standard O, K, and H cultures are summarized for comparative purposes in Table 27 and in consolidated form in Table 28. Also for comparative purposes, data obtained from the examination of 390 additional cultures of *E. coli* are listed in Table 29.

Since the biochemical reactions of isolates of bioserotype 0127a:B8:NM are somewhat atypical as compared to those given by most other cultures of *E. coli*, the most important reactions given by 194 strains of this bioserotype are summarized in Table 30. The data in this table may be compared with those in Tables 27 and 28.

Certain anaerogenic, nonmotile bioserotypes that belong to specified O antigen groups of *E. coli* occur with relative frequency in various kinds of specimens. Members of O antigen group 1 are particularly common. These particular variants formerly were classified as shigellae and as the Alkalescens-Dispar (A-D) bacteria (v. inf.) and they are dealt with separately solely for purposes of discussion. These variants should be reported as *E. coli* with an appropriate note regarding the O antigen group to which they belong and their aberrancies. Summaries of the biochemical reactions given by more than 800 strains of these bacteria are given in Table 31.

The data given in Tables 27 to 31[1] are self-explanatory and no further comment is required. For further information on biochemical reactions, various biotypes within several O antigen groups and sub-O groups, etc., the reader is referred to the publications of Kauffmann and Perch (1943, 1948), Kauffmann and duPont (1950), Kauffmann (1954, 1966), Kauffmann and Ørskov (1956), Ørskov (1951, 1954a, 1954b, 1954c), Ewing et al. (1955), and to references cited in Chapter 3.

THE ANTIGENS OF *E. COLI*

In 1947, Kauffmann published an antigenic schema for *E. coli* which was based upon his earlier work and that of his collaborators, Knipschildt and Vahlne. This schema, which consisted of 25 O antigen groups, 55 K antigens, and 20 H antigens, was an extension of an earlier one devised by Kauffmann in 1944. Since 1947, many investigators have studied serotypes of *E. coli*, and as a result of their efforts, the numbers of delineated O, K, and H antigens have been extended. One hundred forty-eight O antigen groups have been characterized, 92 K antigens have been recognized and given numbers, and 51 H (flagellar) antigens are

---

[1]The data presented in Tables 27 to 31 are from a publication by Ewing et al. (in preparation).

TABLE 27

Summary of the Biochemical Reactions of *Escherichia coli*

Test or Substrate	1021 (1)			210 (2)		
	Sign	%+	(%+)*	Sign	%+	(%+)*
Hydrogen sulfide	−	0		−	0	
Urease	−	0		−	0	
Indol	+	99.2		+	95.7	
Methyl red (37 C)	+	99.1		+	100	
Voges-Proskauer (37 C)	−	0		−	0	
Citrate (Simmons')	−	0.3	(0.3)	−	0	(0.9)
KCN	−	2.7		−	0.9	
Motility	+ or −	68.9		+ or −	70.5	
Gelatin (22 C)	−	0		−	0	
Lysine decarboxylase	d	87.9	(1.2)	+	90.5	
Arginine dihydrolase	d	17.2	(44.8)	d	18.6	(5.2)
Ornithine decarboxylase	d	63.4	(7.1)	d	66.2	(1)
Phenylalanine deaminase	−	0		−	0	
Malonate	−	0		−	0	
Glucose acid	+	100		+	100	
gas	+	90.7		+	92.9	
Lactose	+ or (+)	89.6	(5.5)	+	96.7	(2.8)
Sucrose	d	51.3	(5.3)	d	37.1	(7.2)
Mannitol	+	96.2		+	93.8	
Dulcitol	d	38.1	(17.9)	d	59.5	(12.4)
Salicin	d	37.1	(13.2)	d	53.8	(17.7)
Adonitol	−	4.7	(0.4)	d	10	(1.9)
Inositol	−	1	(0.1)	−	1.4	(1)
Sorbitol	+	93.4	(0.5)	+	93.3	(3.8)
Arabinose	+	100		+	99	(0.5)
Raffinose	d	53.9	(0.5)	d	42.4	(2.4)
Rhamnose	d	74		+ or (+)	88	(5.3)
Jordan's tartrate				+	97.6	(1.9)
Pectate	−	0		−	0	
Sodium alginate	−	0		−	0	
Lipase (Corn oil)	−	0		−	0	
Maltose	d	86.3	(2.5)	+	98.1	(1.9)
Xylose	d	69.7	(11.5)	+	95.1	(2)
Trehalose	+	98.4		+	99	(1)
Cellobiose	−	3.3	(3.3)	d	1.9	(9.1)
Glycerol	+ or (+)	77.8	(19.6)	+	95.7	(2.4)
Alpha methyl glucoside	−	0		−	0	
Erythritol	−	0		−	0	
Esculin	d	30.9	(19.7)			
Nitrate to nitrite	+	99.4		+	100	
Oxidation − fermentation	Ferm	100		Ferm	100	
Oxidase	−	0		−	0	
HCA	−	6.7		−	6.2	

*Figures in parentheses indicate percentages of delayed reactions (3 or more days).

(1) 1021 cultures taken at random from materials submitted for identification.
(2) 210 standard O, K, and H antigen strains.

+, 90% or more positive within one or two days' incubation.
(+), positive reaction after 3 or more days.
−, no reaction (90% or more).
+ or −, majority of strains positive, some cultures negative.
− or +, majority of cultures negative, some strains positive.
(+) or +, majority of reactions delayed, some occur within 1 or 2 days,
d, different reactions: +, (+), −
w, weakly positive reaction.

TABLE 28

Summary of the Biochemical Reactions of *Escherichia coli*

Test or Substrate	Sign	%+	(%+)*
Hydrogen sulfide (TSI agar)	–	0	
Urease	–	0	
Indol	+	98.6	
Methyl red (37 C)	+	99.9	(0.1)
Voges-Proskauer (37 C)	–	0	
Citrate (Simmons')	–	0.2	(0.4)
KCN	–	2.4	
Motility	+ or –	69.1	
Gelatin (22 C)	–	0	
Lysine decarboxylase	d	88.7	(0.8)
Arginine dihydrolase	d	17.6	(33.2)
Ornithine decarboxylase	d	64.2	(5.3)
Phenylalanine deaminase	–	0	
Malonate	–	0	
Glucose acid	+	100	
gas	+	91.1	
Lactose	+	90.8	(5.1)
Sucrose	d	48.9	(5.6)
Mannitol	+	96.8	
Dulcitol	d	49.5	(15)
Salicin	d	40	(14)
Adonitol	–	5.6	(0.7)
Inositol	–	1.1	(0.2)
Sorbitol	+	93.4	(1.1)
Arabinose	+	99.4	(0.3)
Raffinose	d	50.9	(1)
Rhamnose	d	82.3	(2.5)
Mucate	+	91.6	(1.4)
Christensen's citrate	d	15.8	(18.6)
Jordan's tartrate	+	97.6	(1.9)
Pectate	–	0	
Sodium acetate	+ or (+)	83.8	(9.7)
Ammonium salts glucose agar	+	94.5	(1.7)
Sodium alginate	–	0	
Lipase Corn Oil	–	0	
Maltose	+ or (+)	89.9	(2.3)
Xylose	d	81.1	(7.2)
Trehalose	+	98.8	(1.2)
Cellobiose	–	2.4	(6.9)
Glycerol	+ or (+)	89.3	(8.6)
Alpha methyl glucoside	–	0	
Erythritol	–	0	
Esculin	d	30.9	(19.7)
Nitrate to nitrite	+	99.7	
Oxidase	–	0	

*Figures in parentheses indicate percentages of delayed reactions (3 or more days).

+, positive within one or two days' incubation (90% or more).

(+), positive reaction after 3 or more days.

–, no reaction (90% or more).

+ or –, majority of strains positive, some cultures negative.

– or +, majority of cultures negative, some strains positive.

(+) or +, majority of reactions delayed, some occur within 1 or 2 days.

d, different reactions: +, (+), –.

TABLE 29

Summary of Biochemical Reactions of 390 Cultures
Taken at Random During One Year

Test or Substrate	Sign	%+	(%+)*
Hydrogen sulfide	–	0	
Urease	–	0	
Indol	+	98.5	
Methyl red (37 C)	+	100	
Voges-Proskauer (37 C)	–	0	
Citrate (Simmons')	–	0	
KCN	–	0	
Motility	+ or –	77.4	
Lysine decarboxylase	+	93.3	(1.5)
Arginine dihydrolase	d	21.3	(47.3)
Ornithine decarboxylase	d	62.2	(8.7)
Phenylalanine deaminase	–	0	
Glucose acid	+	100	
gas	+	94.6	
Lactose	+	88.7	(1.3)
Sucrose	d	54.4	(1.1)
Mannitol	+	99.8	
Salicin	d	30.7	(8.6)
Adonitol	–	3.6	(0.2)
Inositol	–	1	
Sorbitol	+	98.1	
Raffinose	d	59.3	(0.3)
Maltose	+	93	(0.3)
Oxidase	–	0	

*Figures in parentheses indicate percentages of delayed reactions (3 or more days).

+, positive within one or two days' incubation (90% or more).
(+), positive reaction after 3 or more days.
–, no reaction (90% or more).
+ or –, majority of strains positive, some cultures negative.
– or +, majority of cultures negative, some strains positive.
(+) or +, majority of reactions delayed, some occur within 1 or 2 days.
d, different reactions: +, (+), –.

known. In addition, Ørskov (1956a) determined the nature of the K antigens of 16 standard O group strains, as well as of a number of other cultures of *E. coli*. K antigen numbers have not been assigned to these, however. It should be recalled that O groups 31 and 47 were cancelled because the 031 culture was found to belong to O antigen group 1 and 047 because the standard strain was lost. Further, O antigen groups 67, 72, 94, and 122 were deleted from the list after it was learned that the original cultures were members of the genus *Citrobacter* (Kauffmann, 1954). Most of the descriptive literature regarding the characterization of additional O, K, and H antigens of *E. coli* was cited by Ewing et al. (1956, 1963). Many of the O, K, and H antigens of *E. coli* are related to similar antigens possessed by members of other genera of Enterobacteriaceae (e.g., see Ewing et al., 1956; Kampelmacher, 1959; Guinée et al., 1962; Davis and Ewing, 1963; Grados and Ewing, 1969, 1970; and Chapters 6 and 17).

Serological typing of an unknown culture of *E. coli* is dependent upon the determination of its O antigen group, O antigen factors, K antigens, and H antigen. It is important to note that 0111, for example, is *not* a serotype. It is an O antigen group. Further, 0111:B4 is *not* a serotype; it is an OB serogroup. Complete serotyping involves identification of all of the antigenic components about which information is available. When this has been done the antigenic formula of a culture may be written, e.g., 0111a, 111b:B4:H2. Complete serotyping is essential for meaningful epidemiological investigations.

TABLE 30

The Biochemical Reactions of 194 Cultures of *E. coli* Bioserotype 0127a:B8:NM in Selected Tests

Test or Substrate	Sign	%+	(%+)*
Hydrogen sulfide	–	0	
Urease	–	0	
Indol	+ or –	86.6	
Methyl red (37 C)	+	100	
Voges-Proskauer (37 C)	–	0	
Citrate (Simmons')	–	0	
KCN	–	0	
Motility	–	0	
Lysine decarboxylase	–	0	(1.9)
Arginine dihydrolase	(+)	3.8	(90.6)
Ornithine decarboxylase	– or (+)	0	(47.2)
Phenylalanine deaminase	–	0	
Glucose acid	+	100	
gas	+	100	
Lactose	+	100	
Sucrose	+ or (+)	88.1	(7.1)
Mannitol	+	100	
Salicin	–	0	
Adonitol	–	0	
Sorbitol	–	0	

*Figures in parentheses indicate percentages of delayed reactions (3 or more days).

+, 90% or more, positive within one or two days' incubation.
(+), positive reaction after 3 or more days.
–, no reaction (90% or more).
+ or –, majority of strains positive, some cultures negative.
– or +, majority of cultures negative, some strains positive.
(+) or +, majority of reactions delayed, some occur within 1 or 2 days.
d, different reactions: +, (+), –.

TABLE 31

Summaries of the Biochemical Reactions of Certain Biotypes of *E. coli.*
(The Alkalescens-Dispar Bacteria)

Test or substrate	O1 bioserotype Sign	%+	(%+)*	O2 bioserotype Sign	%+	(%+)*	O3 bioserotype Sign	%+	(%+)*	O4 to O8 bioserotypes Sign	%+	(%+)*
Hydrogen sulfide	–	0		–	0		–	0		–	0	
Urease	–	0		–	0		–	0		–	0	
Indol	+	98.8		+	100		+	100		+	100	
Methyl red (37 C)	+	100		+	100		+	100		+	100	
Voges-Proskauer (37 C)	–	0		–	0		–	0		–	0	
Citrate (Simmons')	–	0	(2w)	–	0		–	0		–	0	(6.9)
KCN	–	0		–	0		–	0		–	0	
Motility	–	0		–	0		–	0		–	0	
Gelatin (22 C)	–	0		–	0		–	0		–	0	
Lysine decarboxylase	– or (+)	78.7	(12.7)	d	47.7	(6.8)	d	52.9	(11.8)	d	48.3	(13.8)
Arginine dihydrolase	d	12	(44.7)	d	6.7	(62.2)	(+) or –	0	(58.8)	(+) or –	0	(51.7)
Ornithine decarboxylase	–	3.3	(4.7)	– or +	13.3		+	94.1		d	51.7	(24.1)
Phenylalanine deaminase	–	0		–	0		–	0		–	0	
Glucose acid	+	100		+	100		+	100		+	100	
gas	–	0		–	0		–	0		–	0	
Lactose	–	1.9	(6.7)	d	2.3	(33.1)	d	22.5	(25)	d	21.4	(39.3)
Sucrose	–	2.7	(2.4)	–	0.8	(4.6)	d	5	(30)	d	30.8	(15.4)
Mannitol	+	99.3		+	97.7		+	100		+	93.1	
Dulcitol	d	79.3	(6.6)	–	0	(7.4)	d	25	(30.6)	+ or –	52.2	
Salicin	–	0.6	(1.3)	d	7	(53.9)	–	0	(2.9)	–	7.7	
Adonitol	–	0		–	0		–	0		–	0	
Inositol	–	0		–	0		–	0		–	0	
Sorbitol	+ or (+)	88	(9.5)	d	48.1	(32.1)	+ or (+)	58.1	(38.7)	d	68.8	(6.2)
Arabinose	+ or (+)	88.9	(5.9)	+	97.9	(1)	+	93.8	(6.2)	+	100	
Raffinose	–	0		–	0		d	44.8	(10.4)	d	38.9	(5.5)
Rhamnose	d	63.2	(19.5)	d	18.7	(42.9)	+	91.2		+	92.3	
Maltose	+ or (+)	88.5	(9)	+	94.5	(1.1)	+	90	(10)	+	90	
Xylose	d	80.8	(3.8)	d	76.8	(5.4)	d	41.7	(44.4)	+	100	
Cellobiose	d	7.7	(5.1)	– or +	33		d	4.2	(12.5)			
Glycerol	+	91.7	(5.6)	d	50	(25)	– or (+)	0	(35)	+	100	

* Figures in parentheses indicate percentages of delayed reactions (3 or more days).

+ Positive within one or two days' incubation (90% or more).
(+) Positive reaction after 3 or more days.
– No reaction (90% or more).
+ or – Majority of strains positive, some cultures negative.

– or + Majority of cultures negative, some strains positive.
(+) or + Majority of reactions delayed, some occur within 1 or 2 days.
d Different reactions: +, (+), –.
w Weakly positive reaction.

THE O ANTIGENS OF *E. COLI*

The O antigens are the heat-stable somatic antigens that are not inactivated by heat at 100 or 121C. The general characteristics of O antigens and variational phenomena that affect them are discussed in Chapter 4.

O antigen suspensions that are to be used for identification may be prepared from agar slant or infusion broth cultures inoculated from smooth (S) colonies selected from platings. For slide agglutination tests it is advisable to use infusion agar slant cultures that have been inoculated over the entire surface of the agar and incubated at 35 to 37 C for a maximum of 18 hr. (A shorter incubation period is preferable). After incubation the growth is removed and suspended evenly in 0.5 ml of sodium chloride (0.5%) solution to form a *dense* suspension. The suspension then is heated at 100 C for 10 or 15 min, cooled and tested. Heat treatment for this period usually suffices for preliminary work. However, it may be expected that there will be instances in which a suspension will require additional heat treatment at 100 C to inactivate the inhibitory properties of the K antigens, and there may be other instances in which it will be necessary to heat a suspension at 121 C for 2 hr. O antigen suspensions prepared as outlined above may be diluted and used for tube agglutination tests in O antisera. However, infusion broth cultures incubated in a water bath at 35 to 37 C for 4 to 6 hr are preferable for the latter purpose. Such broth cultures should be heated for 1 hr at 100 C. When cool, heated antigen suspensions may be preserved by the addition of 0.5% formalin. It is emphasized that most of the periods of heat treatment mentioned above *are not* adequate for antigens that are to be used for antiserum production (v. inf.).

The standard O antigen group cultures are listed in Table 32. These are employed for production and absorption of antisera and for determination of antigenic relationships, etc.

Relationships of O antigens. The results of cross agglutination tests in which O antigen suspensions prepared with each of the standard O group strains were tested in each of the O antisera are given in Table 33. The tests were made to ascertain the number and extent of the interrelationships between the various O antigen groups of *E. coli*, as determined by reactions obtained with antisera prepared by the author. The results subsequently were compared with those obtained by Dr. Ørskov at the International Escherichia Center, Copenhagen, and it was found that while there were a few deviations with respect to minor relationships, the results with regard to major antigenic relationships were similar.

The data contained in Table 33 reveal that although cultures belonging to certain O antigen groups react specifically, there are many strong reciprocal O antigenic relationships among cultures that belonged to other O antigen groups. Also, a number of extensive unilateral or nonreciprocal relationships are evident. Since cultures that belong to many of the O antigen groups may react to high titer in several antisera, it often is impossible to determine the O group of such strains by titration alone. Further, numerous intra-O-antigenic relationships are known (v. inf.). Therefore numerous reciprocally absorbed O antisera are required for complete antigenic characterization of cultures of *E. coli*.

O ANTIGEN DETERMINATION

O Antiserum Pools. Heated O antigen suspensions prepared from unknown cultures of *E. coli* are tested first in O antiserum pools in single tube tests in which the final dilution of each component O antiserum is 1 to 1000. Eight pools, organized according to close antigenic relationships were used by Ewing et al. (1956) for this preliminary procedure, but these were modified (Ewing and Davis, 1961b) as a result of studies on the incidence of members of the various O antigen groups among materials examined over an eleven-year period (Ewing and Davis, 1961a). For example, it was found that about 26 per cent of the total number of cultures examined reacted in O antiserum pool No. 1, which contained antisera for 22 O antigen groups (Table 34). However, 82.5 per cent of the strains that reacted in pool No. 1 belonged to 8 O antigen groups, while 14.5 per cent belonged to 7 other O groups, and 3 percent belonged to the remaining 7 O antigen groups, for which antisera were present in the pool. Hence, pool No. 1 was divided into 1A (8 antisera), 1B (7 antisera) and 1C (7 antisera). Pools 2 to 7 inclusive were divided into 2A, 2B, 2C, etc. on the same basis, since in every instance from 78 to 92 per cent of cultures examined belonged to 6 to 8 O antigen groups represented in the A pools. Since the majority of cultures reacted in only one of the subdivided pools (e.g., 1A), further examination of a strain usually required the use of 6 to 8 individual O antisera instead of 13 to 22 or more as was the case with the original pools. Some cross reactions occurred in the above-mentioned subdivided pools, but these generally formed patterns that are easily recognized after some experience with the use of the pools was acquired.

The above-mentioned percentage figures are based upon data obtained during the examination of materials from human and animal sources submitted over a period of years. Hence, the figures are believed to be fairly indicative of the reactions that may be expected.

TABLE 32

Standard O Group Strains

O Group	K Antigen (Variety)	H Antigen	Culture No.	O Group	K Antigen (Variety)	H Antigen	Culture No.
1	1 (L)	7	U5-41	37	·	10	H 510c
2	1 (L)	4	U9-41	38	·	26	F 11621-41
3	2a,2b (L)	2	U14-41	39	·	NM	H 7
4	3 (L)	5	U4-41	40	·	4	H 316
5	4 (L)	4	U1-41	41	·	40	H 710c
6	2a,2c (L)	1	Bi 7458-41	42	·	37	P 11a
7	1 (L)	NM	Bi 7509-41	43	·	2	Bi 7455-41
8	8 (L)	4	G 3404-41	44	74 (L)	18	H 702c
9	9 (L)	12	Bi 316-42	45	·	10	H 61
10	5 (L)	4	Bi 8337-41	46	·	16	P 1c
11	10 (L)	10	Bi 623-42	48	·	NM	U8-41
12	5 (L)	NM	Bi 626-42	49	·	12	U12-41
13	11 (L)	11	Su 4321-41	50	·	NM (4)	U18-41
14	7 (L)	NM	Su 4411-41	51	·	24	U19-41
15	14 (L)	4	F 7902-41	52	·	10	U20-41
16	1 (L)	NM	F 11119-41	53	·	3	Bi 7327-41
17	16 (L)	18	K 12a	54	·	2	Bi 3972-41
18	76 (B20)	14	F 10018-41	55	59 (B5)	NM	Su 3912-41
19a	·	·	F 8858-41	56	·	NM	Su 3684-41
19b	·	7	F 8188-41	57	·	NM	F 8198-41
20	17 (L)	NM	P 7a	58	·	NM (27)	F 8962-41
21	20 (L)	NM	E 19a	59	·	19	F 9095-41
22	13 (L)	1	E 14a	60	·	NM (33)	F 10167a-41
23	18 (L)	15	E 39a	61	·	19	F 10167b-41
24	·	NM	E 41a	62	·	30	F 10524-41
25	19 (L)	12	E 47a	63	·	NM	F 10598-41
26	60 (B6)	11	H 311b	64	·	NM	K 6b
27	·	NM	F 9884	65	·	NM	K 11a
28	·	NM	K 1a	66	·	25	P 6a
29	·	10	Su 4338-41	68	·	4	P 7d
30	·	NM	P 2a	69	·	38	P 9b
32	·	19	P 6a	70	·	38	P 9c
33	·	NM	E 40	71	·	12	P 10a
34	·	10	H 304	73	·	31	P 12a
35	·	10	E 77a	74	·	39	E 3a
36	·	9	H 502a	75	·	5	E 3b

·, Nature of K antigen undetermined; H undetermined.
NM, Nonmotile, (   ), H antigen of motile variant.
K–, K antigen lacking.
*, Unnumbered K antigen.

TABLE 32 (Continued)
Standard O Group Strains

O Group	K Antigen (Variety)	H Antigen	Culture No.	O Group	K Antigen (Variety)	H Antigen	Culture No.
76	·	8	E 5d	113	75 (B19)	21	6182-50
77	·	NM	E 10	114	·	32	W 26
78	·	NM	E 38	115	·	18	W 27
79	·	40	E 49	116	·	10	W 28
80	·	26	E 71	117	·	4	W 30
81	·	NM	H 5	118	·	NM	W 31
82	·	NM	H 14	119	·	27	W 34
83	·	31	H 17a	120	·	6	W 35
84	·	21	H 19	121	·	10	W 39
85	·	1	H 23	123	·	16	W 43
86	K-	25	H 35	124	72 (B17)	32	227
87	—	12	H 40	125	70 (B15)	19	Canioni (2745-53)
88	·	25	H 53	126	71 (B16)	2	E 611 (6021-50)
89	·	16	H 68	127	63 (B8)	NM	Holcomb (4932-53)
90	·	NM	H 77	128	68 (B12)	2	Cigleris (56-54)
91	·	NM	H 307b	129	(B)*	11	178-54 (1986-54)
92	·	33	H 308a	130	(B)*	9	4866-53
93	·	8	H 308b	131	·	26	HW 27
95	·	NM (33)	H 311a	132	·	28	HW 30
96	·	19	H 319	133	·	29	HW 31
97	·	NM	H 320a	134	·	35	4370-53
98	·	8	H 510d	135	·	NM	Coli pecs
99	·	33	H 504c	136	78 (B22)	NM	1111-55
100	·	2	H 509a	137	79 (L)	41	RVC 1787
101	·	33	H 501a	138	81 (B)	14	62-57
102	·	8	H 511	139	82 (B)	1	63-57
103	·	8	H 515b	140	·	43	149-51
104	·	12	H 519	141	85 (B)	4	RVC 2907
105	·	8	H 520b	142	86 (B)	6	C 771
106	·	33	H 521a	143	·	NM	4608-58
107	·	27	H 705	144	·	NM	1624-56
108	·	10	H 708b	145	·	NM	E 1385 (3)
109	·	19	H 709c	146	·	21	2950-54
110	·	39	H 711c	147	89 (B)	19	G 1253
111	58 (B4)	NM	Stoke W	148	·	28	E 519-66
112	68 (B13)	18	1411-50				

TABLE 33

O Antigenic Relationships of *E. coli*

O Antiserum	Homologous Titer	Titer With Other O Antigens				
1	20,480	2:5,120	10:1,280	14:320	50:1,280	
		64:640	70:320	117:2,560		
2	20,480	1:640	50:320	74:10,240		
3	20,480	13:640	23:640	51:320	70:640	73:640
4	20,480	7:1,280	12:320	13:1,280		16:1,280
		18:5,120	19a,2,560	19a,19b:2,560		25:640
		26:1,280	36:320	51:640	62:320	102:2,560
5	20,480	7:160	70:1,280	114:1,280		
6	20,480	*				
7	20,480	5:640	19a,320	19a,19b:320		25:320
		36:320	71:2,560	116:10,240		141:320
8	20,480	60:160	93:160			
9	20,480	*				
10	20,480	51:640				
11	20,480	*				
12	20,480	4:160	15:640	16:320	51:320	
13	20,480	1:320	3:640	16:160	18:320	19a:1,280
		19a,19b:2,560		25:160	50:320	53:1,280
		62:640	69:320	73:640	99:320	129:1,280
		133:1,280	135:640	147:160		
14	20,480	*				
15	10,240	12:160	51:640	60:640		
16	20,480	3:320	4:320	13:640	94:320	
17	20,480	51:640	62:2,560	73:5,120	77:20,480	106:640
18	20,480	4:2,560	13:2,560	16:320	19a:2,560	19a,19b:640
		25:1,280	62:320	102:640		
19a	5,120	13:20,480	19a,19b:20,480	33:2,560	39:640	
		51:1,280	102:320	147:160		
19a,19b	20,480	13:2,560	19a:2,560	48:1,280	86:320	102:320
20	10,240	*				
21	10,240	32:320	51:2,560	83:640		
22	20,480	3:320	51:320	76:2,560	83:2,560	130:160
23	20,480	3:640	38:1,280		51:640	
24	5,120	*				
25	20,480	4:2,560		7:5,120	13:5,120	18:10,240
		19a:10,240		19a,19b:5,120		26:1,280
		36:160		62:160	68:640	79:320
		102:1,280				
26	20,480	4:320		13:1,280	19a,19b:640	
		25:640		35:320	62:320	68:640
		102:1,280		117:640		
27	20,480	46:640		64:640		
28	10,240	*				
29	20,480	28:320				
30	10,240	21:640		51:640		
32	20,480	21:1,280		83:640		
33	20,480	19a:160				
34	10,240	19a:640		85:320	140:1,280	

Insignificant cross reactions, indicated by a titer of <320, are not recorded unless they happen to be reciprocal.

*Indicates no significant cross reactions.

TABLE 33 (Continued)

O Antigenic Relationships of *E. coli*

O Antiserum	Homologous Titer	Titer With Other O Antigens				
35	20,480	*				
36	20,480	7:160		11:640	25:1,280	102:160
		109:1,280		68:160		
37	5,120	48:5,120		71:320	81:640	
38	10,240	23:2,560		80:640		
39	10,240	14:320		91:320		
40	10,240	*				
41	10,240	*				
42	10,240	62:640				
43	20,480	36:640		118:5,120		
44	10,240	13:1,280		62:2,560	68:640	73:1,280
		77:1,280		106:640		
45	20,480	46:320		62:640	66:20,480	
46	20,480	18:460		102:640		
48	20,480	19a,19b:5,120	37:640	54:640	59:2,560	128:1,280
49	20,480	117:640				
50	20,480	1:1,280	2:320	13:5,120		18:320
		19a:640	19a,19b:2,560			24:320
		25:640	26:320	41:640		107:1,280
		117:10,240	135:320			
51	20,480	117:320	147:640			
52	20,480	87:640				
53	20,480	34:320	108:320			
54	20,480	48:2,560	59:1,280	105:160		
55	20,480	*				
56	10,240	62:640				
57	10,240	6:2,560	41:640			
58	20,480	16:640	45:640	76:320		
59	20,480	48:2,560	54:1,280			
60	20,480	8:640	73:1,280	93:1,280		
61	20,480	108:320	136:1,280			
62	20,480	4:640	13:1,280	16:10,240		17:640
		18:320	25:320	40:5,120	60:640	68:5,120
		70:1,280	73:1,280	106:160	108:640	137:640
63	20,480	50:1,280	54:640	116:320		
64	20,480	1:640	51:640	74:640		
65	20,480	5:10,240	7:640	46:320	70:640	71:10,240
		74:320	116:640	119:640		
66	20,480	13:640	45:2,560	46:320		
68	20,480	7:320	13:640	18:320	19a,19b:320	
		25:5,120	26:1,280	36:10,240	44:160	62:10,240
		102:640				
69	20,480	13:160	19a,19b:320		51:5,120	62:320
						147:640
70	20,480	3:160	5:10,240	13:640	19a,19b:320	
			44:320	45:320	65:5,120	71:1,280
			73:640	74:2,560	103:320	114:320
			116:2,560			

TABLE 33 (Continued)

O Antigenic Relationships of *E. coli*

O Antiserum	Homologous Titer	Titer With Other O Antigens				
71	20,480	5:10,240	7:640	26:1,280	38:640	65:20,480
		70:640	102:640	114:160	116:1,280	117:320
73	20,480	3:1,280	4:1,280	13:2,560	17:1,280	23:640
		44:5,120	52:2,560	53:1,280	60:1,280	62:1,280
		68:1,280	70:1,280	77:2,560	88:320	99:1,280
		106:160	108:320	125:1,280		
74	20,480	2:5,120	40:2,560	50:320	73:320	
75	20,480	126:1,280				
76	20,480	22:20,480	30:1,280	41:640	101:640	
77	20,480	17:10,240	44:2,560	66:640	73:320	
78	20,480	38:1,280	42:1,280	59:640	62:320	66:640
		69:320	71:640	73:1,280	92:640	103:320
		115:1,280	116:2,560	123:640	137:160	
79	20,480	14:320	19a,19b:640	25:320	32:320	41:2,560
		68:320	83:640			
80	20,480	41:1,280	79:10,240			
81	20,480	37:2,560	51:5,120	61:1,280	100:320	
82	10,240	116:320				
83	20,480	21:320	22:1,280	32:2,560	46:1,280	
84	10,240	*				
85	10,240	34:5,120	140:640			
86	20,480	19a,19b:20,480	23:320	48:10,240	90:1,280	127:2,560
87	10,240	41:2,560	48:2,560	76:2,560	96:320	115:1,280
		116:2,560	117:1,280		119:320	128:320
88	20,480	59:320	78:320		116:640	
89	5,120	78:1,280	93:1,280		115:5,120	116:320
90	5,120	19a,19b:320	48:640		86:640	127:1,280
91	10,240	95:1,280				
92	20,480	19a:2,560	63:1,280	78:1,280	81:320	91:2,560
		97:640				
93	20,480	6:320	48:2,560			
95	20,480	2:320	39:640	50:1,280	74:640	91:640
		103:320				
96	10,240	41:2,560	115:1,280	117:320		
97	20,480	*				
98	20,480	*				
99	20,480	13:320	44:320	56:640	63:2,560	73:640
		92:320	97:640			
100	10,240	10:640	37:320	87:640		
101	20,480	38:320	49:320	76:640	116:640	117:2,560
		122:1,280				
102	20,480	4:640	13:320	18:320	19a,19b:640	
		25:10,240	26:5,120	36:2,560	46:640	
103	20,480	*				
104	20,480	*				
105	20,480	54:640	96:320			

TABLE 33 (Continued)

O Antigenic Relationships of *E. coli*

O Antiserum	Homologous Titer	Titer With Other O Antigens				
106	10,240	17:10,240 73:5,120	40:640	44:10,240	62:5,120	72:320
107	10,240	50:160 95:320	56:640 102:640	60:640 117:5,120	62:320 121:640	87:640 123:1,280
108	20,480	53:1,280	61:640			
109	20,480	1:320	73:320			
110	20,480	*				
111	20,480	43:320	66:640	80:320	116:1,280	119:320
112	20,480	48:640	81:640	113:640		
113	20,480	112:640	116:640	117:5,120		
114	20,480	48:2,560	70:640	71:640	82:320	
115	20,480	94:640	109:320			
116	20,480	7:640	38:320	63:160	123:2,560	
117	20,480	50:160	76:2,560	101:640	122:2,560	135:640
118	20,480	121:5,120	123:5,120		43:160	
119	20,480	48:10,240				
120	10,240	53:2,560 117:1,280	102:2,560	105:5,120	115:1,280	
121	20,480	26:320 87:640	34:640 101:2,560	41:1,280 116:320	48:320 78:320 117:1,280	81:640 123:10,240
123	10,240	12:640	78:160	121:5,120		
124	20,480	*				
125	20,480	11:1,280	17:640	73:1,280		
126	5,120	68:160	75:1,280			
127	20,480	86:2,560	90:2,560	128:640		
128	5,120	48:320	127:160			
129	20,480	13:5,120	133:1,280	135:640	141:640	
130	10,240	22:320				
131	5,120	*				
132	10,240	140:640				
133	20,480	13:2,560	129:80	135:320	147:1,280	
134	10,240	*				
135	10,240	13:5,120	129:160	133:5,120	147:320	
136	5,120	*				
137	10,240	*				
138	20,480	39:320	148:640			
139	5,120	102:640				
140	5,120	34:640				
141	5,120	*				
142	5,120	102:320				
143	2,560	*				
144	10,240	112:320				
145	2,560	15:640				
146	5,120	*				
147	5,120	13:1,280	19:1,280	102:1,280	133:5,120	135:640
148	20,480	1:640	138:320			

TABLE 34

Composition of Pools of O Antisera

		Pool Number						
	1	2	3	4	5	6	7	8
A	1a, 1b 2a, 2b 3 4 6 7 25 26 50	8 9 11 15 73 125a, 125b 125a, 125c	20 21 22 28a, 28b 28a, 28c 33 76 81 83	39 44 45 51 54 55 66 108	62 75 78 86a 126 127a 128a, 128b 128a, 128c	88 89 91 93 98 99 104 105 117	111a, 111b 111a, 111c 112a, 112b 112a, 112c 113 119	A-D0: X1* X2 X3 X5 X6 X8 X9 X1: X1:
B	18a, 18b 18a, 18c 23 70 71 114 138 139	12 13 16 17 60 77	24 27 32 34 36 37 85	40 41 46 49 118 145	61 68 69 80 92 124	96 97 100 101 103 109	110 120 121 123 130 131 132 134 136 137 140 141 142 143 144	
C	5 38 64 65 74 102 116	10 14 19a, 19b 129 133 135 147	29 30 35 42 48 4838-69	43 52 53 56 59 106	58 63 79 82 84 90 115 146			

*O antigen groups to which numbers have not been assigned.

However, it should be remembered that these figures may be expected to fluctuate somewhat depending upon whether the cultures of *E. coli* under examination are from more, or from less, highly selected sources than those studied by the author. For example, Ewing and Davis (1961a) reported that about 70 per cent of the cultures of *E. coli* that they examined from urinary tract infections in the human were members of nine O antigen groups (01, 2, 4, 6, 7, 11, 15, 62, and 75). Since members of at least three of these O antigen groups (02, 7, and 62) may react in two or more of the subdivided pools, one would expect a greater percentage of cross reactions than that indicated above if one were dealing only with cultures of *E. coli* from this source. However, investigators interested in *E. coli* from urinary tracts infections in the human might find it profitable to employ a pool composed of O antisera for the above-mentioned nine O antigen groups for preliminary examination of their strains. Other special purpose pools may be devised as needed.

The subdivided O antiserum pools may be prepared in the following manner. To prepare pool 1A, for example, 1 ml of each of the eight constituent O antisera (Table 34, part 1) is placed in a graduated cylinder and the volume is brought to 25 ml by the addition of phenolized (0.5%) physiological saline solution (in this instance 17 ml are added). Since the antisera are preserved with an equal volume of glycerol, the dilution of each component O antiserum is doubled and hence is 1:50 in the diluted pool. The other O antiserum pools are prepared in a similar manner except that the amount of phenolized physiological saline solution added varies in accordance with the number of individual O antisera contained in each.

O antiserum pools diluted in the above-mentioned manner may be used in a final dilution of 1:1000 depending upon the individual titers of the constituent O antisera (see Table 35). Five one hundredths (0.05) ml of each of the diluted pools is placed in a separate tube in a series and 0.95 ml (1.0 ml may be used) of heated, appropriately diluted, broth culture of the strain under examination is added to each tube in the series and to a control tube. The tests are incubated at 48 to 50 C for 6 to 8 hr or overnight. Broth cultures that have been heated at 100 C for 1 hr are tested first in the various pools of O antisera, but if results are negative the tests are repeated using broth cultures that have been heated at 121 C for 2 hr and 1:500 dilutions of antisera. Subsequently, heated antigen suspensions are tested in single tube tests at a dilution of 1:500 in each of the individual O antisera that comprise the pool, or pools, in which a reaction occurred, by titration in O antisera, and by slide tests in absorbed O antisera as indicated.

TABLE 35

Example of Tests with O Antigen Suspensions Prepared with an Unknown Culture of *E. coli*

O Antiserum Pools (1:1000)

1A	2A	3A	4A	5A	6A	7A	1B	2B	3B	4B	5B	6B	7B	1C	2C	3C	4C	5C
—	—	—	—	++++	—	—	—	—	—	—	—	—	—	—	—	—	—	—

Individual O Antisera (Components, Pool 5A)

062	075	078	086	0126	0127	0128
—	—	—	++++	—	—	—

Titration in O antiserum for 086 and slide tests in absorbed O antisera

The subdivided O antiserum pools may be employed in two or more ways, but at least until some experience with their use has been acquired, it is advisable to test all cultures in all three of the components of each pool (i.e., in 1A, 1B, 1C; 2A, 2B, 2C, etc.). Alternatively, O antigen suspensions may be tested first in all of the A pools (1A, 2A, 3A, etc.) and, if a reaction occurs in one of these, one may proceed to single tube agglutination tests in the individual O antisera contained in that pool and, subsequently, to titration, etc., for confirmation of the O antigen group. If this is done, however, it must be remembered that agglutination in a dilution of 1:500 in one of the individual O antisera does not in itself constitute proof that the culture is a member of that O antigen group. The strain may belong to a related O antigen group, hence titrations and other confirmatory tests never should be omitted. Antigen suspensions that are not agglutinated by the O antisera contained in the A pools then are tested in the B and C pools.

When cross reactions occur in two or more of the subdivided pools, steps to be taken in the further examination of the antigen suspension may be indicated by the degree or extent of the agglutination reactions obtained. Suspensions may be examined first in the component O antisera of the pool or pools in which complete agglutination occurred and partial agglutination reactions (1+ or 2+) may be investigated later, if necessary. In the event that only weakly positive agglutination is apparent in one or more of the pools, all such reactions should be investigated. Also, it may be necessary to reinvestigate cultures that give weak reactions, using O antigen suspensions that have been heated at 121 C for 2 hr.

In the case of many cultures of *E. coli* the above-mentioned procedures for O grouping may suffice and further examination of the O antigens may not be warranted. However, several of the serotypes associated with diarrheal disease, for example, contain O antigenic factors that are present in other O antigen groups.[2] In such instances further examination of the O antigen components is necessary if they are to be identified. Also, many of the O groups may be subdivided into two or more subgroups which may be identified by analysis of the component O antigen factors. For

analysis of the factors involved in such relationships, absorbed O antisera may be employed in slide agglutination tests (v. inf.).

Absorption of O Antisera. Cell suspensions (antigen) for absorption of agglutinins may be prepared by inoculating moist, thickly poured (about 30 ml per plate) 90 mm infusion agar plates each with about 0.5 ml of an 18- to 24-hr culture of the selected strain(s). The innoculum should be spread over the entire surface of the agar to assure confluent growth. The plates are incubated in an upright position for 18 to 24 hr, after which the growth is removed. If each plate is flooded with 2 or 3 ml of phenolized (0.5%) physiological saline solution, the growth may be removed easily with the aid of a bent wire or glass rod scraper. The growth from the plates is pooled and the heavy suspension is centrifuged at high speed in an angle centrifuge. The clear supernatant fluid is decanted and diluted antiserum is added to the packed cells. Ordinarily at least 5 ml of antiserum diluted to 1 to $10^{(3)}$ is used, but larger volumes may be absorbed if desired (v. inf.). The bacterial cells should be suspended evenly in the antiserum. The mixture then is incubated at 48 to 50 C (preferably in a waterbath) for two to four hours. The mixtures should be agitated at frequent intervals during incubation. Following this, absorbing suspensions may be stoppered and refrigerated if desired. After incubation the mixture is centrifuged at high speed and the clear absorbed antiserum is removed and preserved by the addition of a small amount (about 2 drops per ml) of chloroform or with another suitable preservative.

The amount of bacterial cells used in an absorption must be sufficient to remove all agglutinins for the absorbing culture(s) from the antiserum. This amount varies with different antisera and with the agglutinin titer of an antiserum for the homologous and related microorganisms. The growth from five 90 mm plate cultures, as described above, ordinarily is sufficient to remove heterologous agglutinins from one ml of glycerolated antiserum. However, it may be expected that there will be instances in which the growth from a larger number of plates will be required.

Although single absorption (addition of all of the absorbing cells to the antiserum at one time) of an antiserum often is effective, the double or two-step absorption procedure is recommended because it is more effective than single absorption. In a double absorption the cell suspension is divided into two equal parts prior to centrifugation. After centrifugation the

[2]It should be emphasized that the reactions and relationships mentioned herein are those obtained with lots of antisera prepared by the writer and that some of the relationships may not be apparent in antisera prepared in other laboratories. It is a recognized fact that antisera prepared in different laboratories with the same cultures may vary in their agglutinin content for minor factors. It is believed that this may be caused by differences in the titers of such antisera, to variations in the manner in which individual animals respond to certain agglutinogens, and to unknown variations in the minor factors.

[3]If antisera are preserved by the addition of an equal volume of glycerol, the preserved antiserum is considered to be diluted 1 to 2. Thus, a 1 to 10 dilution is made by addition of 1 ml of glycerolated antiserum to 4 ml of diluent.

antiserum is added to half of the absorbing cells and the mixture is incubated at 48 to 50 C for 1 to 2 hr then the mixture is centrifuged and the supernatant antiserum is added to the remaining portion of absorbing cells. This is followed by a second period of incubation (1 to 2 hr) and final centrifugation. The need for thorough emulsification of the absorbing antigen (cells) is emphasized. If an antiserum is to be absorbed with two or more different cultures it is advisable to pool all of the growth harvested from the plates, mix it, and then divide it into two approximately even portions.

When relatively large amounts of antiserum are to be absorbed it is expedient to cultivate the bacteria on infusion agar in 150 mm petri dishes (200 to 250 ml of medium per plate) or other suitable large containers. The diameter of the bottom of a so-called 150 X 25 mm plastic petri dish actually is about 142 mm, while that of a 90 mm plate is 88 mm. Therefore the area (d^2 X 0.7854) of the larger plate is about 2.6 times that of the smaller 90 mm plate.

Absorption of antiserum may be regarded as an agglutination reaction performed for the selective removal of unwanted agglutinins. These unwanted (heterologous) agglutinins are bound by the reaction between the antigens of the cells and the agglutinins present in the antiserum. After the reaction takes place, the agglutinins bound to the cells are removed with the cells by centrifugation.

Absorbed O antisera should be tested with heated antigen suspensions prepared from both the homologous and absorbing culture(s). These initial tests should be performed by means of serial dilutions, e.g. 1 to 40 to 1 to 5,120, and antigens prepared from the absorbing culture(s) should not react at 1 to 40 dilution. If the results of titrations are satisfactory, absorbed O antisera are used in slide agglutination tests at the highest dilution that yields rapid complete agglutination of the homologous culture.

Since many investigators may not be interested in complete characterization of the O antigens of cultures of *E. coli* from all sources, a list of all of the absorbed O antisera required for this is not given.[4] It appears that the majority of investigators are interested primarily in cultures from certain sources, e.g., from urinary tract infections, diarrheal disease in the human, intestinal and other diseases in animals, etc. and the number of absorbed antisera required in these specialty laboratories is smaller than the number needed in a reference laboratory. The data given in Table 33 may be used as a guide to the absorption of O antisera, but

in the last analysis all newly prepared antisera must be tested for relationships. Parenthetically, there is a definite need for additional reference centers in which strains from any source may be characterized and it is hoped that as antisera become more readily available from commercial sources more such laboratories will be established.

The absorbed antisera that are required for characterization of the O antigens and O antigen factors of the serotypes of *E. coli* that are generally accepted as being associated with diarrheal disease in children are listed in Table 36. In addition the several kinds of O antigenic relations that may exist between and within these particular O antigen groups of *E. coli* are illustrated in Tables 37 to 41. The data given in these tables are analogous to those obtained with O groups other than those listed and the principles involved are applicable to all O antigen groups. The procedures employed are based upon those of Kauffmann (1947) and collaborators but have been modified in certain respects (Ewing et al., 1956; Ewing, 1956, 1963, 1969).

TABLE 36

Absorption of *E. coli* O Antisera

E. coli O Antiserum 1.0 ml	Absorbed by *E. coli* O group
26	4+ 18+ 25+ 102
86a, 86b	86a+ 90
86a	90
90	127a
127a	90
127a, 127b	127a+ 90
111a, 111b	111a, 111c
111a, 111c	111a, 111b
112a, 112b	112a, 112c
112a, 112c	112a, 112b
119	48
125a, 125b	73
125a, 125b	125a, 125c
125a, 125c	125a, 125b
126	75
75	126
128a, 128b	87
87	128a, 128b
48	128a, 128b+ 87
128a, 128b	128a, 128c
128a, 128c	128a, 128b

[4]A list of most of the absorbed O antisera that are employed by the writer and colleagues was given by Ewing et al. (1956).

O Groups 25 and 26. The O antigens of cultures that belong to O groups 25 and 26 are related to each other (Table 37) and to those of a number of other O antigen groups as well. The relationship shown in Table 37 is known as an a,b - a,c relationship. The factor that is common to the two cultures and which accounts for the cross agglutination in unabsorbed antisera is labeled a, while the factors that remain after reciprocal absorption are labeled b and c, respectively. However, these two absorbed antisera are not entirely specific because cultures of O groups 4, 18, 62, and 102 may react in O 25 antiserum absorbed with O 26 and cultures of O groups 4, 18, and 102 may react in O 26 antiserum absorbed by O 25. Hence, *E. coli* O 26 antiserum should be absorbed by suspensions of O groups 4, 18, and 25, as shown in Table 36. *E. coli* O group 25

antiserum may be absorbed by O groups 4, 18, 26, and 102 to render it more specific.

O Group 55. The O antigens of cultures that belong to *E. coli* O group 55 are not related significantly to those of other *E. coli* O antigens but the O group has been subdivided. Ørskov and Fey (1954) described cultures of O group 55 that differed from the type culture in that they contained an O antigenic factor that was not present in the type culture (i.e., O 55a and O 55a,55b).[5] Although no absorption is listed in Table 36 for the preparation of antiserum for differentiation of cultures that belong to these subgroups, such antiserum could be prepared by absorption of O 55a,55b antiserum with a suspension of O 55a. It should be noted that some *E. coli* O 55 antisera, prepared in this

TABLE 37

Relationship of the O Antigens of *E. coli* O 25 and O 26 Cultures

O Antigen Suspensions (100 C, 1 hr.)	*E. coli* O Antisera			
	25		26	
	Unabsorbed	Absorbed by *E. coli* O 26	Unabsorbed	Absorbed by *E. coli* O 25
E. coli O 25	20,480	10,240	1,280	0
E. coli O 26	1,280	0	20,480	20,480

TABLE 38

The Interrelationship of the O Antigens of *E. coli* Cultures of O Groups 86, 90, and 127.

O Antigen Suspension (100 C, 1 hr.)	*E. coli* O Antisera (unabsorbed)				
	86a	86a, 86b	90	127a	127a, 127b
86a	5,120	5,120	640	0	5,120
86a, 86b	20,480	20,480	640	0	10,240
90	2,560	160	5,120	1,280	5,120
127a	1,280	0	640	20,480	5,120
127a, 127b	10,240	640	1,280	10,240	20,480

[5]See Table 39 for an example of the a-a,b variety of relationship.

and other laboratories, react with *E. coli* O 22 cultures. The reason for this phenomenon is unknown but O 55 antiserum should be tested with a suspension of *E. coli* O 22 and absorbed with it if necessary.

O Group 86. The important O antigenic relationships of *E. coli* O 86 cultures are listed in Table 38 and the relationship between the O antigens of O 86a and O 86a,86b cultures are given in Table 39 (Ewing, Tanner and Tatum, 1955). It will be noted that in the type of relationship (a - a,b) shown in Table 39, only one of the cultures contains a special factor (b). The absorbed O antisera advised for the differentiation of *E. coli* O 86 cultures are given in Table 40 together with the reactions that may be expected.

O Group 111. The O antigens of *E. coli* O 111 cultures are not related significantly to those of other *E. coli* O antigen groups but the O group may be subdivided into at least three factors. Ewing, Galton, and Tanner (1955) described the relationships extant between

cultures of *E. coli* O group 111 from cases of infantile diarrhea (O 111a,111b: B4) and strains from monkeys (O 111a,111c: B4). Reciprocally absorbed O antisera may be used to differentiate these subgroups. Also, Ørskov (1954b) described a culture of *E. coli* O 111 which contained a factor not present in strains from cases of infantile diarrhea.

O Group 112. As mentioned by Ewing and Kauffmann (1950), this O antigen group is divisible into O 112a, 112b and O 112a,112c and these subgroups may be differentiated by the use of cross absorbed O antisera. *E. coli* O 112a,112c cultures are not related to other O antigen groups in any important way but there is a significant relationship between the O antigens of O 112a,112b and O 113 strains (Ewing et al., 1952; Ewing et al., 1956).

O Group 119. Cultures of O group 119 are not known to react significantly in O antisera for other *E. coli* O antigen groups but O 48 suspensions react in O 119

TABLE 39

O Antigen Relationship of *E. coli* O 86 Cultures

O Antigen Suspensions (100 C, 1 hr.)	O Antisera			
	86a		86a, 86b	
	Unabsorbed	Absorbed by 86a, 86b	Unabsorbed	Absorbed by 86a
86a	5,120	0	5,120	0
86a, 86b	20,480	0	20,480	5,120

TABLE 40

Differentiation of *E. coli* O 86 and O 127 Cultures (Slide Tests)

O Antisera	Absorbed by	*E. coli* O Antigen Suspensions (100 C, 1 hr.)				
		O 86a	O 86a, 86b	O 90	O 127a	O 127a, 127b
O 86a, 86b	O 86a + O 90	0	+ + + +	0	0	0
O 86a	O 90	+ + + +	+ + + +	0	0	0
O 90	O 127a	0	0	+ + + +	0	0
O 127a, 127b	O 127a + O 90	0	0	0	0	+ + + +
O 127a	O 90	0	0	0	+ + + +	+ +

antiserum prepared by the writers. This unilateral relationship necessitates the absorption of O 119 antiserum by *E. coli* O 48.

O Group 124. The O antigens of *E. coli* O 124 cultures are not significantly related to those of other *E. coli* O antigen groups.

O Group 125. There is a reciprocal (a,b - a,c) relationship between the O antigens of *E. coli* O 125 and those of *E. coli* O 73 which necessitates cross absorption of the O antisera. Also there are unilateral relationships which cause agglutination of *E. coli* O 7 and O 11 suspensions in O antiserum prepared with *E. coli* O ·125. The latter relationships generally are slight but may require absorption of O 125 antiserum with O group 7 and 11 cultures. Further, O antigen group 125 is divisible into two subgroups, O 125a,125b and O 125a,125c (Ewing and Tanner, unpublished data).

O Group 126. There is a unilateral relationship between *E. coli* O groups 126 and 75 which produces agglutination of O 75 in O 126 antiserum. The writer employs reciprocally absorbed antisera for the differentiation of these O groups. The O group 126 is not significantly related to other known O groups of *E. coli.*

O Group 127. The relationships between *E. coli* O groups 86, 90, and 127 were mentioned. Furthermore, O group 127 may be divided into subgroups O 127a and O 127a,127b (Tables 38 and 41). The absorbed O antisera advised for differentiation of cultures that belong to these subgroups are given in Table 40. A relatively minor relationship is known which causes agglutination of *E. coli* O 128 cultures in O 127a and O 127a,127b O antisera but this usually does not cause difficulty.

O Group 128. The O antigens of cultures of this *E. coli*

O group are related reciprocally to those of strains that belong to O groups 48 and 87 but the relationships are minor and generally are not detected in O antiserum pools used in a dilution of 1:1000. However, if tests indicate that absorption of O 128 antiserum is necessary, the antiserum may be absorbed by O 48 or O 87 because either of these O group cultures removes agglutinin for the other. O antisera for O groups 48 and 87 should be absorbed in the manner given in Table 36. Further, Taylor and Charter (1955) indicated that the O antigens of all *E. coli* O group 128 cultures are not identical and that the group may be subdivided. This observation was confirmed by Ewing and Tanner (unpublished data) who demonstrated subgroups O 128a,128b and O 128a,128c within O group 128. Reciprocally absorbed O antisera (Table 36) may be used to differentiate cultures that belong to these subgroups.

In addition to serotypes of *E. coli* that belong to the above-mentioned O antigen groups certain others have been associated with cases of diarrheal disease. Among these are serotypes that belong to O group 25 (see Braun 1953, 1954, and Ørskov 1954a). It should be recalled that the isolation of A-D 02 (*S. tiete* or Alkalescens II) from cases of diarrhea was described by De Assis (see Weil and Slafkovsky, 1948), and that A-D 02 cultures are anaerogenic, nonmotile varieties of *E. coli* O group 25. Since their description, A-D 02 cultures have been reported from many parts of the world and it is felt that they, as well as biochemically typical O 25 strains of *E. coli,* should be considered in any assessment of the association of these microorganisms with diarrheal disease. Similarly, certain serotypes within several other *E. coli* O antigen groups (e.g. 018, 028, 0114) are under assessment in this connection (for data and references, see Charter, 1956; Ewing et al., 1957; Tatum et al., 1958; Ewing and Davis, 1961a; Ewing et al., 1963). A serotype of O group 73 (73a, 73c:K92(L):34) has been associated with multiple cases of diarrheal disease in infants in

TABLE 41

Relationship of the O Antigens of *E. coli* O 127 Cultures

O Antigen Suspensions (100 C, 1 hr)	O Antisera			
	127a		127a, 127b	
	Unabsorbed	Absorbed by 127a, 127b	Unabsorbed	Absorbed by 127a
127a	20,480	0	5,120	0
127a, 127b	20,480	0	20,480	10,240

two separate outbreaks (Ewing, Davis, and Armstrong, unpublished data, 1967). Further, serotypes of *E. coli* have been reported as being the probable causative agent in outbreaks and sporadic cases of diarrheal disease in children and adults (McNaught and Stevensen, 1953; Ketyi et al., 1957; Costin, 1962, 1966; Costin et al., 1960, 1964, 1965; Stanholtz et al., 1967; and Rowe et al., 1970. See also Ewing et al. 1963 for other references).

THE K ANTIGENS OF *E. COLI*

Characteristics of K Antigens

The K antigens are the heat labile somatic antigens of the bacteria. As pointed out by Kauffmann (1947), the term K is merely a symbol used for a class of antigens composed of several varieties, all of which are envelope or sheath antigens that inhibit agglutination of living enterobacteria in O antisera. The K antigens are divisible into at least three varieties, L, A, and B, based on physical behavior. Some of the more important differences between L, A, and B antigens are given in Table 42. The most striking difference between L and B antigens is the fact that the binding power of the former is inactivated by heat at 100 C for 1 hour. Thus, one may prepare pure L antiserum by absorption of an OL antiserum with a heated suspension of the homologous strain which removes the O agglutinin but leaves L agglutinin. This cannot be done with B antiserum since the binding power of B antigen is not inactivated by heat at 100 C and if one absorbs an OB antiserum with a heated suspension of the homologous strain, agglutinins for both O and B antigens are absorbed from the antiserum. For additional data on the differences between the various kinds of K antigens the reader is referred to Kauffmann, 1947 or 1966. The K antigens and variational phenomena that affect them also are mentioned in Chapter 4.

Interrelationships of K Antigens

A list of the standard K antigen cultures is given in Table 43 and the relationships of the K antigens of *E. coli* are shown in Table 44. Examination of the data given in Table 44 shows that many of the K antigens of

TABLE 42

The K Antigens of *E. coli*

Variety of K Antigen	Characteristics
L	1. Agglutinability of L antigen in L antiserum inactivated by heat at 100 C, 1 hr. 2. Suspensions rendered agglutinable in O antiserum by heat at 100 C, 1 hr. 3. Antibody binding power inactivated by heat at 100 C, 1 hr. 4. Antigenicity inactivated by heat at 100 C, 1 hr. 5. Occur as envelope or sheath, occasionally as a capsule.
A	1. Agglutinability of A antigen in A antiserum inactivated by heat at 120 C, 2½ hr. 2. Suspensions rendered agglutinable in O antiserum by heat at 120 C, 2½ hr. 3. Antibody binding power not inactivated by heat at 100 C, 2½ hr. or at 121 C, 2 hr. 4. Antigenicity inactivated by heat at 120 C, 2½ hr. 5. Occur as capsules.
B	1. Agglutinability of B antigen in B antiserum inactivated by heat at 100 C, 1 hr. 2. Suspensions rendered agglutinable in O antiserum by heat at 100 C, 1 hr. 3. Antibody binding power not inactivated by heat at 100 C, 2½ hr. or at 121 C, 2 hr. 4. Antigenicity inactivated by heat at 100 C, 1 hr. 5. Occur as envelopes or sheaths or as capsules.

TABLE 43

Standard K Antigen Strains

K Antigen No.	Culture	Variety of K	O Group	H Antigen	K Antigen No.	Culture	Variety of K	O Group	H Antigen
1	U9-41	L	2	4	47	A 282a	A	8	2
2	U 14-41	L	3	2	48	A 290a	A	8	9
3	U 4-41	L	4	5	49	A 180a	A	8	21
4	U 1-41	L	5	4	50	PA 80c	A	8	NM (9)
5	Bi 8337-41	L	10	NM (4)	51	A 183a	L	1	NM (1)
6	Bi 7457-41	L	4	5	52	A 103	L	4	NM (4)
7	Pus 3432-41	L	7	4	53	PA 236	L	6	NM
8	G 3404-41	B	8	4	54	A 12b	L	6	10
9	Bi 316-42	B	9	12	55	N 24c	A	9	NM
10	623-42	B	11	NM (10)	56	H 17b	B 1	2	7
11	Su 4321-41	L	13	11	57	H 909d	B 3	9	32
12	Su 65-42	L	4	NM (5)	58	Stoke W	B 4	111	NM
13	Su 4344-41	L	6	1	59	5624-50	B 5	55	NM
14	F 7902-41	L	15	4	60	E 893	B 6	26	NM
15	F 8316-41	L	6	16	61	E 990	B 7	86a	NM
16	K 12a	L	17	18	62	F 1961	L	86	2
17	P 7a	L	20	NM	63	4932-53	B 8	127a	NM
18	E 39a	L	23	15	64	5017-53	B 9	86a,86b	36
19	E 47a	L	25	12	65	2160-53	B 10	127a,127b	4
20	E 19a	L	21	NM (4)	66	1685	B 11	112a,112c	NM
21	H 38	L	23	15	67	56-54	B 12	128	2
22	H 67	L	23	15	68	1411-50	B 13	112a,112b	18
23	H 54	L	25	1	69	34 W	B 14	119	27
24	H 45	L	22	31	70	2745-53	B 15	125	19
25	Bi 7575-41	B 2	8	9	71	E 611	B 16	126	2
26	Bi 449-42	A	9a	NM	72	227	B 17	124	32
27	E 56b	A	8	NM (10)	73	909-51	B 18	28	NM
28	K 14a	A	9a,9b	NM	74	4354-53	L	44	18
29	bi 161-42	A	9	NM	75	6182-50	B 19	113	21
30	E 69	A	9	1	76	F 10018-41	B 20	18a,18b	14
31	Su 3973-41	A	9	NM (4)	77	3219-54	B 21	18a,18c	7
32	H 36	A	9	19	78	1111-55	B 22	136	NM
33	Ap 189	A	9	NM	79	RVC 1787	L	137	41
34	E 75	A	9	NM	80	E 38	B	78	NM
35	A 140a	A	9	NM	81	62-57	B	138	NM
36	A 198a	A	9	19	82	63-57	B	139	NM
37	A 84a	A	9	NM	83	134-51	B	20	26
38	A 262a	A	9	NM	84	2292-55	B	20	26
39	A 121a	A	9	9	85	RCV 2907	B	141	4
40	A 51d	A	8	9	86	C 771	B	142	6
41	A 433a	A	8	11	87	CG 9	B	141	NM
42	A 295b	A	8	NM (4)	88	E 68	L	141	4
43	A 195a	A	8	11	89	D 357	B	147	19
44	A 168a	A	8	NM	90	K10	–	114	32
45	A 169a	A	8	9	92	6181-66	L	73	34
46	A 236a	A	8	30					

TABLE 44

K Antigen Relationships

K Antiserum	Homo-logous Titer	Titer With Other K Antigens	K Antiserum	Homo-logous Titer	Titer With Other K Antigens
1 (L)	320	*	47 (A)	640	72,80
2 (L)	640	*	48 (A)	640	30,80
3 (L)	320	*	49 (A)	640	*
4 (L)	320	*	50 (A)	320	*
5 (L)	640	*	51 (L)	640	*
6 (L)	320	*	52 (L)	640	*
7 (L)	320	*	53 (L)	640	3,40; 18,80; 21,160
8 (L)	1280	13,80; 15,80	54 (L)	320	*
9 (L)	320	*	55 (A)	640	*
10 (L)	640	*	56 (B1)	320	45,40
11 (L)	160	2,80; 73,40	57 (B3)	320	7,80; 73,320
12 (L)	320	*	58 (B4)	640	38,40
13 (L)	640	21, 160;73,40	59 (B5)	640	18,40; 21,320
14 (L)	640	20,40; 52,80	60 (B6)	640	*
15 (L)	320	*	61 (B7)	320	*
16 (L)	160	*	62 (L)	320	*
17 (L)	320	73,320	63 (B8)	320	*
18 (L)	320	2,80	64 (B9)	320	61,40
19 (L)	320	20,320	65 (B10)	640	25,80; 61,40, 63,80
20 (L)	640	38,40	66 (B11)	640	21,80
21 (L)	320	2,160; 15,80; 73,80	67 (B12)	320	*
22 (L)	160	2,40; 73,80	68 (B 13)	640	9,40
23 (L)	640	*	69 (B14)	640	*
24 (L)	320	48,80	70 (B15)	640	*
25 (B2)	320	61,40; 63,40	71 (B16)	640	*
26 (A)	640	*	72 (B17)	320	*
27 (A)	160	*	73 (B18)	640	*
28 (A)	160	15,80; 24,40; 53,40	74 (L)	640	19,40; 28,80
29 (A)	160	*	75 (B19)	320	53,40
30 (A)	160	*	76 (B20)	640	*
31 (A)	640	*	77 (B21)	640	*
32 (A)	640	18,80; 21,320; 41,40	78 (B22)	160	*
33 (A)	640	*	79 (L)	320	*
34 (A)	320	2,80; 41,160; 48,80; 72,80	80	**	
35 (A)	640	*	81	160	
36 (A)	160	*	82	640	
37 (A)	320	2,80	83	320	*
38 (A)	320	*	84	320	*
39 (A)	320	*	85	160	
40 (A)	320	73,40	86	**	
41 (A)	640	23,40	87	**	
42 (A)	640	*	88	320	
43 (A)	320	38,40	89	**	
44 (A)	640	*	90	**	
45 (A)	320	1,80; 56,160	92	640	
46 (A)	640	*			

* Indicates no significant cross reactions of K antigens.

** Antisera unsatisfactory: to be replaced.

E. coli are specific and are not significantly related to any other. Most of the relationships listed in Table 44 are relatively low in titer and unilateral and do not appear to be important.

In working with the K antigens of *E. coli* one must select colonies for K antigen preparation with the utmost care and adequate control work must be done in conjunction with each determination in order to be certain that the agglutination obtained actually is K agglutination and not O, or in some instances, H agglutination. For additional information on the methodology of K antigen determinations, the reader is referred to Kauffmann (1966) and to the publications of the author and colleagues.

K antigen suspensions that are to be used for identification may be prepared from infusion agar slant or infusion broth cultures inoculated from smooth, opaque, rather mucoid K+ forms selected from platings. K antigen suspensions for slide agglutination tests may be prepared by suspending the growth from an agar slant culture in about 0.5 ml of sodium chloride (0.5%) solution or in mercuric iodide solution (Chapter 18). The suspensions should be dense and homogenous. Such suspensions may be diluted and employed in tube agglutination tests but infusion broth cultures that have been incubated for 4 to 6 hr in a waterbath at 35 to 37 C are preferable. Further information regarding K antigen preparation and use, titrations, etc. may be found in the sections entitled Serotypes of *E. coli* Associated with Diarrheal Disease and Antiserum Production (v. inf.).

K Antigen Determination

Pooled K Antisera. Preliminary examination of the K antigens of cultures of *E. coli* that are inagglutinable, or only slightly agglutinable, in O antisera may be made by means of slide agglutination tests in pooled KO antisera. For this purpose the writer and colleagues employ several pools, each of which contain five or occasionally six antisera. No further dilution of the pooled K antisera is made when they are employed in slide agglutination tests. The composition of the pools is as follows:

1. K1 to K5 inclusive	10. K41 to K45
2. K6 to K10	11. K46 to K50 + K55
3. K11 to K15	12. K56 to K59
4. K16 to K20	13. K60 to K65
5. K21 to K25	14. K66 to K70
6. K51 to K54 + K74	15. K71 to 73, 75 to 77
7. K26 to K30	16. K62, 78 to 81
8. K31 to K35	17. K82 to K86
9. K36 to K40	

With a few exceptions the K antisera contained in pools 1 through 6 are L antisera, while those in pools 7 through 11 are A antisera, and with a few exceptions those in pools 12 through 17 are B antisera. A number of other antisera, e.g., K91, have not been incorporated into pools and are used individually.

Investigators who are interested in serotyping of cultures of *E. coli* from particular sources may find it expedient to prepare pools of K antisera that are more suitable for their purposes than those mentioned above. For example, other pools have been devised for use in preliminary examination of strains isolated from patients with diarrheal disease (see Ewing, 1969). The composition of these pooled antisera is as follows:

Polyvalent A
O26:B6 O55:B5 O111:B4 O127:B8
Polyvalent B
O86:B7 O119:B14 O124:B17
O125:B15 O126:B16 O128:B12
Polyvalent C
O18:B21 O20:B7 O20:K84(B)
O28:B18 O44:K74(L) O112:B11

These are prepared by mixing equal quantities of the individual OK antisera. If glycerolated antisera are employed, further dilution of the pools should be minimal. Less is known about the frequency and importance of the serotypes of *E. coli* that belong to the serogroups represented in polyvalent C antiserum than is the case with those represented in A and B antisera. Hence, use of polyvalent C antiserum is recommended, since nothing will be learned about the incidence and relative importance of these microorganisms if they are not looked for.

If a K antigen suspension prepared as outlined above reacts in one or more of the pooled K antisera, it then is tested in the individual antisera that comprise that pool or pools. This is done first by means of slide agglutination tests, the results of which are confirmed by titration of the K antigen suspensions in the indicated K and O antisera. Methods for these procedures are given in the section that deals with the identification of serotypes in diarrheal disease (v. inf.).

In the determination of A antigen (and occasionally the L and B antigens as well) in cultures of *E. coli* it is possible to use the quellung reaction since these antigens occur as capsules. Kauffmann (1954) recommends this method if cultures are not entirely inagglutinable in O antisera. Methods for the performance of such tests are given in Chapter 12.

Absorption of K Antisera. Since the majority of the K antisera used for the delineated K antigens of *E. coli* do not react significantly with K antigens other than the homologous, absorption ordinarily is not required. Should absorption of a K antiserum be necessary it may be done in the manner outlined for absorption of O antisera (v. sup.) except that K forms selected from platings are employed and the cultures are not heated. The H antibody content of all K antisera prepared with motile cultures must be determined, and if the titers are significantly high (e.g. > 1 to 160), the H agglutinins should be removed by absorption.

The manner in which K antisera may be absorbed for certain purposes is indicated in Table 45. As pointed out by Kauffmann (1954), K antisera should not be diluted more than 1 to 5 for absorption.

THE H ANTIGENS OF *E. COLI*

The H antigens are the heat labile protein antigens contained in the flagella of the bacteria. The general characteristics of this class of antigens and variational phenomena that affect them are discussed in Chapter 4.

A list of the standard strains used for production and absorption of H antisera is given in Table 46 and the relationships of the H antigens of *E. coli* are recorded in Table 47.

H Antigen Determination

Suspensions for use in identification of the H antigens of cultures of *E. coli* are prepared from broth cultures of actively motile bacteria. Since the majority of *E. coli* strains are poorly motile when first isolated, it usually is necessary to passage them serially through several tubes of semisolid agar motility medium to

TABLE 45

Absorption of K Antisera

KO Antisera*	Absorbing Suspensions Prepared from	Purpose
OL	Homologous strain, heated 100 C, 2½ hr. and washed twice with saline. Or, another strain of same O group containing a different L antigen, formalinized or heated 100 C, 1 hr.	Pure L antiserum Pure L antiserum
OL (or OB)	Homologous culture, heated 100 C, 2½ hr. and washed twice.	Determination of nature of unknown K antigen (L or B).**
OL	Heterologous culture, formalinized.	Determination of identity or relationship of K antigen (L or B).
OA	K minus or O form isolated from homologous culture, formalinized. Or, another culture of same O group containing an L antigen, heated 100 C, 1 hr.	Pure A antiserum. Pure A antiserum.
OA	K form of heterologous culture formalinized.	Determination of identity or relationship of unknown A antigen.

* K antisera prepared with motile cultures should be tested for H agglutinin content. If such tests indicate the presence of H agglutinin, the antiserum should be absorbed with an H antigen suspension in addition to the absorbing suspensions listed. Formalinized H antigen suspensions prepared from cultures that contain different K antigens must be used.

** Because of the physical properties of B antigens, there is at present no practical method for the preparation of pure B antisera.

TABLE 46

Standard H Antigen Strains

H antigen	O antigen	K antigen	Culture No.
1	2	2	Su 1242
2	43	•	Bi 7455-41
3	53	4	Bi 7327-41
4	2	1	U9-41
5	4	3	U4-41
6	2	1	A20a
7	1	1	U5-41
8	2	•	AP 320c
9	8	25	Bi 7575-41
10	11	10	Bi 623-42
11	13	11	Su 4321-41
12	9	9	Bi 316-42
14	18	76 (B20)	F 10018-41
15	23	18	E 39a
16	6	15	F 8316-41
17	15	•	P 12b
18	17	16	K 12a
19	9	•	A 18d
20	8	•	H3306
21	8	•	U11a-44
23	45	•	HW23
24	51	•	HW25
25	15	•	HW26
26	131	•	HW27
27	15	•	HW28
28	132	•	HW30
29	133	•	HW31
30	38	•	HW32
31	3	•	HW33
32	114	•	HW34
33	11	•	HW35
34	86	61 (B7)	BP 12665
35	135	•	4370-53
36	86	64 (B9)	5017-53
37	42	•	P11a
38	69	•	P9b
39	74	•	E3a
40	79	•	E49
41	137	79	RVC 1787
42	70	•	P9c
43	140	•	149-51
44	3	•	781-55
45	52	•	4106-54
46	26	60 (B6)	5306-56
47	86	•	1755-58
48	16	•	P4
49	6	13	2147-59

* Indicates antigen undertermined.

NOTE: The designations 13 and 22 were assigned to the H antigens of cultures that later were identified as *Citrobacter*. These were deleted and no substitutions have been made.

TABLE 47

H Antigen Relationships

H antigen	Homologous Titer	Titer of other H Relationships		
1	12,800	12,800	16,100	
2	25,600			
3	12,800			
4	6,400	17,400		
5	12,800			
6	6,400			
7	12,800			
8	25,600	40,800		
9	12,800			
10	6,400	9,200		
11	25,600	21,400	40,400	
12	25,600	1,800		
13	12,800			
14	25,600			
15	25,600			
16	25,600			
17	25,600			
18	12,800			
19	25,600			
20	12,800			
21	25,600	11,6400		
22	12,800			
23	12,800			
24	25,600			
25	25,600			
26	25,600			
27	25,600			
28	25,600			
29	25,600			
30	25,600	32,3200		
31	25,600			
32	12,800	30,1600		
33	25,600			
34	12,800	24,100	31,100	
35	12,800			
36	25,600			
37	25,600	41,100		
38	25,600			
39	25,600			
40	25,600	8,200	11,400	37,200
41	25,600	37,100	39,400	
42	12,800	6,200		
43	25,600	37,100		
44	25,600	39,400		
45	25,600	20,100		
46	12,800	17,800		
47	25,600			
48	12,800			
49	25,600	39,400		

Blanks indicate no significant relationships.

enhance motility and H antigen development.[6] The number of passages required cannot be predicted since this varies with individual cultures. However, the average is perhaps two or three passages. When a culture migrates from the point of inoculation at the top of the column of medium to the bottom, or near the bottom, of the tube during overnight incubation it usually is satisfactory. Inoculum taken from the spreading growth at the farthest point of migration then is transferred to infusion broth and the broth cultures are incubated at 37 C for 18 to 20 hr and formalinized by the addition of an equal volume of 0.5% sodium chloride solution to which 0.6% formalin has been added. Incubation for 4 to 6 hr in a waterbath also is satisfactory. Preparation of H antigen suspensions also is discussed in the section on Production of Antisera (v. inf.) and in Chapter 8.

H Antiserum Pools. After passage in semisolid agar medium, formalinized infusion broth antigens prepared in the manner outlined above may be tested in pooled H antisera. If these preliminary tests are negative, further passages in semisolid agar are indicated. The author and colleagues employ ten pools, each of which contains five or six H antisera with the exception of pool 10 which contains four antisera. Because of important reciprocal relationships (Table 47), antisera for several H antigens are paired in the pools. The composition of these pools is as follows:

```
Pool  1   H1,2,3,12,16
      2   H4,5,6,9,17
      3   H8,11,21,40,43
      4   H7,10,14,15,18
      5   H19,20,23,24,25
      6   H26,27,28,29,31
      7   H30,32,34,36,41
      8   H33,35,37,38,45
      9   H39,42,44,46,47
     10   H48,49,50,51
```

One ml of the unknown H antigen suspension is added to an amount of each pooled H antiserum calculated to give a final dilution of 1:1000 of each constituent antiserum contained in a pool. Tests for H agglutination are incubated in a water bath at 48 to 50

C are are read after periods of 15 min, 30 min, and 1 hr. If characteristic H agglutination occurs in one of the H antiserum pools, the suspension then is tested in each of the H antisera represented in that particular pool. These also may be single tube tests in which the final antiserum dilution is 1:1000. Results obtained in the latter tests are confirmed by titration and, if necessary, by tests in absorbed antisera (v. inf.).

Absorption of H Antisera. Although there are several minor relationships among the recognized H antigens, most of these do not appear in tests made at the 1:1000 level of antiserum dilution. However, important relationships exist between the pairs of *E. coli* H antigens listed in Table 48 and absorbed H antisera are required for their exact identification. Bacterial suspensions for absorption of H antisera may be prepared by inoculation of passaged, motile cultures onto plates of infusion agar in the manner described for absorption of O antisera. However, absorbing suspensions are not heated. For these absorption tests one-half of the absorbing suspension may be added to H antiserum diluted 1:50 with phenolized isotonic saline solution. The mixture is placed in a waterbath at 48-50 C for 1 hr and then centrifuged. The diluted antiserum is then added to the second half of the absorbing suspension,

TABLE 48

Absorption of H Antisera

H Antiserum	Absorbing Cultures
1	12
12	1
8	40
40	8+
	11
11	21+
	40
21	11
37	41
41	37+
	39*
49	39*

* These absorptions may not be necessary.

[6]The phase reversal medium recommended for use with cultures of *Salmonella* is not recommended for use with cultures of *E. coli*. Since many cultures of *E. coli* are more actively motile at room temperature than at 35 to 37 C, it may be necessary to incubate cultures being passaged in semisolid agar medium at the lower temperature. In many laboratories the ambient temperature may be low enough to cause the gelatin in phase reversal medium to solidify. If this happens the bacteria will not migrate through it. Therefore a simple semisolid motility medium is recommended for serial passages of cultures other than *Salmonella* and *Arizona*.

reincubated for 1 hr and then centrifuged. Absorbed H antisera may be used in single tube tests in a dilution of 1:1000 or more.

Pure H antisera for use in slide agglutination tests may be prepared by appropriate absorption methods (Ewing et al., 1956). Further, the H antigens of cultures of *E. coli* may be determined by means of microscopic immobilization tests as mentioned by Ørskov (1956b).

IDENTIFICATION OF SEROTYPES OF *E. COLI* ASSOCIATED WITH DIARRHEAL DISEASE

General Considerations. There are certain principles that may be applied to the investigation of any case of diarrheal disease. Materials for culture should be collected before antibiotic therapy is begun. Specimens may be taken from freshly soiled diapers or collected by means of rectal swabs. If stools or other materials must be held several hours before inoculation of media or are to be sent through the mails, specimens should be emulsified in a preservative medium. Samples of stools to be examined for virus should be frozen. Such specimens may be placed in resistant glass or plastic tubes, ointment boxes, or other small containers, then quickly frozen in dry ice or in a deep freeze box and maintained at low temperature (-40 C to -70 C). Specimens taken at autopsy from cases and materials collected for control purposes may be treated in the manner outlined above. If specimens are collected at autopsy, samples of the intestinal contents and tissue specimens should be taken at various levels, such as ileum, ileocecal junction, transverse colon, and descending colon.

Since salmonellae and shigellae are incriminated in certain epidemics of infantile diarrhea, thorough search for members of these groups always should be made. Methods for the isolation and preliminary identification of these microorganisms may be found in Chapter 2.

Isolation. Since strains of *E. coli* serotypes usually do not grow on the highly selective media in general use for the isolation of salmonellae and shigellae, it is imperative that less inhibitory differential media such as MacConkey agar, EMB agar, or other similar media be employed (Figure 2). In addition to the aforementioned plating media, a blood agar plate or infusion agar plate should be streaked with a very small inoculum of stool specimen because it has been noted that, in a few instances, serotypes of *E. coli* associated with infantile diarrhea appeared on blood agar plates and did not grow on MacConkey agar. Blood agar is recommended also because it is easier to avoid contaminants when picking colonies from this medium. In addition, cultures of serogroups 0111:B4 and 055:B5

rarely are hemolytic, a fact that may aid in the recognition of colonies. However, undue reliance should not be placed upon this observation because many other strains of *E. coli* are nonhemolytic.

Preliminary Examination. After incubation for 16 to 20 hr, the primary plating media may be examined for colonies of *E. coli*. No specific directions can be given as to the selection of colonies for transfer because colonies of the various serotypes of *E. coli* all appear quite similar. Portions of ten or more individual colonies that appear on the blood agar or plain infusion agar plates should be tested directly in antisera. If desired polyvalent OB (OB, OL) antisera (v. sup., Pooled K Antisera) may be used for preliminary tests. If this is done the preliminary tests should be followed by tests in the individual O and K antisera for the serogroups included in the polyvalent antisera in which reactions are obtained. If strongly positive slide agglutination tests are obtained by this procedure, it may be considered presumptive evidence that one of the particular serotypes is present in the specimen. This procedure not only affords a rough estimate of the presence or absence of serotypes for which antisera are available, but it also indicates the prevalence of a given serotype in the specimen. Examinations of this sort should be made with colonies on blood or infusion agar plates rather than colonies from MacConkey agar plates since it has been shown that the presence of bile salts may cause confusing agglutination reactions, which are not confirmed in subsequent work (McNaught and Stevenson, quoted by Stevenson, 1956). Three or more entire, smooth, opaque, escherichia-like colonies from each plating medium should be transferred to infusion agar slants (long slants prepared from plain infusion agar, without added sugar). Additional transfers should be made if possible. The slants should be inoculated over the entire surface in order to obtain maximum growth. After incubation these slants are used for biochemical and serological studies.

A generous portion of the growth from each of the agar slant cultures should be emulsified in about 0.5 ml of 0.5% sodium chloride solution. Droplets of the heavy suspensions may be tested for agglutination on slides with droplets of *E. coli* O and OB antisera, first as living suspensions and again after being heated. Agglutination of a living antigen in an OB antiserum and lack of a reaction in the corresponding O antiserum are indicative of the presence of B antigen in the strain. If such a reaction occurs in one of the OB antisera, the suspension is heated at 100 C for about 15 min,[7] cooled, and retested in the indicated OB and O

[7]Occasionally it may be necessary to reheat a suspension for an additional period of time (up to a total of one hour).

Figure 2

**OUTLINE OF METHODS FOR EXAMINATION OF STOOL SPECIMENS FROM CASES
OF DIARRHEAL DISEASE**
(See text for details)

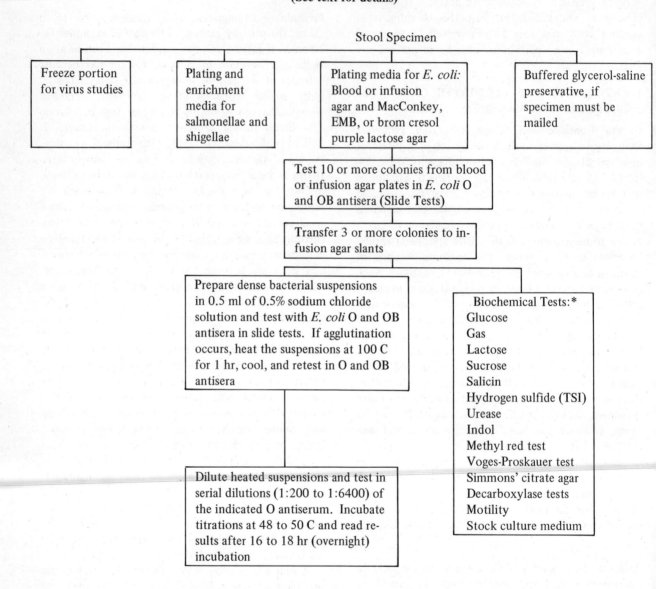

*See Tables 27-30.

antisera. If the culture belongs to the O antigen group, the heated antigen may be expected to react in both OB and O antisera. An example of results that may be anticipated with freshly isolated cultures is given in Table 49. In the example given in Table 49, a K antigen suspension prepared from an unknown culture was agglutinated by O 55:B5 antiserum and reacted only slightly in O 55 antiserum. When the suspension was retested after being heated, it reacted in both O 55:B5 and O 55 antisera. It should be emphasized that in these slide tests, *cognizance should be taken only of the strong, characteristic agglutination reactions* and weak delayed reactions should not be given undue attention. Further, if living suspensions react strongly in O antiserum as well as in KO antiserum, the culture should be plated and K forms isolated for reexamination. If a culture of *E. coli* contains K antigen in considerable quantity, the agglutination observed in

TABLE 49

Preliminary Examination of the K Antigens of Cultures of *E. coli* from Cases of Diarrheal Disease

Antigen Suspensions	Antisera for *E. coli*							
	026:B6	026	055:B5	055	0111:B4	0111	0127:B8	0127
K Antigen (Living)	–	–	+ + + +	± (–)	–	–	–	–
O Antigen (100 C, 1 hr.)	–	–	+ + + +	+ + + +	–	–	–	–

KO antiserum is caused largely by the interaction of K antigen and K antibody and O agglutination is inhibited. However, a suspension prepared from a culture in which the K antigen is poorly developed may be expected to be O agglutinable and to react in both KO and O antisera. Hence, reselection of K forms for antigen preparation is necessary.

Next, the heated suspensions prepared from agar slant cultures are diluted and titrated in serial dilutions (e.g., 1:200 to 1:6,400) of the indicated O antiserum. If desired, broth cultures that have been incubated for 4 to 6 hr and then heated at 100 C for 1 hr may be used in the titrations and, in the absence of O antisera, OB antiserum may be employed. Such tests are read after 16 to 18 hours' incubation in a water bath at 48 to 50 C. If the results of slide agglutination tests performed in the first steps were clear-cut and the culture was inagglutinable in O antisera in the living state, confirmation by agglutination to or near the titer of the antiserum may be expected in the titrations. Other cultures of *E. coli*, however, are related to certain of the serotypes associated with infantile diarrhea through possession of common O antigenic fractions. A strain that belongs to one of these related O groups may be expected to cross-react in the titrations. Also, it should be remembered that living suspensions of cultures of *E. coli* that belong to related O antigen groups may react in OB and O antisera in slide tests if they are not O-inagglutinable, and heated suspensions may be expected to cross-react. As an example of this, the O antigens of O group 25 strains are related to those in 026, and an O group 25 suspension may react in 026:B6 and 026 antisera in slide tests, but when tested by titration in 026 antiserum, only a relatively low-titered reaction is obtained (see Table 37).

After preliminary serological tests are completed, the biochemical reactions of the cultures should be determined. If all of the bacterial suspensions prepared with strains from a particular specimen are agglutinated by one of the antisera, one of these strains may be selected for biochemical studies. If the *Escherichia* present in the specimens are not agglutinated by any of the antisera available, the biochemical reactions of three or more strains from each specimen should be determined. All strains which are examined biochemically also should be placed in stock culture medium (Chapter 18).

Although any desired number of biochemical tests may be made, there are certain tests (Figure 2) that may be regarded as essential for the determination of whether or not a culture belongs to the genus *Escherichia*. In many instances it may be advantageous to determine whether the epidemic strain involved in a particular outbreak ferments sorbitol, dulcitol, salicin, adonitol, or other substrates since several biotypes of *E. coli* 0111:B4, etc., were described by Kauffmann and Ørskov (1956) and others.

For further information concerning isolation and preliminary identification of serotypes of *E. coli* associated with infantile diarrhea, the reader is referred to the publications of Ørskov and of Kauffmann, cited in the foregoing paragraphs, and to Ewing (1963, 1969).

Further Serological Examination of Cultures from Cases of Diarrheal Disease

When indicated O antigen suspensions should be tested (slide tests) in absorbed O antisera (Table 36) to differentiate the O antigens of a culture from those of related O groups and to determine the sub-O group.

The results of K antigen determinations made by slide agglutination should be confirmed by titration of living or formalinized antigen suspensions in KO antisera. *Such tests should be controlled by titration of the same K antigen suspension in the appropriate O anti-*

serum in order to demonstrate inhibition of O agglutination. Control tests are incubated in the same manner as K agglutination tests at 37 C for 2 hr, followed by overnight incubation in a refrigerator. Confirmation of the results of slide tests is obtained when the K antigen suspension is agglutinated in characteristic manner to, or near, the K titer of the KO antiserum and no agglutination or minimal reactions are seen in the O antiserum control. If a number of similar strains are recovered from different patients in connection with an outbreak of diarrheal disease, it is advisable to absorb the KO antiserum in which the cultures reacted with a heated suspension prepared from at least one of the cultures. This procedure aids in the identification of both the K and O antigens of the epidemic strain. For example, if the B and O antigens of the epidemic strain are identical with those of the culture with which the antiserum was prepared, a heated suspension prepared from the former may be expected to remove both O and K agglutinins from the antiserum. It is felt that all cultures isolated from sources other than epidemic diarrhea which are agglutinated by antisera for diarrheal strains should be examined by absorption tests as outlined above. It is known that many cultures from diverse sources are not identical with serotypes isolated from infantile diarrhea cases although they may exhibit varying degrees of relationship. Cultures of *E. coli* from cases of diarrheal disease that do not react in KO antisera prepared with those serotypes known to be associated with diarrheal disease, as well as strains from other diseases, must be examined in antisera prepared with the other K antigens of *E. coli*. Slide agglutination tests with pooled K antisera may be employed for preliminary examinations (v. sup.).

Finally, the H antigens of the cultures should be identified. H antigen suspensions first are tested in pooled H antisera, then in the component antisera (Table 47) that comprise the pool in which a reaction occurred. Lastly, the antigen is tested in absorbed H antisera (Table 48) if required for accurate identification. Methods for these examinations are outlined in the section entitled "The H Antigens of *E. coli*" (v. sup.).

More than 130 serotypes were delineated by Ewing et al. (1963) among cultures that belonged to the nine OB serogroups that usually are accepted as containing serotypes enteropathogenic for the human infant. However, all of these serotypes did not occur frequently among cultures submitted for serological examination. These investigators reported that 81.7 per cent of cultures examined during a twelve-year period belonged to 20 serotypes and bioserotypes. These were as follows:

26:60(B6):NM	119:69(B14):6
26:60(B6):11	125a,125c:70(B15):21
55:59(B5):NM	126:71(B16):NM
55:59(B5):6	126:71(B16):27
55:59(B5):7	127a:63(B8):NM
86a:61(B7):34	127a:63(B8):9
111a,111b:58(B4):NM	127a:63(B8):21
111a,111b:58(B4):2	128a,128b:67(B12):2
111a,111b:58(B4):12	128a,128b:67(B12):7
111a,111b:58(B4):21	128a,128c:67(B12):12

The above-mentioned investigators also listed a number of serotypes belonging to other serogroups of *E. coli* that have been associated with diarrheal disease in the human. These serotypes were members of serogroups 18:76(B20):14; 18:77(B21):7; 20a,20b:84(B):26; 28a,28c:73(B18):NM; 44:74(L):18; 44:74(L):34; 112a,112c:68(B11); 124:72(B17):NM; and 124:72(B17):30.

Descriptive literature regarding serotypes and bioserotypes of *E. coli* that belonged to all of the serogroups mentioned above, and a number of others as well, was reviewed by Ewing et al. (1963).

Stages at which the personnel of laboratories of various kinds logically may terminate their work with enteropathogenic serotypes of *E. coli* were outlined by Ewing (1969). Among other recommendations made was one to the effect that characterization of the O and H antigens of cultures of *E. coli* other than those currently recognized as enteropathogenic should be instituted wherever possible. This would aid in the delineation of additional possible or potential enteropathogenic serotypes and in investigations of nosocomial infections, studies on interhospital spread of infection, etc.

THE ALKALESCENS-DISPAR (A-D) BIOSEROTYPES

In reality the microorganisms included in the Alkalescens-Dispar (A-D) group are anaerogenic, nonmotile, biotypes of *E. coli*, which belong to certain specified O antigen groups and they may be reported as such. Admittedly, the group is an artificial one and it is retained simply as a matter of expediency. Some of the types occur very commonly and must be differentiated from shigellae in particular. Hence, the nomenclature and abbreviations used in connection with these microorganisms in the first edition of this book are retained, since they are well known to most workers and are useful in laboratory practice when referring to these particular *E. coli*.

Frantzen (1950, 1951), utilizing methods recommended by Kauffmann (1947), proposed the antigenic schema given in Table 50 for the A-D group. The schema is based upon study of the relationships of the included types to each other and to *E. coli* O groups. Kauffmann (1949) reported that certain types now included in the A-D group contain K antigens. Frantzen (1950, 1951) confirmed these findings and reported the presence of K antigens of the L and A varieties in certain A-D cultures. Ewing et al., (1950) confirmed the results obtained by Frantzen and reviewed some of the earlier work on this group of microorganisms. These investigators also advocated the adoption and use of the Frantzen schema for the A-D group. The schema is a practical one which affords an accurate and definite means of identifying these bacteria.

A brief review of the terms applied to the bacteria by investigators in the past is required for clarification and reference. In 1918, Andrewes described *Bacillus alkalescens* as well as *Bacillus dispar* and *Bacillus ambiguus (S. dysenteriae 2).* Studies on the antigenic structure and relationships of *S. alkalescens* (A-D O1) were made by De Assis (1939, 1947), Stuart et al. (1943), Neter (1944), Wheeler et al. (1946), and many others. As pointed out by Stuart and his co-workers (1943) and Wheeler and collaborators (1946), there are a large number of biochemical varieties of bacteria that contain all, or a part, of the antigenic complex of A-D O1. These biotypes range from the typical anaerogenic, lactose negative, nonmotile bacterium through intermediate forms to typical cultures of *E. coli.*

De Assis (1939) described a bacterium that was similar to *S. alkalescens* as regards its biochemical reactivities but which contained different O antigens. This serotype was called *S. alkalescens* II but was later named *S. tieté* by Weil and Slafkovsky (1948). Neter (1944) proposed a classification which contained four serotypes of *S. alkalescens.* These were: type I, the

TABLE 50

The O Antigenic Schema for the Alkalescens-Dispar Group*

O Groups	O Antigen	Relationship to *E. coli* O Groups	Earlier Designations
1	1a	identical with 1a 1a,1b	*B. alkalescens* or Alkalescens Type I
2	2	strong relationship with 25 and other groups	Alkalescens type II or *S. tieté*
3	3	strong relationship with 25 and other groups	*S. ceylonensis B* or *S. dispar* Type II or Alkalescens Type III (2-193)
4	4	strong relationship with 4	*S. madampensis* or S. *dispar* I
5	5	identical with 2a	None
6	6	identical with 9	*S. dispar* Type III
7	7	identical with 7	None
8	8	identical with 81	None

* From Frantzen (1950), and Ewing, et. al., 1950.

original *B. alkalescens* of Andrewes; type II, *S. alkalescens* II of De Assis; type III, 2-193 (2372) isolated by Ewing in Italy; and type IV, previously undescribed. Type 2-193 cultures now are regarded as lactose negative variants of A-D O3. They are included in the Frantzen (1950) schema (Table 50) as members of O group 3. A review of the literature on the subject of type 2-193 cultures is given by Ewing et al. (1950). Available cultures of type IV are rough (Frantzen, 1950).

In his publications of 1907 and 1912, Castellani described two bacterial types which were designated *B. ceylanensis* B and *B. madampensis.* A third type, which was described by Castellani (1907) as *B. ceylanensis* A, later was identified as *S. sonnei.* Later, Castellani (1927, 1932) classified these lactose fermenting bacteria as metadysentery bacilli. Andrewes (1918) described a species called *B. dispar* which consisted of a mixed collection of bacterial types, including both indol positive and indol negative microorganisms. Levine (1920) pointed out that the indol negative *B. dispar* cultures of Andrewes were the same as Sonne's bacterium. Welch and Mickle (1932), Carpenter (1943), and Carpenter and Stuart (1946) adopted the name *S. dispar* for the anaerogenic, nonmotile, indol positive serotypes that required 48 hours or longer to produce acid from lactose. Carpenter and Stuart (1950) employed the term *Proshigella dispar* of Borman et al. (1944) to designate the microorganisms.

Carpenter (1943) and Carpenter and Stuart (1946) studied the relationships of *S. dispar* I *(S. madampensis), S. dispar* II *(S. ceylonensis B),* and *S. dispar* III. *S. dispar* II was subdivided into three subtypes, IIa, IIb, and IIc. Later Carpenter (1946) added a fourth subtype, IId. Frantzen (1950) employed antisera prepared with boiled cultures and reported that cultures of *S. dispar* IIa, IIb, and IIc all contained identical O antigens as demonstrated by reciprocal absorption tests. The writer confirmed these results. Ewing (1949) found that subtype IId was serologically identical with *S. alkalescens* II or *S. tiete.* During the study of a large number of cultures that contained O antigens identical with those A-D O2 *(S. alkalescens* II*)* the writer found a series of biotypes comparable to that described by Stuart et al. (1943) for A-D O1 *(B. alkalescens,* of Andrewes*).* Frantzen (1950) reported that the heat-stable and heat-labile antigens of *S. dispar* III were identical with those of certain *E. coli. S. dispar* III was added to the A-D group as O group 6 (Table 50). In 1942, Roelcke recorded the isolation of a new type which he called *B. paradysenteriae palatinense,* Seeliger (1950) studied this culture and found that it was identical with *S. madampensis* and *S. dispar* I.

With few exceptions the biochemical reactions given by the Alkalescens-Dispar biotypes are similar to those given by typical cultures of *E. coli* (Table 31). Lactose, sucrose, and salicin are not utilized by members of the A-D group as frequently as they are by typical *E. coli,* but positive strains occur commonly, particularly among cultures that belong to O groups 2 to 8, incl. Mannitol is fermented by A-D group cultures and exceptions to this apparently are rare. Jeffries and Okabe (1958) reported three cultures of A-D O1 that failed to utilize mannitol, and which they referred to as the Koji biotype of A-D O1. Also, mannitol negative strains of A-D O2 and O5 are known (Ewing et al., 1958, and unpublished data).

Production and Use of Antisera. Cultures to be used for the preparation of vaccines are selected and inoculated into broth in the same manner as cultures of *E. coli.* After incubation for 6 to 8 hr the broth cultures are heated (100 C, 2 hr) to inactivate the L antigens. Broth cultures of A-D O6 must be heated at 121 C for 2 hours to inactivate K antigen A26 contained in cultures of this type. After this treatment 0.5 per cent of formalin is added to each vaccine. The vaccines are then injected into rabbits using the inoculation schedule outlined for production of antisera for *E. coli* (v. inf.).

Preliminary identification of A-D cultures may be made by slide tests with heated suspensions and confirmation of the results obtained by tests in absorbed antisera and by titration in tube tests. Antigen suspensions for slide tests may be prepared by removing the growth from an infusion agar culture and suspending it in 0.5 ml of physiological saline solution. The suspensions then are heated at 100 C for at least one-half hour, cooled, and tested in the antisera. For titrations it is advisable to use heated broth cultures. If no reaction occurs in any of the antisera, the biochemical reactions of the culture must be observed to determine its identity. Such cultures often are anaerogenic *Escherichia* that belong to other O antigen groups.

Certain absorbed antisera are required if one wishes to differentiate members of this group by means of slide agglutination tests. Such absorbed antisera may be prepared in the manner outlined in Table 54. However, it should be pointed out that the use of the absorbed antisera listed in Table 54 does not differentiate members of the A-D group from members of the genus *Shigella.* For example, A-D O1 antiserum absorbed by A-D O5 still reacts with cultures of *Shigella boydii* 1 and 4.

The results of agglutination tests which reveal the O antigen relationships of members of the A-D group to each other, to shigellae, and to known *E. coli* O groups are given in Tables 51, 52, and 53.

TABLE 51

**Relationships of the O Antigens
of the Alkalescens-Dispar Bacteria**

Alkalescens-Dispar O Antigen Suspensions	A-D Group O Antisera								Reactions in antisera for *Shigella*
	01	02	03	04	05	06	07	08	
01	20480	0	0	0	1280	0	0	0	*S. dysenteriae* 1, 320. *S. flexneri* 1a, 320; 2a, 80: 4a, 160; X variant, 640; Y variant 320. *S. boydii* 1, 2560; 4, 320; 11, 320.
02	0	10240	80	40	0	0	80	0	*S. flexneri* 2a, 320; 2b, 320; 3, 160; 4a, 160; 4b, 320; 5, 320; 6, 40; X variant, 320; Y variant, 320. *S. boydii* 5, 640; 9, 640; 12, 80.
03	0	2560	20480	640	0	0	320	0	*S. flexneri* 3, 160; 5, 320; X, 160; Y, 320; *S. boydii* 4, 640; 5, 1280; 9, 640; 13, 1280; 12, 640.
04	0	320	40	20480	0	0	0	0	*S. flexneri* 2a, 80; 2b, 640; 3, 320; 4a, 160; 5, 320; X, 320; Y, 320.
05	2560	0	0	0	20480	0	0	0	*S. flexneri* 1a, 320; 4b, 320; 5, 160; X, 640; Y, 640. *S. boydii* 1, 640; 4, 640; 11, 640
06	0	0	0	0	0	5120	0	0	
07	0	80	80	0	0	0	20480	0	*S. boydii* 12, 5120.
08	0	0	0	0	0	0	0	20480	

SEROTYPES OF *E. COLI* FROM INFECTIONS IN ANIMALS

Cultures of *E. coli* isolated from various pathologic processes in animals may be examined by the methods outlined in the foregoing sections. However, O and KO antisera for the serotypes frequently found in swine edema disease, in calf scours, or in infections in poultry or other animals will be required, depending upon the interests of the investigator. Slide agglutination tests in O and KO antisera and other preliminary steps may be carried out in the manner described for presumptive identification, and final identification may be made by

TABLE 52

Results of Agglutination Tests with Alkalescens-Dispar Antisera and Cultures of *E. coli*

E. coli O Antigen Suspensions	A-D Group O Antisera							
	01	02	03	04	05	06	07	08
01	10,240	0	0	0	2,560	0	0	0
02a	0	0	0	0	10,240	0	0	0
02a, 2b	1,280	0	0	0	5,120	0	160	0
04	0	160	1,280	5,120	0	0	0	0
07	0	1,280	0	0	0	0	20,480	0
09	0	0	0	160	0	5,120	0	0
025	0	20,480	1,280	0	80	0	320	0
081	0	0	0	0	0	0	0	10,240

TABLE 53

Results of Agglutination Tests with *E. coli* Antisera and Alkalescens-Dispar Cultures

Alkalescens-Dispar O Antigen Suspensions	*E. coli* O Antisera						
	01	02	04	07	09	025	081
01	20,480	1,280	0	0	0	0	0
02	0	0	640	0	0	20,480	0
03	0	0	5,120	1,280	0	10,240	0
04	0	0	20,480	320	0	2,560	0
05	1,280	20,480	0	0	0	0	0
06 (121 C)	0	0	0	0	20,480	0	0
07	0	0	1,280	20,480	0	1,280	0
08	0	0	0	0	0	0	20,480

means of complete biochemical and serologic characterization as outlined. Further information on complete antigenic analysis of *E. coli* strains may be found in the publications of Kauffmann (1954), Ørskov (1951 et seq.), and Ewing et al. (1956) and some data on the occurrence of members of the various *E. coli* O antigen groups among materials from animal sources were recorded by Ewing and Davis (1961) and Ewing (1962).

The following references are cited for the purpose of providing sources for further studies by readers who may be concerned with veterinary bacteriology and the importance of *E. coli* serotypes to it. Many of the publications contain extensive bibliographies.

Diarrheal disease and enteritis: Marsh and Tunnicliff (1938), Wramby (1948), Lovell (1955), Bokari and Ørskov (1952), Ulbrich (1954a, 1954b), Wood (1955), Fey (1956, 1957a, 1957b), Charles (1957), Rees

TABLE 54

Preparation of Absorbed Antisera for Differentiation of A-D Cultures

A-D O Antiserum	Absorbing Culture
01	05
02	03 +
	04
03	02 +
	04
04	02 +
	03
05	01
07	03 +
	04

(1958a, 1958b), Roberts (1958), Glantz et al (1959), Fey and Margadant (1961), Smith (1962), Gay (1965), and Glantz (1970a, 1970b).

Mastitis: Fey (1958), Murphy and Ryan (1958), and Dunn (1959).

Swine edema disease: Timony (1949, 1950, 1956), Gregory (1958, 1960), Ewing et al. (1958), Kelen et al. (1959), Campbell (1959), Mushin and Basset (1964), and Gossling and Rhoades (1966).

Avian disease: Hjarre and Wramby (1945), Hjarre (1949), Ulbrich (1951), Edwards and Ewing (1954), Gross (1956, 1957), Bankowski (1961), Glantz et al. (1962), and Takahashi and Miura (1968).

Experimental: Taylor et al. (1961) and Smith and Halls (1968).

General: Taylor (1961), Guinée (1963), Barnum et al. (1967), Smith and Halls (1967), and Glantz (1968).

ANTISERUM PRODUCTION

In all instances cultures to be employed in antiserum production should be plated on infusion agar plates and smooth (S) colony forms selected for use. Selected colonies are transferred to infusion agar slants or into semisolid medium if H antisera are to be produced.

O Antisera. After plating and selection of S colonies,

subsequent treatment of antigen suspensions to be used for O antiserum production varies according to the variety of K antigen possessed by a particular *E. coli* O group culture, as follows:

(a) Cultures known to contain K antigens of the L or B varieties. S colonies are inoculated into infusion broth and incubated for 6 to 8 hr at 37 C, and then heated at 100 C for 2½ hr. This procedure inactivates the L or B antigen as well as the H antigen of the cultures. Only suspensions that are homogeneous and not autoagglutinable upon heating are selected for inoculation.

(b) Cultures known to contain K antigen of the A variety. When possible, smooth translucent O forms, or K minus forms, are selected from platings. Such O forms may be inoculated into broth and heated at 100 C for 2½ hr as outlined in (a) above. When it is necessary to employ the more opaque KO, or K plus, colony forms, the suspensions prepared from them should be autoclaved at 121 C for 2 hr.

(c) If the nature of the K antigen contained in a culture is not known with certainty, antigen suspensions to be used for O antiserum production should be treated in the manner outlined in (b). In other words, it must be assumed that the culture might contain an A antigen and should be so treated.

(d) Vaccines for O antiserum production also may be prepared from agar slant cultures instead of from 6 to 8 hour broth cultures. Smooth colonies are selected from platings and inoculated over the entire surface of infusion agar slants. After incubation for 12 to 16 hr, the growth is removed from the agar slant cultures and emulsified in 0.5% sodium chloride solution. After being subjected to heat, as in a, b, or c, above, the suspensions may be diluted to a density approximating that of a 6- to 8-hr broth culture and preservative added. Also, O antisera may be produced according to the method of Roschka (1950) which has been used for the preparation of O antisera for members of various groups of enteric bacteria, e.g., see Edwards (1951).

Two or more rabbits should be injected with each *E. coli* O antigen and all inoculations are given in the marginal ear vein. The first dose is 0.5 ml, the second, 1.0 ml, third, 2.0 ml, and the fourth and fifth, 4.0 ml each. The interval between injections should be 4 to 5 days and the animals may be exsanguinated 6 to 8 days after the last inoculation. After the sera are separated, they may be preserved by the addition of an equal volume of glycerol or with another suitable preservative.

Antigen suspensions to be used for testing O antisera are prepared in the manner outlined above (a, b, c, or d), except that O antigens prepared with cultures known to contain L or B antigens may be heated at 100 C for only 1 hr. Agglutination tests used in the study of *E. coli* O antigens should be incubated in a water bath at 48 to 50 C for 16 to 18 hr.

K Antisera. Although there are several methods by which antigens may be prepared for K antiserum production, those given below may be expected to give good results. Regardless of which method is chosen, nonmotile strains should be used for K antiserum production whenever possible. This obviates concern over the presence of H agglutinin in K antisera. If nonmotile cultures are not available, freshly isolated strains that *have not been passaged in semisolid medium* should be employed for this purpose. The cultures should be plated and several colonies of the smooth, more opaque, rather mucoid, K forms selected for transfer. The medium known commercially as blood agar base is excellent for such platings since it appears to enhance K antigen production. Occasionally, K forms are not clearly discernable on plates incubated for only 24 hr at 37 C. If such is the case, the plates may be allowed to remain at room temperature for an additional 24 hours. This usually enhances K antigen production and facilitates colony selection. If antisera are available, it is advisable to test portions of colonies in O and KO antisera on slides prior to transfer. Colonies that react rapidly and completely in KO antisera and are inagglutinable or only slightly agglutinated in O antisera should be selected for transfer and subsequent use. If KO antiserum is not available, inagglutinability in O antiserum will prove a valuable aid in selection of K forms. After selection, each colony should be transferred to infusion broth and an infusion agar slant. Following overnight incubation, the broth culture is heated at 100 C for 1 hr. This procedure is a control of the stability of the O antigen. If any culture is autoagglutinable after this heat treatment, the corresponding agar slant should be discarded and should not be employed for K antiserum production. The heated broth cultures are not needed further and may be discarded.

Method 1. A tube of infusion broth is inoculated from the selected agar slant culture. The broth culture is incubated at 37 C for 4 or 5 hr and then formalinized by the addition of sufficient formalin to give a final concentration of 0.5%. Tubes containing 10 ml of broth are convenient for this purpose and 0.05 ml of formalin may be added to each culture after incubation. The formalinized broth cultures are placed in a refrigerator overnight; then the animals are given the first injection. For the second and third injections, fresh 4- to 5-hr broth cultures are used but the rabbits may be inoculated an hour or two after the formalin is

added. For subsequent injections, 4- to 5-hr broth cultures also are used but formalin treatment is omitted and living bacteria are injected. The agar slant culture used for inoculation of the broth should be transferred twice a week so that actively growing bacteria are present in the inoculum.

Method 2. This method is similar to the first except that the cells to be injected are grown on infusion agar slants. The growth from slants that have been incubated for 12 to 14 hr is suspended in formalinized saline solution for the first few injections and in isotonic saline for subsequent injections. The density of the suspensions should approximate that of a broth culture of 4 to 5 hr incubation.

Method 3. This procedure is a modification of the Roschka (1950) method and is similar to methods described for Vi antiserum production. It is mentioned because it is useful when nonmotile strains are not available for OB antiserum production. Its use is advised only in the case of cultures which possess A or B antigens. Selected cultures are inoculated over the surface of agar plates or onto several agar slants and after incubation the growth is removed and placed directly into absolute ethyl alcohol. The bacteria should be emulsified and suspended as well as possible in the alcohol. The suspension is placed in the incubator at 37 C for 2 hr, centrifuged, and the supernatant alcohol discarded. The sedimented bacteria are then resuspended in absolute alcohol and placed in the incubator for an additional 2 hours. After this the suspension is again centrifuged and the bacteria resuspended in acetone. The bacteria should be washed twice in acetone, then resuspended in a small amount of acetone, and placed in an incubator overnight or until the actone is evaporated. The dried bacteria are ground in a mortar until powdered and stored in an airtight moisture proof container. When animals are to be injected, a small amount of the powder is suspended in isotonic saline solution. The density of the suspension may approximate that of a 24-hour broth culture. Although this alcohol treatment may not completely inactivate the flagellar antigens of the bacteria, the production of H agglutinin is minimized, especially when poorly motile strains are selected.

The same schedule for inoculation of animals may be used with antigens prepared by method 1, 2, or 3 and all injections are made in the marginal ear vein. The first dose given may be 0.3 ml, the second 0.5 ml, the third 1.0 ml, and the fourth, fifth, and sixth, 2.0 ml each. The intervals between inoculations and the time for bleeding the animals is the same as that given for O antiserum production. If several animals are inoculated with one antigen, the animals should be bled into separate containers. The sera are not pooled until after tests are completed.

Antigens for titration of K antibody content of antisera are selected and prepared in the manner outlined above for rabbit inoculation. Broth cultures incubated 4 to 5 hours and then formalinized may be used as antigens and added to serial dilutions of antisera ranging from 1:20 to 1:2560. The O inagglutinability of the antigen should be controlled by titration of the same formalinized antigen suspension in serial dilutions of the appropriate O antiserum. The tests are placed in an incubator at 37 C for 2 hours and then removed to a refrigerator for overnight incubation. The type of agglutination obtained is fairly characteristic if K antigen suspensions which contain adequate amounts of antigen are used. The cells are agglutinated in the bottom of the tube in the form of a disc or pellicle. Titers are recorded as the highest dilution that gives good disc formation and antisera that have a K titer of 1:160 to 1:640 are satisfactory. Such antisera may be used in dilutions of 1:5 or 1:10 for slide agglutination tests. The O antibody content of KO antisera should be determined by titration of a heated O antigen suspension prepared from a smooth culture.

H Antisera. In order to prepare H antigen suspensions for production of H antisera or for testing antisera, it is necessary to promote motility and H antigen development by passage of the cultures through a semisolid agar medium. A medium containing 0.1% to 0.2% agar is recommended for the first few passages and one with 0.3% to 0.4% agar may be used for subsequent transfers. Usually about five passages are necessary before satisfactory H antigens are obtained but occasionally more are required. After enhancement of motility, flasks of infusion broth are inoculated and incubated at 37 C for 15 to 18 hr. The broth cultures then are preserved by the addition of 0.5% formalin. Rabbits are inoculated with H antigen suspensions prepared in this manner using the same route and schedule of injections as for O antiserum production. After immunization of rabbits is completed, the H antigen suspensions may be kept at room temperature, under which conditions they remain usable for agglutination tests[8] for several years. The H antibody content of antisera is determined by titration in serial dilutions of 1:100 to 1:25,600. Tests for H agglutination should be incubated in a water bath at 48 to 50 C and read after 15 min, 30 min, and 1 hr.

---

[8]The agglutinogenic properties of H antigen suspensions are greatly diminished after storage for about one month, and suspensions should not be used for antiserum production after that period.

REFERENCES

Andrewes, F. W. 1918. Lancet i, 560 (April 20).

Bankowski, R. A. 1961. Brit. Vet. J., **117**, 306.

Barnum, D. A., P. J. Glantz, and H. W. Moon. 1967. Colibacillosis. Summit, N. J.: Vet. Sales Div., CIBA Pharmaceutical Co.

Bokari, S. M. H., and F. Ørskov. 1952. Acta Pathol. Microbiol. Scand., **30**, 87.

Borman, E. K., C. A. Stuart, and K. M. Wheeler. 1944. J. Bacteriol., **48**, 351.

Braun, O. H. 1953. Das Problem der Pathogenitat von *Escherichia coli* im Saeuglingsalter. Heidelberg: Springer.

————. 1954. Z. f. Hyg., **139**, 478.

Campbell, S. G. 1959. Vet. Rec., **71**, 909.

Carpenter, P. L. 1943. Proc. Soc. Exper. Biol. Med., **53**, 129.

Carpenter, P. L., and C. A. Stuart. 1946. Ibid., **61**, 238.

————. 1950. J. Immunol., **64**, 237.

Castellani, A. 1907. J. Hyg., **7**, 1.

————. 1912. Zentralbl. f. Bakt. I. Orig., **65**, 262.

Charles, G. 1957. Aust. Vet., **33**, 329.

Charter, R. E. 1956. J. Pathol, Bacteriol., **72**, 33.

Costin, I. D. 1962. Microbiol., Parasitol., Epidemiol., **7**, 335.

————. 1966. Pathol. Microbiol., **29**, 214.

Costin, I. D., P. David, M. Dinculescu, A. Weiszberger, G. Olariu, N. Schiften, and T. Marton. 1960. Microbiol., Parasitol., Epidemiol., **6**, 531.

Costin, I. D., D. Voiculescu, and V. Gorcea. 1964. Pathol. Microbiol., **27**, 68.

Costin, I. D., and N. Olinici. 1965. Microbiol., Parasitol., Epidemiol., **10**, 49.

Davis, B. R., and W. H. Ewing. 1958. Can. J. Microbiol., **4**, 517.

————. 1963. Amer. J. Clin. Pathol., **39**, 198.

De Assis, A. 1939. O Hospital, **15**, 455.

Dunne, H. W. 1959. Can. J. Comp. Med., **23** 101.

Dupont, A. 1955. Epidemic infantile diarrhea with special reference to *Escherichia coli*. Copenhagen: Munksgaard.

Edwards, P. R. 1951. Pub. Health. Repts., **66**, 837.

Edwards, P. R., and W. H. Ewing. 1954. Cornell Vet., **44**, 50.

Ewing, W. H. 1953. J. Bacteriol., **66**, 333.

————. 1956. Ann. N. Y. Acad. Sci., **66**, 61.

————. 1962. J. Infect. Dis., **110**, 114.

————. 1963a. Isolation and identification of *Escherichia coli* serotypes associated with diarrheal diseases. CDC Publ.*

---

*CDC Publ., Publication from Center for Disease Control, (formerly the Communicable Disease Center), Atlanta, Ga. 30333.

————. 1963b. Studies on the occurrence of *Escherichia coli* serotypes associated with diarrheal disease. CDC Publ.*

————.1967. Revised definitions for the family Enterobacteriaceae, its tribes and genera. CDC Publ.*

————. 1969. Pub. Health Lab., ** **27**, 19.

Ewing, W. H., M. W. Taylor, and M. C. Hucks. 1950. Pub. Health Repts., **65**, 1474.

Ewing, W. H., and F. Kauffmann. 1950. Ibid., **65**, 1341.

Ewing, W. H., M. C. Hucks, and M. W. Taylor. 1952. J. Bacteriol., **63**, 319.

Ewing, W. H., K. E. Tanner, and H. W. Tatum. 1955. Pub. Health Repts., **70**, 107.

Ewing, W. H., M. M. Galton, and K. E. Tanner. 1955. J. Bacteriol., **69**, 549.

Ewing, W. H., H. W. Tatum, B. R. Davis, and R. W. Reavis. 1956. Studies on the serology of the *Escherichia coli* group. CDC Publ.*

Ewing, W. H., K. E. Tanner, and H. W. Tatum. 1956. Pub. Health Lab.,** **14**, 106.

Ewing, W. H., H. W. Tatum, and B. R. Davis. 1957. Ibid., **15**, 118.

————. 1958. Corriell Vet., **48**, 201.

Ewing, W. H., R. W. Reavis, and B. R. Davis. 1958. Can. J. Microbiol., **4**, 89.

Ewing, W. H., and B. R. Davis. 1961a The O antigen groups of *Escherichia coli* from various sources. CDC Publ.*

————. 1961b. O antiserum pools for preliminary examination of *Escherichia coli* cultures. CDC Publ.*

Fey, H. 1956. Schweitz. Z. f. Allg. Path. u. Bakt., **19**, 543.

————. 1957a. Zbl. f. Veterinarmed., **4**, 309.

————. 1957b. Ibid., **4**, 447.

————. 1958. Schweiz. Z. f. Allg. Path. u. Bakt., **19**, 926.

Fey, H., and Margadant, A. 1961. Zentralbl. f. Bakt. I. Orig., **188**, 71.

Frantzen, E. 1950. Acta Pathol. Microbiol. Scand., **27**, 236.

————. 1951. Ibid., **28**, 103.

Gay, C. C. 1965. Bacteriol. Revs., **29**, 75.

Glantz, P. J. 1968. Appl. Microbiol., **16**, 435.

————. 1970a.Can. J. Comp. Med., **34**, 47.

————. 1970b. Ibid., **34**, 101.

Glantz, P. J., H. W. Dunne, C. E. Heist, and J. F.

Hokanson. 1959. Bull. 645, Penna. State Univ. Coll. Agric., Agric. Exper. Sta., University Park, Penna.

————. 1962. Avian Dis., **6**, 322.

Gossling, J., and H. E. Rhoades. 1966. Cornell Vet., **56**, 344.

Grados, O., and W. H. Ewing. 1969. Technics for characterization of soluble antigens of Enterobacteriaceae. CDC Publ.*

————. 1970. J. Infect. Dis., 122, 100.

Gregory, D. W. 1958. Vet. Med., **53**, 77.

————. 1960. Amer. J. Vet. Res., **21**, 88.

Gross, W. B. 1956. Poultry Sci., **35**, 765.

————. 1957. Amer. J. Vet. Res., **18**, 724.

Guinée, P. A. M. 1963. Zentralbl. f. Bakt. I. Orig., **188**, 201.

Guinée, P. A. M., E. H. Kampelmacher, and H. M. C. C. Willems. 1962. Antonie v. Leeuwenhoek, **28**, 17.

Hjarre, A. 1949. Gastvorlesung an der Vet. Med. Fakültat Berlin, Dec. 10.

Hjarre, A., and G. Wramby. 1945. Scand. Vet. Tidskr., **35**, 449.

Report, 1958. International Enterobacteriaceae Subcommittee. Int. Bull. Bacteriol., Nomen. Tax., 8, 25.

Jeffries, C. D., and K. Okabe. 1958. U. S. Armed Forces Med. J., **9**, 965.

Kampelmacher, E. H. 1959, Antonie v. Leeuwenhoek, **25**, 289.

Kauffmann, F. 1943. Acta Pathol. Microbiol. Scand., **20**, 21.

————. 1944a. Ibid., **21**, 20.

————. 1944b. Ibid., **21**, 46.

————. 1947. J. Immunol., **57**, 71.

————. 1952. Acta Pathol. Microbiol. Scand., **31**, 355.

————. 1954. Enterobacteriaceae, 2nd Ed. Copenhagen: Munksgaard.

————. 1966. The bacteriology of Enterobacteriaceae. Copenhagen: Munksgaard.

Kauffmann, F., and B. Perch. 1943. Acta Pathol. Microbiol. Scand., **20**, 201.

Kauffmann, F., and G. Vahlne. 1944. Ibid., Suppl., **54**, 80.

Kauffmann, F., and B. Perch. 1948. Ibid., **25**, 507.

Kauffmann, F., and A. Dupont. 1950. Ibid., **27**, 552.

Kauffmann, F., F. Ørskov, and W. H. Ewing. 1956. Int. Bull. Bacteriol., Nomen. Tax., **6**, 63.

Kauffmann, F., amd F. Ørskov. 1956. Die Bakteriologie der *Escherichia coli* Enteritis. *In* Säuglings-Enteritis, ed. A. Adam. Stuttgart: Thiem.

Kelen, A. E., S. G. Campbell, and D. A. Barnum. 1959. Can. J. Comp. Med., **23**, 216.

Kétyi, I., P. Kneffel, and J. Domjän. 1957. Zentralbl. f. Bakt. I. Orig., **70**, 423.

*CDC Publ., Publication from Center for Disease Control, (formerly the Communicable Disease Center), Atlanta, Ga. 30333.
**Public Health Laboratory: Journal of the Conference of Public Health Laboratory Directors.

Levine, M. 1920. J. Infect. Dis., **27**, 31.

Lovell, R. 1955. Vet. Rev. and Annot., **1**, 1.

Marsh, H., and E. A. Tunnicliff. 1938. Dysentery of newborn lambs. Bull. No. 361, Montana State Coll. Agric. Exper. Sta., Bozeman, Mont.

Murphy, J., and M. A. Ryan. 1958. Irish Vet. J., **12**, 51.

Mushin, R., and C. R. Basset. 1964. Aust. Vet. J., **40**, 315.

McNaught, W., and J. S. Stevenson. 1953. Brit. Med. J., ii, 182.

Neter, E. 1944. Proc. Soc. Exper. Biol. Med., **57**, 200.

Neter, E., R. F. Korns, and R. E. Trussell. 1953. Pediatrics, **12**, 377.

Olarte, J., and G. Varela. 1952. J. Lab. & Clin. Med., **40**, 252.

Ørskov, F. 1951. Acta Pathol. Microbiol. Scand., **29**, 373.

———. 1952. Ibid., **31**, 51.

———. 1953. Ibid., **32**, 241.

———. 1954a. Ibid., **35**, 187.

———. 1954b. Ibid., **35**, 72.

———. 1954c. Ibid., **35**, 179.

———. 1955. Ibid., **36**, 375.

———. 1956a. Ibid., **39**, 147.

———. 1956b. Escherichia coli. Om type-bestemmlse, specielt med pa stammer fra infantil diarrhoe (English Summary). Copenhagen: Nyt. Nordisck Forlag, Arnold Busck.

Ørskov, F., and H. Fey. 1954. Acta Pathol. Microbiol. Scand., **35**, 165.

Ørskov, F., and I. Ørskov. 1960. Ibid., **48**, 47.

Ørskov, F., I. Ørskov, T. A. Rees, and K. Sahab. 1960. Ibid., **48**, 48.

Rees, T. A. 1958a. J. Comp. Pathol. Thera., **68**, 388.

———. 1958b. Ibid., **68**, 399.

Roberts, P. S. 1958. Aust. Vet. J., **34**, 152.

Roelke, K. 1942. Z. f. Hyg., **124**, 356 (Quoted by Seeliger, 1950).

Roschka, R. 1950. Klin. Med., **5**, 88.

Rowe, B., J. Taylor, and K. A. Bettelheim. 1970. Lancet, i, 1.

Seeliger, H. 1950. Z. f. Hyg., **130**, 490.

Stanholtz, M. I., P. White, and S. Granger. 1967. Morbidity and Mortality Weekly Repts., **16**, 246.

Smith, J. 1949. J. Hyg., **47**, 221.

Smith, H. W. 1962. J. Pathol. Bacteriol., **84**, 147.

Smith, H. W., and S. Halls. 1967. Ibid., **93**, 499.

———. 1968. J. Med. Microbiol., **1**, 61.

Stevenson, J. S. 1956. Bull. Hyg., **31**, 570.

Takahashi, K., and S. Miura. 1968. Jap. J. Vet. Res., **16**, 65.

Tatum, H. W., W. H. Ewing, and B. R. Davis. 1958. Pub. Health Lab.,** **16**, 8.

Taylor, J. 1961. J. Appl. Bact., **24**, 316.

Taylor, J., B. W. Powell, and J. Wright. 1949. Brit. Med. J., ii, 117.

Taylor, J., and R. E. Charter. 1952. J. Pathol. Bacteriol., **64**, 715.

———. 1955. J. Clin. Pathol., **8**, 276.

Taylor, J., M. P. Wilkins, and J. M. Payne. 1961. Brit. J. Exper. Pathol., **42**, 43.

Timoney, J. F. 1949. Vet. Rec., **61**, 710.

———. 1950. Ibid., **62**, 748.

———. 1956. Irish Vet. J., **10**, 233.

Ulbrich, F. 1951. Wien. tierarztl. Wchnschr., **37**, 395.

———. 1954a. Zentral. Veterinarmed., **1**, 603.

———. 1954b. Zentralbl. f. Bakt. I. Orig., **160**, 506.

Weil, A. J., and H. Slafkovsky. 1948. J. Bacteriol., **55**, 759.

Welch, H., and F. L. Mickle. 1932. J. Infect. Dis., **50**, 524.

Wheeler, K. M., C. A. Stuart, and W. H. Ewing. 1946. J. Bacteriol., **51**, 169.

Wood, P. C. 1955. J. Pathol. Bacteriol., **70**, 179.

Wramby, G. 1948. Investigations into the antigenic structure of *Bact. coli* from calves. Upsala: Appelsbergs.

Chapter 6
The Genus *Shigella*

Chantemesse and Widal (1888) reported the isolation of a bacterium from feces in five cases of acute dysentery, and from the large intestine and mesenteric glands as well as the stools of a soldier dead of the disease. These workers studied some of the cultural characteristics of the bacterium and with it produced ulcerative lesions in the large intestine of animals. Available literature to that date was reviewed. Cultures isolated by Chantemesse and Widal were studied by Vaillard and Dopter (1903) who found them to be the same as those described by Shiga (1898). Grigorieff (1891) isolated cultures from eleven cases of dysentery in Russia; he considered these to be identical with the bacterium described by Chantemesse and Widal (see Shiga, 1898). According to both Shiga (1898) and Flexner (1900), Klebs (1886) described a small bacterium as the cause of dysentery but whether this microorganism was compared to Shiga's bacillus or that of Chantemesse and Widal is not known. Therefore, it is clear that Chantemesse and Widal (1888) may be regarded as the discoverers of dysentery bacteria.

In the last decade of the nineteenth century severe and extensive epidemics of bacillary dysentery occurred in Japan. According to Shiga (1898), 89,400 cases with 22,300 deaths were reported within one short period. During this period Shiga (1898) reported the isolation of an identical bacterium from 34 of 36 persons afflicted with the disease. The bacterium was reported to be similar to the typhoid bacterium in morphological and staining characteristics. Shiga's microorganism was selected subsequently as the type species of the genus *Shigella* (Castellani and Chalmers, 1919), since the specific epithet employed by Shiga had priority.

Other reports followed that of Shiga. Kruse isolated bacteria that resembled those of Shiga from dysentery patients in Westphalia in 1900. Flexner, in the same year, reported the recovery of microorganisms which he regarded as the same Shiga's bacillus from an American soldier in the Philippines. Strong and Musgrave (1900) confirmed the work of Flexner and produced dysentery with the bacteria in a monkey and in a condemned criminal. There followed in rapid succession a large number of publications from virtu-

ally all parts of the world which recorded the isolation of these and similar microorganisms from acute and chronic cases of dysentery.

More detailed discussion of the work of earlier investigators may be found in the publications of Gardner (1929), Dudgeon (1929), Felsen (1945), Madsen (1949), and in the other references cited.

CLASSIFICATION AND NOMENCLATURE WITHIN THE GENUS *SHIGELLA*

The genus *Shigella* was defined by Ewing (1949) as follows: "gram-negative bacteria that are aerobic, non-sporulating, nonmotile, and, with a few exceptions, nonproductive of gas from fermentable substances. They do not utilize salicin, adonitol, or citrate, or hydrolyze urea, liquefy gelatin, or form acetylmethylcarbinol. Lactose is utilized by only two recognized species *(Shigella sonnei* and *Shigella dispar)* and by these only upon prolonged incubation." Following the suggestion of Kaufmann (1949) and others (e.g., Frantzen, 1950; Ewing et al., 1950) *Shigella alkalescens* and *Shigella dispar* were deleted from the genus by the Shigella Commission of the International Enterobacteriaceae Subcommittee in 1950.

The above-mentioned Shigella Commission included definitions of the genus *Shigella* in its recommendations (see Reports, International Enterobacteriaceae Subcommittee, 1954, 1958). However, the definition given by Ewing (1967) has been adopted for use herein. It is as follows:

The genus *Shigella* is composed of nonmotile bacteria that conform to the definitions of the family ENTEROBACTERIACEAE and the tribe ESCHERICHIEAE. With the exception of certain biotypes of *Shigella flexneri* 6, visible gas is not formed from fermentable carbohydrates. Compared with *Escherichia*, *Shigella* are less active in their utilization of carbohydrates. Salicin, adonitol, and inositol are not fermented. *Shigella sonnei* strains ferment lactose upon extended incubation, but other species do not utilize this substrate in conventional medium. Lysine is not decarboxylated, the majority of strains do not possess a demonstrable arginine dihydrolase system, and ornithine is decarboxylated only by *S.*

Sonnei and *Shigella boydii* 13. The type species is *Shigella dysenteriae* (Shiga) Castellani and Chalmers.

When used in conjunction with the definition of the tribe Escherichieae (Chapter 1), this definition provides a convenient starting point for a discussion of the characteristics of shigellae.

Although agglutinin techniques were mentioned at an early date in connection with studies on shigellae (e.g., Park et al., 1904), earlier classifications and methods for differentiation of members of the genus *Shigella* were based upon biochemical tests or upon biochemical tests combined with agglutination in unabsorbed antisera. While the classification of a microorganism in the genus *Shigella* and in a particular species within the genus is based upon certain physiological reactions, most bacteriologists today do not place great reliance in carbohydrate fermentation tests for *type determination* within a subgroup or species. However, certain tests are of assistance in this regard (v. inf.).

The *nomenclature* and taxonomic schema of the genus *Shigella* adopted herein (Figure 3) is that proposed by Ewing (1949) as modified by the Shigella Commission in 1950 and as amended and extended in subsequent reports by the International Enterobacteriaceae Subcommittee (1954, 1958). The Shigella schema is based in part upon biochemical characteristics, in part upon intragroup antigenic relationships, and in part upon tradition. Arabic instead of roman numerals are employed to designate serotypes. This change was made to avoid possible confusion between the letters V and X and the romans numerals V and X. However, roman numerals are employed still to designate the type of specific antigen when writing the formula for a serotype of *Shigella*. For comparative purposes, older nomenclature and designations are shown in Figure 3. The nomenclature of the four species (or subgroups) of the genus *Shigella* and the use of arabic numerals for the designation of individual serotypes within each species, as employed in the International Shigella Schema, has been acted upon and approved by the Judicial Commission of the International Nomenclature Committee (see Int'l. Bull. Bact. Nomen. Tax., 10: 85, for references). In 1966 Ewing and Carpenter recommended certain changes in the designations of subserotypes 3a and 3b of *S. flexneri* 3. These recommendations have been adopted, and the serotypes of *S. flexneri*, abbreviated antigenic formulae for the subserotypes, and the manner in which cultures may be expected to react absorbed antisera are given in Table 55.

Since translations of publications on shigellae and bacillary dysentery by Russian investigators have become more readily available, a comparison of the designations used by many Soviet writers with those employed in the International Shigella Schema is given in Figure 4 for use by interested investigators. The data in the figure are based upon comparative studies made by Ewing et al. (1959) and by Ewing and Trabulsi (1962). Similar investigations were made by Olinici et al. (1958) and by Novgorodskaja (1959, 1960). In 1963 Soviet investigators adopted a schema that is more like the International Shigella Schema than that given in Figure 4 (see Zhadanov and Andreeva, 1963). For example, the method of designating serotypes of *S. flexneri* used in the International Shigella Schema was adopted. However, *S. flexneri* 6 and *S. boydii* were regarded as subspecies of *S. flexneri*.

Subgroup A is made up of those shigellae that characteristically do not utilize mannitol, and which, with few exceptions, do not bear close seriological relationships to members of other subgroups of *Shigella*. These are *Shigella dysenteriae* 1 (Shiga), *Shigella dysenteriae* 2 *(Shigella ambigua* or *Shigella schmitzii),* the members of the so-called Large-Sachs group, *Shigella dysenteriae* serotypes 3 to 7 inclusive (Archer, 1933; Large and Sankaran, 1934; Boyd, 1935; Sachs, 1943), *Shigella dysenteriae* 8 (Ewing et al., 1952), and two serotypes that were added more recently, *Shigella dysenteriae* 9 and *Shigella dysenteriae* 10. The last two serotypes named, formerly labeled 58 (Cox and Wallace, 1948) and 2050-52 (Ewing, 1953) were *sub judice* for several years but were added to subgroup A in the 1958 Report of the International Enterobacteriaceae Subcommittee (see also Ewing et al., 1958). Several provisional serotypes of *Shigella* are under consideration for possible addition to subgroup A but still are *sub judice*. These are serotypes 3873-50, 2000-53, and 3341-55 (Ewing and Hucks, 1952; Ewing et al., 1958). The microorganisms described under the designations A12 (Sachs, 1943), 1831 (Wheeler and Stuart, 1946) and H62 (Piechaud and Rubensten, 1953) formerly were considered as possible members of subgroup A but are no longer so considered. The A12 biotypes are aerogenic, utilize citrate mutatively in Christensen's citrate agar medium (1949) and some cultures yield mutative fermentation of lactose (Ewing and Hucks, 1950; unpublished data). They are serologically related to *Escherichia Coli* O groups 32 and 83, as well as to *S. boydii* 14 and certain O antigen groups of *Salmonella* and *Arizona*. Serotype 1831 cultures, some of which utilize mannitol, usually utilize mucate rapidly, some strains yield mutative fermentation of salicin, and some utilize citrate in Christensen's medium (Ewing et al., 1958). The 1831 strains are anaerogenic varieties of *E. coli* O group 22. The H62 strains are mannitol negative cultures of *E. coli* O group 25, or Alkalescens-Dispar O group 2

TABLE 55

Serotypes and Subserotypes of *S. flexneri*

Serotype	Subsero-type	Abbreviated antigenic formula	Absorbed antisera								
			Type specific						Subserotype complexes		
			I	II	III	IV	V	VI	3, 4	6	7, 8
1	1a	I:4 . . .	++++	-	-	-	-	-	++++	-	-(+)
	1b	I:6 . . .	++++	-	-	-	-	-	-	++++	-
2	2a	II:3,4 . . .	-	++++	-	-	-	-	++++	-	-
	2b	II:7,8 . . .	-	++++	-	-	-	-	-	-	++++
3	3a	III:6:7,8 . . .	-	-	++++	-	-	-	-(+)	++++	++++
	3b	III:3,4:6 . . .	-	-	++++	-	-	-	+	++++	-
	3c	III:6 . . .	-	-	++++	-	-	-	-	++++	-
4	4a	IV:3,4 . . .	-	-	-	++++	-	-	++++(-)	-	-
	4b	IV:6 . . .	-	-	-	++++	-	-	-	++++	-(+)
5	5	V:7,8 . . .	-	-	-	-	++++	-	-	-	++++
6	6	VI:- . . .	-	-	-	-	-	++++	-(+)	-	-
X variant		-:7 . . .	-	-	-	-	-	-	-	-	++++
Y variant		-:3,4 . . .	-	-	-	-	-	-	++++	-	-

Degrees of agglutination: + + + +, complete reaction; + (to ++) weakly positive reaction.

Symbols in parenthese indicate occasional reactions.

Adapted from Ewing and Carpenter (1966).

Figure 3
SHIGELLA NOMENCLATURE AND TAXONOMY

Species and Subgroup	Subcommittee ① 1958	Shigella Commission 1953	Shigella Commission 1950	Ewing 1949 (with additions) Type	Ewing Abbreviated Antigenic formula	Kauffmann & Ferguson 1947	Wheeler 1944a*	Boyd 1940* 1946	Boyd 1938	Weil, Black, and Farsetta 1944	English (older)	German	Other
A. *S. dysenteriae*	1	1	1	I								Shiga-Kruse I	*Bacterium shigae*
	2	2	2	II									*S. ambigua, S. schmitzii, B. ambiguus*
	3	3	3	III									Q771, type 8524†, *S. arabinotarda* A††
	4	4	4	IV									Q1167, *S. arabinotarda* B
	5	5	5	V									Q1030
	6	6	6	VI									Q454
	7	7	7	VII									Q902
	8	8	8										Serotype 599-52
	9												Serotype 58
	10												Serotype 2050-50
B. *S. flexneri*	1a	1a	1a	I	I:4	1b	I	I	V	I	V	B,C	Flexner
	1b	1b	1b	I	I:4,6	1a	I	II	W	I, III	VZ	A	
	2a	2a	2a	II	II:4	2a	IIa	III	Z	II, VII	W	D	Strong, Hiss-Russell
	2b	2b	2b	II	II:7,8,9	2b	IIb	IV	103	II, VII	WX	DX	
	3a	3	3	III	III:6,7	3	III	IV	103Z	III	Z	H	
	3b			III	III:4,6,7			V	p.119	IV		F	Lentz Y2
	3c			III	III:(4),6			VI	88	III,IV		J	*S. saigonensis, S. rio* ②
	4a	4a	4a	IV	IV:4	4a	IV			V(V,VII)		F	*S. newcastle*
	4b	4b	4b	IV	IV:6	4b	IV			VII		G	
	5	5	4	V	V:7	5	VI			VIII	X	L	Hiss-Russell
	6	6	5	VI	VI:(4)	6	X				Y	Y	
	X	X	6	—	—:7,8,9		Y						
	Y	Y	X	—	—:3,4								
			Y										
C. *S. boydi*	1	1	1	I			I		170	IX		R	Type T, Lavington, *S. etousa*
	2	2	2	II			II		p.288	X			Serotype 112
	3	3	3	III			III		D.1	XI			Serotype 1296/7
	4	4	4	IV					p.274	XIV			Serotype 430
	5	5	5	V					p.143	XIII			Serotype 34
	6	6	6	VI					D19	XII		N	Serotype 123
	7	7	7	VII								P	Serotype 425
	8	8											Serotype 2770-51
	9	9											Serotype 703
	10	10											
	11	11											
	12												
	13												
	14												
	15												
D. *S. sonnei*	*S. sonnei*	*s. sonnei*	*S. sonnei*	*S. sonnei*								E	Sonne-Duval, Sonne III, *S. ceylonensis* A

① International Enterobacteriaceae Subcommittee, Report, 1958
② Mannitol negative biotypes of *S. flexneri* 4, usually 4a
* Antigenic formulas proposed by these investigators are given in Table 75.
† Gober et. al. (1944).
†† Christensen and Gowan (1944).

English: Andrewes and Inman (1919), Murray (1918), Gettings (1919), Boyd (1938). Boyd (1938).
German: Kruse (1900), Sartorius and Reploh (1932), Winkle (1949), Seeliger (1949).

Figure 4

IDENTIFICATION OF CULTURES OF *SHIGELLA* FROM THE USSR

Culture No. (Original)	Classification according to International Schema as determined at International Center, Atlanta		Soviet Method of Designation √[1]		
			Species	Subspecies	Serotype
	Subgroup A				
*	*S. dysenteriae*	1	Grigorjeff-Shiga		
*		2	Stutzer-Schmitz		
4662		3	*Bact. dysenteriae*	Novgorodskaja-Semenova	Roman
2291		4	do	do	1618
1676		5	do	do	819
443		6	do	do	Tjacht
2435		7	do	do	2435
*		8			
*		9			
*		10			
	Subgroup B				
2303	*S. flexneri*	1a	Flexner	Flexner	$f(f_2)$
767		1b	do	do	$f(f_1)$
729		2a	do	do	c
726		2b	do	do	b
737		3a	do	do	e
*		3b			
414		3c	do	do	d
752		4a	do	do	$a(a_2)$
718		4a	do	do	$a(a_1)$
2465		4b	do	do	$a(a_3)$
1307		5	do	do	$g(g_1)$
1007		5	do	do	$g(g_2)$
347		6**	Flexner	Newcastle	
339		6	do	do	
358		6	do	do	
271		6	do	do	
790		6	do	do	
808		6	do	do	
	Subgroup C				
439	*S. boydii*	1	Flexner	Boyd-Novgorodskaja	I
2012		2	do	do	V
*		3			
4322		4	do	do	III
2619		5	do	do	VII
*		6			
4830		7	do	do	II
*		8			
12181		9	do	do	IV
78		9	do	do	IV
*		10			
*		11			
2469		12	do	do	VI
*		13			
*		14			
*		15			
	Subgroup D				
946	*S. sonnei*		Sonne		
523			do		
2374			do		

√[1] Designations that occurred on the culture labels and in data that accompanied the cultures.

* Cultures of these serotypes were not received.

** The six *S. flexneri* 6 cultures were of three biotypes: Boyd 88 (347, 339), Manchester (358, 271), and Newcastle (790, 808).

(Ewing et al., 1958). None of these three biotypes (A12, 1831, H62) is regarded as a member of the genus *Shigella* by the author. It may be added that the exclusion of these bacterial types from the genus *Shigella* by the author is without reference to the possible pathogenic propensities of the strains, either in man or in experimental animals.

Members of the genus *Providencia* (Chapter 17) sometimes are confused with shigellae because they fail to utilize mannitol and may be anaerogenic. *Providencia*, previously known as the 29911 group of Stuart (Stuart et al., 1946), may be excluded from the genus *Shigella* on the basis of motility, sodium citrate utilization, adonitol or inositol fermentation, and by production of phenylpyruvic acid from phenylaninine.

Subgroup B is composed of those microorganisms formerly referred to as *Shigella paradysenteriae* Flexner, and is now named *S. flexneri,* serotypes 1 through 6 (Figure 3). Members of this subgroup are related one to another through the possession of common group antigens but each serotype contains a type or main antigen by which it may be identified. Mannitol usually is fermented by members of subgroup B but exceptions are known. They may be differentiated from members of subgroups A and C by means of additional biochemical reactions as well as by serological analysis (v. inf.).

Subgroup C of the genus *Shigella* is composed of the serotypes of *Shigella boydii*, of which fifteen are recognized (Figure 3). Biochemically members of subgroup C are somewhat similar to members of the *S. flexneri* subgroup but differ in several important aspects; furthermore, members of subgroup C lack close serological relationship to the *S. flexneri* serotypes. *S. boydii* serotypes 12, 13, 14, and 15 were maintained as provisional serotypes for many years (Ewing and Gravatti, 1947; Ewing and Hucks, 1952; Ewing, 1953; Ewing et al., 1958), but were added to the schema (Figure 3) in the 1958 Report of the International Subcommittee on Enterobacteriaceae. Several other serotypes (e.g., 3615-53, 1621-54, and 2710-54) remain *sub judice* (see Ewing et al. 1958 for references). The bacterium described by Cefalu and Gulotti (1953) as a provisional member of the *S. boydii* subgroup may be classified as *E. coli* on the basis of its biochemical reactions and its serological relationships permitted classification as *E. coli* 028a, 28c, K73 (B), the Katwijk type (see Chapter 5). The microorganisms described by Manolov (1959) under the designation of "Shigella 13" are not members of the genus *Shigella* but anaerogenic members of *E. coli* O group 25. Bacteria of this sort were described earlier as type 145-46 by Dr. K. P. Carpenter of the International Shigella Center (London) and by Ewing et al. (1958), but were not called shigellae.

Subgroup D contains only *S. sonnei*. This microorganism resembles members of subgroups B and C as regards many of its biochemical reactions but differs in that *S. sonnei* ferments lactose upon extended incubation, decarboxylates ornithine, and produces a color reaction in phenyl-proprionic acid medium (v. inf.). The O antigens of *S. sonnei* cultures are not significantly related to those of other described shigellae.

BIOCHEMICAL REACTIONS OF CULTURES OF *SHIGELLA*

The biochemical reactions given by members of the genus *Shigella* are summarized in Table 56. The data listed in Table 56 are based upon the reactions given by more than 5,000 cultures representative of each of the four species. Further, the reactions given by members of each species are summarized separately in Table 57. From the data presented in the above-mentioned tables it may be seen that members of the genus *Shigella* do not produce hydrogen sulfide in triple sugar iron (TSI) agar nor in media that possess a similar level of sensitivity, they do not grow on Simmons' citrate agar, do not hydrolyze urea, do not produce detectable amounts of acetylmethylcarbinol or 2,3-butyleneglycol, do not decarboxylate lysine, and do not utilize sodium malonate. Shigellae do not produce phenylalanine deaminase nor lipase. They are nonmotile and fail to produce acid from salicin, adonitol, inositol, alpha methyl glucoside, erythritol, or esculin. Only *S. sonnei* and some strains of *S. boydii* 9 ferment lactose. Sodium mucate is utilized by some cultures of *S. sonnei* but not by other shigellae (v. inf.) and only *S. sonnei* and cultures of *S. boydii* 13 are known to possess an ornithine decarboxylase system. Sodium acetate is utilized as a sole source of carbon by some biotypes of *S. flexneri* 4a but not by other shigellae (v. inf.) The data given in the above-mentioned tables are in general agreement with those reported by Boyd (1940), Weil (1944), Wheeler (1944a), Madsen (1949), and others.

The biochemical reactions given by provisional *(sub judice)* serotypes of *Shigella* are shown in Table 58. Since serotype 2000-53 possesses an ornithine decarboxylase system and is able to utilize sodium acetate (Trabulsi and Ewing, 1962) it is probable that this serotype should no longer be considered as a possible *Shigella*.

In addition to those contained in Tables 56 and 57 the data given in Tables 3, 5, and 10 (Chapter 3) should be helpful in the differentiation of members of the genus *Shigella*.

In practice, the biochemical tests given in Tables 4 and 5 (Chapter 3) may be used for general work. Evidence of hydrogen sulfide formation in TSI agar medium and urease production aid in the elimination

TABLE 56

The Biochemical Reactions of Shigellae

Test or substrate	Sign	%+ (%+)*	Test or substrate	Sign	%+ (%+)*
Hydrogen sulfide	–	0	Rhamnose	d	16.6 (6.1)
Urease	–	0	Malonate	–	0
Indol	– or +	37.8	Mucate	–	3.5 (6.5)**
Methyl red (37 C)	+	100	Christensen's citrate	–	0
Voges-Proskauer (37 C)	–	0	Jordan's tartrate	– or +	29.4
Citrate (Simmons')	–	0	Sodium pectate	–	0
KCN	–	0	Sodium acetate	–	0
Motility	–	0	Sodium alginate	–	0
Gelatin (22 C)	–	0	Lipases		
Lysine decarboxylase	–	0	Corn oil	–	0
Arginine dihydrolase	d	7.6 (5.6)	Triacetin	–	0
Ornithine decarboxylase	d	20 (0.3)	Tributyrin	–	0
Phenylalanine deaminase	–	0	Maltose	d	26.6 (57)
Glucose acid	+	100	Xylose	d	4.3 (11.1)
gas	–	2.1	Trehalose	d	76.4 (19.3)
Lactose	d	0.3 (11.4)**	Cellobiose	–	2.9 (0.4)
Sucrose	d	0.9 (31.1)**	Glycerol	d	13.8 (39.6)
Mannitol	+ or –	80.5	Alpha methyl glucoside	–	0
Dulcitol	d	5.4 (12.7)	Erythritol	–	0
Salicin	–	0	Esculin	–	0
Adonitol	–	0	Beta galactosidase	– or +	13.7**
Inositol	–	0	Nitrate to nitrite	+	99.8
Sorbitol	d	29.1 (21.9)	Oxidation – fermentation	+	100 F
Arabinose	d	67.8 (11.2)	Oxidase	–	0
Raffinose	d	20.7 (19.1)			

[1] Based upon the examination of 5166 cultures representative of all four species.

* Figures in parentheses indicate percentages of delayed reactions (3 or more days)

** See text.

+ 90% or more positive within one or two days' incubation.
(+) Positive reaction after 3 or more days.
 – No reaction (90% or more –).
+ or – Majority of strains positive, some cultures negative.
 – or + Majority of cultures negative, some strains positive.
(+) or + Majority of reactions delayed, some occur within 1 or 2 days.
 d Different reactions: +, (+), –.

of many microorganisms. *Any microorganism that blackens triple sugar iron (TSI) agar, is urease positive, or is able to grow on Simmons' citrate medium, or decarboyxlates lysine, produces phenylpyruvic acid from phenylalanine, or is motile does not belong to the genus Shigella.*

All shigellae produce acid from glucose within 24 hours and, with the exception of two biotypes of *S. flexneri* 6, (Table 59), they are anaerogenic. The Newcastle bacillus described by Clayton and Warren (1929) and the Manchester bacillus (Downie et al. 1933) are biotypes of *S. flexneri* 6 which produce gas

TABLE 57

Biochemical Reactions of Serotypes of *Shigella*[1]

Subgroup and Serotype	Mannitol	%+	Dulcitol	%+	Xylose	%+	Rhamnose	%+	Raffinose	%+	Glycerol	%+	Indol	%+	Ornithine Decarboxylase	%+
Subgroup A *S. dysenteriae*																
1	−	0	−	0	−	0	−	0	−	0	+ or (+)	100	−	0	−	0
2	−	0	−	0	−	0	+	98	−	0	(+) or +	98	+	100	−	0
3	−	0	−	0	−	0	−	0	−	0	(+) or +	100	−	0	−	0
4	−	0	−	0	−	0	−	0	−	0	(+) or +	100	−	0	−	0
5	−	0	+ or (+)	100	−	0	−	0	−	0	+ or (+)	100	−	0	−	0
6	−	0	−	0	−	0	−	0	−	0	− or (+)	38	−	0	−	0
7	−	0	−	0	−	0	(+) or +	90	−	0	−	0	+	100	−	0
8	−	0	−	0	+ or (+)	96	−	8	−	0	+ or (+)	100	+	100	−	0
9	−	0	−	0	−	0	−	0	−	0	+ or (+)	100	−	0	−	0
10	−	0	−	0	+	100	−	0	−	0	+ or (+)	100	−	0	−	0
Subgroup B *S. flexneri*																
1	+	95	−	0	−	0	−	0	d	89	−	0	− or +	35	−	0
2	+	99	−	0	−	0	−	0	d	77	−	0	− or +	44	−	0
3	+	98	−	0	−	0	d	12	d	88	−	0	+ or −	88	−	0
4	+	99	−	0	−	0	d	23	d	82	−	0	+ or −	55	−	0
4	−	0	−	0	d	71	− or +	48	−	3	−	0	+	98	−	0
5	+	99	−	0	−	0	d	5	d	72	−	0	+	95	−	0
6	+	>99	d	80	−	4	−	6	−	0	d	88	−	0	−	0
6ᵃ	+	100	d	86	d	75	−	0	−	0	+ or (+)	100	−	0	−	0
6ᵃ	−	0	+ or (+)	100	−	0	−	0	−	0	(+)	100	−	0	−	0
Subgroup C *S. boydii*																
1	+	100	−	1	+ or (+)	97	−	0	−	0	(+) or +	96	−	0	−	0
2	+	100	−	1	−	0	−	0	−	0	+ or (+)	100	−	0	−	0
3	+	100	d	75	d	86	−	0	−	0	+ or (+)	91	−	0	−	0
4	+	99	− or (+)	28	−	0	−	0	−	0	+ or (+)	100	−	0	−	0
5	+	100	−	0	(+)	94	−	0	−	0	d	61	+	100	−	0
6	+ or (+)	100	(+) or +	100	+	100	−	0	−	0	(+) or +	100	−	0	−	0
7	+	100	−	0	+	98	−	0	−	0	+ or (+)	98	+	100	−	0
8	+	100	−	0	+	94	d	80	−	0	(+) or +	100	−	0	−	0
9	+	95	−	0	−	0	−	0	−	0	(+) or −	82	+	100	−	0
10	+	94	+	100	d	84	d	0	−	0	+ or (+)	100	−	0	−	0
11	+	100	− or (+)	34	+ or (+)	100	−	0	−	0	− or +	14	+	100	−	0
12	+	100	− or (+)	14	−	0	−	0	−	0	(+) or −	63	−	0	−	0
13	+	100	−	0	−	0	−	0	−	0	+ or (+)	100	+	100	+	100
14	− or +	29	−	0	(+) or +	100	−	0	−	0	+ or (+)	100	+	100	−	0
15	+	90	−	0	−	0	−	0	−	0	(+) or −	64	+	100	−	0
Subgroup D *S. sonnei*	+	99	−	1	−	1	+ or (+)	98	d	84	d	46	−	0	+	>99

+ 90% or more positive in 1 or 2 days. − 90% or more negative. + or − Majority positive. − or + Majority negative. (+) Delayed positive. d Different reactions [+, (+), −].

ᵃSome *S. flexneri* 6 cultures (the Newcastle and Manchester biotypes) produce gas from fermentable substrates; other shigellae are anerogenic.

NOTE: In this table percentages of + and (+) reactions are combined.

TABLE 58

Biochemical Reactions of Described Provisional Serotypes of *Shigella* (Sub Judice)[a]

Serotype	Mannitol	Maltose	Dulcitol	Arabinose	Sorbitol	Xylose	Rhamnose	Raffinose	Indol
3873-50	–	– or (+)	–	+	(+)	(+)	–	–	–
2000-53	–	+	–	+	–	–	+	–	+
3341-55	–	–	–	+	(+)	(+)	+	–	–
3615-53	+	– or (+)	–	+	+	+	–	–	+
2710-54	+	– or (+)	–	+	+	+	–	–	+
1621-54	+	+	–	– or (+)	(+) or –	–	– or (+)	(+)	+

[a] All available strains of the serotypes listed fermented glucose without gas production, all produced nitrite from nitrate, and all gave positive methyl red tests. Cultures of serotype 1621-54 fermented sucrose and raffinose mutatively, all others were negative on these substrates. All of the cultures failed to produce acid from lactose, salicin, adonitol, and inositol, were Voges-Proskauer negative, and failed to utilize citrate in either Simmons' or Christensen's medium. All strains gave negative results in tests for gelatin liquefaction, hydrogen sulfide production, mucate and malonate utilization, urease and phenylalanine deaminase production, all failed to grow in KCN medium, all failed to decarboxylate lysine, and all were nonmotile.

+   Positive in 1 or 2 days.

(+)   Delayed positive reaction.

–   Negative test.

TABLE 59

Biotypes of *S. flexneri* 6

Biotypes	Glucose	Mannitol	Dulcitol
Boyd 88	A	A	–
Boyd 88	A	A	(A)
Manchester	AG	AG	(AG)
–	AG	AG	–
Newcastle	AG	–	(AG)
–	A	–	–
–	A	–	(A)

A, acid reaction   AG, acid and gas   ( ), delayed

from fermentable substances. Lactose is fermented by *S. sonnei* but with very rare exceptions fermentation is delayed several days. Also, some cultures of *S. boydii* 9 utilize lactose but this serotype apparently is relatively rare. Thus, cultures that produce acid from lactose within 24 hours need not be examined further, as a rule. Gas production in lactose medium indicates that the culture in question is not a member of the genus *Shigella* since the only lactose fermenting *Shigella* known is anaerogenic. Occasionally, strains of *S. sonnei* that do not ferment lactose are reported (Rubinsten and Piechaud, 1952) and strains of this sort were isolated in Italy by the author during 1943-1945. It should be noted that occasional cultures of *S. dysenteriae* 1 which have been cultivated on artificial media for some time are able to produce acid from lactose. This observation was reported by Winter (quoted by Andrewes, 1918), Wheeler and Stuart (1946), Madsen (1949), and also was made by the writer. Cultures of this serotype examined by the writer also produced acid from maltose.

The beta-D-galactosidase activities of Enterobacteriaceae and certain other bacteria have been reported upon by several investigators (e.g., Le Minor and Ben Hamida, 1962; Szturm-Rubinsten and Piechaud, 1963a, 1963b; Buelow, 1964; Lubin and Ewing, 1964; and Costin (1966). The ONPG test (Chapter 18) is used for this determination. While the use of this test is not regarded as essential for differentiation of members of the genera of Enterobacteriaceae, the test can be of value, especially with members of the tribe Salmonelleae, when the results are interpreted judiciously. As noted above, some cultures of *S. dysenteriae* 1 ferment lactose slowly. However, all strains of this serotype tested gave positive results in the ONPG test (see Lubin and Ewing, 1964). In addition, a few cultures of several of the other serotypes of *S. dysenteriae* and several of the serotypes of *S. boydii* yielded positive results in the ONPG test (loc. cit.). Of 490 strains of *S. flexneri* tested only three gave positive results in this test and none of these fermented lactose in conventional medium. Since the ONPG test may be regarded as a measure of latent ability to ferment lactose, it would be expected that the majority of strains of *S. sonnei* would yield positive results when tested by this technique, and indeed they do. However, some cultures of *S. sonnei* fail to ferment lactose and yield negative results in the ONPG test. Among others Rubinsten and Piechaud (1952) and Szturm-Rubinsten and Piechaud (1957), and Szturm-Rubinsten (1963) mentioned two biotypes of *S. sonnei* based upon utilization of lactose and xylose (see also Tee, 1952). Szturm-Rubinsten (1963) confirmed the existence of two biotypes with the aid of the ONPG test. These biotypes are as follows:

a, ONPG +, lactose + (slowly), xylose -
d, ONPG -, lactose -,   xylose +

It is possible to delineate other biotypes of *S. sonnei* (v. inf.).

Cultures of some bioserotypes of the Alkalescens-Dispar (A-D) group (Chapter 5) yield positive results in the ONPG test (see references cited above). However, the majority of strains of A-D 01, which occurs more commonly than other members of this group, are ONPG negative and must be differentiated from shigellae by other means.

Of the shigellae described, only *S. sonnei* is able to utilize sucrose regularly. As in the case of lactose, fermentation of sucrose occurs only after continued incubation. Older classifications of *S. flexneri* were based upon sucrose fermentation by some *S. flexneri* serotypes and not by others, but it was found by Boyd (1940), by Bridges and Taylor (1946), and by Ewing (unpublished data) that sucrose is not fermented by freshly isolated cultures of *S. flexneri*. However, cultures may develop the ability to attack this substance in the laboratory.

The use of mannitol is not essential but the carbohydrate is included because of the information provided when a culture is classified as a member of the genus *Shigella*. The entire *S. dysenteriae* subgroup is mannitol negative. The only known exception to this is a mannitol-positive variant of *S. dysenteriae* 3, described by Dr. K. P. Carpenter (personal communication, 1956) and mentioned by Ewing et al. (1958). The *S. flexneri* subgroup characteristically is mannitol positive but varieties of each serotype are known which do not utilize mannitol. The subject of mannitol-negative varieties of *S. flexneri* serotypes 1, 2, 3, 4, and 6 is reviewed by Ewing (1954) and Seeliger (1955) described a culture of *S. flexneri* 5 which failed to ferment mannitol. Mannitol-negative varieties of *S. flexneri* 4a and 6 appear to be the most common of the mannitol-negative varieties of *S. flexneri* but apparently they do not occur as frequently as their mannitol-positive counterparts. The microorganisms variously known as *Shigella rio* (Weil et al., 1948), *Shigella saigonesis* (Pacheco and Rodriques, 1930), and *Bacillus rabaulensis* (Mumford and Mohr, 1944) are considered to be cultures of *S. flexneri* 4a. The author (Ewing 1954) has examined many mannitol-negative cultures of *S. flexneri* 4a from various parts of the world and they have been studied by many other investigators as well, e.g., MacLennan (1945), Nelson (1947), Seeliger (1950). The various biotypes of *S. flexneri* 6, including those that do not ferment mannitol, studied by the writer are given in Tables 57 and 59. The different biotypes of *S. flexneri* 6 also were reported upon by Ewing and Taylor, 1945. Of the *S. boydii* serotypes, certain strains that belong to serotypes 3, 4, 6, 9, 10, 14, and 15 are known to be mannitol negative or to utilize that substrate slowly (MacLennan, 1945; Vandepitte et al., 1956; and unpublished data). *S. sonnei* cultures utilize mannitol within 24 hours' incubation with rare exceptions. Three strains of *S. sonnei* that failed to ferment mannitol were identified among cultures isolated in Italy by the author during 1943-1945 and since 1953 a number of such cultures have been recognized among strains isolated in the United States. The first of these was received in 1953 from Dr. G. F. Forster of the Illinois State Health Department. In 1956, a culture of *S. sonnei* that failed to utilize mannitol was received from Dr. Omori of Osaka, Japan, and Graham (1958) reported than an outbreak of dysentery in England was caused by such a strain. Also Rose (personal communication, 1959) reported the isolation of a strain of this biotype in New Zealand. It should be noted that in comparison to the incidence of its mannitol positive counterpart, the mannitol negative biotype of *S. sonnei* is very rare. However, it is well to remember that an epidemic strain of this sort might become prevalent in a given locality or region. Also, in Castellani's descriptions of *B. ceylanensis* A (1907, 1912) the microorganism is described as mannitol negative. In subsequent publications up to 1937, *B. ceylonensis* A is listed as mannitol positive or negative, while in 1937 and subsequent papers only mannitol positive cultures of this serotype are mentioned by Castellani. Cerruti (1930) demonstrated that *B. ceylonensis* A cultures were agglutinated by antiserum prepared with the bacillus of Sonne (1915) and in 1934 Cerruti concluded that the types were identical. Similar conclusions were reached by Welch and Mickle (1932) and by Soule and Heyman (1933).

The addition of xylose and dulcitol broths often aids in the differentiation of shigellae; results obtained with them should not be judged individually, however, but rather should be assessed along with those obtained with other carbohydrate media. Many cultures that belong to the *S. boydii* subgroup or species utilize xylose or dulcitol (Table 57), whereas their utilization by members of subgroups A, B, and D is rare. Of the subgroup B members only cultures of the mannitol negative variety of *S. flexneri* 4 commonly utilize xylose and only certain biotypes of *S. flexneri* 6 are known to produce acid from xylose or dulcitol. Cultures of *S. sonnei* usually do not ferment xylose (v. sup.) and only rarely do they produce acid from dulcitol. Xylose fermentation by certain strains of *S. sonnei* was reported by Roelke (1943) and by Rubinsten and Piechaud (1952), and Tee (1952) reported that of 812 cultures of *S. sonnei* examined about 5 percent utilized this substance. An occasional culture of *S. sonnei* that utilized xylose was recognized among strains isolated by the author in Italy during 1943-1945 and a few strains were identified among cultures isolated in the United States since 1948, but the percentage of occurrence of such cultures was very small when compared to *S. sonnei* strains that failed to utilize xylose. Fermentation of raffinose by *S. sonnei* cultures has been reported by several investigators (e.g., Nelson, 1930; Welch and Mickle, 1932; Sears and Schoolnik, 1936). Reavis and Ewing (1958) reported that *S. sonnei* cultures usually utilized raffinose rapidly and that this characteristic was an aid to their differentiation from members of subgroups A and C, as well

as in the differentiation of members of subgroups B and C (Table 57).

The indol reactions recorded in Tables 56 and 57 are based upon tests made with Kovacs' reagent on peptone water cultures incubated at 37 C for 2 days. Members of the species *S. dysenteriae*, *S. boydii* and *S. sonnei* give consistent results when tested for indol production after 24 hours' incubation. Cultures of *S. flexneri* serotypes 1 to 4 vary considerably and may give no evidence of indol production if tested at 24 hours. However, many such strains form detectable amounts of indol upon continued incubation. Of numerous cultures of *S. flexneri* 5 examined by the author, only two failed to produce indol and all cultures of *S. flexneri* 6 failed to do so.

The author has found that Christensen's citrate (Christensen, 1949), sodium acetate, and sodium mucate media are of considerable value in the differentiation of members of the genus *Shigella* from *Escherichia* particularly from anaerogenic nonmotile biotypes of *E. coli* (Chapter 5), and from members of certain other genera as well. These observations have been confirmed by other investigators (e.g., Costin, 1965). The reactions of representative cultures of *Shigella* and *E. coli* in these media are recorded in Tables 60a, and 60b.

Salicin, adonitol, and inositol are not utilized by members of the genus *Shigella* as defined above and any organism that utilizes these substances may be excluded from the genus. As regards salicin there may be some exceptions, since Sereny (1959b) has reported a strain of *S. sonnei* that produced acid from that substrate after 9 days' incubation. (See episomal transmission of aberrant characters, Chapter 4.)

Members of the genus *Shigella* are nonmotile. If a semisolid motility medium is employed, many organisms which do not belong to the genus can be eliminated. Bacteria that might otherwise be confused with shigellae often can be eliminated by this test alone. Microscopic methods should not be relied upon to determine lack of motility.

Two methods of approach for identifying unknown cultures are open to the laboratory worker. The choice of which to follow is a problem which must be solved in each laboratory. In the first, outlined in Figure 1 (Chapter 2), urease-negative cultures are tested in antisera for *Shigella* by the slide agglutination technique. A heavy suspension of the microorganisms may be made in physiological saline or mercuric iodide solution from the growth on the TSI agar slants. Or, an infusion agar slant may be inoculated and incubated for about 6 hr and a suspension made from this. Cultures are tested first in polyvalent and then in

TABLE 60a

The Reactions of Shigellae and *E. coli* in Acetate,
Christensen's Citrate, and Mucate Media

Genera and species	Sodium acetate				Christensen's citrate				Sodium mucate			
	No. tested	Sign	%+	(%+)*	No. tested	Sign	%+	(%+)*	No. tested	Sign	%+	(%+)*
dysenteriae	50	–	0	0	294	–	0	0	63	–	0	0
flexneri	100[1]	–	0	0	1375	–	0	0	423	–	0	0
boydii	50	–	0	0	442	–	0	0	123	–	0	0
sonnei	100	–	0	0	209	–	0	0	165	d	6.4[W]	(30.3[W])
coli	186	+ or (+)	83.8	(9.7)	423	d	15.8	(18.4)	344	+	91.6	(1.4)
"Alkalescens-Dispar" biotypes	238	+ or (+)	89.6	(4.7)	200	d	75	(12.5)	61	d	29.5	(27.9)

Adapted from Trabulsi and Ewing (1962), and unpublished data.

Includes all *S. flexneri* serotypes (see also table 4b).

Figures in parentheses indicate percentages of delayed reactions (3 to 7 days).

Weakly positive reaction.

TABLE 60b

Reactions of Cultures of *S. flexneri* Serotype 4 in Sodium Acetate Medium

Subserotype or biotype of *S. flexneri* 4	No. tested	%+	(%+)*
4a	52	0	(8[W])
4a (mannitol negative)	50	0	(43[W])
4b	52	0	0

Adapted from Ewing (1966).

* Positive in 2 to 7 days.

[W] Weakly positive reaction.

monospecific antisera. A positive reaction in one of the polyvalent antisera is *presumptive* evidence that the culture is a *Shigella*. Similarly, agglutination in one of the monovalent antisera is a *presumptive* indication of the serotype to which the culture belongs.

After the slide agglutination tests are completed, cultures must be confirmed as members of the genus *Shigella* by a study of their biochemical and physiological reactions, since it has been found that bacteria of other genera are related to certain *Shigella*. Furthermore, any culture that gives a typical reaction on TSI medium, is negative on urea medium, and fails to agglutinate in any of the antisera for *Shigella* must be studied by means of biochemical tests. Such a culture may be one of the less common shigellae, a new *Shigella* serotype, or a member of another genus.

Cultures that appear to be shigellae, but which agglutinate poorly or not at all, should be tested for the presence of heat labile inhibitory substances. This may be done by heating a suspension in a water bath at 100 C for 15 to 60 min. After such treatment, the suspension is cooled and retested on a slide for agglutination. Substances that inhibit O agglutination are present commonly in cultures of *E. coli* but also occur in serotypes of *Shigella*.

The second avenue of approach open to the investigator is to test urease negative cultures for their biochemical reactions before doing serologic tests. The list of test substances given in Tables 3, 5, and 10 is recommended for this purpose. This method effects a considerable saving of antiserum.

SEROLOGIC IDENTIFICATION OF CULTURES OF *SHIGELLA*

Production of Antisera. The methods used for the production of antisera for *Shigella* are similar to those outlined for the production of O antisera for *E. coli* or *Salmonella*. Cultures to be used for immunization are selected for specificity, smoothness, and agglutinability. Some cultures of certain *S. flexneri* react much more specifically than others and show less cross agglutination. Selected cultures are plated on infusion agar and smooth colonies transferred to infusion agar slants. After incubation for 16 to 18 hr, the growth may be tested in 1:10 or 1:20 dilution of homologous antiserum for agglutinability. The importance of the use of smooth cultures cannot be overemphasized because the occurrence of rough agglutinins in antisera leads to confusing cross reactions. The selected culture is used to inoculate a flask of meat infusion broth (pH 7.4). The flasks are incubated for 6 to 8 hr and then heated at 100 C for 2 hr and centrifuged. After centrifugation the supernatant fluid is discarded and the bacteria are resuspended in physiological saline solution containing 0.5% formalin. The density of the suspension should approximate that of the 6- to 8-hr broth cultures. If so desired, the broth cultures may be formalinized and used as vaccines in the unheated state. However, in recent years the author has used only heated antigens to prepare typing antisera for shigellae. Fewer animals are lost during the course of immunization when heated antigens are used and the resulting antisera are free of K agglutinins (v. inf.). The antigens are given by intravenous inoculation and

injections of 0.5, 1.0, 2.0, 4.0, and 4.0 ml amounts are administered at 4 to 5 day intervals. The animals may be test bled 5 to 7 days after the last injection. If titers are satisfactory the animals are exsanguinated. Such antisera are preserved by the addition of an equal volume of glycerine. Each newly prepared antiserum should be tested with a living suspension of a culture known to contain the alpha antigen of Stamp and Stone (1944). If alpha agglutinin is present in an antiserum, it should be absorbed with a suspension of the alpha culture. Antisera for shigellae may also be produced with antigens prepared according to the technique of Roschka, 1950 (see Chapter 5 or Chapter 8).

Polyvalent grouping antisera may be produced for each of the species or subgroups listed in Figure 3. The composition of the grouping sera for subgroups A, B, and C employed by Ewing (1950) is advised for general laboratory work. Subgroup A polyvalent antiserum is produced by the injection into two or more rabbits of a mixed vaccine containing equal amounts of antigen for each of the first seven members of the species *S. dysenteriae*. Similarly, polyvalent antisera for subgroups B and C are produced by injection of mixed vaccines composed of the six members of subgroup B and of the first seven members of suborup C, respectively (Figure 3). In addition a mixed antiserum for *S. sonnei* is recommended. This is produced by the injection of a vaccine made by pooling broth cultures of *S. sonnei,* form I(S) and form II(R). A polyvalent O antiserum for those anaerogenic, nonmotile biotypes of *E. coli* referred to as the Alkalescens-Dispar (A-D) group (see Chapter 5) may be used in conjunction with antisera for the subgroups of *Shigella*. Satisfactory polyvalent antisera for *Shigella* also may be produced according to the Roschka (1950) method or by use of chrome alum treated vaccines as directed by Ando et al. (1952). Grouping antisera should be tested for the presence of alpha agglutinins as advised in the preceding paragraph.

After production, polyvalent antisera should be tested with suspensions of each of the serotypes that were used in the vaccine. Slide tests may be used for this purpose and each suspension should be tested in antiserum dilutions of 1:5, 1:10, and 1:20. For use, a dilution should be selected which causes rapid and complete agglutination of each component serotype. Such grouping antisera may be used in the unabsorbed state providing certain cross agglutination reactions are borne in mind. Results that may be expected are listed in Table 61. If necessary, the grouping antisera may be absorbed in such a manner as to free them of cross reactions. The manner in which these absorptions may be made in is given in Table 62 and results obtained

TABLE 61

Reactions Obtained with Polyvalent
Shigella Antisera (Slide Tests)

Living Cultures		Unabsorbed			Absorbed		
		A	B	C	A	B	C
S. dysenteriae	1	4	1s	—	4	—	—
	2	4	—	—	4	—	—
	3	4	d	—	3	—	—
	4	4s	d	—	4	—	—
	5	4	—	2	3	—	—
	6	4	—	—	4	—	—
	7	4	d	—	4	—	—
S. flexneri	1a	1s	4	—	—	4	—
	1b	1s	4	1vs	—	4	—
	2a	—	4	1vs	—	4	—
	2b	—	4	—	—	4	—
	3	—	4	—	—	4	—
	4a	—	4	2vs	—	4	—
	4b	ds	4	1vs	—	4	—
	4	—	4	—	—	4	—
	5	—	4	1vs	—	4	—
	6	—	4	—	—	4	—
	var. "X"	—	4	—	—	4	—
	var. "Y"	3s	4	1s	1s	4	1s
S. boydii	1	—	1s	4	—	—	4
	2	ds	1s	4	—	—	4
	3	—	2s	4	—	—	4
	4	d	1s	4	—	—	4
	5	—	ds	4	—	—	4
	6	—	—	4	—	—	4
	7	—	—	4	—	—	4
S. sonnei	I	d	1s	—	—	—	—
	II	d	1s	1	—	—	—
A-D*	01 L–	3s	4s	4s	—	—	—
	L+	d	d	d	—	—	—
	02	—	—	—	—	—	—
	03	d	3s	4s	—	—	—
	04	d	4s	4s	—	—	—

1, 2, 3, and 4 indicate degrees of agglutination.

d, doubtful. s, slow. vs, very slow.

TABLE 62

Absorption of *Shigella* Polyvalent Antisera

Antisera	Absorbing Cultures
Polyvalent A (*S. dysenteriae* 1-7)	S. *dysenteriae* 10* + S. *boydii* 1* + S. *boydii* 15 + A-D 01
Polyvalent A-1 (S. *dysenteriae* 8-10)	S. *dysenteriae* 2 + S. *boydii* 1* + S. *boydii* 15*
Polyvalent B (*S. flexneri* 1-6)	A-D 01 + A-D 03 + A-D 04
Polyvalent C (*S. boydii* 1-7)	A-D 01 + A-D 03 + A-D 04 + S. *sonnei*, R (II)
Polyvalent C-1 (*S. boydii* 8-11)	S. *boydii* 1 + S. *boydii* 4 + A-D 01 + A-D 02
Polyvalent C-2 (*S. boydii* 12-15)	S. *dysenteriae* 2 + S. *dysenteriae* 8* + A-D 01 + A-D 02 + A-D 07

*Absorption with cultures of serotypes marked with an asterisk may not be required. Antisera should be tested and absorbed as required.

with some such absorbed antisera are listed in Table 61. Techniques for agglutinin absorption tests are discussed below.

In addition to the polyvalent grouping antisera mentioned above, some workers may wish to produce similar antisera for the serotypes of *Shigella* added to the schema since 1950. The writer produces a polyvalent antiserum A-1 by injection of a mixed vaccine composed of heated (100 C. 2 hrs.) suspensions of S. dysenteriae 8, 9, 10, and provisional serotype 3873-50. A polyvalent C-1 grouping antiserum may be produced by injection of a mixed vaccine composed of heated cultures of S. *boydii* 8, 9, 10, and 11 and polyvalent C-2 by use of a mixed vaccine containing heated cultures of S. *boydii* 12, 13, 14, and 15. While the above-mentioned grouping antisera often may be used in the unabsorbed state, it is advisable to absorb them in the manner given in Table 62.

Technique of Agglutinin Absorption. The techniques used for absorption of agglutinins from antisera vary considerably from one laboratory to another (see Boyd, 1936, 1938; Wheeler, 1944b; Weil et al., 1944). The methods outlined here are satisfactory for the preparation of small quantities of typing fluids. (See also Chapter 5.)

Smooth cultures of the microorganisms to be used for absorption are tested for agglutinability and then inoculated into tubes of infusion broth. After incubation for 6 to 8 hr at 37 C, the broth cultures are used to seed infusion agar plates. Standard 90 mm petri dishes containing about 40 ml of infusion agar are employed. Each plate is seeded with 0.3 to 0.4 ml of broth culture and the inoculum is spread over the entire surface of the agar. Such plates are incubated in an upright position for 18 to 20 hr at 37 C. If the cultures to be used for absorption are O-agglutinable, i.e., react well in antiserum in the unheated state, the growth from the plates may be harvested in 0.5 formalinized physiological saline solution. However, if the cultures contain antigens that inhibit O agglutination, the growth from the plates should be harvested in 0.5% sodium chloride solution and the tubes containing the harvested growth heated at 100 C for 1 hr. The heat treatment inactivates labile (K) antigens and the procedure usually results in a more satisfactory absorption of O agglutinin. The microorganisms then are sedimented by centrifugation, the supernatant fluid is discarded, and diluted antiserum is added to the packed cells. The mixture is shaken to resuspend the sedimented bacteria. Antisera to be absorbed are diluted 1:5 to 1:10 in 0.5% phenolized saline solution and added to an absorptive dose calculated to be in excess of that required. Better results are obtained if the cells to be used for absorption are divided into two portions. Diluted antiserum is added to one portion which is incubated at 48 to 50 C for 4 hr, and then centrifuged. The antiserum is added to the second portion of packed bacterial cells. The mixtures again are incubated at 48-50 C for an hour or two and then placed in a refrigerator (about 4 C) overnight. Then the mixtures are centrifuged and the absorbed antiserum tested.

In most cases the growth from one 90 mm plate of each bacterial type effectively removes the heterologous agglutinins from 0.1 ml. There are exceptions, of course, in which antisera must be reabsorbed by additional bacteria to remove all heterologous agglutinins.

Absorbed antisera should be tested in serial dilution (1:40 to 1:10240) with dense suspensions prepared from several strains of the homologous and the absorbing bacteria. Absorbed antisera should be free of agglutinins for the absorbing culture or cultures at a dilution of 1:40 or less. Also, absorbed antisera should be tested on slides in dilutions of 1:5 or 1:10, 1:15, and 1:20. A dilution of antiserum that causes rapid and complete agglutination of the homologous serotype should be selected for use.

Use of Antisera. The methods used for serotyping cultures of *Shigella* are based upon the fact that each serologic type contains a specific or major somatic antigen which is characteristic of the serotype. Certain shigellae also contain group or minor antigens, which in many cases are shared by other serotypes. For example, serotypes of *S. flexneri* contain common group antigens and, consequently, each type of this species reacts to some extent in antisera prepared against other serotypes of the species. Such relationships make the use of absorbed antisera a necessity when dealing with *S. flexneri* as well as with certain other serotypes. The cross agglutination reactions of shigellae are listed in Tables 63, 66, and 70. From these tables it may be deduced which antisera must be absorbed before they can be used for identification.

Slide tests may be used for most serotyping of cultures of *Shigella*. For these tests glass slides 2 X 3 inches in size are convenient. Such slides may be divided into a number of sections with a wax pencil. Or, if desired, a larger glass plate may be employed. Thick suspensions of the bacteria to be tested are prepared by removing the growth from a 16- to 20-hr agar slant culture and emulsifying it in about 0.5 ml of physiological saline solution. A droplet of the suspension (3 mm loop) is placed on the slide and then a similar droplet of antiserum is placed below the antigen suspension. The droplets of antiserum and antigen are mixed with the loop in such a way as to make a narrow track of mixture (about 5 mm wide and 15 mm long). Mixing may then be continued by tilting the slide back and forth.

Subgroup A, *S. dysenteriae*. With certain exceptions antisera produced with serotypes of *S. dysenteriae* usually may be used in the unabsorbed state if the cross agglutination reactions shown in Tables 61 and 63 are kept in mind. Minor relationships, indicated by titers of 1:160 or less, usually do not interfere with slide agglutination when antisera are diluted 1:10 or more. However, it is advisable to prepare absorbed antisera for the subgroup A serotypes listed in Table 64 and it always is advisable to test newly prepared antisera for all known relationships, minor or not, and to absorb them as required. The dose of heterologous bacteria required to rid a given antiserum of agglutinins for it may be expected to vary somewhat with each lot of antiserum.

TABLE 63

Agglutination Reactions of *S. dysenteriae* Cultures*

| Antigen Suspensions (100 C, 1 hr.) | Unabsorbed Antisera | | | | | | | | | | | | | Reactions in Antisera for other shigellae and for the A-D group |
| | *S. dysenteriae* | | | | | | | | | | Provisional Serotypes | | | |
	1	2	3	4	5	6	7	8	9	10	3873-50	2000-53	3341-55	
S. dysenteriae 1	10240	0	0	0	0	0	0	0	0	0	0	0	0	A-D 01, 320
2	0	10240	0	0	0	0	0	40	0	640	0	0	0	*S. boydii* 1,640; *S. boydii* 15,640
3	0	0	20480	0	0	0	0	0	0	0	0	0	0	Serotype 3615-53, 5120
4	0	0	0	10240	0	0	0	0	0	0	0	0	0	
5	0	0	0	0	5120	0	0	0	0	0	0	0	0	
6	0	0	0	0	0	10240	0	0	0	0	0	0	0	
7	0	0	0	0	0	0	5120	0	0	0	0	0	0	
8	0	0	0	0	0	0	0	10240	0	0	0	0	0	*S. boydii* 15, 320
9	0	0	0	0	0	0	0	0	5120	0	0	0	0	
10	0	640	0	0	0	0	0	0	0	2560	0	0	0	*S. boydii* 1,640; *S. boydii* 4,40
Provisional Serotypes 3873-50	0	0	0	0	0	0	0	0	0	0	5120	0	0	
2000-53	0	0	0	0	0	0	0	0	0	0	0	5120	0	
3341-55	0	0	0	0	0	0	0	0	0	0	0	0	5120	

Titers are expressed as the reciprocal of the highest dilution that gave agglutination.

O, negative at lowest dilution tested, 1:40.

*Adapted from Ewing et al. (1958) and Ewing and Johnson (1961).

TABLE 64

Absorption of Antisera for *S. dysenteriae*

Antisera	Absorbing Cultures
S. dysenteriae 1	Alkalescens-Dispar 01
S. dysenteriae 2	*S. boydii* 1 *S. boydii* 15
S. dysenteriae 8a, 8b	*S. dysenteriae* 8a, 8c
S. dysenteriae 8a, 8c	*S. dysenteriae* 8a, 8b + *S. boydii* 15
S. dysenteriae 10	*S. dysenteriae* 2 + *S. boydii* 1 + Alkalescens-Dispar 01

TABLE 65

Antigens of S. *flexneri* Comparison of the Results of Boyd and Wheeler*

Andrewes and Inman Boyd Types	Type Specific Antigen	Group Antigens	
		Boyd, 1938	Wheeler, 1944
V, VZ	I	1, 2, 3, 4, 5, 6	1, 2, 4, 5, 6, 9 . . .
W	II	1, 2, 4,	1, 3, 4, . . .
WX	II		1, 7, 8, 9 . . .
X		1, 2, 6 (?)	1, 7, 8, 9 . . .
Y			1, 3, 4 . . .
Z	III	1, 2, 6	1, 6, 7, 8, 9 . . .
103	IV	1, 2, 3	1, 6, . . .
P119	V	1, 5,	1, 5, 7, 9 . . .
88	VI	1, 2, 4,	1, 2, 4, . . .

* Adapted from Boyd (1938) and Wheeler (1944) with modifications.

Subgroup B, *S. flexneri*. The work of Boyd (1936, 1938, 1940) with the serotypes of *S. flexneri* laid the foundation for the development of serologic typing methods for exact differentiation of these micro-organisms. Previous investigators failed to recognize the fact that qualitative, as well as quantitative, differences exist between the various serotypes of *S. flexneri*. For example, Murray (1918), Gettings (1919), Andrewes and Inman (1919), Davison (1920), and others described what were called races of *S. flexneri* which were said to contain a spectrum of antigens, V to Z. Each of the races contained all of the antigens of the spectrum, and differences between the races were explained on the basis of varying quantities of the antigens of the spectrum. A type V was called a V because it contained a large quantity of that antigen and only a little of each of the other antigens. Boyd was able to prove that there is a specific or type antigen in each of the serotypes of *S. flexneri*, and this antigen is *qualitatively* different in each serotype. There also are common group factor antigens present in serotypes of *S. flexneri* which account for the complex intragroup relationships among serotypes of *S. flexneri*. A group factor antigen is so named because it occurs in more than one serotype of *S. flexneri* in contrast to a specific or type antigen which character-izes only one of the several serotypes. The group factor antigens are of considerable importance for a number of reasons. For example, consideration of group factor antibody in an antiserum and in the cultures selected to absorb such an antiserum is impor-tant in the production of absorbed specific typing antisera without which serotyping of *S. flexneri* is impossible. Also, determination of certain group factor antigens is essential to the determination of the sub-serotype of cultures of *S. flexneri* that belong to serotypes 1, 2, 3, and 4. Further, it is known that there often is quantitative variation in the amount of group antigen present in cultures of the same serotype. A thorough knowledge of the group factor antigens of serotypes of *S. flexneri* and the variational phenomena that affect them is very helpful in the selection of cultures for antiserum production and is essential to any critical study of *S. flexneri*.

In addition to the group factors listed in Table 65, cultures of *S. flexneri* 1b contain a minor antigenic factor that is not present in other known serotypes of *S. flexneri*. Repeated absorption of antiserum for *S. flexneri* 1b by cultures of all of the other serotypes including *S. flexneri* 1a does not remove the agglutinins for this minor factor. Cultures of *S. flexneri* 1b react in a dilution of 1:80 in such an absorbed antiserum (author's unpublished data). This factor was studied by other workers, e.g., Madsen (1949), Seeliger (1953), and by Formal and Baker (1953).

Other workers confirmed Boyd's findings regarding serotypes of *S. flexneri* and Wheeler (1944a) and the author studied the group or minor antigens in greater detail. The results of Boyd and of Wheeler are compared in Table 65. Kauffmann (1942, 1948) studied the antigens of *S. flexneri* 1, 2, and 3 and was able to derive formulas based upon antigens of diagnos-tic importance.

The interrelationships of the various serotypes of *S. flexneri* are shown in Table 66. A study of the data contained in the table reveals why it is necessary to prepare absorbed antisera in order to identify cultures of *S. flexneri* by serologic methods. The data given in Table 66 were obtained with lots of antisera produced with heated antigens and differ somewhat from results obtained with antisera prepared with ordinary unheated formalinized antigens (e.g., compare Table 20, Edwards and Ewing, 1951).

Methods for preparing absorbed antisera are given by Wheeler (1944b), Weil et al. (1944), Ewing (1944, 1950), Seeliger (1953), and others. The procedures outlined here are based upon the investigations of Boyd (1938, 1940), Wheeler (1944a, b), and Ewing. Results obtained are comparable to those outlined by Wheeler (1944b).

The antisera and cultures employed in preparing type specific antisera are listed in Table 67. The number of plates of absorbing bacteria required varies with the quantity of heterologous agglutinins in the antiserum. Therefore, it may be necessary to reabsorb some antisera with the growth from additional plates of each culture.

Newly prepared absorbed typing fluids must be tested with known cultures of the *S. flexneri* subgroup. Properly prepared fluids react specifically and do not agglutinate serotypes of *S. flexneri* other than the homologous ones. Such type specific antisera may be diluted considerably for diagnostic use. The highest dilution that gives rapid, specific agglutination on a slide is the dilution of choice. This dilution generally is between 1:10 and 1:20, but may be higher.

Similar methods for the preparation of antisera for identification of some of the group antigenic factors of cultures of *S. flexneri* are given in Table 68. Absorbed antisera for identification of group factors 3,4;6, and 7,8 are necessary to separate the subserotypes of *S. flexneri* 1,2,3, and 4. Aside from ordinary interest in the bacteria, determination of the subserotype is of value in epidemiology. Newly absorbed factor antisera are tested in the same way as type specific antisera and diagnostic dilutions are determined in the same way.

The results to be expected when testing suspensions of cultures of *S. flexneri* in properly absorbed type specific and group factor antisera are given in Table 69. (See also Table 55.)

TABLE 66

Agglutination Reactions of Cultures of *S. flexneri*

Antigen Suspensions (100 C, 1 hr.)	Unabsorbed Antisera													Reactions of antisera for other shigellae and for the A-D group
	1a	1b	2a	2b	3 (III:6:7,8)	3 (III:3,4:6)	3 (III:6)	4a	4b	5	6	X	Y	
S. flexneri 1a	10240	2560	320	640	160	160	40	320	160	80	80	5120	640	*S. boydii* 2, 20: *S. boydii* 11, 80; *S. boydii* 12,160;A-D 01;640; A-D 05, 160.
1b	5120	10240	320	640	2560	1280	1280	160	80	80	160	640	160	*S. boydii* 2, 20:*S. boydii* 12, 160; A-D 01, 80; A-D 02, 40; A-D 04, 160; A-D 05, 160.
2a	160	40	10240	5120	80	80	160	1280	80	80	320	160	640	*S. boydii* 12, 80.
2b	320	0	5120	10240	5120	80	80	160	80	0	320	10240	320	*S. boydii* 11, 160;*S. boydii* 12, 80,
3 (III:6:7,8)	160	2560	80	1280	10240	5120	2560	160	160	40	40	2560	320	*S. boydii* 13, 20;*S. boydii* 12, 80.
3 (III:3,4:6)	640	5120	160	320	20480	10240	5120	320	160	40	40	1280	640	*S. boydii* 13, 20:*S. boydii* 12, 160.
3 (III:6)	640	10240	640	640	20480	20480	10240	1280	640	160	320	1280	2560	
4a	320	40	320	40	0	0	40	5120	2560	40	40	320	320	
4b	80	640	80	80	1280	1280	1280	5120	5120	80	40	320	640	*S. boydii* 5, 160.
5	160	0	40	160	160	320	40	0	160	2560	0	640	160	
6	640	320	80	80	40	320	40	160	0	0	20480	160	320	*S. boydii* 5, 160;*S. boydii* 12, 80; A-D 02, 40.
X Variant	640	0	40	1280	1280	80	80	160	640	320	40	20480	320	*S. boydii* 2, 320;*S. boydii* 5, 320; *S. boydii* 9, 640;*S. boydii* 11, 640;A-D 01, 1280;A-D 02, 1280; A-D 03, 160;A-D 04, 640;A-D 05, 320;A-D 07, 160.
Y Variant	5120	640	10240	2560	1280	5120	2560	5120	5120	1280	1280	20480	20480	

Except for the antisera for X and Y variants, these antisera were produced with heated (100 C, 2 hrs.) antigens prepared from freshly isolated cultures.

TABLE 67

Preparation of Specific Typing Fluids for *S. flexneri*

S. flexneri Antiserum	Absorbing Cultures	Specific Factor
1a	2a or Y (–:3, 4) + 5 or X (–:7, 8)+ 6	I
1b	2a or Y (–:3, 4) + 3 + 5 or X (–:7, 8) + 6	I
2a	1a or 1b + Y (–:3, 4)	II
2b	1a or 1b + X (–:7, 8)	II
3	1b + 2 a or Y (–:3, 4) + 2b or X (–:7, 8)	III
4a	1a or 1b or X (–:7, 8) + 2a or Y (–:3, 4)	IV
4b	1b + 2a or Y (–:3, 4) + 3	IV
5	1a or 1b + 2b or X + 3	V
6	1a or 1b	VI

Variation in serotypes of *S. flexneri*. Two kinds of modified form variation are known to occur in cultures of *S. flexneri* and a knowledge of them is particularly important in the selection of cultures to be used for antiserum production. The first of these results in the loss of the type specific antigen of the bacteria and gives rise to an X-like or a Y-like variant, depending on the nature of the parent strain. This loss variation long has been known and has been studied by many investi-gators among whom are Boyd (1938), Veasie, (1949), Seeliger (1950), and Ewing (1954). Andrewes and Inman (1919) also investigated X and Y variant cultures but employed different terminology in their discussions. Loss variation of this sort has been studied particularly in *S. flexneri* 2a and 4a cultures but also has been observed in *S. flexneri* 1a, 2b, 3, and 5 (Ewing, 1954 and unpublished data). In one instance, a culture of *S. flexneri* 2a lost its specific antigen and

TABLE 68

Antisera for Group Factors of *S. flexneri*

Antiserum for type 1.0 ml		Absorbed by		Factor(s) Remaining
2a	II: 1, 3, 4 . . .	2b	II: 1, 7, 8, 9 . . .	3, 4
1b	I: 1, 2, 4, 5, 6, 9	1a	I: 1, 2, 4, 5, 9 . . .	6
2b	II: 1, 7, 8, 9 . . .	1a or 1b 2a		7, 8

became a Y-variant. It was possible to prepare a specific typing antiserum by absorbing an antiserum prepared before the variation occurred with the homologous culture after it had undergone loss variation. Most cultures labeled X or Y received for examination by the writer actually were *S. flexneri* 2a or 4a but a few cultures were *S. flexneri* 1a, 2b, 3, or 5. The methods used by the writer in the examination of such cultures were mentioned by Ewing (1954). A culture is first examined as received and then plated and numerous individual colonies are examined in specific and group factor antisera. As an example, a culture was received in 1953 labeled as *S. flexneri,* X variant. When received, the parent culture did not react in any of the absorbed antisera containing specific factors I to VI, inclusive. However, when this culture was plated and a large number of individual colonies examined, colonies were found that contained the specific antigen of *S. flexneri* 5 and others were identified that lacked the specific antigen and contained only group factor 7,8 as follows:

	V	3,4	6	7,8	
Col. 1	++++	-	-	++++	*S. flexneri* 5
Col. 2	-	-	-	++++	X variant

It should be mentioned that in some cultures examined by the writer, the loss variation had progressed to a point where it was no longer possible to identify a single colony that contained one kind of specific antigen even though many individual colonies were examined. In these instances the cultures were considered to be true X or Y variants. It will be apparent from the discussion above that all X and Y variants studied by various investigators are not necessarily the same, but rather that these variants may be derived from several sources. If a culture labeled X or Y can be typed in specific typing antisera prepared from known serotypes, then the culture is not a true X

or Y variant, but instead it is a culture containing type specific antigen. If a strain that behaves like an X or Y variant cannot be serotyped, but is found to contain a specific antigen (not I to VI or R), then it is not an X or Y variant, but a new serotype of *S. flexneri.*

The second variety of modified form variation involves changes in the group factor antigens of serotypes of *S. flexneri* and may occur simultaneously with, or independently of, the loss of specific antigen. This sort of variation is known to occur particularly in cultures of *S. flexneri* 4 (Ewing, 1954) but may be observed in *S. flexneri* 1, 3, and 5 as well. In cultures of these serotypes variations may occur in the 3,4;6; or 7,8 group factor antigens. In general, it may be said that if a large amount of the 3,4 factor is present in a colony, factor 6 is absent or present in lesser amounts, as measured by the degree and rapidity of the reactions in factor antisera. In some instances both factors may be absent. One may study variation in group factor antigens by plating the culture and then examining a number of individual colonies in absorbed antisera. It often is advantageous to do so when selecting cultures to be used for antiserum production. In the following example, Colony 4 would be selected for antigen preparation if one wished to produce an antiserum for *S. flexneri* 4 because experience has shown that such an antiserum requires less absorption to render it specific.

	Specific	Absorbed Group Factor Antisera		
Colony	IV	3,4	6	7,8
1	-	++++	+++	+
2	++++	++++	++	-
3	++++	++++	-	-
4	++++	-	-	-

Thus, the quantity of agglutinin for factor 3,4 or for 6 in certain antisera largely can be controlled by

TABLE 69

S. flexneri: **Agglutination in Typing Antisera**
(Slide Tests)

Test Antigens	Absorbed Antisera								
	Type Specific						Group Factors		
	I	II	III	IV	V	VI	3, 4	6	7, 8
I: 1,2,4,5,9 …	+ + + +	–	–	–	–	–	+ + + +	–	–(++)
I: 1,2,4,5,6,9 …	+ + + +	–	–	–	–	–	++(–)	+ + + +	–
II: 1,3,4 …	–	+ + + +	–	–	–	–	+ + + +	–	–
II: 1,7,8,9 …	–	+ + + +	–	–	–	–	–	–	+ + + +
III: 1,6,7,8,9 …	–	–	+ + + +	–	–	–	–	+ + + +	+ + + +
III: 1,3,4,6,7,8,9 …	–	–	+ + + +	–	–	–	++	+ + + +	+ + + +
III: 1,3,4,6 …	–	–	+ + + +	–	–	–	++	+ + + +	–
III: 1,6 …	–	–	+ + + +	–	–	–	–	+ + + +	–
IV: 1,3,4 …	–	–	–	+ + + +	–	–	+ + + +(–)	–	–(++)
IV: 1,6 …	–	–	–	+ + + +	–	–	–(++)	+ + + +	–
V: 1,5,7,9 …	–	–	–	–	+ + + +	–	–	–	+ + + +(–)
VI: 1,2,4 …	–	–	–	–	–	+ + + +	+(–)	–	–

Degrees of agglutination: + + + +, complete reaction; + (to ++) relatively weaker reactions. Symbols in parentheses indicate occasional reactions.

selection of colonies for antigen preparation. Similarly, the nature of the cultures selected to test and absorb such antisera may be determined accurately. Colonies selected for either purpose should be smooth as judged by appearance, should not become autoagglutinable when heated, and preferably, they should be derived from recently isolated strains. Frequently cultures of *S. flexneri* 4 that possess very little group factor 3,4 antigen are encountered. Such strains may fail to react in slide tests with antiserum for the 3,4 factor, but when antisera for them are produced some agglutinin for that factor appears in the antisera. Occasionally, strains of other serotypes of *S. flexneri* that possess lesser quantities of one or another of the group factors are encountered. For example, strains of *S. flexneri* 1, 2, and 5 that appear to be devoid of group factor antigens occasionally are seen. Such cultures are useful for antiserum production since they generally require less absorption to render them type specific.

It is known that variant cultures sometimes undergo other changes in group factor antigen content. For example, cultures of *S. flexneri* 1a, 2a, and 4a may become somewhat agglutinable in 7,8 group factor antiserum. This change often takes place in cultures in which loss of the specific antigen is in progress but may occur in freshly isolated cultures that contain a normal amount of specific antigen. It seems probable that variational phenomena such as those mentioned may account for the appearance of strains which differ in group antigen content from the known and accepted subserotypes. For example, both Boyd (1938) and Wheeler (1944a) mentioned the occurrence of cultures of *S. flexneri* 3 which contained the Y (3,4) component, and the writer characterized the following varieties of *S. flexneri* 3 (abbreviated formula):

    III: 3,4 6 7,8
    III: (3,4) 6 7,8
    III: 6 7,8 (classical "Z" form)
    III: 6

Similarly, a number of varieties of *S. flexneri* 1a and 4a are known. The writer believes that while the characterization of an epidemic strain is of considerable importance to epidemiology, it is neither advantageous nor desirable to attempt to label each such strain with a different letter of the alphabet if it happens to differ slightly from the ordinary or known subserotypes. The use of an abbreviated antigenic formula, as suggested by Ewing (1949), appears to be a convenient and practical way to characterize such cultures. Parenthetically, lysogenic conversion may be involved in some of the above-mentioned variational phenomena. This subject is discussed briefly in Chapter 4.

The fimbrial antigens of *S. flexneri*. Bacterial fimbriae are filamentous appendages, which are smaller and more numeous than flagella, and which have nothing to do with motility in a culture. Such structures have been demonstrated in a variety of bacterial types such as strains of *E. coli, Salmonella, Klebsiella* and in certain cultures of *S. flexneri* (Duguid et al., 1955; Duguid and Gillies, 1957; Gillies and Duguid, 1958; and Duguid, 1959). Of the shigellae, fimbriae' have been demonstrated only in strains of *S. flexneri* 1 to 5 inclusive and have not been demonstrated in cultures of *S. flexneri* 6 or in members of subgroups A, C, and D (Duguid and Gillies, 1957; Gillies and Duguid, 1958).

Fimbriae possess adhesive qualities, cause haemagglutination in the absence of antiserum, and are antigenic. Cross agglutination and absorption studies carried out by Gillies and Duguid (1958) indicated that the fimbrial antigens of *S. flexneri* serotypes 1 to 5 were identical, regardless of serotype. These investigators reported that the antigenicity of the fimbriae was destroyed by heat at 100 C for 2½ hr, but the fimbrial agglutinin binding power of detached fimbriae was not. Antiserum for the fimbriate forms of *S. flexneri* strains may be prepared by absorption of an antiserum prepared by injection of a fimbriate strain with a culture of the homologous serotype in the nonfimbriate form (loc. cit.). The aggregates formed during agglutination of fimbriate cultures by antiserum that contains agglutinins for these structures resemble those seen in H (flagellar) agglutination. Fimbriae are discussed in some detail in Chapter 4.

These structures are mentioned here only because improper cultivation and subsequent treatment of strains to be used for antiserum production, and the consequent occurrence of antibody for fimbriae in the antisera, could lead to serious error.

Subgroup C, *S. boydii*. The species *S. boydii* (Figure 3) is composed of fifteen recognized serotypes. The first six of these originally were isolated in India and were described by Boyd (1931, 1932, 1936, 1938, 1940). They since have been recognized in various parts of the world (e.g., see Ewing and Gravatti 1947, Vandepitte, 1950, and Olarte and Varela, 1953). Most of these six serotypes have been reported from the United States (Morris et al., 1944a; Morris et al., 1944b; Ferguson and Carlson, 1945; Nelson et al., 1946; and Fulton and Curtis, 1946) and several have been identified in the USSR (Novgorodskaja 1959, 1960; Ewing et al. 1959; Ewing and Trabulsi, 1962). Apparently *S. boydii* 7 first was isolated in North Africa in 1943 by Stock (Ernst and Stock, 1943) who referred to the organism as type T. Cultures of this serotype were isolated in Algeria and in Italy by Ewing (1946) who found that Stock's type T was identical with cultures recovered in Italy and with cultures isolated in England and France (Lavington et al., 1946; Heller and Wilson, 1946; Stock et al., 1947). Cultures of *S. boydii* 7 were isolated in Finland during World War II by Gildemeister (Winkle, 1949; Gildemeister, 1956) and these cultures were designated type N in the German literature. Seeliger (1949) reported that the German type N was identical with *S. boydii* 7 and also reported the isolation of this type in Germany during the war. The writer confirmed Seeliger's observation regarding the identity of type N and *S. boydii* 7. DuBois and Ferguson (1950) reported the isolation of a culture of *S. boydii* 7 in the United States from a traveler recently returned from Spain and Portugal. Since 1948, the writer has identified several cultures of *S. boydii* 7 that were isolated from cases of dysentery in the United States. The first of these was isolated in 1951 from an American Indian in Arizona. The second and third cultures were received in 1953 from California where they were recovered from two children in the same family. Another culture originated in an outbreak of dysentery in an institution and was representative of a number of similar cultures isolated during the epidemic (Cares and Goldman, 1955).

S. boydii 8 first was described by Cox and Wallace (1948) under the designation type 112. The original cultures of this bacterial type were isolated in India and Burma. Between the years 1948 and the early part of 1950 Courtois and Vandepitte (1950) isolated and studied eleven cultures of serotype 112 in the Congo. The author compared strains of serotype 112 from Cox and from Courtois and Vandepitte and found them to be identical. The cultures are not significantly related to other shigellae. Courtois and Vandepitte (1950) proposed the designation provisional *S. boydii* 8 for serotype 112 of Cox and Wallace (1948) and it was accepted (Reports, International Enterobacteriaceae Subcommittee, 1954).

The serotype now labeled *S. boydii* 9 was isolated in Finland by Gildemeister in 1944 (see Winkle, 1949, and Gildemeister, 1956) and its isolation in the Middle East was mentioned by Boyd (1946). Gildemeister's cultures became known as type P in the older system of German taxonomy, while the cultures mentioned by Boyd (1946) were labeled 1296/7. Seeliger (1949) found that the German type P was the same as type 1296/7, an observation which was confirmed in the writer's laboratories. Francis (1946) and Madsen (1949) included cultures of type 1296/7 in their studies and Ewing et al. (1951a) proposed the name *S. boydii* 9 for type 1296/7 cultures after an investigation of their interrelationships and biochemical reactions. Courtois and Vandepitte (personal communication,

1950) isolated three cultures in the Congo during 1949 which they stated were antigenically identical with type 1296/7 *(S. boydii 9)* but which produced acid from lactose. These observations were confirmed by Ewing et al. (1951a) and Ewing et al. (1958) suggested that these cultures should be regarded as aberrant, lactose-positive strains of *S. boydii* 9, since they differed from typical shigellae only as regards their fermentation of lactose.

The names *S. boydii* 10 and *S. boydii* 11 were proposed by Ewing and Taylor (1951) for two serotypes isolated in 1943 and 1944 in Algeria and in Italy. Cultures which later were shown by Ewing and Taylor (1951) to be identical with the serotype now called *S. boydii* 10 were isolated in Madagascar and described by Szturm et al. (1950) under the designation D15. Cultures of *S. boydii* 10 and 11 differ in their biochemical reactions (Table 57) and although they are related antigenically each contains a specific factor by which it may be recognized (Ewing and Taylor, 1951). These two serotypes were accepted (Reports, International Enterobacteriaceae Subcommittee 1954).

Additional cultures of four provisional *S. boydii* serotypes were studied by Ewing et al. (1958) and it was recommended that they should be added to subgroup C as *S. boydii* 12 (serotype 123), *S. boydii* 13 (serotype 425), *S. boydii* 14 (serotype 2770-51), and *S. boydii* 15 (serotype 703). This recommendation was adopted by the International Enterobacteriaceae Subcommittee in the 1958 Report. The four serotypes were characterized by Ewing et al. (1952) and Ewing and Hucks (1952). The biochemical and serological reactions of cultures of *S. boydii* 12, 13, 14, and 15, are given in Tables 57 and 70.

Three other serotypes have been described, which may or may not be added to the schema for *Shigella* in future. One of these is serotype 3615-53, which is related (Table 70) to *S. dysenteriae* 3 (Ewing et al., 1956). However, the biochemical reactions given by strains of serotype 3615-53 (Table 58) are similar to those of members of subgroup C. Ewing et al. (1958) suggested that this serotype might be considered a member of subgroup C having intersubgroup antigenic relationships with *S. dysenteriae* 3, but recommended that its status should remain *sub judice* for the time being. Similar recommendations were made regarding the status of serotype 2710-54. The O antigens of this serotype are not significantly related to those of other shigellae (Table 70) but they are related to those of *E. coli* O group 41 cultures. The biochemical reactions given by cultures of this serotype are similar to those of members of subgroup C (Table 58). The *sub judice* serotype, 1621-54, was described by Sahab (1953) and was studied further by Ewing et al. (1956) and by

TABLE 70

Agglutination Reactions of Cultures of *S. boydii*

Antigen Suspensions (100 C, 1 hr.)	\<Unabsorbed Antisera — *S. boydii*\> 1	2	3	4	5	6	7	8	9	10	11	12	13	14	15	\<Provisional Serotypes\> 3615-53	2710-54	Reactions in Antisera for other shigellae and for the A-D group
S. boydii 1	20480	0	0	640	0	0	0	0	0	0	640	0	0	0	0	0	0	*S. dysenteriae* 2, 640; *S. dysenteriae* 10. 640; A-D 01, 1280; A-D 05, 160.
2	0	20480	0	0	0	0	0	0	0	0	0	0	0	0	0	0	0	
3	0	0	20480	0	0	0	0	0	0	0	0	0	0	0	0	0	0	
4	2560	0	0	10240	0	0	0	0	160	0	1280	0	0	0	0	0	0	A-D 01, 2560; A-D 05, 640; *S. flexneri* 5, 320.
5	0	0	0	0	10240	0	0	0	320	0	0	0	0	0	0	0	0	A-D 02, 1280; A-D 03, 320; A-D 04, 320; A-D 07, 40.
6	0	0	0	0	0	20480	0	0	0	0	0	0	0	0	0	0	0	*S. sonnei* (see text)
7	0	0	0	0	0	0	20480	0	0	0	0	0	0	0	0	0	0	
8	0	0	0	0	0	0	0	5120	0	0	0	0	0	0	0	0	0	
9	0	0	0	320	320	0	0	0	20480	0	0	0	1280	0	0	0	0	A-D 02, 1280; A-D 03, 320; A-D 04, 320; *S. flexneri* 2, 160; *S. flexneri* 4, 160; *S. flexneri* 5, 160.
10	0	0	0	0	0	0	0	0	0	20480	1280	0	0	0	0	0	0	A-D 01, 320; A-D 05, 640.
11	320	0	0	1280	0	0	0	0	0	2560	20480	0	0	0	0	0	0	
12	0	0	0	0	0	0	0	0	0	0	0	20480	0	0	0	0	0	A-D 02, 160; A-D 07, 640; *S. flexneri* 6, 1280.
13	0	0	0	320	0	0	0	0	320	0	0	0	20480	0	0	0	0	*S. dysenteriae* 2, 640;
14	0	0	0	0	0	0	0	0	0	0	0	0	0	20480	0	0	0	*S. dysenteriae* 8, 320.
15	0	0	0	0	0	0	0	0	0	0	0	0	0	0	20480	0	0	
Provisional Serotype 3615-53	0	0	0	0	0	0	0	0	0	0	0	0	0	0	0	5120	0	*S. dysenteriae* 3, 10, 240.
2710-54	0	0	0	0	0	0	0	0	0	0	0	0	0	0	0	0	10240	

Ewing et al. (1958) who reported that the O antigens of cultures of this serotype were identical with those of *E. coli* 07 strains and related to those of *S. boydii* 12. The biochemical reactions of cultures of serotype 1621-54 were somewhat similar to those given by members of subgroup C but they differed in certain important ways.

The serological interrelationships of the serotypes of *S. boydii* and of the provisional serotypes are given in Table 70. Unabsorbed antisera may be used in dilutions of 1:5 or 1:10 for the identification of cultures of *S. boydii* 2, 3, 7, 8, and 14. Antisera for the others should be absorbed in the manner shown in Table 71. The reciprocal relationships between cultures of *S. boydii* serotypes 1 and 4 and A-D 01 are particularly important to diagnostic work because these serotypes of *S. boydii* are among the more common members of the subgroup C and A-D 01 is of common occurrence. The

multiple absorption of antisera for *S. boydii* 1 and *S. boydii* 4 shown in Table 71 is necessary because the relationships are such that single absorption of antiserum *S. boydii* 1 by A-D 01 alone does not remove agglutinin for *S. boydii* 4, and conversely, absorption of antiserum for *S. boydii* 4 by A-D 01 alone does not remove agglutinin for *S. boydii* 1 from the antiserum. The other multiple absorptions listed in Table 71 are necessary because of analogous complex relationships.

It will be noted that in Table 70 *S. boydii* 6 is listed as having no important cross reactions, while in Table 71 an absorption of antiserum for *S. boydii* 6 by a *S. sonnei* II (R) strain is listed as being necessary. The reason for this apparent discrepancy lies in the cultures selected for antiserum preparation. If antiserum for *S. boydii* 6 is prepared with a culture that has been maintained in the laboratory for some time, it may be expected to agglutinate cultures of *S. sonnei* II (R) to

TABLE 71

Preparation of Absorbed Antisera for the Identification of Cultures of *S. boydii* and Provisional Serotype 3615-53

Antiserum	Absorbing Cultures
S. boydii 1	*S. dysenteriae* 2 + *S. boydii* 4 + A-D 01
S. boydii 4	*S. boydii* 1 + A-D 01
S. boydii 5	*S. boydii* 9 + A-D 03
S. boydii 6	*S. sonnei* R (II)
S. boydii 9	*S. boydii* 13 + A-D 02 + A-D 03
S. boydii 10	*S. boydii* 11
S. boydii 11	*S. boydii* 10 + A-D 01
S. boydii 12	A-D 07 +
S. boydii 13	*S. boydii* 9 + A-D 01
S. boydii 15	*S. dysenteriae* 2 + *S. dysenteriae* 8a, 8c
Provisional Serotype 3615-53	*S. dysenteriae* 3

75 percent or more of the homologous titer. This reaction is reciprocal, i.e., antiserum for *S. sonnei* II (R) reacts to about 50 percent of the titer with antigens prepared with such a culture of *S. boydii* 6. However, freshly isolated strains of *S. boydii* 6 did not react in antisera for *S. sonnei* and an antiserum prepared with a recently isolated culture of *S. boydii* 6 gave only minimal reactions at 1:40 and 1:80 dilutions with cultures of *S. sonnei* II (R). Since *S. boydii* 6 is not a common serotype, antisera generally must be produced with stock strains and absorbed in the manner given in Table 71. This absorption has little effect upon the homologous titer of antiserum for *S. boydii* 6. There is no relationship between *S. boydii* 6 (S) and *S. sonnei* I (S) and the reactions mentioned above between *S. boydii* 6 (SR or R) and *S. sonnei* II (R) are believed to be caused by common rough antigens. These reactions are depicted in Table 72 and are discussed further in the section on subgroup D. The statement has been made that it is impossible to differentiate cultures of *S. boydii* 6 and *S. sonnei* II (R) by serological methods and that antigenically *S. boydii* 6 is identical with *S. sonnei* "phase" II. This is incorrect; the bacterial forms are not antigenically identical and it is possible to differentiate them by serological methods, as shown by Wheeler and Mickle (1945) and by the author (Table 72). The R form of *S. boydii* 6 is identical with that of *S. sonnei* II (R) (v. inf. and Table 72).

Subgroup D, *S. sonnei.* The O antigens of *S. sonnei* (form I or S) are not significantly related to those of other shigellae and the relationships of form II or R cultures generally may be ascribed to roughness. The phenomenon of smooth (S) to rough (R) variation in *S. sonnei* strains has been studied extensively by many investigators (for references, see Neter, 1942; Roelke, 1943; De Blasi and Gabrelli, 1954, and Chapter 4) and various terms have been applied to the cultural forms involved. Ørskov and Larson (1925) apparently were first to demonstrate the serological differences between the various forms of *S. sonnei*. Wheeler and Mickle (1945) described two forms of *S. sonnei* which they called phases I and II. The author prefers the terms form I or S and form II or R to the term phase for use in this connection. The term phase should be reserved for the flagellar variation of Andrewes when used in connection with Enterobacteriaceae. The variation in

TABLE 72

Slide Agglutination Reactions of *S. sonnei* and *S. boydii* 6

Antisera	Absorbed by	Suspensions				
		S. sonnei			*S. boydii* 6	
		I (S)	I, II (SR)	II (R)	S	SR
S. sonnei I (S)	Unabsorbed	+ + + +	+ +	–	–	–
S. sonnei I, II* (SR)	Unabsorbed	+ +	+ +	+ + + +	–	+ + + +
	S. sonnei I (S)	–	+ +	+ + + +	–	+ + + +
	S sonnei II (R)	+ +	–	–	–	–
	S. boydii 6 (SR)	+ +	–	–	–	–
S. sonnei II (R)	Unabsorbed	–	+ + + +	+ + + +	–	+ + + +
	S. sonnei I (S)	–	+ + + +	+ + + +	–	+ + + +
	S. boydii 6 (SR)	–	–	–	–	–
S. boydii 6 (S)	Unabsorbed	–	–	–	+ + + +	+ + +
S. boydii 6 (SR)	Unabsorbed	–	+ + + +	+ + + +	+ + + +	+ + + +
	S. sonnei II (R)	–	–	–	+ + + +	+ + + +

* Produced with an unselected culture. Such an antiserum usually contains more agglutinin for *S. sonnei* II (R) than for I.

S. sonnei cultures that is frequently referred to as phase is S-R variation of Arkwright, and it probably would be better to call it that. In addition to the S and R forms there is a third form, which probably is a rho (ρ) form in the terminology of Bruce White (see Chapter 4). Transitional forms are encountered and it is not uncommon to find cultures that are mixtures of S and R forms (Table 72). However, the third form occurs very infrequently in untreated cultures. It is possible to produce a pure S or form I antiserum by proper selection of freshly isolated smooth cultures for injection. Such an antiserum does not react with the R or form II or with *S. boydii* 6 antigens. Similarly, an R or form II antiserum may be produced which will not react with the S form of *S. sonnei*. However, if antisera for the two forms are prepared without particular attention to the state of cultures used as antigens, considerable cross agglutination may be expected between the two. If desired, such antisera may be cross absorbed but this is not essential in ordinary laboratory practice.

The relationship between cultures of *S. boydii* 6 and *S. sonnei* was studied by Wheeler and Mickle (1945). These investigators analyzed the antigenic components of cultures of *S. sonnei* II (R), *S. boydii* 6, and an untyped microorganism called Harris and assigned letters to the major factors. Factor A was found in *S. sonnei* II (R) and the Harris culture and factor B was contained in cultures of *S. sonnei* II (R) and in *S. boydii* 6. A third factor, C, was common to all three types of these microorganisms. In addition to the above, the specific antigen of *S. boydii* 6 may be identified after absorption with *S. sonnei* II (R) (Table 72). Except for the special factor A which the writer did not identify in cultures of *S. sonnei* II (R), the relationships described by Wheeler and Mickle (1945) were confirmed. The Harris bacterium is anaerogenic and nonmotile but fermented inositol. Data indicate that the microorganisms contain antigens that are very closely related to the O antigens of *E. coli* O 14. Absorption experiments with cultures heated at 100 C for 1 hr revealed that *E. coli* O 14 suspensions extracted all agglutinins from antiserum prepared with the Harris culture, but in the reciprocal absorption test, a minor fraction remained in antiserum for *E. coli* O 14 following thorough absorption with the Harris culture.

The reactions of cultures of *S. sonnei* and *S. boydii* 6 in absorbed and unabsorbed antisera are given in Table 72. It should be mentioned that absorption of antiserum for *S. sonnei* II (R) by *S. boydii* 6 (SR form) results in the removal of all agglutinin (Table 72). An antiserum prepared with a suspension of the C27 type that has been heated at 100 C for 2½ hr may be employed in the examination of *S. sonnei* cultures in the S (I) form, as suggested by Bader (1954). However,

the R antigens of *S. sonnei* cultures apparently are unrelated to those of the C27 type, hence rough cultures of *S. sonnei* do not react in such an antiserum. Further, *S. boydii* 6 cultures do not react in O antiserum prepared with a C27 culture. The C27 type was described by Ferguson and Henderson (1947) and by Bader (1954) and was one of several O antigen groups within *Aeromonas shigelloides* (see Ewing and Johnson, 1960, and Ewing et al., 1961, for further references).

Variational phenomena in cultures of *S. sonnei* were reinvestigated by Slopek et al. (1960) and Rauss et al. (1961) with results not unlike those reviewed above. Sereny (1959a, 1960) studied dissociation in subgroup D strains with the aid of lighting methods similar to those reported by Cooper et al. (1957).

THE K ANTIGENS OF *SHIGELLA*

It has been known for many years that some strains of certain serotypes of *Shigella* were inagglutinable in antisera and that if the cultures were subjected to heat they became agglutinable. In 1917 Seligmann (quoted by Dudgeon, 1929) mentioned inagglutinable strains of *S. dysenteriae* 1 and in 1918 Bauch (see Kauffmann, 1954) reported that such cultures became agglutinable after boiling. Benians (1920) reported studies upon two variants of *S. dysenteriae* 1 derived from a single culture and although there is some indication that roughness may have been involved, there also is evidence that Benians' inagglutinable variant may have been what we now would call a K+ form. The work of Scheutze (1944) appears in retrospect to be an early study of the K antigens of shigellae, since his work suggested the presence of a Vi-like substance in cultures of *S. dysenteriae* 1 which inhibited O agglutination. In absorption experiments, Scheutze noted that heated suspensions of the inagglutinable strain removed all agglutinin for the inagglutinable form from the homologous antiserum. Such results would now be interpreted as evidence that the inhibitory antigen was of the B variety of K antigens. Shelubsky and Olitzki (1947) demonstrated a precipitinogen in smooth cultures of *S. dysenteriae* 1, the reactivity of which was inactivated by heat. The results of their absorption studies indicated that all the precipitin was removed from an antiserum produced with living bacteria by absorption with a heated (100 C, 2 hrs.) suspension. This precipitinogen also was compared to the factor which inhibited O agglutination. As regards other shigellae, Dudgeon (1929) referred to the work of Gehrmann (1918) who reported that it was necessary to heat cultures of *S. dysenteriae* 2 for an hour at 100 C in order to render them agglutinable and Ornstein (1921) made similar observations. Inagglutinable forms of many other shigellae since have been reported. For example, inagglutinable strains of *S. flexneri* 6 frequently are encountered and in his studies Madsen (1949) noted inhibition of O agglutination in several subgroup A and C serotypes, as well as in *S. flexneri* 6 cultures. Further, Bader and Kleinmaier (1952) reported the presence of a thermolabile antigen in strains of *S. flexneri* 6.

It is now known that the inhibition of O agglutination is caused by the presence of envelope antigens, the masking effects of which are inactivated by heat. The envelope antigens of shigellae belong to the general class of somatic antigens called K, characterized by Kauffmann (see Chapter 5). Ewing et al. (1951b) demonstrated the presence of K antigens of the B variety in cultures of *S. boydii* 1 and 2 and reported that the B antigens of *S. boydii* serotypes 1 and 2, respectively, were the same in encapsulated and nonencapsulated strains. The B antigens of *S. boydii* 1 were not related to those of *S. boydii* 2. Ewing and Tanner (1955) reported that cultures of *S. dysenteriae* 2 contained a B antigen which was closely related to the B antigen of *E. coli* 0112a,112c: B11 (Chapter 5).

The work on the K antigens of shigellae reported in the first edition of this book was continued and extended, but remained unpublished (Ewing and coworkers). K antigens of the B variety were demonstrated in all of the serotypes of *Shigella* listed in Table 73. The designations employed in this table were those proposed earlier as a practical system, which may be expanded without alteration of designations already made.

A number of strains of each serotype (Table 73) were tested in the K antisera and it was found that in each instance individual cultures of each serotype possessed the same K (B) antigen. However, additional cultures of each serotype must be tested before definite conclusions can be drawn regarding the similarity of the K antigens of members of the various individual serotypes. No relationship has as yet been detected between the K antigens of the subgroup A serotypes listed in Table 73. Further, the K antigens of provisional serotypes (e.g., 3873-50, 2000-53, and 3341-55) appeared to be unrelated to those of recognized shigellae. Nor was any relationship apparent between the K antigens of serotypes of *S. dysenteriae* and those of serotypes of *S. boydii*. With few exceptions, the K antigens of individual subgroup C serotypes (Table 73) were unrelated. Reciprocal relationships were noted between the K antigens of *S. boydii* 4 and 11 and between *S. boydii* 10 and 11. While the former relationship appeared to be a minor one, the latter was fairly extensive but in each instance reciprocal absorption tests demonstrated the individuality of the K antigens of these serotypes. In those instances in which the O antigens of serotypes of

Shigella are identical with those of an *E. coli* O antigen group (v. inf.), the K antigens also are identical or are closely related. This was demonstrated by the fact that absorption of K antiserum produced with *Shigella* with heated suspensions of the appropriate culture of *E. coli*

TABLE 73

Designations for the K Antigens of Shigellae

Serotype	K Antigen Designation
A1*	KA1
A2	KA2
A3	KA3
A4	KA4
A5	KA5
A6	KA6
A7	KA7
A8	KA8
A9	KA9
A10	KA10
B2a	KB2a
B6	KB6
C1	KC1
C2	KC2
C3	KC3
C4	KC4
C5	KC5
C6	KC6
C7	KC7
C8	KC8
C9	KC9
C10	KC10
C11	KC11
C12	KC12
C13	KC13
C14	KC14
C15	KC15
D	KD

*These designations are merely a convenient way of labeling cultures in the laboratory. The letter A refers to subgroup A and the figure indicates the serotype. Thus, A1 is *S. dysenteriae* 1, B2a is *S. flexneri* 2a, C1 is *S. boydii* 1, etc.

TABLE 74

Relationship of the Heat Stable Antigens of Serotypes of *Shigella* to Those of Coliform Cultures. (from Ewing, 1953, and Ewing et al., 1956, modified)

Shigella		Related to *Escherichia coli* O group (s) or to coliform or other cultures
Subgroup A		
Shigella dysenteriae	1	1, reciprocal, a,b-a,c*; 120, unilateral
	2	112a,112c, identical
	3	124, identical
	4	88, unilateral; coliform 3588-51, identical
	5	58, identical
	6	130, reciprocal, a,b-a,c
	7	121, reciprocal, a-a,b
	8	38, reciprocal, a,b,c-a,b; 23,a,d
	9	No relationship, 1-148
	10	144, reciprocal, a,b-a,c
Subgroup B		
Shigella flexneri	1a	1,19a,62,69,73
	1b	1,16,19a,62,69,73
	2a	13, reciprocal, a,b-a,c
	2b	13, reciprocal, a,b-a,c; 73; 147a,147b, identical
	3 (III: 6,7,8)	13,16 (identical culture 5444-80)[1]
	3 (III: 3,4,6,7,8)	13,16
	4a	1,13
	4b	135, identical[2], 13
	4	1,17,19a,73
	5	129, identical[3]
	6	19a; coliform 3438-51 reciprocal, a,b-a,c
	X variant	1,2,13,19
	Y variant	1,2,13,19
Subgroup C		
Shigella boydii	1	2 (50), reciprocal, a,b-a,c
	2	87a,87b, identical; 87a,87b,reciprocal, a,b-a,c; 96, receiprocal, a,b-a,c
	3	85, reciprocal, a,b-a,c
	4	53, identical
	5	79, identical
	6	76, reciprocal, a,b-a,c
	7	4838-69, identical
	8	143, identical 114 minor
	9	102, reciprocal, a,b-a,c
	10	105a,105c, reciprocal, a,b-a,c
	11	105a,105b, identical
	12 (and M)	7, reciprocal, a,b,c-a,d (M = a,b)
	13	28a,28c,reciprocal, a,b-a,c; 98 reciprocal a,b-a,c
	14	32, identical; 83, reciprocal, a,b-a,c
	15	112a,112b, identical
Subgroup D		No relationship 1-148
Shigella sonnei		Identical, type C27[4] *(Aeromonas shigelloides)*

*These arbitrary formulas are used merely as an aid to make the relationships clearer and they should not be interpreted as permanent designations for the antigens involved.

[1] Slopec and Dabrowshi (1957a). [2] Rauss and Vertenyi (1956).

[3] Seeliger (1955); Slopec and Dabrowski (1957b).

[4] Ferguson and Henderson (1947); Bader (1954); Ewing et al. (1961).

usually resulted in removal of all agglutinin. Also, absorption of antiserum for *S. sonnei* with a heated suspension of a C27 *(Aeromonas shigelloides)* culture resulted in the removal of both O and B agglutinins.

Antisera for the B antigens discussed in the foregoing paragraphs were produced in the same manner as antisera for the B antigens of *E. coli* (Chapter 5). The methods used to study these antigens in shigellae and to determine the variety of K to which they belonged also were the same as those used in work with *E. coli.*

EXTRAGENERIC RELATIONSHIPS OF SHIGELLAE

Although it is known that certain serotypes of *Shigella* are related to members of other genera within the family Enterobacteriaceae, only the relationships known to exist between shigellae and *Escherichia* will be mentioned here. This subject was reported upon by Ewing (1953) and by Ewing et al. (1956) and a table based primarily upon those reports is presented (Table 74). Also, the relationships of the O antigens of certain *sub judice* serotypes to those of *E. coli* are given in Table 75.

The O antigens of most serotypes of *Shigella* are either identical with or related to those of *E. coli* O antigen groups or those of intermediate bacteria. Only two serotypes are unrelated to the O antigens of *E. coli* O groups 1 to 148. For details on the nature of the relationships involved here, the reader is referred to Ewing (1953) and Ewing et al. (1956) and for other data on extrageneric relationships, to Kampelmacher (1959) and Kauffmann (1966).

It should be stressed that knowledge of relationships such as those mentioned above is of importance from the practical point of view as well as of academic interest. In the indentification of an unknown enteric bacterium one cannot rely entirely on serology, nor can complete characterization be made on the basis of biochemical reactions alone. Both methods must be employed for accurate identification. Biochemical reactions are used to determine the genus and species to which a culture belongs, and serology is employed to determine the serotype. Quite often anaerogenic and microaerogenic cultures of *E. coli* are isolated which do not utilize lactose or sucrose promptly and hence give the appearance of shigellae on triple sugar iron agar and similar media. If such cultures are examined only by serological methods, the chance of error is great.

TABLE 75

The O Antigenic Relationships of Certain *Sub Judice* Serotypes to *E. coli* O Antigen Groups

Serotype	*E. coli* O Antigen Group and Variety of Relationship
3873-50	29, reciprocal, a-a,b
2000-53	6, identical
3615-53	124, reciprocal, a,b-a,c
1621-54	7, identical
2710-54	41, reciprocal, a,b-a,c
3341-55	3, reciprocal, a,b-a,c

REFERENCES

Ando, K., H. Shimojo, and I. Tadokoro. 1952. Jap. J. Exper. Med., **22**, 485.

Andrewes, F. W. 1918. Lancet, i. 560 (April 20).

Andrewes, F. W., and A. C. Inman. 1919. Med. Res. Council Gt. Brit., Spec. Rep. Ser. No. 42.

Bader, R. E. 1954. Z. f. Hyg., **140**, 450.

Bader, R. E., and H. Kleinmaier. 1952. Ibid., **135**, 27: 82.

Benians, T. H. C. 1920. J. Pathol. Bacteriol., **23**, 171.

Boyd, J. S. K. 1931. J. Roy. Army Med. Corps, **57**, 161.

————. 1932. Ibid, **59**, 241.

————. 1936. Ibid. **66**, 1.

————. 1938. J. Hyg., **38**, 477.

————. 1940. Tr. Roy. Sec. Trop. Med. Hyg., **33**, 553.

————. 1946. J. Pathol. Bacteriol., **58**, 237.

Bridges, R. F., and J. Taylor. 1946. J. Hyg., **44**, 346.

Buelow, P. 1964. Acta Pathol. Microbiol. Scand., **60**, 376.

Cares, R., and J. Goldman. 1955. Amer. J. Clin. Pathol., **25**, 629.

Castellani, A. 1907. J. Hyg., **7**, 1.

————. 1912. Zentralbl. f. Bakt., **65**, 262.

————. 1927. Am. J. Trop. Med., **7**, 199.

————. 1929. Lancet i, 370 (August 24).

————. 1930. Ann. Med. navale e coloniale, **1**, 1.

————. 1932. Zentralbl. f. Bakt., **125**, 37.

————. 1937. J. Trop. Med. Hyg., **40**, 197.

Castellani, A., and A. J. Chalmers. 1919. Manual Trop. Med. London: Bailliere, Tyndall, and Cox.

Cefalu, M., and A. Gulotti. 1953. Nuovi Ann. d'Ig. et Microbiol., **4**, 237.

Cerutti, C. F. 1930. Quoted by Soule and Heyman 1933.

————. 1934. Quoted by Kauffmann 1954.

Chantemesse, A., and F. Widal. 1888. Gazette Medicale de Paris **19**, 522.

Christensen, W. B. 1949. Res. Bull. Weld. Co. Colo. Health Dept., **1**, 3.

Christensen, W. B., and G. H. Gowan. 1944. J. Bacteriol., **47**, 171.

Clayton, F. H. A., and S. H. Warren. 1929. J. Hyg., **28**, 355: **29**, 191.

Cooper, M. L., H. M. Keller, and E. W. Walters. 1957. J. Immunol., **78**, 160.

Costin, I. D. 1965. J. Gen. Microbiol., **41**, 23. .

————. 1966. Zentralbl. f. Bakt. I. Orig., **200**, 49.

Courtois, G., and J. Vandepitte. 1950. Ann. Soc. Belge Med. Trop., **30**, 149.

Cox, C. D., and G. L. Wallace. 1948. J. Immunol., **60**, 465.

Davison, W. C. 1920. Bull. Johns Hopkins Hosp., **31**, 225.

DeBlasi, R., and G. Gabrelli. 1954. Riv. Ital. d'Ig., N.9-10, 3.

Downie, A. W., E. Wade, and J. A. Young. 1933. J. Hyg., **33** 196.

DuBois, M., and W. W. Ferguson. 1950. Am. J. Clin. Pathol., **20**, 491.

Dudgeon, L. S. 1929. A system of bacteriology. Med. Res. Council, London, **9**, 184.

Duguid, J. P. 1959. J. Gen. Microbiol., **21**, 271.

Duguid, J. P., I. W. Smith, G. Dempster, and P. N. Edmunds. 1955. J. Pathol. Bacteriol., **70**, 335.

Duguid, J. P., and R. R. Gillies. 1957. Ibid., **74**, 397.

Edwards, P. R., and W. H. Ewing. 1951. A manual for enteric bacteriology. Gov't. Printing Office. Washington, D.C.

Ernst, K. F., and A. H. Stock. 1943. Report to the Base Surgeon, Atlantic Base Section, U. S. Army.

Ewing, J. O., and J. Taylor. 1945. Month. Bull. Ministry of Health, **4**, 130.

Ewing, W. H. 1944. J. Bacteriol., **48**, 703.

————. 1946. Ibid., **51**, 433.

————. 1949. Ibid., **57**, 633.

————. 1950. J. Lab. & Clin. Med., **36**, 471.

————. 1953. J. Bacteriol., **66**, 333.

————. 1954. J. Immunol., **72**, 404.

————. 1966. Preliminary serologic examination of *Salmonella* and *Shigella* cultures. CDC Publ.*

————. 1967. Revised definitions for the family Enterobacteriaceae, its tribes and genera. CDC Publ.*

Ewing, W. H., and J. L. Gravatti. 1947. J. Bacteriol., **53**, 191.

Ewing, W. H., and M. C. Hucks. 1950. Ibid., **60**, 367.

Ewing, W. H., and F. Kauffmann. 1950. Pub. Health Rep., **65**, 1341.

Ewing, W. H., M. W. Taylor, and M. C. Hucks. 1950. Pub. Health Repts., **65**, 1474.

Ewing, W. H., M. C. Hucks, and M. W. Taylor. 1951. Ibid., **66**, 1579.

Ewing, W. H., P. R. Edwards, and M. C. Hucks. 1951. Proc. Soc. Exper. Biol. Med., **78**, 100.

Ewing, W. H., M. C. Hucks, and M. W. Taylor. 1952. J. Bacteriol., **63**, 319.

Ewing, W. H., and M. C. Hucks. 1952. J. Immunol., **69**, 575.

Ewing, W. H., J. Vandepitte, A. Fain, and M. Schoetter. 1952. Ann. Soc. Belge. Med. Trop., **32**, 585.

Ewing, W. H., and K. E. Tanner. 1955. J. Bacteriol., **69**, 89.

Ewing, W. H., H. W. Tatum, B. R. Davis, and R. W. Reavis. 1956. Studies on the serology of the *Escherichia coli* group. CDC Publ.*

Ewing, W. H., R. W. Reavis, and B. R. Davis. 1958. Can. J. Microbiol., **4**, 89.

Ewing, W. H., J. G. Johnson, and B. R. Davis. 1959. Int. Bull. Bacteriol. Nomen. Tax., **9**, 177.

Ewing, W. H., and J. G. Johnson. 1960. Int. Bull. Bacteriol. Nomen. Tax., **10**, 223.

Ewing, W. H., R. Hugh, and J. G. Johnson. 1961. Studies on the *Aeromonas* group. CDC Publ.*

Ewing, W. H., and L. R. Trabulsi. 1962. Int. Bull. Bacteriol. Nomen. Tax., **12**, 1.

Ewing, W. H., and K. P. Carpenter. 1966. Int. J.

*CDC Publ., Publication from the Center for Disease Control (formerly Communicable Disease Center), Atlanta, Ga. 30333.

**Public Health Laboratory: Journal of the Conference of Public Health Laboratory Directors.

System. Bacteriol., **16**, 145.

Felsen, J. 1945. Bacillary dysentery, colitis, and enteritis. Philadelphia. Saunders.

Ferguson, W. W., and M. J. Carlson. 1945. J. Bacteriol., **49**, 526.

Ferguson, W. W., and N. D. Henderson. 1947. Ibid., **54**, 179.

Flexner, S. 1900a. Phil. Med. J., **6**, 414.

_____. 1900b. Bull. Johns Hopkins Hosp., **11**, 231.

Formal, S. B., and E. E. Baker. 1953. J. Immunol., **70**, 260.

Fulton, M., and S. F. Curtis. 1946. J. Infect. Dis., **78**, 198.

Francis, A. E. 1946. J. Pathol. Bacteriol., **58**, 320.

Frantzen, E. 1950. Acta Pathol. Microbiol. Scand., **27**, 236.

Gardner, A. D. 1929. A system of Bacteriology. Med. Res. Council, London, **9**, 161.

Gettings, H. S. 1919. Med. Res. Council Gt. Brit. Spec. Rep. Ser. No. 30.

Gildemeister, H. 1956. Zentralbl. f. Bakt. I. Orig., **166**, 225.

Gillies, R. R., and J. P. Duguid. 1958. J. Hyg., **56**, 303.

Gober, M., V. Stacy, and M. Woodrow. 1944. Am. J. Hyg., **40**, 209.

Graham, J. M. 1958. J. Pathol. Bacteriol., **76**, 291.

Grigorieff, A. W. 1891. Woennomedicinsk Journal, T1, 71, 73. (Abstr. 1892, Zentralbl. f. Bakt., 12, 876.)

Gullotti, A. 1953. Nuovi Ann. d'Ig. e Microbiol., **5**, 325.

Heller, G., and S. G. Wilson. 1946. J. Pathol. Bacteriol., **58**, 98.

International Enterobacteriaceae Subcommittee. 1954. Reports. Int. Bull. Bacteriol. Nomen. Tax., **4**, 1.

_____. 1958. Ibid., **8**, 25; 93.

Kampelmacher, E. H. 1959. Antonie Leeuwenhoek, **25**, 289.

Kauffmann, F. 1942. Acta pathol. microbiol. Scand., **19**, 53.

_____. 1948. Ibid., **25**, 619.

_____. 1949. Ibid., **26**, 879.

_____. 1954. Enterobacteriaceae, 2nd ed. Copenhagen: Munksgaard.

Kruse, W. 1900. Deutsche Med. Wchnschr. **26**, 237.

Large, D. T. M., and O. K. Sankaran. 1934. J. Roy. Army Med. Corps, **63**, 231.

Lavington, R. J., A. J. Matheson, J. Taylor, and W. J. D. Flemming. 1946. J. Pathol. Bacteriol., **58**, 101.

LeMinor, L., and F. Ben Hamida. 1962. Ann. Inst. Pasteur, **102**, 267.

Lubin, A. H., and W. H. Ewing. 1964. Pub. Health Lab., ** 22, 83.

MacLennan, J. D. 1945. J. Pathol. Bacteriol., 57, 307.

Madsen, S. 1949. On the classification of the Flexner types. Copenhagen: Munksgaard.

Manolov, D. G. 1959. J. Hyg., Epidemiol., Microbiol., Immunol., **3**, 184.

Morris, J., A. Brim, T. F. Sellers, and R. D. Payne. 1944. J. Infect. Dis., **75**, 106.

Morris, J., A. Brim, and T. F. Sellers. 1944. Am. J. Pub. Health, **34**, 1277.

Mumford, E., and J. Mohr. 1944. Am. J. Trop. Med. Suppl. pp. 9-10.

Murray, E. G. D. 1918. J. Roy. Army Med. Corps, **31**, 257.

Nelson, M. G. 1947. J. Pathol. Bacteriol., **59**, 316.

Nelson, R. L. 1930. J. Bacteriol., **20**, 183.

Nelson, C. T., A. Berg, J. Spizizen, and M. J. Barnes. 1946. Am. J. Pub. Health, **36**, 51.

Neter, E. 1942. Bacteriol. Revs., **6**, 1.

Novgorodskaja, E. M. 1959. Zh. Microbiol. Epidemiol., Immunobiol., **30**, 9.

_____. 1960. Int. Bull. Bacteriol. Nomen. Tax., **10**, 239.

Olarte, J., and G. Varela. 1953. Rev. Inst. Salubridade Enfermedades, **12**, 65.

Olinici, N., S. Lescinski, and S. Busila. 1958. Microbiol., Parasitol., Epidemiol., No. 5, p. 453.

Ørskov, J., and A. Larson. 1925. J. Bacteriol., **10**, 473.

Pachecho, G., and C. Rodrigues. 1930. Arch. Inst. Biol., **3**, 145.

Park, W. H., K. R. Collins, and M. E. Goodwin. 1904. J. Med. Res., **11**, 553.

Piéchaud, D., and S. Rubinsten. 1953. Ann. Inst. Pasteur, **84**, 654.

Rauss, K., and Vertényi, A. 1956. Acta Microbiol. (Hung.), **8**, 53.

Reavis, R. W., and W. H. Ewing. 1958. Int. Bull. Bacteriol. Nomen. Tax., **8**, 75.

Roelcke, K. 1943. Die Kruse-Sonne (E) Ruhr. Stuttgart: Stahle & Friedel.

Roschka, R. 1950. Klin. Med., **5**, 88.

Rubinsten, S., and D. Piéchaud. 1952. Ann. Inst. Pasteur, **82**, 770.

_____. 1954. Ibid., **86**, 250.

Sachs, A. 1943. J. Roy. Army M. Corps, **80**, 92.

Sahab, K. 1953. Documenta Med. Geograph. Trop., **5**, 361.

Sartorius, F., and H. Reploh. 1932. Zentralbl. f. Bakt., I Orig., **126**, 10.

Schuetze, H. 1944. J. Pathol. Bacteriol., **56**, 250.

Sears, H. J., and Schoolnik. 1936. J. Bacteriol., **31**, 309.

Seeliger, H. 1949a. Z. f. Hyg. u. Infektionskr., **129**, 379.

_____. 1949b. Ibid., **129**, 444.

_____. 1950. Ibid., **131**, 509.

_____. 1951. J. Bacteriol., **62**, 243.

_____. 1953. Die Laboratoriumsdiagnostik der Bakterienruhr. Heft. 6. Beiträge zur Hygiene und Epidemiologie. Leipzig: Barth.

_____. 1954. Z. f. Hyg. u. Infektionskr., **139**, 55.

_____. 1955. Zentralbl. f. Bakt. I. Orig., **163**, 7.

Sereny, B. 1959a. Acta Microbiol. (Hung.) **6**, 179.

_____. 1959b. Ibid., **6**, 217.

_____. 1960. Ibid., **7**, 51.

Shelubsky, M., and L. Olitzki. 1947. J. Hyg., **45**, 123.

Shiga, K. 1898. Zentralbl. f. Bakt. I. Abt., **24**, 817, 870, 913.

Slopec, S., and L. Dabrowski. 1957a. Schweiz. Z. Path. Bakt., **20**, 330.

_____. 1957b. Ibid., **20**, 337.

Slopec, S., M. Mulczyk, T. M. Lachowicz, and A. Krukowska. 1960. Arch. Immunol. & Therap. (Poland), **8**, 593.

Sonne, C. 1915. Zentralbl. f. Bakt. I. Orig., **75**, 408.

Soule, M. H., and A. M. Heyman. 1933. J. Lab. Clin. Med., **18**, 549.

Stamp, L., and D. M. Stone. 1944. J. Hyg., **43**, 266.

Stock, A. H., I. Eisenstadt., G. W. Triplett, and A. Cato. 1947. J. Infect. Disc., **81**, 59.

Strong, R. P., and W. E. Musgrave. 1900. J. A. M. A., **35**, 498.

Stuart, C. A., K. M. Wheeler, and V. McGann. 1946. J. Bacteriol., **52** 431.

Szturm, S., M. Piéchaud, and R. Neel. 1950. Ann. Inst. Pasteur, **78**, 146.

Szturm-Rubinsten, S. 1963. Ibid., **104**, 423.

Szturm-Rubinsten, S., and D. Piéchaud. 1957. Ibid., **92**, 335.

_____. 1963a. Ibid., **104**, 284.

_____. 1963b. Ibid., **103**, 935.

Tee, G. H. 1952. Month. Bull. Med. Res. Council (Gt. Brit.), **11**, 68.

Trabulsi, L. R., and W. H. Ewing. 1962. Pub. Health Lab., ******, **20**, 137.

Vaillard, L., and Gh. Dopter. 1903. Ann. Inst. Pasteur, **131**, 463.

Vandepitte, V. 1950. Ann. Soc. Belge Med. Trop., **30**, 1567.

Vandepitte, J. M., W. H. Ewing, R. W. Reavis, and N. Marrecau. 1956. Ann. Soc. Belge Med. Trop., **36**, 117.

Veasie, L. 1949. J. Immunol., **61**, 307.

Weil, A. J., J. Black, and K. Farsetta. 1944. J. Immunol., **49**, 321.

Weil, A. J., A. De Assis, and H. Slafkovsky. 1948. Ibid., **58**, 23.

Welch, H., and F. L. Mickle. 1932. Am. J. Pub. Health, **22**, 263.

Wheeler, K. M. 1944a. J. Immunol., **48**, 87.

_____. 1944b. Am. J. Pub. Health, **34**, 621.

Wheeler, K. M., and F. L. Mickle. 1945. J. Immunol., **51**, 257.

Wheeler, K. M., and C. A. Stuart. 1946. J. Bacteriol., **51**, 169.

Wheeler, K. M., C. A. Stuart, and W. H. Ewing. 1946. J. Bacteriol., **51**, 169.

Winkle, S. 1949. Zur Diagnostik und Epidemiologie der Paradysenterie. Jena: Gustav Fischer.

Zhadanov, V. M., and Z. M. Andreeva. 1963. Int. Bull. Bacteriol, Nomen. Tax., **13**, 59.

Chapter 7
The Genus *Edwardsiella*

The genus *Edwardsiella* is composed of bacteria that yield uniform biochemical reactions but which may be separated into a large number of serotypes based upon their O and H antigens. It has been said that *Edwardsiella* resemble *Escherichia* but this resemblance is very superficial since many of their biochemical reactions are different. The genus was defined as follows by Ewing and McWhorter (see Ewing et al., 1967).

The genus *Edwardsiella* is composed of motile bacteria that conform to the definitions of the family ENTEROBACTERIACEAE and the tribe EDWARDSIELLEAE. Lysine and ornithine are decarboxylated, but neither malonate nor mucate is utilized. Glucose and maltose are fermented promptly, and with rare exceptions gas is formed from these substrates. Glycerol is utilized slowly by the majority of strains, but lactose, sucrose, mannitol, dulcitol, salicin, inositol, sorbitol, raffinose, rhamnose, xylose, cellobiose, and alpha methyl glucoside are not attacked. The type species is *Edwardsiella tarda*. Ewing and McWhorter.

A definition of the tribe Edwardsielleae is given in Chapter 1.

The bacteria known as *Edwardsiella tarda* first were recognized as an entity early in 1959 when culture No. 1483-59 and others like it were characterized. The designation biotype 1483-59 was used for these microorganisms and cultures of the biotype were included in work with sodium acetate medium (Trabulsi and Ewing, 1962), the ONPG test (Lubin and Ewing, 1964), and on the lipolytic, pectolytic, and alginolytic activities of Enterobacteriaceae (Davis and Ewing, 1964). Further, the biochemical reactions given by cultures of biotype 1483-59 were reported upon by Ewing et al. (1964). A search of the literature failed to reveal a description of a microorganism that resembled the 1483-59 bacteria. However, when Dr. R. Sakazaki of the National Institute of Health, Tokyo, learned of the above-mentioned work through correspondence, he wrote (1964) that in 1962 he had presented a paper entitled "The New Group of Enterobacteriaceae, the Asakusa Group" at the 1962 meeting of the Japan Bacteriological Society and that a summary of the presentation (Japanese text) was published in 1962. Dr. Sakazaki kindly furnished a translation of the above-mentioned abstract. From this it appeared that the majority of the cultures were isolated from snakes and that the strains described were similar to biotype 1483-59. Further, King and Adler (1964) described the isolation of a culture of bacterium 1483-59, which they labeled the Bartholomew group.

In 1965 Ewing et al. gave a detailed description of 37 cultures of biotype 1483-59 and compared them with members of other genera of Enterobacteriaceae. They concluded that the bacteria were *sui generis* and proposed the species name *Edwardsiella tarda* (tribe Edwardsielleae) for them.

Sakazaki (1967) gave a report of the biochemical reactions given by 256 cultures of *E. tarda*. Of these strains, 248 were isolated from intestinal contents of reptiles, two were from seals, and five were isolated from stool specimens of humans with acute gastroenteritis. Cultures of *E. tarda* have been isolated from a variety of sources including pig bile (Arambulo et al., 1967), gastroenteritis in humans (Bhat et al., 1967), and meningitis (Sonnenwirth and Kallus, 1968). Some investigators have stressed the occurrence of these bacteria in cold-blooded animals, but the majority of the cultures examined by the writer and colleagues have been from human sources.

Biochemical Reactions of *E. tarda*

A summary of the biochemical reactions given by 394 cultures of *E. tarda* is given in Table 76. These data are revised from the publication of Ewing et al. (1969). Further, the reactions given by strains of *E. tarda* are compared with those given by members of other genera of Enterobacteriaceae in Tables 5 and 11 (Chapter 3). The data recorded in the aforementioned tables are self-explanatory and should be sufficient to enable bacteriologists to identify the bacteria with ease. A culture that forms abundant hydrogen sulfide from inorganic sulfur compounds in a medium such as TSI agar and produces indol but which fails to ferment mannitol and fails to hydrolyze urea or to deaminate phenylalanine should be suspected of being *E. tarda*.

143

TABLE 76

Summary of Biochemical Reactions of *E. tarda*

Test or substrate	Sign	%+	(%+)*	Test or substrate	Sign	%+	(%+)*
Hydrogen sulfide	+	99.7	(0.3)	Jordan's tartrate	–	0	
Urease	–	0		Sodium pectate	–	0	
Indol	+	99	(0.5)	Sodium acetate	–	0	
Methyl red (37 C)	+	100		Lipases			
Voges-Proskauer (37 C)	–	0		Corn oil	–	0	
Citrate (Simmons')	–	0		Triacetin	–	0	
KCN	–	0		Tributyrin	–	0	
Motility	+	98		Maltose	+	99.1	(0.3)
Gelatin (22 C)	–	0	(0.3)	Xylose	–	0	
Lysine decarboxylase	+	100		Trehalose	–	0.3	
Arginine dihydrolase	–	0	(0.3)	Cellobiose	–	0	
Ornithine decarboxylase	+	99.7	(0.3)	Glycerol	$(+^W)$ or $+^W$	35.8^W	(60.4^W)
Phenylalanine deaminase	–	0		Alpha methyl glucoside	–	0	
Glucose acid	+	100		Erythritol	–	0	
gas	+	99.2		Esculin	–	0	
Lactose	–	0		Beta galactosidase	–	0	
Sucrose	–	0.3		Nitrate to nitrate	+	100	
Mannitol	–	0		Oxidation – fermentation	+ (Ferm)	100	
Dulcitol	–	0		Oxidase	–	0	
Salicin	–	0	(0.3^W)	HCA	–	0	
Adonitol	–	0		Cetrimide	–	0	
Inositol	–	0					
Sorbitol	–	0.3		Organic acids **			
Arabinose	–	9.4	(0.3)				
Raffinose	–	0		Citrate	(+)	100	
Rhamnose	–	0		D-tartrate	–	0	
Malonate	–	0		i -tartrate	–	0	
Mucate	–	0	(0.3)	l -tartrate	–	0	
Christensen's citrate	+	99	(0.7)				

* Figures in parentheses indicate percentages of delayed reactions (3 or more days).

+ 90% or more positive within one or two days' incubation.
(+) Positive reaction after 3 or more days.
– No reaction (90% or more).
+ or – Majority of strains positive, some cultures negative.
– or + Majority of cultures negative, some strains positive.
(+) or + Majority of reactions delayed, some occur within 1 or 2 days.
d Different reactions: +, (+), –.
w Weakly positive reaction.

Serological Reactions of *E. tarda*

A provisional antigenic schema for these bacteria was presented by McWhorter et al. (1967). Work on this schema has continued and 49 O antigens, 37 H antigens, and 148 serotypes have been characterized among the aforementioned 394 cultures of *E. tarda*. However, it has been possible to determine the serotype (i.e., both O and H antigens) of only 296 (75 per cent) of the cultures with the available antisera, although these antisera have permitted partial characterization (i.e., either O group or H antigen determined) of many additional strains. Clearly additional O and H antisera are required to attain a 90 percent (or greater) level of coverage. Since the provisional schema currently is regarded as inadequate, it is not given in detail.

The above-mentioned 394 cultures of *E. tarda* included 17 received from Dr. Sakazaki and were derived from 385 diverse sources or foci. There are some indications that some serotypes, at least, may be associated with intestinal disease but their relative importance in this regard cannot be assessed at present because of the inadequacies of the data concerning the sources of the cultures examined and of the provisional schema. However, their importance in extraintestinal infections hardly can be disputed.

REFERENCES

Arambulo, P. V., N. C. Westerlund, R. V. Sarmiento, and A. S. Abaga. 1967. Far East Med. J., **3** 385.

Bhat, P., R. M. Myers, and K. P. Carpenter. 1967. J. Hyg., **65**, 293.

Davis, B. R., and W. H. Ewing. 1964. J. Bacteriol., **88**, 16.

Ewing, W. H., A. C. McWhorter, M. R. Escobar, and A. H. Lubin. A new group of Enterobacteriaceae: Biotype 1483-59. CDC Publ.*

_____. 1965. Int. Bull. Bacteriol. Nomen. Tax., **15**, 33.

Ewing, W. H., A. C. McWhorter, M. M. Ball, and S. F. Bartes. 1967. The biochemical reactions of *Edwardsiella tarda*, a new genus of Enterobacteriaceae. CDC Publ.*

_____. 1969. Pub. Health Lab.,**27**, 129.

King, B. M., and D. L. Adler. 1964. Amer. J. Clin. Pathol., **41**, 230.

Lubin, A. H., and W. H. Ewing. 1964. Pub. Health Lab.,**22**, 83.

McWhorter, A. C., W. H. Ewing, and R. Sakazaki. 1967. Bacteriol. Proc., p.89.

Sakazaki, R. 1967. Japan. J. Med. Sci. Biol., **20**, 205.

Sonnenwirth, A. C., and B. A. Kallus. 1968. Amer. J. Clin. Pathol., **49**, 92.

Trabulsi, L. R., and W. H. Ewing. 1962. Pub. Health Lab.,**20**, 137.

*CDC Publ., Publication from the Center for Disease Control (formerly the Communicable Disease Center), Atlanta, Ga., 30333.
**Public Health Laboratory: Journal of the Conference of Public Health Laboratory Directors.

Chapter 8
The Genus *Salmonella*

In order to identify members of the genus *Salmonella*[1] it is necessary to consider the differential characters of the genus and to define it. Although many definitions of this genus are extant in the literature, the definition given by Ewing and Ball (1966) and Ewing (1967) has been adopted. It is as follows:

The genus *Salmonella* is composed of motile bacteria that conform to the definitions of the family ENTEROBACTERIACEAE and the tribe SALMONELLEAE. Urease is not produced, sodium malonate is not utilized, gelatin is not liquefied, and growth does not occur in medium containing potassium cyanide. Lysine, arginine, and ornithine are decarboxylated. Acid is produced in Jordan's tartrate medium. Dulcitol is fermented and inositol is utilized by numerous strains. Sucrose, salicin, raffinose, and lactose are not fermented.

This definition should be used in conjunction with that for the tribe Salmonelleae (Chapter 1).

Occasionally an aberrant strain of one of the commonly occurring salmonellae may be seen. Such a culture may ferment lactose, sucrose, or salicin, or it may form indol. These are exceptional, and if all the other biochemical reactions are like those given by salmonellae, then these abberant strains are members of the genus *Salmonella*. In other words, the determination as to whether a microorganism is a member of the genus is made on the basis of *all* its biochemical characteristics and not by means of a single test such as lactose utilization or lack of it. For example, a strain that ferments lactose may occur in a product and give rise to a number of infections, but it should be recalled that it is still a single strain represented by multiple isolants. Or, to put it another way, the aberrant strain represents one focus of infection with multiple cases (and isolates). This fact should be taken into account

in attempts to determine the frequency of occurrence of aberrant strains of salmonellae. Further, it should be recalled that with few exceptions (Ewing and Ball, 1966; Martin et al., 1969), strains seen in daily practice are perfectly typical, a fact that is more remarkable perhaps than the occurrence of an occasional aberrant culture. Nevertheless, aberrant strains should be recognized when they occur, and this is one of the reasons that the use of bismuth sulfite agar, along with other plating media, is recommended for the isolation of salmonellae.

The nomenclatural system used in connection with the genus *Salmonella* is discussed in some detail in Chapter 1. Accordingly, the three species of *Salmonella* are *Salmonella cholerae-suis*, *Salmonella typhi*, and *Salmonella enteritidis*. The numerous serotypes (ser) and bioserotypes (bioser) of *S. enteritidis* are written as follows: *S. enteritidis* ser Enteritidis, *S. enteritidis* bioser Paratyphi-C, *S. enteritidis* ser Typhimurium etc.

BIOCHEMICAL IDENTIFICATION OF *SALMONELLA*

Introduction. In a series of papers published 1960 through 1965, Kauffmann proposed a subdivision of the genus *Salmonella* into four subgenera which were labeled I, II, III, IV (see Kauffmann, 1966, p. 57). Subgenus I contained salmonellae that were typical biochemically, subgenera II and IV consisted of serotypes that yielded aberrant reactions in certain biochemical tests, and the genus *Arizona* was incorporated into the genus *Salmonella* as subgenus III.

The following is included for those who may be interested in the separation of subgenera I, II, and IV of Kauffmann.

In his book Kauffmann (1966) listed 962 serotypes and bioserotypes of *Salmonella,* of which 155 (16.1 per cent) and 18 (1.9 per cent) were designated as members of subgenera II and IV, respectively. Two other serotypes were listed as atypical; apparently these did not belong to I, II, or IV of Kauffmann (1960, 1965, 1966). Although the percentage of atypical or subgenus II and IV serotypes listed in the antigenic schema for *Salmonella* was relatively great, the incidence of these particular serotypes in man and warm-blooded animals was very low as far as the

---

[1]The generic term *Salmonella* was given to these microorganisms by Lignières in 1900 (see Report, Salmonella Committee, 1934; or Ewing, 1963), in honor of Dr. D. E. Salmon, a codiscoverer of the microorganism now known as *Salmonella cholerae-suis*. This generic term was adopted by international agreement on the basis of priority, in accordance with international rules of nomenclature, and has been employed universally since 1933. It cannot be changed by any legislative body.

Test or Substrate	I	II	III[a]	IV
Lactose	-	-	+ or X	-
Beta-galactosidase	-	- or X	+[b]	-[e]
Dulcitol	+	+	-	-
Salicin	-	-	-[c]	+[f]
D-tartrate	+	- or X	- or X	- or X
Mucate	+	+	d	-
Malonate	-	+	+	-
Gelatin	-	+	+[d]	+
KCN	-	-	-	+

X, late and irregularly positive. d, different types.

Adapted from Kauffmann, 1966.

[a]The genus *Arizona* in this book.

[b]Lubin and Ewing (1964) reported 7.2 per cent negative.

[c]Ewing et al. (1965) found that 10 per cent were positive.

[d]Ewing et al. (1965) reported 4 per cent negative.

[e]Weakly positive reactions sometimes are seen.

[f]Some strains do not ferment salicin.

author was able to determine. A review of several publications on the occurrence of serotypes of *Salmonella* in the United States revealed that only 33 cultures (0.04 per cent) of subgenus II of Kauffmann occurred among more than 86 thousand cultures (see Ewing and Ball, 1966 for references).

The incidence of strains that would belong to subgenus IV of Kauffmann also appeared to be very low. A survey of the literature from other countries also indicated a very low incidence of serotypes of subgenus II of Kauffmann (see Ewing and Ball, 1966, for references).

The several proposals of Kauffmann regarding the above-mentioned subgenera, as well as certain other subjects, were reviewed and discussed by Edwards (1963), Edwards et al. (1965), and by Ewing and Ball (1966) and need not be dealt with in detail here. It is sufficient to state that the late Dr. Edwards did not subscribe to Kauffmann's views and neither does the author. The numerous serotypes and bioserotypes of *Salmonella,* whether named or unnamed, are not

regarded as individual species. On the contrary, they are considered to be infrasubspecific entities and many investigators have expressed views contrary to those of Kauffmann in this regard (see Ewing and Ball, 1966, for references). With respect to the so-called subgenera, the author believes that if a culture possesses the overall biochemical characteristics of members of the genus *Salmonella*, it should be labeled a *Salmonella*. If it is atypical, the aberrant characters should be noted.

Granted atypical strains do occur, but they apparently are infrequent in occurrence among cultures seen in daily practice. As mentioned at the beginning of this chapter atypical cultures of what would be termed subgenus I also occur occasionally. However, the frequency of their occurrence must be low, judging from the fact that few have been detected among more than 80,000 cultures of *Salmonella* examined during the past 22 years. Further, it should be recalled that occasional atypical or borderline strains occur in or between all of the genera of Enterobacteriaceae. The author does not agree to a combination of the genera *Arizona* and *Salmonella*. These are distinct entities which can be separated by biochemical methods. Parenthetically there is no more reason to incorporate the genus *Arizona* into *Salmonella* than there would be to incorporate the genus *Shigella* into the genus *Escherichia*.

Biochemical Reactions of Salmonella. Tests and media that are helpful in preliminary examination of isolants are given in Chapter 2. For example, the use of TSI agar and lysine iron agar media are discussed there, and the reactions given by the majority of Enterobacteriaceae are tabulated (Tables 1 and 2).

Members of the genus *Salmonella* conform to the definition of the family Enterobacteriaceae and the tribe Salmonelleae (Chapter 1). With a few exceptions hydrogen sulfide is produced abundantly in triple sugar iron (TSI) or Kligler's iron agars as well as in media of similar sensitivity (e.g., peptone iron agar). The notable exceptions include some strains of *S. cholerae-suis* (usually, but not always, the diphasic forms), most cultures of *S. enteritidis* bioser Paratyphi-A, and *S. enteritidis* bioserotypes Sendai, Abortusequi, Gallinarum, and Berta. Some strains of *S. typhi* also fail to produce detectable hydrogen sulfide on the above mentioned media, and on lysine iron agar (LIA). For example, the following reactions were obtained with 37 cultures of *S. typhi* isolated from carriers:

	+ 1 da	+ 2 da	+ 3 da	-14 da
TSI agar	30	5		2
PI agar	35			2
LIA	33	1	1	2

The same two strains were negative in all three media. Further, it should be borne in mind that occasional strains of almost any serotype of *Salmonella* may fail to produce hydrogen sulfide in the above-mentioned media.

Salmonellae yield positive methyl red and negative Voges-Proskauer tests. With very few exceptions indol is not produced, and malonate is utilized only by occasional cultures seen in daily practice. The same may be said of gelatin liquefaction and production of beta galactosidase. Exceptional reactions that occasionally may be seen are apparent in the tables that follow.

Cultures may be encountered, which, when inoculated into fluid glucose or other carbohydrate medium, ferment the carbohydrate and then revert to an alkaline condition within a relatively short incubation period (24 to 48 hr). Such strains almost invariably are members of one of the genera that contain bacteria that yield methyl red negative and Voges-Proskauer positive reactions. Provided that they are pure, cultures that revert in this manner can be eliminated if one is interested primarily in recovery of salmonellae and related bacteria.

Differentiation of members of the genus *Salmonella* from related bacteria has been facilitated by the introduction of several tests and methods. Some of these were described as early as 1932, while others were introduced more recently. Since about 1955 these tests have been evaluated thoroughly by a number of investi-

gators in different laboratories and their value has been established. Medium that contains potassium cyanide (KCN) in a definite, prescribed concentration is of considerable value in the differentiation of members of the genera *Salmonella* and *Arizona* from members of the genus *Citrobacter*. This medium is of particular value in the differentiation of cultures of *Citrobacter* that ferment lactose slowly or fail to utilize lactose, i.e., strains formerly classified as Bethesda-Ballerup bacteria. Braun and Guggenheim (1932) and Braun (1938, 1939) reported that growth of cultures of *E. coli* was inhibited by certain concentrations of KCN whereas growth of strains of *Klebsiella* and *Enterobacter* was not. Buttiaux (1952) found that growth of salmonellae was inhibited in Braun's KCN medium while cultures of *Citrobacter* were able to grow in it. These observations were confirmed by Moeller (1954a) and by Kauffmann (1954). Moeller (1954) modified the medium of Braun, and it is the Moeller modification that is recommended. Other investigators confirmed the value of KCN medium for differentiation within the tribe Salmonelleae. The results reported by some of these investigators are recorded in Table 77. It may be seen that with few exceptions members of the genera *Salmonella* and *Arizona* fail to grow in KCN medium, whereas members of the genus *Citrobacter*, with few exceptions, are able to grow in it.

The tests for decarboxylation of diamino acids introduced by Moeller (1954b, 1955) also have proven

TABLE 77

Reactions of Members of the Tribe SALMONELLEAE in KCN Medium As Reported by Several Authors. (Data adapted from publications cited.)

Genus	No. tested	No. positive	(%)	No. negative	(%)	References
Salmonella	78	0	(0)	78	(100)	Moeller, 1954a
Arizona	35		(6)		(94)	
Citrobacter	72	72	(100)	0	(0)	
Salmonella	900	10	(1.1)	890	(98.9)	Edwards and Fife, 1956
Arizona	501	39	(7.8)	462	(92.2)	
Citrobacter	580	574	(98.9)	6	(0.1)	
Arizona	150	13	(8.7)	137	(91.3)	Ewing, Fife, and Davis, 1965
Citrobacter	582	565	(97.1)	17	(2.9)	Davis and Ewing, 1966

NOTE: Numerals in parentheses are percentages.

to be of great value, not only for differentiation of salmonellae, but for differentiation throughout the family (e.g., see Chapter 3). Some of the results of tests for lysine and ornithine decarboxylases and for arginine dihydrolase reported by several investigators are summarized in Table 78. The Moeller media are regarded as standard or recommended, and the use of all three amino acid media in addition to a control tube is recommended.

Similarly, the test for utilization of sodium malonate in a modification of Leifson's (1933) formula is of considerable value for differentiation of members of the genera *Salmonella* and *Arizona* and for differ-

entiation in certain other areas of the family as well (Chapter 3, and Table 79).

The organic acid media of Kauffmann and Petersen (1956) are of value in the differentiation of cultures of *Salmonella* and *Arizona*, and are particularly helpful in determining whether an atypical or aberrant strain is a member of the genus *Salmonella* or of the genus *Arizona*. Some of the organic acids were employed earlier, but the tests were refined by Kauffmann and Petersen (1956). The results obtained by Ellis et al. (1957) in tests for utilization of certain organic acids and for utilization of sodium malonate are given in Table 80.

TABLE 78

The Decarboxylase Reactions Given by Members of the Tribe SALMONELLEAE as Reported by Several Authors. (Data adapted from publications cited.)

Genus	No. tested	Lysine +	Lysine (+)	Lysine –	Arginine +	Arginine (+)	Arginine –	Ornithine +	Ornithine (+)	Ornithine –	References
Salmonella *	74	74	0	0	0	72	2	74	0	0	Moeller, 1954b, 1955
Bioser Paratyphi-A	4	0	0	4[a]	0	4	0	4	0	0	
Bioser Gallinarum	1	1	0	0	0	1	0	0	0	1	
S. typhi	3	3	0	0	0	3	0	0	0	3	
Arizona	36	36	0	0	0	36	0	36	0	0	
Citrobacter	116	0	0	116	0	112	4	0	33	83	
Salmonella *	65	63	0	2	1	63	1	65	0	0	Ewing, Davis and Edwards, 1960
Bioser Paratyphi-A	10	0	0	10	0	7	3	10	0	0	
Bioser Gallinarum	10	10	0	0	0	0	10	0	0	10	
S. typhi	11	11	0	0	0	0	11	0	0	11	
Arizona	50	50	0	0	2	48	0	50	0	0	
Citrobacter	54	0	0	54	0	52	2	3	1	50	
Arizona	150	150	0	0	19	127	4	150	0	0	Ewing and Fife, 1966
Citrobacter	582	0	0	582	254	257	71	100	1	481	Davis and Ewing, 1966

+ Positive reaction within 24 hours.

(+) Positive reaction within 2 to 4 days.

– Negative, final reading 4 days.

* *Salmonella* other than *S. typhi* and the bioserotypes (of *S. enteritidis*) listed.

[a] Moeller reported that four ser Paratyphi-A gave weakly positive reactions in lysine medium. Such reactions were not noted by other investigators, at least not in a four-day incubation period.

TABLE 79

Comparison of Results of Tests for Utilization of Malonate as Reported by Several Investigators. (Data adapted from publications cited.)

Genus	No. tested	Malonate +	Malonate –	References
Salmonella	Unknown	0	all	Schaub, 1948 (Group IIC)
Arizona	Unknown	+	none	
Salmonella	228	11 (4.8)	217 (95.2)	Shaw, 1956
Arizona	67	62 (92.5)	5 (7.5)	
Salmonella	58	4 (6.9)	54 (93.1)	Ewing, Davis, and Reavis, 1957
Arizona	131	123 (94.9)	8 (6.1)	
Salmonella	1136	36 (3.2)	1100 (96.8)	Ellis, Edwards, and Fife, 1957, and unpublished data
Arizona	619	585 (94.5)	34 (5.5)	
Citrobacter	250	30 (12)	220 (88)	
Arizona	150	140 (93.3)	10 (6.7)	Ewing and Fife, 1966
Citrobacter	582	127 (22.5)	451 (77.5)	Davis and Ewing, 1966

NOTE: Figures in parentheses are percentages.

The ONPG test for detection of beta galactosidase activity is an aid in the differentiation of cultures of *Salmonella* and *Arizona*, particularly strains of *Arizona* that ferment lactose slowly or fail to utilize it. This test also is helpful in the characterization of cultures of *Citrobacter* that utilize lactose slowly. The results obtained by several investigators with this test are summarized in Table 81.

Judicious use of tests for decarboxylation of lysine, arginine, and ornithine, for utilization of organic acids and sodium malonate, for production of beta galactosidase, and tests for growth in KCN medium may be expected to yield information of value in the differentiation of members of the tribe Salmonelleae. Other media and tests that are helpful in effecting this differentiation are listed in the tables that follow. Percentage data are included in these tables. These data indicate the reactions given by the majority of cultures in the various tests. They also give indications of occasional atypical or aberrant reactions.

The results obtained in a study (Ewing and Ball, 1966) of 371 cultures that belonged to the three species of *Salmonella* are summarized in Table 82.

Since these results were thought to be fairly representative of the genus as a whole they were employed as the basis for the definition of the genus *Salmonella* given at the beginning of this chapter. The above-mentioned 371 strains of *Salmonella* were selected from 1544 cultures submitted for serologic analysis during a seven-month period in 1964. Of the 1544, 160 were strains of *S. typhi*. With very few exceptions these strains were taken at random from the above-mentioned materials. One exception was that when groups of cultures were received from a single outbreak or focus of infection, only one strain from each such group was selected. Also, insufficient numbers of *S. cholerae-suis* and *S. enteritidis* bioserotypes Paratyphi-A, Gallinarum, and Pullorum were augmented with strains received earlier since only a few of these were submitted during the above-mentioned period. The numbers of these strains were approximately proportional to the total. Of the total of 371 cultures examined, 20 were *S. cholerae-suis*, 16 were *S. typhi*, and 335 were serotypes and bioserotypes of *S. enteritidis* (see Ewing and Ball, 1966, for details).

Three cultures gave evidence of hydrogen sulfide

production in TSI agar medium after 3 to 7 days of incubation. Two of these were strains of *S. cholerae-suis* and one was a culture of *S. typhi*. Twenty-eight cultures failed to show evidence of hydrogen sulfide production during 14 days of incubation. These were as follows:

S. cholerae-suis	6 strains
S. typhi	1 strain
S. enteritidis ser Berta	2 strains
S. enteritidis bioser Paratyphi-A	13 strains
S. enteritidis bioser Pullorum	2 strains
S. enteritidis ser Senftenberg	2 strains
S. enteritidis ser Typhimurium	2 strains

Four cultures yielded positive indol reactions. Three of these were strains of *S. enteritidis* serotype Muenchen, ser Javiana, and ser Bareilly. One was a culture of *S. enteritidis* ser Gaminara which also liquefied gelatin after 30 days of incubation.

Growth on Simmons' citrate agar medium was delayed in 26 instances. The majority (18) of these delayed reactions were obtained with cultures of *S. cholerae-suis*. In addition, growth on Simmons' citrate agar occurred slowly (3 to 12 days) with one culture of ser Bareilly, one strain of bioser Gallinarum, one culture of ser Infantis, four strains of bioser Paratyphi-A, and one isolant of ser Pensacola. Forty-eight

TABLE 80

Patterns of Reactions Given by Members of the Tribe SALMONELLEAE in Organic Acid Media *

Substrate				Genus			
Citrate **	D-tartrate **	Malonate	Mucate **	*Salmonella*	*Arizona*	*Citrobacter*	Remarks
–	–	–	–	27 (2.4)	30 (4.8)	30 (12)	13 cultures of ser Paratyphi-A
–	–	+	–	8 (0.7)	449 (72.5)	9 (3.6)	
–	–	+	+	19 (1.7)	134 (21.7)	21 (8.4)	
+	+	–	+	652 (57.4)	0	0	
+	–	–	+	81 (7.1)	1 (0.2)	2 (0.8)	17 cultures ser Paratyphi-B
–	–	–	+	38 (3.3)	3 (0.5)	187 (74.8)	
+	+	–	–	55 (4.8)	0	0	12 stock cultures of *S. typhi* 12 stock cultures of *S. cholerae-suis*
–	+	–	+	185 (16.3)	0	0	
–	+	–	–	53 (4.7)	0	0	26 freshly isolated cultures of *S. typhi*
+	+	+	+	8 (0.7)	2 (0.3)	0	
+	–	–	–	9 (0.8)	0	1 (0.4)	
–	+	+	+	1 (0.1)	0	0	
Total				1136 (100)	619 (100)	250 (100)	

NOTE: Figures in parentheses indicate percentages.

* Results based upon readings made after 20 hours of incubation. (Data adapted from Ellis, Edwards, and Fife, 1957, and unpublished results.)

** Media of Kauffmann and Petersen, 1956.

See also Table 14, Chapter 3.

TABLE 81

Results of Beta Galactosidase (ONPG) tests with Members of the Tribe SALMONELLEAE
(Data adapted from publications cited.)

Genus	No. tested	ONPG +	ONPG –	References
Salmonella	61	0	61	LeMinor and Ben Hamida, 1962
Arizona	40	30	10	
Citrobacter	25	25	0	
Salmonella, subgenus I	140	3 (2.1)	137 (97.9)	Kauffmann, 1963b (This paper should be consulted for details.)
Salmonella, subgenus II	143	59 (41.3)	84 (58.7)	
Arizona (subgenus III)	51	"ca. 100%"		
Salmonella, I[a]	60	0	60	Buelow, 1964
Salmonella, II[b]	12	5	7	
Arizona (III)	38	35	3	
Citrobacter	30	18	12	
Salmonella, I[a]	271	10 (3.7)	261 (96.3)	
Salmonella, II[b]	181	7 (3.9)	174 (96.1)	Lubin and Ewing, 1964
Arizona	446	414 (92.8)	32 (7.2)	
Citrobacter	39	39	0	

NOTE: Figures in parentheses are percentages.

[a] I, Subgenus I of Kauffmann

[b] II, Subgenus II of Kauffmann

cultures failed to show evidence of growth on Simmons' citrate medium. These were as follows:

S. typhi	16 cultures
S. enteritidis bioser Gallinarum	9 cultures
S. enteritidis bioser Paratyphi-A	12 cultures
S. enteritidis bioser Pullorum	10 cultures
S. enteritidis ser Typhimurium	1 culture

These strains yielded typical reactions in other biochemical tests. Note that failure to grow on Simmons' citrate medium is typical of some of the above-mentioned microorganisms. Only two cultures grew in KCN medium. One, a strain of *S. enteritidis* ser Ohio, showed evidence of growth within 24 hours while the other (ser Tennessee) was positive after 4 days of incubation. Both of these isolants were typical in other respects. All of the cultures tested were motile except strains of *S. enteritidis* bioserotypes Gallinarum and Pullorum (see also Table 88). Four isolates liquefied gelatin. One of these, a culture of ser Gaminara, was mentioned above in connection with indol production. A strain of ser Berta liquefied gelatin in 5 to 7 days, and one culture each of ser New Mexico and ser Travis did so after 30 days of incubation.

Lysine was not decarboxylated by one strain of *S. cholerae-suis* and two cultures of bioser Pullorum. Sixteen isolants of *S. enteritidis* bioser Paratyphi-A failed to decarboxylate this amino acid as might be expected (see Table 87). One strain of bioser Paratyphi-A failed to decarboxylate ornithine as well. Twenty-seven cultures failed to decarboxylate ornithine. One of these was the strain of bioser Paratyphi-A just mentioned, and the remainder were cultures of *S. typhi* and bioser Gallinarum (see also

Tables 84, 87 and 88). Sodium malonate was utilized by only two cultures. These were one strain each of ser Bovis-morbificans and ser Albany. Three cultures fermented lactose promptly, i.e., within one or two days. One of these was a strain of ser Typhimurium that also failed to produce hydrogen sulfide and did not grow on Simmons' citrate agar medium (v. sup.). The other two cultures were ser Tennessee strains which also produced acid from sucrose. Only three strains fermented salicin and these did so slowly. The three belonged to different serotypes: ser Florida, ser Indiana, and ser Newmexico.

With the exceptions noted, the above-mentioned 371 cultures yielded typical biochemical reactions. The biochemical reactions given by cultures of *S. choleraesuis* and *S. typhi* are recorded in Tables 83 and 84,

TABLE 82

The Biochemical Reactions of Members of the Genus *Salmonella*

Test or substrate	Sign	%+	(%+)*	Test or substrate	Sign	%+	(%+)*
Hydrogen sulfide (TSI agar)	+	91.6	(0.8)	Raffinose	–	3	(0.3)
Urease	–	0		Rhamnose	+	90.3	(1.1)
Indol	–	1.1		Malonate	–	0.5	
Methyl red (37 C)	+	100		Mucate	d	73.6	(1.3)
Voges-Proskauer (37 C)	–	0		Jordan's tartrate	+ or (+)	89.3	(1.2)
Citrate (Simmons')	d	80.1	(7)	Stern's glycerol	d	81	(0.5)
KCN	–	0.3	(0.3)	Sodium acetate	d	80	(2.7)
Motility	+	94.6		Maltose	+	96	(1.3)
Gelatin (22 C)	–	(1.1)		Xylose	+	94	(0.6)
Lysine decarboxylase	+	94.6		Trehalose	+	93.5	(1.1)
Arginine dihydrolase	+ or (+)	58.5	(34)	Cellobiose	d	6.5	(76.4)
Ornithine decarboxylase	+	92.7		Glycerol	d	4.5	(17)
Phenlalanine deaminase	–	0		Nitrate to nitrite	+	100	
Glucose acid	+	100		Oxidase	–	0	
gas	+	91.9		HCA	–	0	
Lactose	–	0.8		Cetrimide	–	1.4	(1.4)
Sucrose	–	0.5					
Mannitol	+	99.7		Organic acids **			
Dulcitol	d	86.5	(2.7)				
Salicin	–		(0.8)	citrate	+ or (+)	87	(4.6)
Adonitol	–	0		D-tartrate	+ or (+)	84.3	(5.7)
Inositol	d	34.5	(0.8)	i -tartrate	d	3.8	(51.2)
Sorbitol	+	94.1	(4)	l -tartrate	d	9.3	(66.7)
Arabinose	+ or (+)	89.2	(0.8)				

* Figures in parentheses indicate percentages of delayed reactions (3 or more days).

+ Positive within one or two days' incubation.
(+) Positive reaction after 3 or more days.
 – No reaction.
 + or – Majority of strains positive, some cultures negative.
 – or + Majority of cultures negative, some strains positive.
(+) or + Majority of reactions delayed, some occur within 1 or 2 days.
 d Different reactions: +, (+), –.

** Method of Kauffmann and Petersen, 1956.

TABLE 83

The Biochemical Reactions of *S. cholerae-suis*

Test or substrate	Sign	%+	(%+)*	Test or substrate	Sign	%+	(%+)*
Hydrogen sulfide	d	60	(10)	Mucate	–	0	
Urease	–	0		Christensen's citrate	+	90	(10)
Indol	–	0		Jordan's tartrate	d	80	(5)
Methyl red (37 C)	+	100		Stern's glycerol	–	0	
Voges-Proskauer (37 C)	–	0		Sodium acetate	– or (+^{w})		(20)
Citrate (Simmons')	(+)		(90)	Sodium alginate	–	0	
KCN	–	0		Lipases (corn oil)	–	0	
Motility	+	100		Maltose	+ or (+)	100	
Gelatin (22 C)	–	0		Xylose	+	100	
Lysine decarboxylase	+	90		Trehalose	–	0	
Arginine dihydrolase	(+)		(90)	Cellobiose	–	0	
Ornithine decarboxylase	+	100		Glycerol	–	0	
Phenylalanine deaminase	–	0		Alpha methyl glucoside	–	0	
Glucose acid	+	100		Erythritol	(+) or –		(85)
gas	+	90		Esculin	–	0	
Lactose	–	0		Beta galactosidase	–	0	
Sucrose	–	0		Nitrate to nitrite	+	100	
Mannitol	+	100		Oxidation – fermentation	F	100	
Dulcitol	d	5	(15)	Oxidase	–	0	
Salicin	–	0					
Adonitol	–	0		Organic acids **			
Inositol	–	0					
Sorbitol	+ or (+)	85	(15)	citrate	+	90	(5)
Arabinose	–	0		D-tartrate	+	95	
Raffinose	–	0		i -tartrate	(+) or –		(85)
Rhamnose	+	100		l -tartrate	– or (+)		(35)
Malonate	–	0					

* Figures in parentheses indicate percentages of delayed reactions (3 or more days).

** Method of Kauffmann and Petersen, 1956.

F Fermentation utilization of glucose.

+ Positive within one or two days' incubation.

(+) Positive reaction after 3 or more days.

– No reaction.

+ or – Majority of strains positive, some cultures negative.

– or + Majority of cultures negative, some strains positive.

(+) or + Majority of reactions delayed, some occur within 1 or 2 days.

d Different reactions: +, (+), –.

TABLE 84

The Biochemical Reactions of *S. typhi*

Test or substrate	S. typhi Sign	%+	(%+)*	Test or substrate	S. typhi Sign	%+	(%+)*
Hydrogen sulfide	+	94.3		Raffinose	–	0	
Urease	–	0		Rhamnose	–	0	
Indol	–	0		Malonate	–	0	
Methyl red (37 C)	+	100		Mucate	–	0	
Voges-Proskauer (37 C)	–	0		Christensen's citrate	d		(12.5)
Citrate (Simmons')	–	0		Jordan's tartrate	+ or –	87.5	
KCN	–	0		Stern's glycerol	–	0	
Motility	+	100		Sodium acetate	–	0	
Gelatin (22 C)	–	0		Maltose	+	100	
Lysine decarboxylase	+	100		Xylose	+	93.7	(6.3)
Arginine dihydrolase	(+) or –		(81.3)	Trehalose	+	100	
Ornithine decarboxylase	–	0		Cellobiose acid	d	6^w	(31)
Penylalanine deaminase	–	0		Glycerol	(+^{w})		(93.7)
Glucose acid	+	100		Beta galactosidase	–	0	
gas	–	0		Nitrate to nitrite	+	100	
Lactose	–	0		Oxidation – fermentation	F	100	
Sucrose	–	0		Oxidase	–	0	
Mannitol	+	100					
Dulcitol	d		(31.3)	Organic acids **			
Salicin	–	0					
Adonitol	–	0		citrate	+	100	
Inositol	–	0		D-tartrate	+ or (+)	87.5	(6.3)
Sorbitol	+	100		i -tartrate	–	0	
Arabinose	–		(6.3)	l -tartrate	–	0	

* Figures in parentheses indicate percentages of delayed reactions (3 or more days).

** Method of Kauffmann and Petersen, 1956

F Fermentative utilization of glucose.

+ Positive within one or two days' incubation.
(+) Positive reaction after 3 or more days.
– No reaction.
+ or – Majority of strains positive, some cultures negative.
– or + Majority of cultures negative, some strains positive.
(+) or + Majority of reactions delayed, some occur within 1 or 2 days.
d Different reactions: +, (+), –.

respectively, and those given by *S. enteritidis* are given in Table 85. Although the numbers of strains of *S. cholerae-suis* and *S. typhi* examined by Ewing and Ball (1966) were not large, the results recorded in Tables 82 and 83 correlated very closely with those reported in other studies (Kauffmann, 1966; Martin et al., 1969; and others).

Tests that are of particular value in the differentia-

TABLE 85

The Biochemical Reactions of *S. enteritidis*[1]

Test or substrate	Sign	%+	(%+)*	Test or substrate	Sign	%+	(%+)*
Hydrogen sulfide (TSI agar)	+	93.7		Raffinose	–	3.3	(0.3)
Urease	–	0		Rhamnose	+	94	(1.2)
Indol	–	1.2		Malonate	–	0.6	
Methyl red (37 C)	+	100		Mucate	d	81.5	(1.5)
Voges-Proskauer (37 C)	–	0		Jordan's tartrate	+	90	(1)
Citrate (Simmons')	+ or (+)	88.7	(2.1)	Stern's glycerol	+	98.2	(0.6)
KCN	–	0.3	(0.3)	Sodium acetate	d	86.9	(2.1)
Motility	+	94		Maltose	+	96.4	(0.6)
Gelatin (22 C)	–		(1.1)	Xylose	+	93.7	(0.3)
Lysine decarboxylase	+	94.9		Trehalose	+	98.8	(1.2)
Arginine dihydrolase	+ or (+)	65.4	(27.4)	Cellobiose	(+) or –	7	(83.5)
Ornithine decarboxylase	+	96.7		Glycerol	d	5.1	(14.1)
Phenylalanine deaminase	–	0		Nitrate to nitrite	+	100	
Glucose acid	+	100		Oxidase	–	0	
gas	+	96.1		HCA	–	0	
Lactose	–	0.9		Cetrimide	–	1.5	(1.5)
Sucrose	–	0.6					
Mannitol	+	100		Organic acids **			
Dulcitol	+	95.7	(0.3)				
Salicin	–		(0.9)	citrate	+ or (+)	86.2	(4.8)
Adonitol	–	0		D-tartrate	d	83.5	(6)
Inositol	d	38.2	(0.9)	i -tartrate	d	4.2	(54.6)
Sorbitol	+	94.3	(3.6)	l -tartrate	d	10.1	(73.5)
Arabinose	+	98.8	(0.6)				

* Figures in parentheses indicate percentages of delayed reactions (3 or more days).

** Method of Kauffmann and Petersen, 1956.

[1] Exclusive of cultures of *S. cholerae-suis* and *S. typhi*.

+ Positive within one or two days' incubation.

(+) Positive reaction after 3 or more days.

– No reaction.

+ or – Majority of strains positive, some cultures negative.

– or + Majority of cultures negative, some strains positive.

(+) or + Majority of reactions delayed, some occur within 1 or 2 days.

d Different reactions: +, (+), –.

tion of the three species of salmonellae *(S. cholerae-suis, S. typhi,* and the commonly occurring forms of *S. enteritidis)* are given in Table 86. In this table the hydrogen sulfide reactions of cultures of *S. cholerae-suis* are recorded as "d." The reason for this is that the

most commonly occurring biotype (Kunzendorf) of this species produces hydrogen sulfide abundantly. However, it should be recalled that not only is *S. cholerae-suis* the type species of the genus but also that the original cultures of this species failed to form

TABLE 86

Differentiation of species of *Salmonella*[1]

Test or substrate	S. cholerae-suis			S. typhi			S. enteritidis[1]		
	Sign	%+	(%+)*	Sign	%+	(%+)*	Sign	%+	(%+)*
Hydrogen sulfide (TSI)	d	60	(10)	+^{w}	94.3		+	98	
Citrate (Simmons')	(+)		(90)	–	0		+	99.3	(0.7)
Ornithine decarboxylase	+	100		–	0		+	100	
Gas from glucose	+	100		–	0		+	97.7	
Dulcitol	d	5	(15)	– or (+)	0	(31.3)	+	98.3	
Inositol	–	0		–	0		d	42.8	(1)
Trehalose	–	0		+	100		+	100	
Arabinose	–	0		–		(6.3)	+	99.3	
Rhamnose	+	100		–	0		+	95	
Cellobiose	–	0		d	37.5		(+)	5	(92.8)
Erythritol	(+^{w}) or –		(85)	–	0		–	0.6	
Sodium acetate	– or (+^{w})		(20)	–	0		+	92.4	(2.2)
Mucate	–	0		–	0		+ or (+)	88.3	(1.7)
Stern's glycerol fuchsin	–	0		–	0		+	98.2	(0.6)
Organic acids **									
citrate	–	10		–	10		+	96	(4)
D-tartrate	+	95		+	87.5	(6.3)	+	91	(5.3)
i -tartrate	–	0	(85)	–	0		d	4.7	(57.5)
l -tartrate	–	0	(35)	–	0		d	11.8	(75.2)

[1] Adapted from Ewing and Ball (1966). The reactions listed for *S. enteritidis* are those obtained with 299 cultures of commonly occuring serotypes.
*Figures in parentheses indicate percentages of delayed reactions (3 days or more).
**Method of Kauffmann and Petersen, 1956. (Twenty hour readings except where indicated.)

^{w} Weakly positive reactions.
+ Positive within one or two days' incubation.
(+) Positive reaction after 3 or more days.
 – No reaction.
+ or – Majority of strains positive, occasional cultures negative.
 – or + Majority of cultures negative, occasional strains positive.
(+) or + Majority of reactions delayed, some occur within 1, or 2 days.
 d Different reactions: +, (+), –.

hydrogen sulfide. Hence the neotype strain of the species and others like it do not produce this substance or do so slowly a in small amounts. This point may be of importanc i framing a definition of the species *S. cholerae-suis.*

The differe ation of strains of *S. typhi* poses no particular prc m since the reactions of these microorganisms ar inique (Table 86), a fact that has been known for r iy years.

Culture *S. cholerae-suis* and *S. enteritidis* may be differentia d by the delayed and usually weak reactions ven by strains of the former species in Simmon citrate medium as contrasted with the rapid growth nd abundant utilization of citrate in this mediu seen with cultures of the latter (Table 86). All of th isolants of *S. enteritidis* that failed to grow in Sim ns' citrate agar were strains of bioserotypes Par yphi-A, Pullorum, and Gallinarum (v. inf.). The re ions given by cultures of *S. cholerae-suis* and *S. e eritidis* in dulcitol medium are of assistance in the ç ferentiation of these species. Only 5 per cent of the rains of the former species tested fermented dulcitol within one or two days (Table 86) whereas 98.3 per cent of isolates of the latter produced acid rapidly from this substrate. Similarly inositol is useful in the differentiation of cultures of these two species since 42.8 per cent of the strains of *S. enteritidis* fermented inositol. Thus a strain that utilized inositol would not be expected to be *S. cholerae-suis.* As was pointed out many years ago by Krumweide et al., 1916, 1918; Jordan et al., 1917; and by Koser, 1921 (see Bruner and Edwards, 1940, for references), arabinose and trehalose media are of particular value in differentiating members of the two species mentioned above. Neither of these substances is fermented by cultures of *S. cholerae-suis* while 99.3 per cent of the strains of *S. enteritidis* examined utilized arabinose and all of them fermented trehalose (Table 86). Cellobiose and erythritol media also appeared to be useful. Cultures of *S. cholerae-suis* failed to ferment the former, but the majority of strains produced delayed weak reactions in the latter medium. Conversely, the strains of *S. enteritidis* tested produced acid slowly from cellobiose but with rare exceptions failed to ferment erythritol. Likewise, sodium acetate agar medium appeared to be of considerable value (Table 86). Only four isolants of *S. cholerae-suis* grew on this medium and in these instances the growth was scanty and did not appear until after five to seven days of incubation. In contrast, the cultures of *S. enteritidis* tested grew well and produced strong reactions in sodium acetate medium, as follows: 79.9 per cent + on the first day; 17.4 per cent + on the second day; 0.7 per cent + on the third day; and 1.3 per cent + after 5 to 7 days. Only two of

the 299 strains of *S. enteritidis* failed to grow on this medium. Sodium mucate, Stern's glycerol fuchsin medium, and the organic acid media of Kauffmann and Petersen, 1956, (especially sodium citrate and D-tartrate) also were useful in the differentiation of *S. cholerae-suis* and *S. enteritidis.*

The biochemical reactions of representative cultures of *S. enteritidis* bioser Paratyphi-A are given in Table 87. Although this important microorganism no longer is common in the United States, it is seen occasionally and bacteriologists should be able to recognize it and make a report as soon as possible, as would be done with a culture of *S. typhi.* Also, bioser Paratyphi-A strains occur more frequently in other parts of the world than in the United States. For these reasons and because some of the biochemical reactions of this bioserotype are slightly atypical its reactions are listed separately in Table 87.

The author's reasons for listing separately the biochemical reactions given by representative strains of *S. enteritidis* bioser Pullorum and bioser Gallinarum (Table 88) are analogous to those given in the preceding paragraph. The microorganisms are of particular importance to veterinary bacteriologists, but those working with materials from humans also should be familiar with them. This is especially true of bioser Pullorum. Larger numbers of cultures of bioser Pullorum and bioser Gallinarum were studied by Trabulsi and Edwards (1962) and by Costin et al. (1964). These investigators listed tests that were of particular value in the differentiation of these two bioserotypes, namely production of gas from glucose and reactions in maltose, dulcitol, Jordan's tartrate medium, mucate, ornithine, and cysteine gelatin medium. From the data given in Table 88 it appeared that tests for fermentation of cellobiose and glycerol, and the organic acid media of Kauffmann and Peterson (1956) also were useful for this purpose.

In a separate investigation of the biochemical reactions given by members of the genus *Salmonella,* Martin et al. (1969) reported the results obtained in the examination of 562 cultures. These were selected from a total of 876 received for serologic identification during a three-month period in 1968. In this study cultures were examined as received each day except that when groups of cultures from a single focus of infection were received only representative isolates from each such group were examined. Of the 562 cultures, 518 (92.2 per cent) were members of serogroups A to E and G of *Salmonella.* This figure is in good agreement with data reported by Edwards (1962), Ewing and Ball (1966), and Kelterborn (1967). However, 71 (12.6 per cent) of the 562 cultures belonged to 41 serotypes that usually occur uncom-

TABLE 87

Biochemical Reactions of *S. enteritidis* Bioserotype Paratyphi-A[1]

Substrate or test	Bioser Paratyphi-A			*S. enteritidis*		
	Sign	%+	(%+)*	Sign	%+	(%+)*
Hydrogen sulfide (TSI)	– or +^{W}	12.5		+	98	
Citrate (Simmons')	– or (+)	0	(25)	+	99.3	(0.7)
Lysine decarboxylase	–	0		+	99.7	
Jordan's tartrate	–	0		+	92.5	
Inositol	–	0		d	42.8	(1)
Xylose	–	0		+	99	
Cellobiose	d	12.5	(6,2)	(+)	5	(92.8)
Glycerol	(+)	0	(100)	d	5.7	(7.2)
Stern's glycerol fuchsin medium	–	0		+	98.2	(0.6)
Sodium acetate	–		(6.2)	+	92.4	(2.2)
Mucate	–	0		+ or (+)	88.3	(1.7)
Organic Acids **						
citrate	–	0		+	96	(4)
D-tartrate	–	0		+	92.3	(4)
i -tartrate	–	0		d	4.7	(57.5)
l -tartrate	–	0		d	11.8	(75.2)

[1] Adapted from Ewing and Ball (1966).

*Figures in parentheses indicate percentages of delayed reactions (3 or more days).
**Method of Kauffmann and Petersen, (1956).

NOTE: Bioser Paratyphi-A does not occur commonly in the United States, but investigators should be able to recognize it.

+ Positive within 1 or 2 days' incubation (90% or more).
(+) Positive reaction after 3 or more days.
– No reaction (90% or more).
+ or – Majority of strains positive, some cultures negative.
– or + Majority of cultures negative, some strains positive.
(+) or + Majority of reactions delayed, some occur within 1 or 2 days.
d Different reactions: +, (+), –.
w Weakly positive reaction.

monly and would not be expected to be encountered frequently in most laboratories. Nevertheless, the results of this investigation confirmed those reported by Ewing and Ball (1966), i.e., the *Salmonella* are remarkably consistent as regards their biochemical reactions. This observation applied particularly to the commonly occurring serotypes, but it also applied to the majority of cultures of the less common serotypes encountered during the investigation.

The above-mentioned work (Martin et al., 1969) also assisted in the selection of tests and substrates that are of particular value in the differentiation of the species *Salmonella enteritidis, Arizona hinshawii,* and *Citrobacter freundii.* These are summarized in Table

TABLE 88

Differentiation of *S. enteritidis* bioserotypes Pullorum and Gallinarum[1]

Substrate or test	Bioser Pullorum			Bioser Gallinarum		
	Sign	%+	(%+)*	Sign	%+	(%+)*
Jordan's tartrate	–	0		+	100	
Ornithine decarboxylase	+	100		–	0	
Mucate	–	0		+	90.3	
Gas from glucose	+	95.1		–	0	
Dulcitol	–	0		+	99	(1)
Maltose	– or (+)		(35)	+	98.1	(1.9)
Cellobiose	–	0		d	60	(30)
Glycerol	–	0		(+)		(90)
Cysteine-gelatin	–	0		+	98.1	(1.6)
Organic acids **						
citrate	–	0		– or (+)		(40)
D-tartrate	–	0		+	92.2	(3.9)
i -tartrate	–	0		(+)		(100)
l -tartrate	–	0		(+) or –		(80)

[1] Modified from Trabulsi and Edwards (1962) and Ewing and Ball (1966).

*Figures in parentheses indicate percentages of delayed reactions (3 days or more).
**Method of Kauffman and Petersen, 1956.

+ Positive within 1 or 2 days' incubation (90% or more).
– No reaction (90% or more).
+ or – Majority of strains positive, occasional cultures negative.
– or + Majority of cultures negative, occasional strains positive.
(+) or – Majority of reactions delayed, some occur within 1 or 2 days.
d Different reactions: +, (+), –.

89. Readers also are referred to Table 12 (Chapter 3) and to Chapters 10 and 11 for additional information on the identification of members of the genera *Salmonella*, *Arizona*, and *Citrobacter*.

Differentiation of Certain Antigenically Related Salmonellae

Several species and bioserotypes of *Salmonella* are closely related antigenically, and for accurate characterization of these a few substrates must be used in addition to those normally employed in many laboratories. These microorganisms are listed in Table 90 together with their antigenic formulas. Parenthetically, bioser Paratyphi-A is included because of its impor-

tance when it occurs, not because of antigenic relationships. The biochemical reactions obtained with cultures that belonged to the species and bioserotypes listed in Table 90 are summarized in Tables 91, 92, and 93 inclusive. With the exception of bioser Paratyphi-A (Table 91), the bacteria are arranged in the tables according to their antigenic relationships. The percentages of positive and positive delayed reactions are included in the tables for all of the salmonellae tested except bioser Decatur and bioser Sendai. In these instances the numbers of strains available were two small for percentage data to be meaningful.

The data presented in Tables 91, 92, and 93 are clear and little comment is needed. The majority of

TABLE 89

Differentiation of *Salmonella*, *Arizona*, and *Citrobacter*

Test or substrate	*Salmonella enteritidis*[1]			*Arizona hinshawii*[2]			*Citrobacter freundii*[3]		
	Sign	%+	(%+)*	Sign	%+	(%+)*	Sign	%+	(%+)*
Urease	–	0		–	0		d	69.4	(6.9)
KCN	–	0.1	(0.1)	–	8.7		+	96.2	(0.9)
Gelatin (22C)	–	0.4[4]		(+)		(92)	–		(0.9)
Lysine decarboxylase	+	94.4	(0.1)	+	100		–	0	
Ornithine decarboxylase	+	100		+	100		d	17.2	(0.2)
Lactose	–	0.3		d	61.3	(16.7)	(+) or +	39.3	(50.9)
Sucrose	–	0.2		–	4.7		d	15.3	(9.4)
Dulcitol	+	97.7		–	0		d	59.8	(0.7)
Inositol	d	43.8	(2)	–	0		–	3.3	(1.9)
Malonate	–	1	(0.1)	+	92.6	(0.7)	d	21.8	(0.7)
Jordan's tartrate	d	84.2	(1)	–	5.3	(19.3)[5]	+	100	
Beta galactosidase	–	2.1		+	92.8		+ or –	74.4	
D-tartrate **	+	92.3	(4)	(+) or –		(83.3)	(+)		(90.9)

+   90% or more positive within 1 or 2 days.

–   90% or more, no reaction.

(+)   Delayed positive 3 or more days.

d   Different biochemical reactions, +, (+), –.

+ or – Majority of strains positive, some cultures negative.

– or + Majority of cultures negative, some strains positive.

*Figures in parentheses indicate percentages of delayed reactions (3 or more days).

[1] Summary of reactions given by 787 cultures: Ewing and Ball, 1966 (tables 8 and 9) and Martin et al. (1969).

[2] Adapted from Ewing and Fife, (1966) (specific epithet emended).

[3] Adapted from Davis and Ewing (1966).

[4] Two cultures among 526 strains examined by Martin et al. (1969).

[5] Cognizance should not be given to delayed reactions. Final readings of reactions in Jordan's tartrate should be made at 48 hours.

**Method of Kauffmann and Petersen (1956).

cultures that yielded positive reactions on a given substrate did so after 1 or 2 days of incubation, usually within 24 hr. With very few exceptions delayed reactions occurred within 3 to 7 days. Gas volumes produced by cultures of bioser Paratyphi-A were comparatively small (a bubble to 5 or 10%), and production of hydrogen sulfide was weak when it occurred. Cultures of bioser Typhisuis, as well as bioser Paratyphi-A, failed to decarboxylate lysine. These two are the notable exceptions to the rule that salmonellae decarboxylate this amino acid. Cultures of *S. typhi* did not decarboxylate ornithine. This also is exceptional (along with bioser Gallinarum) as reported earlier, e.g., see Ewing et al., 1960. Since most salmonellae ferment mannitol, this substrate is not listed. However, bioser Typhisuis is an exception, as noted by White (1926, p.

TABLE 90

Antigenic Formulas of Certain Salmonellae

Microorganism	Antigenic formula
S. enteritidis bioser Paratyphi-A	1,2,12:a:-
S. enteritidis ser Paratyphi-B	1,4,5,12:b:1,2
S. enteritidis ser Paratyphi-B, bioser Java	1,4,5,12:b:[1,2]
S. cholerae-suis	6,7:[c]:1,5
S. enteritidis bioser Paratyphi-C	6,7:[Vi]:c:1,5
S. enteritidis bioser Decatur	6,7:c:1,5
S. enteritidis bioser Typhisuis	6,7:c:1,5
S. typhi	9,12:Vi:d:-
S. enteritidis bioser Sendai	1,9,12:a:1,5
S. enteritdis bioser Miami	1,9,12:a:1,5

27, Glaesser-Voldagsen type). None of the twelve strains tested utilized mannitol. Therefore, this substrate is of value in the differentiation of bioser Typhisuis. Further, isolants of this bioserotype failed to produce hydrogen sulfide in triple sugar iron agar within 24 hours, but all strains produced trace to weakly positive reactions at 48 hours (Table 92). With very few exceptions the microorganisms listed in Tables 91, 92 and 93 utilized D-tartrate in Jordan's medium within 48 hours.

Many publications dealing with the biochemical reactions of salmonellae have appeared since the original isolation of *S. typhi* and *S. cholerae-suis*. Persons interested in pursuing this subject further are advised to consult Winslow et al. (1919), in which publication much of the early work is reviewed, and the bibliographies in the publications cited in the foregoing paragraphs.

SEROLOGICAL IDENTIFICATION OF *SALMONELLA*

Of the microorganisms now included in the genus *Salmonella*, the first to be described was *S. typhi* (Eberth, 1880; Gaffky, 1884). Soon after, Salmon and Smith (1885) isolated *S. cholerae-suis*, and this discovery was followed by the isolation and description of a number of other microorganisms belonging to the genus (e.g., *S. enteritidis*, Gaertner, 1888). Isolation of closely related bacteria from diverse sources inevitably led to confusion. In their discussion of a bacterium (probably ser Typhimurium), Smith and Stewart (1897) considered this confused situation and stated that the microorganisms "... belong to one great group (or species) in virtue of the identity of their morphological and biological characters." This statement fairly accurately described the state of the knowledge of the genus until Schuetze (1921) published a short paper which gave an insight into the numerous serotypes within the genus and the possibility of distinguishing them by using absorbed antisera. However, it was not until the appearance of the work of White (1925, 1926), who recognized the necessity of considering important discoveries concerning antigenic variation such as those of Andrewes (1922) in differentiation of the bacteria, that serological classification of salmonellae was placed upon a firm basis. This work was confirmed by Kauffmann (e.g. 1930, 1941), who modified, systematized, and greatly extended White's work to form the basis for the present serological classification within the genus. Thus rapid and accurate identification of serotypes was made possible. (See also Chapter 9)

VARIATIONAL PHENOMENA THAT AFFECT SEROTYPES OF *SALMONELLA*

Since successful serological classification of *Salmonella* is dependent upon the understanding and recognition of the variational changes to which the organisms are subject, it is necessary to enumerate and discuss these phenomena before proceeding with a description of the details of serological identification of salmonellae. Several of these variational phenomena affect members of many or all of the genera of Enterobacteriaceae and for this reason they were discussed in Chapter 4. Those

TABLE 91

Differentiation of Certain Bioserotypes of *Salmonella*

Test or substrate	*S. enteritidis*								
	bioser Paratyphi-A			ser Paratyphi-B			bioser Java		
	Sign	%+	(%+)*	Sign	%+	(%+)*	Sign	%+	(%+)*
Hydrogen sulfide (TSIA)	–	10^W		+	100		+	100	
Simmons' citrate	– or (+)		(24)	+	95		+	100	
Stern's glycerol-fuchsin medium	–	0		+ or –	80		+	95	
Lysine decarboxylase	–	0		+	95	(5)	+	100	
Arginine dihydrolase	(+) or +	10	(81)	+	90	(5)	+ or (+)	85	(15)
Ornithine decarboxylase	+	100		+	95		+	100	
Mucate	–	0		+	90		+ or –	80	
Jordan's D-tartrate	–	0		–	0		+	95	
Gas production	+†	100		+	100		+	100	
Arabinose	+	100		+	100		+	100	
Trehalose	+	100		+	100		+	100	
Dulcitol	+ or (+)	86	(14)	+	100		+	100	
Xylose	–	0		+	90	(5)	+	100	
Sorbitol	+	100		+	100		+	100	
Organic acid media									
Citrate	–	0		+	100		+	100	
D-tartrate	–	0		–	0		+	100	
l -tartrate	–	0		+	95	(5)	+	100	
i -tartrate	–	0		+ or (+)	50	(45)	(+) or +	40	(60)

NOTE: In tables 91-93, symbols represent the following:

  + Positive within one or two days' incubation.
(+) Positive reaction after 3 or more days.
 – No reaction.
+ or – Majority of strains positive, occasional cultures negative.
– or + Majority of cultures negative, occasional strains positive.
(+) or + Majority of reactions delayed, some occur within 1 or 2 days.
 d Different reactions: +, (+), –.
^{W} Weakly positive reaction.

 * Figures in parentheses indicate percentages of delayed reactions (3 or more days).
 † Volumes of gas produced by bioser Paratyphi-A were relatively small (usually a bubble to 5 or 10%).

phenomena that were mentioned in Chapter 4 will be enumerated but not discussed further in this chapter. **HO-O Variation.** Loss variation in motile (HO) cultures that gives rise to nonmotile O variants was discussed in Chapter 4. In serologic examination of strains of salmonellae it is absolutely essential to distinguish between H (flagellar) antigens and O (heat stable somatic) antigens and to distinguish between the reactions of their corresponding agglutinins. This can be done without difficulty (v. inf.).

S-R Variations. Smooth (S) to rough (R) variations were mentioned in some detail in Chapter 4. It is obvious that only smooth forms should be used to prepare O antisera since antisera derived from rough bac-

TABLE 92

Differentiation of *S. cholerae-suis* and Certain Bioserotypes of Salmonellae

Test or substrate	S. cholerae-suis			S. enteritidis							
				ser Paratyphi-C			ser Decatur		ser Typhisuis		
	Sign	%+	(%+)*	Sign	%+	(%+)*	Sign	No.†	Sign	%+	(%+)*
Hydrogen sulfide (TSIA)	d	60	(10)	+	100		+	3/0	+^{W}	92	(8)
Simmons' citrate	+ or (+)	60	(30)	+	100		+	3/0	–	0	
Stern's medium	–	0		–	0		+	3/0	–	0	
Lysine decarboxylase	+	95		+	100		+	3/0	–	0	
Arginine dihydrolase	(+)		(95)	+ or –	65		+	3/0	(+)		(92)
Ornithine decarboxylase	+	100		+	100		+	3/0	+	100	
Mucate	–	0		–	0		+	3/0	–	0	
Jordan's D-tartrate	+ or –	85		+	100		+	3/0	–	0	
Gas production	+	100		+	100		+	3/0	+	100	
Arabinose	–	0		+	94		+	3/0	+	100	
Trehalose	–	0		(+) or +	41	(59)	+	3/0	+	92	(8)
Dulcitol	d	5	(25)	+	100		+	3/0	(+)		(100)
Xylose	+	100		+	100		+	3/0	+	100	
Sorbitol	+ or (+)	85	(15)	+	100		+	3/0	–	0	
Organic acid media											
Citrate	+	90	(5)	+	100		(+)	(3)/0	–	0	
D-tartrate	+	95		+	100		+	3/0	–	0	
l -tartrate	(+) or –		(85)	– or (+)		(24)	(+)	(3)/0	–	0	
i -tartrate	– or (+)		(35)	–	0		–	3/0	–	0	

NOTE: See note to table 91 for key to symbols.

  * Figures in parentheses indicate percentage of delayed reactions (3 or more days).

  †Numerator indicates the number of strains that gave positive reactions; denominator indicates the number of strains that were negative.

teria would have a tendency to agglutinate many rough cultures regardless of their identity. Satisfactory antigens for agglutination tests often can be prepared from autoagglutinable bacteria by the method of White (1926), who suspended the growth from agar cultures in ethyl alcohol (96% or absolute), heated the suspensions at 60 C for 1 hour, sedimented the cells by centrifugation, and resuspended them in sodium chloride solution. Very often stable suspensions may be obtained by this method when ordinary saline suspensions of the bacteria agglutinate spontaneously. Suspensions treated in this way often react to some degree with antiserum prepared from smooth races, so that O antigens of the rough strain can be identified. In other instances, the antigens so prepared react only with antisera derived from rough races. In these cases it is difficult or impossible to identify the O antigens of the cultures. However, if a culture is not completely and irrevocably rough, its O antigens frequently may be determined by means of immunogel diffusion techniques (Grados and Ewing, unpublished data).

T Forms and T Antigens. Kauffmann (1956, 1957) described somatic antigens which were called T antigens, the designation being derived from the word *transient.* The first of these, T_1 was found originally in ser Paratyphi-B and ser Typhimurium. The second T antigen, T_2 was found in ser Bareilly. In all the cultures

TABLE 93

Differentiation of *S. typhi* and Certain Bioserotypes of Salmonellae

| | *S. typhi* | | | *S. enteritidis* | | | | | | | | |
| | | | | bioser Sendai | | bioserotypes of Miami | | | | | | |
Test or substrate	Sign	%+	(%+)*	Sign	No.†	Sign	%+	(%+)*	Sign	%+	(%+)*
Hydrogen sulfide (TSIA)	$+^W$	95		–	0/4	+	100		+	100	
Simmons' citrate	–		(5^W)	–	0/4	+	100		+	100	
Stern's medium	–	0		–	0/4	+	100		+	100	
Lysine decarboxylase	+	100		d	1,(2)/1	+	100		+	100	
Arginine dihydrolase	d	2	(46)	(+)	(4/0)	+ or (+)	57	(43)	(+)		(100)
Ornithine decarboxylase	–	0		+	4/0	+	100		+	100	
Mucate	–	0		–	0/4	–	0		+	100	
Jordan's D-tartrate	+	92		–	0/4	(+) or +	5	(90)	+	100	
Gas production	–	0		+	4/0	+	100		+	100	
Arabinose	–		(6)	+	4/0	–	0		+	100	
Trehalose	+	100		+	4/0	+	100		+	100	
Dulcitol	– or (+)		(35)	(+) or –	(2)/2	–	0		+	100	
Xylose	+	94	(6)	– or (+)	(1)/3	+	100		+	100	
Sorbitol	+	100		+	4/0	+	100		+	100	
Organic acid media											
Citrate	+	100		–	0/4	+	100		+ or (+)	94	(6)
D-tartrate	(+) or +	25	(69)	–	0/4	(+) or –	0	(71)	+	94	(6)
l -tartrate	–	0		–	0/4	–	0		(+) or –		(78)
i -tartrate	–	0		–	0/4	–	0		–	0	

NOTE: See note to table 91 for key to symbols.

  * Figures in parentheses indicate percentage of delayed reactions (3 or more days).

  † Numerator indicates the number of strains that gave positive reactions; denominator indicates the number of strains that were negative.

From Ewing, Ball, Bartes, and McWhorter. 1970. The biochemical reactions of certain species of bioserotypes of *Salmonella*. *Journal of Infectious Diseases,* 121 (3): 288-194.© 1970 by the University of Chicago. All rights reserved. Reprinted with permission.

that contained the T antigens, some rough colonies could be detected but the T forms themselves were morphologically smooth. They did not react with available O antisera used in serotyping nor with antisera derived from R forms. Antisera prepared from antigens T_1 and T_2 each agglutinated its homologous antigen in high dilutions but failed to agglutinate the other or to react with S or R forms. An indication of the true nature of the T antigens is the fact that while no cross agglutination was observed between T and R forms, absorption of T antisera by R forms or R antisera by T forms brought about a marked reduction in titer. No such reduction in titer was observed when S forms and S sera were used in similar experiments.

In the author's experience, colonies of T forms are morphologically smooth; suspensions in saline are stable, do not become agglutinable upon heating at 100 C, and are not agglutinable by trypaflavine. Since they are not readily distinguished by their morphological and physical properties, it is desirable to produce antisera for their detection. These are prepared in the same manner as O antisera. T_1 forms have been found in certain serotypes within serogroups B, E_1, E_4, and G. The second form (T_2) has been found only in ser Bareilly.

V-W (Vi-O) Variation. It has been known for many years that recently isolated cultures of *S. typhi* are likely to be inagglutinable in O antisera. The reason for this was unknown until Felix and Pitt (1934, 1934a) demonstrated that O inagglutinable cultures possessed a special antigen which they called virulence or Vi antigen. They found that cultures possessing this antigen were more virulent for mice than those lacking it. Further, mice were not protected against Vi cultures by vaccination with O forms, although vaccine produced with Vi cultures gave a high degree of resistance to the virulent strains. It is now generally accepted that Vi antigen is not necessarily associated with virulence since Kauffmann (1936) demonstrated that the presence of Vi in cultures ser Paratyphi-C did not parallel the virulence of the cultures for mice (see also Chapter 4). Further, mice are not suitable animals for testing the virulence of *S. typhi* since overwhelming doses of the bacteria must be administered to produce death. The results of Felix and Pitt probably are attributable to an added toxicity in Vi cultures. This conclusion is supported by the results of Archer and Whitby (1967) who studied the action of ser Paratyphi-C in mice vaccinated with suspensions containing O antigen, suspensions containing Vi antigen, and suspensions containing both O and Vi antigens. It was found that the vaccines containing Vi antigen protected against massive doses which produced acute intoxication, but not against smaller doses which produced an infection. Conversely, O vaccines did not protect against intoxication but were just as effective as Vi + O vaccines in protecting mice against infection. Nevertheless, Vi antigen remains an important constituent of *S. typhi* and the recognition of this antigen often aids in the identification of the organism (v. inf.). It also conditions the susceptibility of cultures to the Vi bacteriophages used in phage typing of *S. typhi.*

The agglutinability of Vi strains changes markedly when suspensions are heated in physiological saline solution. After boiling, the organisms no longer are agglutinated in pure Vi antiserum but are agglutinable in O antiserum. The length of time that Vi cultures must be heated to inactivate their Vi agglutinability and render them fully O agglutinable varies with different strains. It probably is roughly proportional to the amount of Vi antigen produced by individual strains and to the density of the suspensions (Spaun, 1952). While heat treatment of Vi cultures alters the manner in which they react in Vi and O antisera, it actually does not destroy the Vi antigen. Stuart and Kennedy (1943) demonstrated that the supernatant fluid from heated Vi suspensions contained Vi antigen which was precipitable by Vi antiserum and was capable of producing antibodies in rabbits. Spaun (1951) sensitized erythrocytes with solutions of Vi

antigen and rendered them specifically agglutinable. Landy (1952), Webster et al. (1951), and Landy and Webster (1952) demonstrated that Vi antigen could be purified by chemical methods (see also Chapter 4). After purification the antigen could be absorbed on collodion particles which then reacted specifically with Vi antisera. The purified antigen also reacted specifically in precipitin and complement fixation tests and was antigenic in mice, rabbits, and man, engendering agglutinins, precipitins, and protective antibodies.

The Vi antigen separates readily from cells when suspensions in physiological saline solution are heated. However, when cells are dried by alcohol or acetone treatment and heated in the dry state their agglutinative properties are unaffected. Peluffo (1941) found that the reactions of cells treated with absolute alcohol or acetone were unchanged and, after being dried, cultures could be heated at 150 C for 1 hr without changing their reactions in Vi and O antisera. Vi antigen may be regarded as one of a group of envelope or sheath antigens, of which many occur among Enterobacteriaceae, and which are designated by the general term K antigens. See Chapters 4 and 5.

Felix and Pitt (loc. cit.) reported that cultures maintained in the usual manner rapidly lost Vi antigen and reverted to the O agglutinable form. Craigie and Brandon (1936) and Giovanardi (1938) noted that colonies that contained Vi antigen differed in morphology from those in which it was lacking. The Vi colonies were slightly more opaque, and when viewed by reflected light presented a ground-glass appearance. Kauffmann (1935) called those colonies that possessed Vi antigen, V colonies, and those that lacked it, W colonies. V colonies were agglutinated by Vi antiserum, but not by pure O antiserum, while W colonies were agglutinated by O antiserum, but not by pure Vi antiserum. He also found that on continued transfer each form gave rise to the other, and he named this variation V-W variation. Later Kauffmann (1936, 1941) found Vi antigen in ser Paratyphi-C and in certain cultures which were called *S. coli*. Kauffmann and Moeller (1940) also described its presence in a new serotype which they labeled *Salmonella ballerup*. These *S. coli* cultures later were identified as members of the genus *Citrobacter*. Later, *S. ballerup* was removed from the antigenic schema. It too, is *Citrobacter freundii.*

The agglutinating properties of Vi antigen and its power to inhibit O agglutination are lost gradually in aqueous suspension. This change is greatly accelerated in phenolized saline solution. That is, a Vi+ culture of *S. typhi* suspended in phenolized physiological saline solution ceases to agglutinate in Vi antiserum and becomes agglutinable in O antiserum after standing for a few hours at room temperature. For this reason, a culture suspected of being *S. typhi* should be suspended in physiological saline solution and examined immediately by slide agglutination (v. inf.)

The recognition of Vi antigen in cultures of *S. typhi* is of the greatest importance. Usually freshly isolated cultures are poorly motile so that they are not flocculated by H antiserum. In such cultures the recognition of Vi antigen aids in distinguishing them from anaerogenic cultures of other serotypes within serogroup D of the genus *Salmonella,* the members of which possess the same O antigens as *S. typhi.* In addition, an occasional culture may be found which is rough when isolated, or becomes rough very shortly thereafter. When such strains are nonmotile or poorly motile recognition of Vi antigen must be relied upon for their serologic identification, since the O antigens may be so changed as to be unrecognizable.

When a Vi strain is encountered and it is wished to determine the O antigens, it should be suspended in physiological saline solution and the tube placed in a boiling water bath for 10 to 20 min. The suspension may then be examined by slide agglutination with O antisera (v. inf.).

M-N Variation. This term was used by Kauffmann (1936) to denote variation between mucoid (M) and nonmucoid (N) forms of salmonellae. It long has been known that occasional cultures of *Salmonella* have a tendency to produce moist, glistening, domed colonies which often were mucoid and resembled colonies of *Klebsiella.* This form of growth is most likely to occur in freshly isolated cultures, and its natural occurrence has been observed more often in ser Paratyphi-B than in other types, particularly when cultures are incubated at about 22 C. Freshly isolated cultures of ser Paratyphi-B, when incubated overnight at 37 C and then left for 1 or 2 days at 22 C, develop a raised, moist rim around the circumference of the colonies, the so-called slime-wall frequently mentioned in the older German literature. See also Chapter 4.

Kauffmann (1935, 1936) demonstrated a special antigen (designated M) in mucoid cultures. Mucoid forms are not agglutinated by O antisera but if living mucoid cultures are injected into rabbits, agglutinins for the M form are produced. M agglutinins usually are low in titer and produce an agglutination which is firm and dislike in character. It is possible also to produce M agglutinins by injection of formalinized cultures and cultures which have been heated at 60 C for one hr. On the contrary, Kauffmann found that mucoid cultures which had been heated at 100 C for 2 hr, or treated with 96% alcohol or N/1 HCl for 20 hr at 37 C did not produce M agglutinins. By absorption of an anti-M serum with a nonmucoid form of the serotype with which the antiserum was produced, it is possible to produce a pure anti-M serum which agglutinates mucoid forms but not normal, nonmucoid cultures (Kauffmann, 1935, 1936).

The M antigen of all *Salmonella* in which it has been examined apparently is the same although the organisms which produce it belong to a variety of serotypes and to different O antigen serogroups. It is of some interest that the M antigen of salmonellae is rather strongly related to the capsular antigen of *Klebsiella* type 13 and, lacking an antiserum for the M of *Salmonella,* mucoid salmonellae may be tested with antiserum for *Klebsiella* type 13 to demonstrate the M antigen. This procedure has been used successfully by the writer.

Definite capsules can be demonstrated in some, but not all, mucoid strains of salmonellae. When present, they easily are demonstrable in moist India ink preparations. Capsulated strains give definite quellung (capsular) reactions with anti-M serum or with antiserum for *Klebsiella* type 13. From the foregoing brief description it is apparent that the M antigen of salmonellae resembles the sheath, envelope, or capsular antigens of other Enterobacteriaceae. For descriptions of these antigens the reader is referred to chapters on the genera *Shigella, Escherichia,* and *Klebsiella* and to Chapter 4.

Since mucoid forms are inagglutinable in O antisera, the identification of their O antigens, and hence of their serotype, presents a problem. As noted by Kauffmann, O-inagglutinability is inactivated by heating at 100 C for 1 to 2 hr. If one is dealing with a perfectly smooth strain which has no tendency to become autoagglutinable when heated, such a procedure is satisfactory. However, in the experience of the writer, production of M antigen often occurs in cultures that are not completely smooth and which become autoagglutinable when heated. In such instances it is necessary to resort to another method. Alcohol treatment of mucoid strains often results in the production of a gummy, rubbery mass and resuspension in physiological saline solution is difficult. A method which usually is successful is passage of the organisms through one or more tubes of semisolid agar medium as explained in the section on preparation of H antisera. Mucoid organisms usually are nonmotile or are sluggishly motile. In the growth of the culture nonmucoid forms appear and these usually are more actively motile and migrate through the medium more rapidly than do the mucoid forms. Hence, one or more passages through semisolid agar medium usually results in the isolation of a motile, nonmucoid form that presents no difficulty in the determination of its O and H antigens.[2] Also, in the author's experience, cultures so treated are likely to have less tendency toward roughness and this is an added advantage in antigenic analysis.

Nonfimbriate to Fimbriate Variation. This subject was mentioned in some detail in Chapter 4, and only a few comments are needed here to reemphasize the importance of this variation in the production of antisera. It is clear that fimbrial agglutinins are unlikely to be stimulated by the injection of the heated suspensions used in the production of O antisera for *Salmonella*, particularly when the antigens are prepared by the method of Roschka (1950). On the contrary, it is possible that such agglutinins might be formed through the injection of unheated, formalinized broth cultures in the production of H antisera. Taylor (personal communication, 1960) observed very high titred agglutinins (1 to 256,000) in certain H antisera which were thought to be antifimbrial or "X" agglutinins. Thus, it becomes even more important to titrate all antisera carefully against homologous and heterologous antigens in order to assure that antifimbrial agglutinins which might interfere with serologic identification are not present.

Antigenic Changes Induced by Bacteriophage. This subject is mentioned in Chapter 4 under the heading of Genetic Recombinations That May Affect Serological Typing. Of the mechanisms mentioned there, lysogenic conversion probably has the greatest effect on serotypic identity and epidemiology. Since certain O antigen factors of salmonellae are controlled by lysogenic conversion, the presence or absence of these particular O antigens in certain serotypes is not necessarily of epidemiological significance. For example, ser Anatum and ser Newington were present in an outbreak of disease from which both serotypes first were isolated in 1920. These two often have been found in association and now it is apparent that the factor that determines their identity is infection of ser Newington by phage epsilon 15. However, it is notable that these two serotypes have been found separately in outbreaks more often than in association. Similarly, the presence or absence of O antigens 1, or 14, or 27 is not necessarily of epidemiological importance. The reader is referred to Chapter 4 for other examples of O

antigens that may be controlled by this mechanism and for references.

Form Variation. The O antigens of the genus *Salmonella* are designated by arabic numerals and certain minor O antigens of the genus overlap; that is, they occur in more than one of the serogroups into which the bacteria are divided on the basis of their major O antigens. Among these overlapping O antigens are 1,6, and 12. Kauffmann (1940, 1941a) found that when certain serotypes of salmonellae that contain antigen 1 were plated, some colonies contained large amounts of 1 (++++) while others contained slight amounts of the antigen. Each form gave rise to the other. Likewise antigen 12 was found to be variable and certain colonies contained larger amounts of 12 than others. In this instance the situation was more complicated than with antigen 1, since Kauffmann subdivided 12 into 12_1, 12_2, and 12_3, and reported that only 12_2 was variable. The type of variation which involved 1 and 12 was designated as form variation. From the work of Levine and Frisch (1935) and Kauffmann (1937) it was known that antigen 6 of serogroups C and H was divisible into two portions, labeled 6_1 and 6_2, and 6_1, like 1 and 12_2, was subject to form variation.

While form variation is known to affect only the so-called minor antigens of the bacteria, it may exert a pronounced effect upon their behavior in agglutination tests. It often has a marked influence on diagnosis of infection by agglutination tests with patients' sera. Further, it is necessary to consider form variation in selecting strains and colonies for the production of O antisera, otherwise the antisera may not differentiate clearly between antigens prepared from the various serogroups.

The effect of form variation on cross agglutination between serogroups C_1 (6,7) and C_2 (6,8) is illustrated in Table 94. The reactions of 6,7 and 6,8 strains that have a large (++++) and small (+) amount of antigen 6_1 with antisera prepared from the respective forms are included in Table 94. It may seem that 6,7 organisms react in high dilution in antisera prepared from 6,7 strains without regard to 6_1 content. However, there is a marked difference in the reactions of the + and ++++ forms in 6,8 antiserum prepared from 6_1++++ forms. Similar reactions are apparent in 6,8 strains which contain large and small amounts of 6_1. From a study of the table it readily can be seen that if antisera for serogroups C_1 and C_2 were prepared from forms which contained large amounts of antigen 6_1, it would be impossible to use them without absorption. On the contrary, use of forms which contain only a *minimal* amount of this antigen results in antisera which are quite specific. Thus, while form variation affects only

---

[2]This method also is helpful for the separation of a serotype of *Salmonella* from a microorganism such as *Pseudomonas* when the two occur in mixed culture. Salmonellae usually are more actively motile than pseudomonads and generally can be isolated in pure culture from the bottom of the tube of semisolid medium.

TABLE 94

Effect of Form Variation on O Agglutination

Antigen suspensions	Antisera			
	6,7 (6_1+)	6,7 (6_1++++)	6,8 (6_1+)	6,8 (6_1++++)
6,7 (6_1+)	1,000*	1,000	20	100
6,7 (6_1++++)	1,000	1,000	40	1,000
6,8 (6_1+)	20	40	1,000	2,000
6,8 (6_1++++)	100	400	1,000	2,000

* Figures indicate the highest dilution at which agglutination occurred.

those components referred to as minor O antigens, it may exert a pronounced effect on the determination of the serogroup to which a culture belongs.

Not only does form variation affect the common, or overlapping antigens which occur in more than one serogroup, but it also occurs in antigens which distinguish different subgroups within a single serogroup. For example, in serogroup G some cultures are characterized by antigens 13,22 while others possess antigens 13,23. The components 22 and 23 are subject to form variation and colonies of a single 13,22 culture may contain 22+ forms and 22++++ forms. Similarly, 13,23 cultures may yield 23+ and 23++++ colonies. Antigens 24 and 25 in O serogroup H also are subject to similar variation. In instances of this sort colonies that have _maximum_ amounts of antigens 22, 23, 24, and 25, respectively, should be selected for antiserum production. If this is done, the resultant antisera will yield stronger single factors after absorption and the subgroups to which unknown cultures belong can be determined more readily. While all the O antigens of the genus have not been examined intensively for form variation, it is probable that other antigens also are subject to form variation.

Thus, it may be seen that form variation may affect the specificity of O antigen grouping antisera or it may determine the effectiveness of antisera in distinguishing between subgroups within a given serogroup. In the case of antigens that are common to two or more serogroups, one should select forms which contain as _little_ of the variable antigen as possible so that grouping antisera will be more specific. In the case of antigens that distinguish subgroups within a given serogroup, one should select forms with _maximal_ development of the distinguishing antigens so that antisera will contain a strong agglutinin for them.

It should be emphasized that S-R, T, and V-W variations, and M-N and form variations affect only the somatic antigens of the bacteria and their reactions in the corresponding antisera. They have no effect on H antigens except as roughness affects the stability of bacterial suspensions and, therefore, their agglutination. _In contrast, phase variation, which is described below, influences only the H antigens and is without effect upon the O antigens._

Phase Variation. Since phase variation affects H antigens, it occurs only in flagellated cultures. It cannot properly be said to occur only in _motile_ cultures since certain nonmotile cultures may be flagellated (Kauffmann, 1941; Edwards et al. 1946). Ordinarily, however, phase variation is associated with flagellated, actively motile cultures. For many years it was known to occur only in salmonellae but Edwards et al. (1947) reported a number of cultures of _Arizona_ that displayed typical phase variation. The term phase has been employed in many ways with diverse connotations such as the phases of _Hemophilus pertussis_ or the growth phases of streptococci. Here, the term phase is used strictly in the sense of Andrewes (1922) and applies _only_ to the antigenic properties of the H components of the flagellated bacteria and their corresponding agglutinins.

For years it was recognized that many cultures of _Salmonella_ contain specific and nonspecific antigenic factors. The nonspecific antigens of different types were closely related and caused the types to cross agglutinate in such high dilution that simple agglutination could not be depended upon for differentiation. Agglutinin absorption was used to fix the identity of the various serotypes as they then were known, and while this method was fairly accurate, it was laborious and time-consuming. It was also subject to error since in its use the complex antigens were considered as a whole and no attempt was made to differentiate between H and O antigens. The classical work of Andrewes (1922, 1925) on phase variation explained many curious observations concerning variability in _Salmonella_, and, together with the observations of Weil and Felix on HO-O variation, served as a foundation for the work of White on classification by antigenic analysis.

Andrewes found that if two serotypes of _Salmonella_ with related H antigens were plated, certain colonies were agglutinated by antisera derived from both types, whereas other colonies were agglutinated only by antiserum derived from the homologous type. Furthermore, subsequent transfers derived from the colonies had a marked tendency to retain these properties but eventually reverted to the mixed characters of the whole culture from which the colonies were derived. The organisms that were agglutinated only by homologous antiserum were called the specific phase while colonies that were agglutinated by both homologous and heterologous antisera were designated as the nonspecific phase. The colonies of the two phases were identical in appearance and physical characters, and differed only in the serologic behavior of their H antigens.

An illustration of the agglutinative behavior of specific and nonspecific phases with antisera derived from homologous and heterologous types is given in Table 95. The antisera used in the illustration (Table 95) were prepared from whole cultures of the bacteria without selection. Hence they contain agglutinins for both phases of the culture from which they were derived. Both specific and nonspecific phases of ser Paratyphi-B were agglutinated in high dilution by antiserum for ser Paratyphi-B since agglutinins for both phases were present in large amounts. Also the nonspecific phase of ser Typhimurium was agglutinated in high dilution because the antigens of that phase were closely related to the nonspecific phase of ser Paratyphi-B. In marked contrast was the action of the antiserum on the specific phase of ser Typhimurium. Agglutination of that phase by ser Paratyphi-B antiserum was negligible since the antiserum contained no agglutinins for the phase. The slight agglutination which occurred (Table 95) was caused by traces of nonspecific components in the antigen suspension. In the case of ser Typhimurium antiserum the situation was reversed. Andrewes demonstrated that specific phases contained traces of nonspecific components and nonspecific phases carried small amounts of specific factors. Antisera produced from specific phases agglutinated the homologous serotype in high dilution but

had very low titers for the heterologous serotypes. Antisera derived from nonspecific phases agglutinated both homologous and heterologous serotypes in high dilution.

As further serotypes were studied it became apparent that the specific phases of Andrewes were not specific for one serotype only, but that they often occurred in other serotypes in combination with different O antigens or different nonspecific antigens. For this reason the Salmonella Subcommittee (1940) substituted the designations phase 1 and phase 2, respectively, for specific phase and nonspecific phase. This change was desirable because Kauffmann and Mitsui (1930) described another sort of phase variation which they called alpha-beta variation. This variation was in no way different from the phase variation of Andrewes except that the antigens found in phase 2 were not related to the nonspecific phases of Andrewes, but formed an agglutinative group peculiar to themselves. Edwards and Bruner (1938) described a kind of phase variation in which the antigens of both phase 1 and phase 2 are antigens that ordinarily occur in phase 1 of a number of serotypes of _Salmonella_. For example, ser Meleagridis (3,10:e,h:l,w) contains antigens in phase 1 and phase 2, both of which occur in other serotypes as phase 1 in combination with typical phase 2 antigens. Such combinations with l,w antigens later were found in a number of serotypes. Rarely, diphasic serotypes have been found in which antigens of both phases ordinarily occur as phase 2 components.

In antigenic analysis the major antigens are assigned symbols and the same antigen always is designated by the same symbol, even though it may appear in a number of different serotypes of _Salmonella_. No attempt is made to delineate the complete antigens of the serotypes; only those antigens that are of significance in identification are expressed in the schema. The O antigens of the bacteria are denoted by arabic numerals, the antigens of phase 1 by small roman letters and the antigens of the nonspecific phases of Andrewes by arabic numerals. The beta phases of Kauffmann and Mitsui, like the antigens of phase 1, are characterized by small roman letters even though they represent phase 2 of the bacteria. This anomalous situation is caused by the fact that the beta phases (e,n) are antigenically related to certain antigens found in phase 1 (e,h) and it was not until after the antigenic schema was established that organisms containing antigens e,n were found to be subject to phase variation. Formerly they were thought to be monophasic and after phase variation was demonstrated in these organisms the designations for the respective antigens remained the same.

In order to illustrate more clearly the application of

symbols to antigens, Tables 96 and 97 are included. It can be seen that the antigenic formula of ser Paratyphi-B is 4,5,12:b:1,2 and that the antigens in phase 1 are 4,512:b while those in phase 2 are 4,5,12:1,2. The antigenic formula of ser Abortusbovis is 1,4,12,27: b:e,n. The antigens of phase 1 are 1,4,12,27:b while those of phase 2 are 1,4,12,27:e,n. Only the H antigens are influenced by phase variation.

The agglutination reactions given in Table 96 are H reactions only, and do not reflect the related O antigens of the bacteria. If H agglutination only is considered, it is obvious that phase 1 of ser Paratyphi-B and phase 1 of ser Abortusbovis will cross agglutinate since each contains the antigen b. It is also clear that phase 2 of the bacteria would not cross agglutinate since one contains antigens 1,2 while the others contains antigens e,n. Phase 2 of ser Paratyphi-B cross agglutinates with phase 2 of ser Saintpaul, because each contains antigens 1,2. Phase 2 of ser Abortusbovis cross agglutinates with phase 2 of ser Sandiego because both contain antigens e,h. Phase 1 of ser Saintpaul and phase 1 of ser Sandiego cross agglutinate to a lesser degree with phase 2 of ser Sandiego since all contain the common minor antigen e. Phases containing antigens e,h and e,n can be differentiated with certainty by using absorbed, pure h and n antisera. Single factor h antiserum can be prepared by absorbing e,h antiserum with an e,n phase, thus removing the agglutinins for the common antigen e. Pure n antiserum is produced by absorption of e,n antiserum with an e,h phase. The agglutination reactions illustrated in Table 97 are typical of those obtained when actively motile cultures of diphasic salmonellae are plated, individual colonies picked to broth, and the resulting broth cultures formalinized and used as antigens in agglutination tests.

Not all serotypes of *Salmonella* are diphasic. Many are monophasic and in these every colony contains identical H antigens. This monophasic state is char-

TABLE 95

Agglutination of Specific and Nonspecific Phases in Antisera of Homologous and Heterologous Types
(Figures indicate highest dilution at which H agglutination occurred.)

H Antigens	Antisera for *S. enteritidis* serotypes:	
	Paratyphi-B	Typhimurium
S. enteritidis		
ser Paratyphi-B, specific phase	10,000	100
ser Paratyphi-B, nonspecific phase	10,000	5,000
ser Typhimurium, specific phase	100	10,000
ser Typhimurium, nonspecific phase	5,000	10,000

TABLE 96

Antigenic Formulas

Serotype	O Antigens	H Antigens	
		Phase 1	Phase 2
Paratyphi-B	4,5,12	b	1,2
Abortusbovis	1,4,12,27	b	e,n
Sandiego	4,5,12	e,h	e,n
Saintpaul	1,4,5,12	e,h	1,2

TABLE 97

Agglutination of Isolated Phases in Phase 1 and Phase 2 Antisera
(Figures indicate highest dilution at which H agglutination occurred.)

H Antigen suspensions	Antisera			
	Paratyphi-B Phase 1 (4,5,12:b)	Paratyphi-B Phase 2 (4,5,12:1,2)	Sandiego Phase 1 (4,5,12:e,h)	Sandiego Phase 2 (4,5,12:e,n)
Paratyphi-B Phase 1 (4,5,12:b)	20,000	500	0	0
Paratyphi-B Phase 2 (4,5,12:1,2)	500	20,000	0	0
Abortusbovis Phase 1 (1,4,12,27:b)	20,000	100	0	0
Abortusbovis Phase 2 (1,4,12,27:e,n)	0	0	2,000	20,000
Sandiego Phase 1 (4,5,12:e,h)	0	0	20,000	2,000
Sandiego Phase 2 (4,5,12:e,n)	0	0	2,000	20,000
Saintpaul Phase 1 (1,4,5,12:e,h)	0	0	20,000	2,000
Saintpaul Phase 2 (1,4,5,12:1,2)	0	20,000	0	0

acteristic of *S. enteritidis* ser Enteritidis (9,12:g,m) and of the majority of presently recognized monophasic serotypes containing antigens g or m. In all instances in which only one flagellar phase is recognized it is necessary to attempt the isolation of a second phase which may be temporarily suppressed in order to be certain that all such cultures are truly monophasic. Methods for isolation of suppressed phases are described below.

Irreversible Variation and the Occurrence of Multiple Phases. Since 1942 a number of salmonellae have been encountered in which complex flagellar antigen phases have been recognized. Some of these are diphasic while others are triphasic or quadriphasic. These complex forms may lose a major antigenic component spontaneously, in which case they become indistinguishable from well-known normal serotypes. Multiphasic forms may lose the ability to produce one or two phases spontaneously and revert to stable diphasic serotypes.

Further, diphasic cultures of serotypes that normally are monophasic have been found. The most commonly encountered example of the latter is ser Senftenberg (1,3,19:g,s,t:-). Cultures of this serotype that possess one or another of five different H antigens ($z_{27}, z_{37}, z_{43}, z_{45}$, or z_{46}) have been encountered repeatedly. The presence of one of these antigens obscures agglutination in g,s,t antiserum, but the antigen may be lost spontaneously and the cultures then are recognizable as typical strains of ser Senftenberg. The phases of some of the multiphasic forms may be reversed with ease (v. inf., Isolation of Phases, Phase Reversal). However, the loss of an antigenic component or of a phase from one of the complex forms referred to above almost invariably is permanent, and the variation therefore is irreversible. The importance of complex phases and of irreversible variations of the sort mentioned lies in the fact that cultures exhibiting these may cause confusion in serotyping, and unless the possibility of the occurrence of such strains is borne in mind, epidemiological relationships between certain strains may not be recognized. For example, a culture characterized as $1,3,19:z_{43}$ or $1,3,19:g,s,t:z_{43}$, isolated with other cultures characterized as $1,3,19:g,s,t$, may not be recognized for what it really is (a diphasic form of ser Senftenberg). Antigens z_{47}, z_{48}, and z_{49} are known to occur in other serotypes (Table 98).

TABLE 98

Examples of Complex Serotypes and Serotypes in Which Irreversible Variation is Known

Serotype	Designation	Reference
$4,12:d,a:d,1,7$	ser Arechavaleta, complex	McWhorter and Edwards, 1963
$4,12:m,t:z_{45}$	ser California, diphasic	Unpublished data
$4,12:d,e,h:d,e,n,z_{15}$	ser Salinatis, complex	Edwards and Bruner, 1942
$1,4,12,27:g,s,t:z_{43}$	ser Kingston, diphasic	Taylor et al., 1960
$6,7:y:e,n,z_{15}:z_{47}:z_{50}$	ser Mikawasima, quadriphasic	Edwards et al., 1962 b
$6,7:z_{49},r:z_{49},1,5$	ser Infantis, complex, triphasic	Edwards et al., 1962 a
$3,10:g,s,t:z_{43}$	ser Westhampton, diphasic	Taylor et al., 1960
$3,10:g,s,t:z_{45}$	do	ibid
$3,10:z_{45}, y:z_{45},1,2$	ser Amager, complex	Unpublished data
$3,10:z_{49},z_{10}:z_{49},1,5$	ser Lexington, complex	McWhorter et al., 1964 b
$3,10:e,h:z_{48}$	ser (Unnamed)	Unpublished data
$3,15:z_{27},e,h:z_{27},1,2$	ser Hamilton = ser Goerlitz	Moran and Edwards 1958
$3,15:g,s,t:z_{45}$	ser Halmstad, diphasic	Taylor et al., 1960
$11:d,r:d,e,n,x$	ser Rubislaw, complex	McWhorter et al., 1964 a
$11:d,a:d,e,n,z_{15}$	ser Montgomery, complex	Edwards et al., 1957
$1,3,19:g,s,t:z_{27}$	ser Senftenberg, diphasic	Edwards et al., 1947
$1,3,19:g,s,t:z_{37}$	do	Unpublished data
$1,3,19:g,s,t:z_{43}$	do	Taylor et al., 1960
$1,3,19:g,s,t:z_{45}$	do	Unpublished data
$1,3,19:g,s,t:z_{46}$	do	do
$1,13,23:z_{29}:z_{37}$	ser Cubana, diphasic	Taylor et al., 1960
$1,13,23:z_{29}:z_{43}$	do	ibid
$1,13,23:d:z_{37}$	ser Wichita, diphasic	Edwards et al., 1954
$1,13,23:z_{43}, (z),:z_{43},l,w$	ser Worthington, complex	Douglas and Edwards, 1962
$1,13,23,36:f,g:z_{45}$	ser Havana, diphasic	Taylor et al., 1960
$6,14,18:z_4,z_{23}:z_{45}$	ser Siegburg, diphasic	Le Minor, 1967, personal communication
$35:z_4,z_{23}:z_{37}$	ser Alachua, diphasic	Edwards and Ewing, 1962 (2nd ed.)
$35:z_4,z_{23}:z_{43}$	do	ibid
$35:f,g:z_{27}$	ser Adelaide, diphasic	Taylor, 1964, personal communication
$39:z_{48}:1,5$	ser Cook, diphasic	McWhorter et al., 1962
$47_1,47_2:z_{45}:z_4, z_{23}:z_6$	ser Bere, triphasic	Unpublished data

The culture of ser Salinatis described by Edwards and Bruner (1942) may be cited as an example of a complex diphasic serotype that possesses a major antigenic component common to both phases. The antigenic structure of ser Salinatis is $4,12:d,e,h:d,e,n,z_{15}$. However, when ser Salinatis was cultivated in semisolid phase reversal medium containing d antiserum, a diphasic serotype indistinguishable from ser Sandiego ($4,12:e,h:e,n,z_{15}$) was produced.

The bacteria (Pc 230 etc.) described by Edwards (1950) are variants of ser Salinatis. These cultures ($4,12:d,e,h$) fermented sucrose rapidly and originally they were called intermediate coliform bacteria. Except for sucrose fermentation, the biochemical reactions of the strains are those of typical salmonellae (unpublished data, 1967). These cultures were unusual in that it was possible to isolate three separate phases (d;e,h; and e,n) from the original monophasic forms by cultivation in the presence of appropriate antisera.

If a culture that possesses O antigens, 1,3,19 is in the z_{27} (or z_{37}, z_{43}, z_{45}, or z_{46}) phase when isolated or received, the possibility that it might be ser Senftenberg should be considered. Such a culture should be inoculated into semisolid phase reversal medium containing appropriate antiserum (z_{27}, z_{43}, etc.) in order to isolate the g,s,t phase. A culture of this sort then becomes stable in the g,s,t phase, since this variation is irreversible, as illustrated below:

$$1,3,19:z_{43} \xrightarrow{z_{43}} g,s,t \xrightarrow{g,s,t} \text{immobilized.}$$

In this diagram the arrows indicate the direction of variation. The symbols on the arrows refer to the antisera incorporated in semisolid medium.

Usually the change from the z_{37} (or z_{43} etc.) to the g,s,t phase can be accomplished without difficulty. However, in the case of ser Simsbury ($1,3,19:z_{27}$) the change from z_{27} to g,s,t sometimes is difficult and may require repeated transfers in medium with z_{27} antiserum (Edwards and Moran, 1947). It would appear that ser Simsbury actually is phase 2 of ser Senftenberg.

Antigens z_{27} and z_{43} also may occur in serotypes that normally are diphasic. Le Minor and Edwards (1960) reported that when ser Hamilton ($3,15:z_{27}$) was cultivated in semisolid medium that contained z_{27} antiserum a culture having the formula $3,15:e,h:1,2$ was recovered. This change since has been observed to occur spontaneously. The $3,15:e,h:1,2$ form recovered from ser Hamilton is biochemically and serologically identical with ser Goerlitz, so it would seem that ser Hamilton actually is a third phase of ser Goerlitz. Here again, while e,h:1,2 easily can be recovered from the z_{27} phase, it is only occasionally and with difficulty

that the z_{27} phase can be recovered from ser Hamilton after it has assumed the $3;15:e,h:1,2$ form.

Additional examples of analogous diphasic and complex forms are given in Table 98. Interested investigators are referred to the publications cited above and in Table 98 for further details.

ANTIGENIC ANALYSIS

Somatic Antigens of *Salmonella*

In the characterization and classification of serotypes of *Salmonella* by antigenic analysis the O, or heat stable somatic, antigens of the bacteria are identified first. This is accomplished by use of O antisera representative of all of the heat stable antigens possessed by members of the genus. Use of the antisera listed in Table 99 permits determination of the serogroup to which a culture belongs. This determination is followed by the use of absorbed single factor antisera when indicated (v. inf.). Methods for production of antisera are given near the end of this chapter.

The O antigens of salmonellae are numbered consecutively from 1 to 64 by means of arabic numerals. However, this series is not completely continuous. That is, there are no O antigens in the schema represented by the numerals 29, 31, 32, 33, 49, 62, and 63. Originally, the numerals 29, 31, and 32 were given to antigens of certain cultures that were related to those of salmonellae (Kauffmann, 1941). These particular cultures were labeled *S. ballerup* and *S. coli* by Kauffmann, but when it was learned that the cultures actually were members of the genus *Citrobacter* the antigens were deleted from the schema and the numbers 29, 31, and 32 were not applied to anything else. Unfortunately, some writers have used the designation *E. coli* for these cultures. They are not *E. coli* and Kauffmann never labeled them as such. Kauffmann (e.g., 1966) combined O antigen 49 with 40 and deleted O49 from the schema. This policy has been retained herein. Kauffmann (e.g., 1966) assigned the numerals 62 to 65 to the O antigens of certain members of the genus *Arizona*. This procedure is not followed by the author. It seems likely that eventually salmonellae will be characterized, the O antigens of which can be designated by the numerals 61, 62, and 65, as happened in the case of the designation O antigen 64. The cultures originally assigned to O33 also are members of the genus *Arizona*.

Some of the O antigens of salmonellae have been subdivided (Kauffmann and Petersen, 1962, 1963) and selectively absorbed antisera are required for exact characterization of the O antigens of serotypes within these groups (v. inf.). However, workers in many laboratories will not find such exact characterization

TABLE 99

Antisera for Identification of O Antigens

Agglutinin content	Cultures used for O antiserum production	Agglutinin content	Cultures used for antiserum production	Agglutinin content	Cultures used for antiserum production
1,2,12	bioser Paratyphi-A	6,14,24	ser Carrau	44	ser Niarembe
1,4,5,12	ser Typhimurium	(1),6,14,25	ser Florida	45a,45b	ser Deversoir
1,4,12,27	ser Schleissheim	16	ser Gaminara	45a,45c	ser Dugbe
4,12	ser Essen	17	ser Kirkee	(9),46	ser Haarlem
6,7	ser Thompson	6,14,18	ser Siegburg	47a,47b	ser Bergen
$6_1,7$	S. cholerae-suis, stock 36	18	ser Cerro	47a,47c	ser Kaolack
$6_2,7$	S. cholerae-suis, stock 37	(8),20	ser Kentucky	48a,48b	ser Dahlem
6,(7),(14)	ser Bornum	21,26	ser Minnesota	48a,48b,48c	ser Djakarta
6,8	ser Newport	3,10,26	ser London	50a,50b,50c	ser Wassenaar
(8)	ser Virginia	28a,28b	ser Pomona	50a,50b,50d	ser Greenside
9,12	bioser Gallinarum	28a,28c	ser Dakar	51	ser Treforest
$9,12_1,12_2,12_3$	S. typhi, 901	30a,30b	ser Urbana	52	ser Utrecht
$9,12_1,12_2,12_3$	bioser Pullorum	35	ser Monschaui	53	ser Humber
3,10	ser Anatum	38	ser Inverness	54	ser Uccle
3,15	ser Newington	39	ser Champaign	55	ser Tranoroa
1,3,19	ser Senftenberg	40a,40b	ser Riogrande	56	ser Artis
(3),(15),34	ser Minneapolis	1,40a,40c	ser Bulawayo	57	ser Locarno
6,14	ser Boecker	41	ser Waycross	58	ser Basel
11	ser Aberdeen	42_1	ser Weslaco	59	ser Betoiky
13,22,36	ser Poona	1,42a,42b	ser Loenga	60	ser Luton
13,23,36	ser Atlanta (Mississippi)	43a,43b,43c	ser Milwaukee	61	ser Eilbek
1,13,23,37	ser Worthington	43a,43c,43d	ser Bunnik	64	ser 64:k:e,n,x,z_{15},z_{16} (unnamed)

necessary or desirable. The O antigens that have been subdivided are 28, 30, 40, 42, 43, 45, 47, 48, and 50.

Use of O Antisera. The O antisera are used in slide agglutination tests to identify the heat stable antigens of the bacteria. Antigen suspensions for use in these tests may be prepared by emulsifying the growth from an agar slant culture in 0.5 ml of physiological saline solution[3] to form a *dense*, homogeneous suspension. Another method of O antigen preparation is a modification of that described by White (1926). For this an infusion agar slant (long slant) should be inoculated over the entire surface. After incubation the growth is removed and suspended as well as possible in 1.0 ml of ethyl alcohol (96% or absolute). The suspension then is placed in a water bath at 60 C for 1 hr, centrifuged, and the alcohol decanted. After the alcohol has drained 0.5 ml of phenolized physiological saline solution is added and the tube is shaken gently. At this point the antigen should be allowed to stand for about 30 min, after which it is shaken vigorously to suspend the bacteria. The latter method is particularly useful in the examination of cultures that are slightly rough. There are two further advantages in the use of this method. First, one may work entirely with killed cultures. This is particularly desirable when comparatively inexperienced persons are involved in the work and when cultures of *S. typhi* are being examined. Second, there is somewhat less cross agglutination between the various serogroups when alcohol-treated organisms are used than when living cultures are employed in slide tests. The author and colleagues use this method routinely in the examination of cultures of *Salmonella*. When the organisms are treated with alcohol they become somewhat more resistant to agglutination and the antisera must be used in lower dilutions. If alcohol-treated O antigen suspensions are not used, plain saline solution should be used, since agents such as phenol and formalin affect the flagella of the bacteria, and the altered flagella may inhibit agglutination of the cells by O antisera.

The antisera are diluted with phenolized physiological saline solution.[4] The dilution employed, usually 1 to 5 or 1 to 10, is dependent upon the strength of the individual antiserum. Each newly produced antiserum should be titrated with its homologous antigen and used in the highest dilution in which it gives a strongly positive slide agglutination within 30 sec to 1 min. If the antisera are used in lower dilutions (i.e., more concentrated), confusing cross reactions may result. In slide tests the antigen is put down on a slide in droplets of approximately 0.005 ml each (or an amount roughly equivalent to a 3 mm loopful). A droplet of equal volume of each of the diluted antisera is mixed with an individual droplet of antigen, and the slide tilted back and forth until agglutination is apparent. Agglutination occurs in one or more antisera that contain agglutinins for the somatic antigens of the organism that is being examined. With diluted unabsorbed antisera the speed of the reactions is important. Homologous reactions occur rapidly and are complete, while heterologous agglutination usually is slower and weaker. Typical reactions of the various antigens with O antisera are given in Table 100. The O antisera represented in the table were prepared by the writer and colleagues and are used daily in the examination of cultures. They are used in 1 to 10 dilution with alcohol-treated antigens. If it is desired to use antigens without alcohol treatment, the dilution of the antisera may be doubled and the results will be essentially the same. Warning should be given against the use of large drops of antiserum and antigen in slide agglutination tests. If large drops are used, confusing cross reactions are likely to develop. These cross reactions are diminished when small amounts of the reagents are used. If a 0.2 ml pipette containing antiserum or antigen is held vertically with the index finger covering the bore at the upper end and the lower end brought in contact with the slide, a sufficient volume of liquid will be deposited upon the slide by capillary action.

Marked cross agglutination occurs between serotypes that possess related O antigens (Table 100). For instance, an organism having the antigens 4,5,12, will be agglutinated not only by antiserum derived from an organism with identical O antigens, but also by an antiserum that contains agglutinins for the antigens 4,12,27, since the antigens 4 and 12 are common to both serotypes. For this reason it is necessary to use absorbed antisera to determine whether an unknown culture contains certain antigens (v. inf.). The reactions shown in Table 100 indicate that anomalous cross agglutination occurs between certain antigens and antisera. While some of these reactions may be the result of incipient roughness in the strains used to prepare the antisera, the majority are caused by antigenic relationships which, for one reason or another, are not expressed in the antigenic schema. It must be remembered that only the antigens that are of major significance in identification are expressed in the antigenic schema for *Salmonella*. No attempt is made to

---

[3]The author routinely uses 0.5% sodium chloride rather than 0.85% in such solutions.

[4]If antisera are preserved by the addition of an equal volume of glycerol, the preserved antiserum is considered to be diluted 1 to 2. Thus, a 1 to 10 dilution is made by addition of 1 ml of glycerolated antiserum to 4 ml of diluent.

TABLE 100

Agglutination Reactions in O Antisera
(Slide Tests)

O Antisera

O Antigen suspensions	1,2,12	4,5,12	4,12,27	6,7	(8)	(8),20	9,12	9,46	3,10	3,15	(3),(15),34	1,3,19	11	13,22	1,13,23	6,14,24	1,6,14,25	16*	17	18
1,12,12	++++	+	+	–	–	–	+	–	–	–	–	+	–	–	+	–	+	–	–	–
4,5,12	–	++++	++	–	–	–	+	–	–	–	–	–	–	–	–	–	–	–	–	–
4,12,27	–	++++	++++	–	–	–	+	–	–	–	–	–	–	–	–	–	–	–	–	–
6,7	–	–	–	++++	–	–	–	–	–	–	–	–	–	–	–	+	+	–	–	–
6,8	–	+	–	+	++++	++	–	–	–	–	–	–	–	–	–	+	+	–	–	–
(8),20	–	–	–	–	++	++++	–	–	–	–	–	–	–	–	–	–	–	–	–	–
9,12	+	+	+	–	–	–	++++	++++	–	–	–	–	–	–	–	–	–	–	–	–
(9),46	–	–	–	–	–	–	++++	++++	–	–	–	–	–	–	–	–	–	–	–	–
3,10	–	–	–	–	–	–	–	–	++++	++++	+	+	–	–	–	–	–	–	–	–
3,15	–	–	–	–	–	–	–	–	++++	++++	++++	+	–	–	–	–	–	–	–	–
(3),(15),34	–	–	–	–	–	–	–	–	–	+	++++	–	–	–	–	–	–	–	–	–
1,3,19	++	–	–	–	–	–	–	–	+	+	–	++++	–	–	–	–	–	–	–	–
11	–	–	–	–	–	–	–	–	–	–	–	–	++++	–	–	–	–	–	–	–
13,22	–	–	–	–	–	–	–	–	–	–	–	–	–	++++	+	–	–	–	–	–
1,13,23	+	–	–	–	–	–	–	–	–	–	–	–	–	+	++++	–	–	–	–	–
6,14,24	–	–	–	+	–	–	–	–	–	–	–	–	–	–	–	++++	++	–	–	–
1,6,14,25	+	–	–	+	–	–	–	–	–	–	–	–	–	–	–	++	++++	–	–	–
16	–	–	–	–	–	–	–	–	–	–	–	–	–	–	–	–	–	++++	+++	–
17	–	–	–	–	–	–	–	–	–	–	–	–	–	–	–	–	–	–	++++	–
18	–	–	–	–	–	–	–	–	–	–	–	–	–	–	–	–	–	–	–	++++

NOTE: Diluted antisera for serogroups 16 to 64 usually do not agglutinate heterologous antigens. However, there may be some exceptions (see text) and newly prepared antisera must be tested.

+, ++++ indicate degrees of agglutination.

enumerate and label all of the antigens of the various serotypes. Certain overlapping O antigens, e.g., 1 and 12, occur in combination with a number of other antigens. These may cause confusing cross agglutination reactions. When doubt exists concerning the O antigens of a culture, absorbed antisera should be used. Conversely, some cross agglutination reactions actually may be helpful in the determination of the serogroup or O antigen group to which a strain belongs. It is emphasized that investigators must test their antisera with cultures of known antigenic composition and learn their characteristics. This applies to all antisera.

Many of the antisera listed in Table 99 may be used in the unabsorbed state for practical work in most laboratories and for preliminary examinations in all laboratories. Further, a number of the absorbed single factor antisera (Table 101) may be needed only in certain specialty and reference laboratories. However, some of the absorbed antisera listed in Table 101 are needed in all laboratories in which complete serotyping is done (v. inf.). In practice, unabsorbed O antisera may be used in predetermined dilutions in the following manner for determination of many O antigens.

01 – Agglutination in 1,2,12 *and* 1,3,19. Antisera diluted 1 to 2 may be needed.

02 – Agglutination in 1,2,12 antiserum.

03 – Agglutination in 3,10 *and* 3,15 antisera.

04 – Agglutination in 4,12 antiserum.

06 – Agglutination in 6,7 *and* 6,8 antisera.

07 – Agglutination in 6,7 antiserum.

08 – Agglutination in (8)[5] antiserum.

09 – Agglutination in 9,12 antiserum.

011 – Agglutination in 11 antiserum.

013 – Agglutination in 13,22,36 *and* 13,23,37 antisera.

016 – Agglutination in 16 antiserum.

017 – Agglutination in 17 antiserum.

021 – Agglutination in 21,26 antiserum; not in 3,10,26.

026 – Agglutination in 21,26 *and* 3,10,26 antisera.

028 – Agglutination in mixed 28a, 28b + 28a, 28c antiserum. Because of a minor relationship it may be necessary to absorb this pooled antiserum with a culture of 039.

030 – Agglutination in 30a, 30b antiserum. Absorption of this antiserum with a culture of 040 may be necessary.

035 – Agglutination in 35 antiserum.

038 – Agglutination in 38 antiserum.

039 – Agglutination in 039 antiserum. It may be necessary to absorb this antiserum with cultures of 028.

040 – Agglutination in mixed 40a, 40b + 40a, 40c antiserum. Absorb antiserum with a culture that possesses 01 if necessary.

041 – Agglutination in 41 antiserum. Some members of this serogroup possess 01. Test in 1,2,12 antiserum.

042 – Agglutination in 42a antiserum. Absorb antiserum with an 054 culture if necessary.

043 – Agglutination in 43a, 43b, 43c antiserum.

044 – Agglutination in 44 antiserum. Absorb antiserum with 051 cultures if necessary.

045 – Agglutination in mixed 45a, 45b + 45a, 45c antiserum. Absorb antiserum with 050 cultures if necessary.

047 – Agglutination in mixed 47a, 47b + 47a, 47c antiserum.

048 – Agglutination in mixed 48a, 48b + 48a, 48b, 48c antiserum.

050 – Agglutination in 50 antisera. Absorb antisera with 045 cultures if necessary.

051 – Agglutination in 51 antiserum. Absorb antiserum with 044 culture if necessary.

052 – Agglutination in 52 antiserum.

053 – Agglutination in 53 antiserum. Absorb antiserum with 017 cultures if necessary.

054 – Agglutination in 54 antiserum. Absorb antiserum with 042 culture if necessary.

055 to 061 and 064 – Agglutination in respective antisera 55, 56, 57, 58, 59, 60, 61, and 64.

In several places in the listing of antisera (above) that generally may be used in the unabsorbed state a notation is included which indicates that certain antisera should be absorbed if necessary. In these instances O antigenic relationships to the indicated O antigen groups are known which may cause cross agglutination. Usually these relationships are minor and are not apparent, or very weak, when the antisera are diluted for use in identification. However, the extent of these relationships can be determined only by tests with known, characterized cultures. If the cross reactions are sufficiently strong to be troublesome, the antisera should be absorbed in the manner indicated (see also Table 101).

Absorption of O Antisera. Cell suspensions (antigen) for absorption of agglutinins are prepared by inoculating moist, thickly poured (about 30 ml per plate) 90 mm infusion agar plates with about 0.5 ml of an 18 to 24 hr broth culture of the selected strain(s). The inoculum should be spread over the entire surface of

---

[5]Note: Enclosure of the designation of an antigen in parentheses indicates that the complete antigen is not present, e.g., (8). Enclosure of an antigen designation in brackets indicates that the antigen or antigen complex may be present or absent, e.g., [1], [1,2] etc.

TABLE 101

Preparation of Absorbed Single O Factor Antisera

Single Factor	Antiserum	Absorbing Culture(s)
1	1,2,12 bioser Paratyphi-A	bioser Paratyphi-A variant Durazzo:2,12
2	1,2,12 bioser Paratyphi-A	*S. typhi*:9,12 + ser Senftenberg:1,3,19
4	4,12 ser Essen	*S. typhi*:9,12(901)
5	1,4,5,12 ser Typhimurium	ser Typhimurium, variant Copenhagen (1,4,12) + ser Bredeney:4,12,27 + bioser Paratyphi-A:1,2,12
6_1	6,7 *S. cholerae-suis* (prepared with stock culture 36)	*S. cholerae-suis*:6_2,7 (stock culture 37)
6_2	6_2,7 *S. cholerae-suis* (prepared with stock culture 37)	*S. cholerae-suis*:6_1,7 (stock culture 36)
7	6,7 ser Thompson	ser Newport:6,8 + ser Onderstepoort: 1,6,14,25
8*	6,8 ser Newport	ser Thompson:6,7 + ser Onderstepoort:1,6,14,25
9	9,12 bioser Gallinarium	bioser Paratyphi-A variant Durazzo:2,12 + Typhimurium variant Copenhagen:4,12
10	3,10 ser Anatum	ser Newington:3,15 + ser Senftenberg:1,3,19
12_2	9,12_1,12_2,12_3 *S. typhi* 901	bioser Paratyphi-A variant Durazzo:2,12_1,12_3 or with *S typhi* str T2:9,12_1,12_3
12_3	9,12_1,12_2,12_3 bioser Pullorum	bioser Pullorum:9,12_1,12_2 (variant strain, 12_2 + + +)
14	6,14 ser Boecker (or 6,14,24 ser Carrau)	ser Thompson:6,7 + ser Cerro:18 (or ser Newport + ser Oranienburg 6,8 6,7
15	3,15 ser Newington	ser Anatum:3,10 + ser Senftenberg:1,3,19

*If desired, a single factor 6 antiserum may be prepared by absorption of 6,8 (ser Newport) antiserum with ser Virginia (8) and ser Kentucky (8,20).

TABLE 101 (Continued)

Preparation of Absorbed Single O Factor Antisera

Single Factor	Antiserum	Absorbing Culture(s)
18	18 ser Cerro	ser Boecker:6,14
19	1,3,19 ser Senftenberg	ser Anatum:3,10 + ser Newington:3,15 + ser Bredeney:1,4,12,27
20	(8),20 ser Kentucky	ser Newport:6,8
21	21,26 ser Minnesota	ser London:3,10,26
22	13,22,36 ser Poona	ser Mississippi (Atlanta): 13,23,36
23	13,23,36 ser Mississippi (Atlanta)	ser Bredeney:1,4,12,27 + ser Poona:13,22,36
24	6,14,24 ser Carrau	ser Onderstepoort:1,6,14,25 + ser Boecker:6,14
25	1,6,14,25 ser Florida	ser Senftenberg:1,13,19 + ser Carrau:6,14,24
26	3,10,26 ser London	ser Anatum:3,10
27	1,4,12,27 ser Schleissheim	ser Reading:4,5,12 + ser Abortusequi:4,12 + ser Typhimurium, variant Copenhagen:1,4,12(273-67)
28b	28a,28b ser Telaviv	ser Dakar:28a,28c
28c	28a,28c ser Dakar	ser Telaviv:28a,28b
30	30a,30b ser Urbana	ser Riogrande:40

TABLE 101 (Continued)

Preparation of Absorbed Single O Factor Antisera

Single Factor	Antiserum	Absorbing Culture(s)
30b	30a,30b ser Urbana	ser Soerenga:30a
34	(3),(15),34 ser Minneapolis	ser Newington:3,15 + *S. typhi*: 9,12 (901)
36	13,23,36 ser Atlanta	ser Worthington:1,13,23,37
.37	1,13,23,37 ser Worthington	ser Senftenberg:1,13,19 + ser Grumpensis:13,23,36 +ser Atlanta: 13,23,36
39	39 ser Champaign	Absorb with 028 cultures if necessary
40b	40a,40b ser Riogrande	ser Bulawayo:1,40a,40c
40c	1,40a,40c ser Bulawayo	ser Riogrande:40a,40b + ser Bukavu:1,40a,40b
42a	42a Weslaco	ser Uccle:54 if necessary
42b	1,42a,42b ser Loenga	ser Weslaco: 42a + ser Senftenberg:1,13,19
43b	43a,43b,43c ser Milwaukee	ser Houten:43a,43c,43d or ser Bunnik
43c	43a,43b,43c ser Milwaukee	ser Kingabwa: 43a,43b
43d	43a,43c,43d ser Bunnik	ser Milwaukee:43a,43b,43c
44	44 ser Niarembe	Absorb with 051 cultures if necessary
45b	45a,45b ser Deversoir	ser Dugbe:45a,45c. Absorb with 050 culture if necessary.

TABLE 101 (Continued)

Preparation of Absorbed Single O Factor Antisera

Single Factor	Antiserum	Absorbing Culture(s)
45c	45a,45c ser Dugbe	ser Deversoir:45a,45b. Absorb with O50 culture if necessary.
46	(9),46 ser Haarlem	*S. typhi*:9,12 + ser Anatum:3,10
47b	47a,47b ser Bergen	ser Kaolack:47a,47c
47c	47a,47c ser Kaolack	ser Bergen:47a,47b
48b	48a,48b ser Dahlem	*Citrobacter*, culture 2624-36
48c	48a,48b,48c ser Djakarta	ser Dahlem:40a,40b
50	50 ser Greenside	Absorb with O45 cultures if necessary
50b	50a,50b,50c ser Wassenaar	ser Hooggraven:50a,50c
50c	50a,50b,50c ser Wassenaar	ser Greenside:50a,50b,50d
50d	50a,50b,50d ser Greenside	ser Wassenaar:50a,50b,50c
51	51 ser Treforest	Absorb with O1 and O44 culture if necessary
53	53 ser Humber	Absorb with O17 culture if necessary
54	54 ser Uccle	Absorb with O42 cultures if necessary

NOTE: All of the single factor O antisera listed may not be required in all laboratories. Some will be needed only in certain specialty and reference laboratories. See text.

the agar in order to assure confluent growth. The plates are incubated in an upright position for 18 to 24 hr, after which the growth is taken up in phenolized physiological saline solution. If each plate is flooded with 2 or 3 ml of the phenolized saline solution, the growth may be removed easily with the aid of a bent wire or glass rod scraper. The growth from the plates is pooled and the heavy suspension is centrifuged at high speed in an angle centrifuge. The clear supernatant fluid is decanted and the bacterial cells are resuspended in 0.5 ml of phenolized physiological saline solution. One ml of the glycerolated antiserum to be absorbed then is added to the cells and *thoroughly mixed.* The mixture is incubated at 48 to 50 C (preferably in a waterbath) for 2 hr. (Such mixtures should be shaken at frequent intervals during incubation.) After incubation the mixture is centrifuged and the clear absorbed antiserum is removed and preserved by the addition of a small amount (two drops per ml) of chloroform or with merthiolate. The dilution of a glycerolated antiserum absorbed in the manner outlined above will be 1 to 3. Ordinarily it is inadvisable to dilute absorbed O antisera further.

The amount of cells used in absorption must be sufficient to remove all agglutinins for the absorbing culture(s) from the antiserum. This amount varies with different antisera and with the agglutinin titer of an antiserum for the homologous and related micro-organisms. The growth from five 90 mm plate cultures, as described above, usually is sufficient to remove heterologous agglutinins from 1.0 ml of antiserum. However, it may be anticipated that there will be instances in which the growth from a larger number of plates (e.g., ten) will be required to clear an antiserum of unwanted agglutinins. In some instances fewer than five plates may be sufficient.

Although single absorption (addition of all of the absorbing cells to the antiserum at one time) of an antiserum is effective in many instances, the double or two-step absorption procedure is recommended because it is more effective than single absorption in removal of heterologous agglutinins. In a double absorption the absorbing suspension is divided into equal parts prior to centrifugation. After centrifugation the antiserum is added to half of the absorbing cells and the mixture is incubated at 48-50 C for 1 hr. Then the mixture is centrifuged and the supernatant antiserum is added to the remaining portion of absorbing cells. This is followed by a second period of incubation (1 hr) and final centrifugation. The need for thorough emulsification of the absorbing antigen (cells) in the antiserum is emphasized. If an antiserum is to be absorbed with two or more different cultures, it is advisable to pool all of the growth harvested from

plates, mix it, and then divide it into two equal portions.

When relatively large amounts of antiserum are to be absorbed it is expedient to cultivate the bacteria on infusion agar in 150 mm petri dishes (200 to 250 ml of medium per plate) or other suitable large containers. The diameter of the bottom of a so-called 150 X 25 mm plastic petri dish actually is about 142 mm, while that of a 90 mm plate is 88 mm. Therefore, the area (d^2 X 0.7854) of the larger dish is about 2.6 times greater than that of the smaller 90 mm dish.

Absorption of antiserum may be regarded as an agglutination reaction performed for the selective removal of unwanted agglutinins. These unwanted (heterologous) agglutinins are bound by the reaction between the antigens of the cells and the agglutinins present in the antiserum. After this reaction takes place, the agglutinins bound to the cells are removed with the cells by centrifugation.

Determination of Single O Antigen Factors. It is unnecessary to titrate absorbed O antisera by means of serial dilutions in tube agglutination tests. In fact, such a procedure would be wasteful. It is necessary only to test the absorbed antiserum by slide agglutination to make certain that it no longer reacts with the absorbing strain(s) but actively agglutinates the homologous culture.

The manner in which O antisera may be absorbed for the production of single factor O antisera is given in Table 101. These single factor antisera may be used for the determination of single O antigens that have significance in serotyping. It is emphasized that investigators should not substitute cultures for those listed for either O antiserum production (Table 99) or for absorption (Table 101) without first having made certain of the *complete* antigenic structure of the proposed substitute cultures. Kauffmann (1966, pp. 97-108) listed other cultures that may be employed for both of the above-mentioned purposes.

The O antigens of serotypes of serogroup G are more complicated than is indicated by the abbreviated formulas given in the antigenic schema. Careful examination of some of these serotypes by Kauffmann (1944) and Kauffmann and Rohde (1961) revealed the following:

ser Poona	13,22,23
ser Willemstad	1,13,22,37
ser Rotterdam	1,13,22,36,37
ser Tanger	1,13,22,36,37
ser Grumpensis	13,23,36
ser Atlanta	13,23,36
ser Worthington	1,13,23,37
ser Mississippi	1,13,23,36,37

TABLE 102

Reactions of *S. typhi* in Slide Agglutination Tests

	Antisera		
Suspension	*Salmonella* polyvalent	*Salmonella* O Group D *(9,12)*	*Vi (Citrobacter 029)*
Living	+ + + +	—	+ + + +
Heated	+ + + +	+ + + +	—

Adapted from Ewing, 1966

As mentioned on page 183, several of the O antigens of salmonellae may be subdivided. Some of the serotypes that are members of these groups and the antigen fractions that they possess are given by Kauffmann and Petersen (1962, 1963). See also Table 101.

Detection of Vi Antigen. Since Vi antigen in aqueous suspension is inactivated by heat, a strong Vi antiserum that is free of H agglutinins cannot be prepared unless a permanently nonflagellated bacterium that possesses Vi antigen is available. Lacking such a culture, the author employs a Vi (V) form of a culture of *Citrobacter* to prepare Vi antiserum. Originally this culture was characterized by Kauffmann and Moeller (1940) as *S. ballerup* ($29:Vi:z_{14}$) but later was excluded from the antigenic schema. Antiserum prepared with this strain contains O, Vi, and H agglutinins, but the O and H agglutinins do not react with the O and H antigens of salmonellae that are known to possess Vi antigen. A pure Vi antiserum may be prepared from this antiserum by absorption with an O (W) form of the homologous culture. However, the unabsorbed antiserum may be used for detection of Vi antigen in cultures of *S. typhi* and ser Paratyphi-C. The manner in which ViO forms of *S. typhi* may be expected to react is depicted in Table 102. The reversal of the reactions in O (9,12) and Vi antisera after the suspension has been heated at 100 C for about 15 min is typical.

THE H ANTIGENS OF *SALMONELLA*

H Antisera. The H antigens of salmonellae are determined by selective use of antisera representative of the H antigens possessed by members of the genus. Determination of the H antigens that are known to be of significance in identification of salmonellae may be accomplished with the antisera listed in Table 103. Production of H antisera is discussed near the end of this chapter. The method and dilutions employed in identification are such that only H agglutination is apparent in the tests.

The H antigens of salmonellae are designated by letters of the alphabet (a, to z, z_1, z_2 etc.) and by the arabic numerals.[6] The series a to z is complete except for j, which is not expressed in the schema. Antigen j is an induced phase of *S. typhi*. This phase never has been reported in a naturally occurring strain of *S. typhi*. The limitations of the alphabet made it necessary to use subscript numbers (z_1,z_2 etc.) to designate antigens that were characterized after the designation z had been applied. Antigens designated z_1, z_2 etc. are not subdivisions of z. On the contrary, each of these antigens is a separate entity, although there are relationships between some of these and other H antigens. For example, antigens z_1,z_2,z_3,z_7,z_8, and z_9 all are related to the g . . . . complex (g,m;g,m,s;etc). The series of antigens labeled in this manner extends from z_1 to z_{59}, but some of these are not expressed in the antigenic schema or are not identified in practice. Further, antigens designated z_{54} to z_{57} and z_{59} by Kauffmann (see Kauffmann, 1966) actually are antigens of serotypes of *Arizona*. The phase 2 antigens of numerous salmonellae are designated by arabic numerals. This series extends from 1 to 12, but antigens 3, 4 and 8 to 12 are not expressed in the antigenic schema. Nevertheless some of these unexpressed antigens are important, particularly in absorption of antisera (v. inf.).

The antigens formerly designated $z_{14},z_{20},z_{21},z_{22},z_{30}$, and z_{31} have been deleted from the antigenic schema. The cultures in which these antigens were characterized were found to be members of the genus *Citrobacter*.

[6]It may be somewhat difficult to distinguish between the roman letter l and the arabic numeral 1 in some instances. If the following is borne in mind, this difficulty may be alleviated. When the character 1 occurs with a roman letter (e.g.l,v;l,w) it is the letter. If the character 1 occurs with an arabic numeral (1,2;1,5 etc.) it is the numeral. These characters never are interchanged; the letter never is used with the numeral or *vice versa*. Also the antigens 1,2;1,5 etc. always occur in phase 2.

Several of the H antigens of salmonellae may be subdivided (see Kauffmann, 1966). For example, antigen b may be subdivided into fractions b_1, b_2, b_3 etc. Similarly, fractions of antigens a, c, d, i, k, r, z, z_{10} and z_{29} may be delineated. While subdivision of H antigens in this manner is not required for identification of salmonellae in most laboratories, such determinations may be found useful in certain reference centers.

TABLE 103

Antisera for Identification of H Antigens

Agglutinin content	Cultures used for H antiserum production	Agglutinin content	Cultures used for H antiserum production
a	bioser Paratyphi-A	z_6	ser Kentucky, ph 2
b	ser Paratyphi-B, ph 1	z_{10}	ser Illinois, ph 1
c	*S. cholerae-suis*, ph 1	z_{27}	ser Simsbury*
d	*S. typhi*	z_{29}	ser Tennessee
e,h	ser Reading, ph 1	z_{35}	ser Chittagong, ph 2
e,n,x	ser Abortusequi	z_{36}	ser Weslaco
e,n,z_{15},z_{17}	ser Salinatis, Pc 230*	z_{37}	ser Wichita, ph 2
f,g,z_8	ser Derby	z_{38}	ser Lille
g,o,m,z_1,z_2	ser Enteritidis	z_{39}	ser Quimbamba, ph 2
g,m,s	ser Montevideo	z_{41}	ser Karamoja, ph 1
g,o,m,q,z_1	ser Blegdam	z_{42}	ser Bunnik, ph 2
g,p	ser Dublin	z_{43}	ser Senftenberg, 2170-58, ph 2*
g,p,u	ser Rostock	z_{44}	ser Quinhon, ph 1
g,o,q,z_3	ser Moscow	z_{45}	ser Senftenberg, 2282-58, ph 2*
g,s,t	ser Senftenberg, ph 1	z_{46}	ser Senftenberg, 840-59, ph 2*
g,[p],z_{51}	ser Wayne, ph 1	z_{47}	ser Mikawasima, 2547-60,z_{47} ph*
i	ser Typhimurium, ph 1	z_{48}	ser Cook, 364-61, ph 1*
k	ser Thompson	z_{49}	ser Infantis, 2783-61,z_{49} ph*
l,v	ser London, ph 1	z_{50}	ser Mikawasima, 2122-59,z_{50}ph*
l,w	ser Daressalaam, ph 1	z_{52}	ser Tulear, ph 2
l,z_{13}	ser Uganda, ph 1	z_{53}	ser Tokai, ph 3
l,z_{28}	ser Javiana, ph 1	z_{59}	ser Sternchanze, ph 2
l,z_{40}	ser Rutgers, ph 1	1,2	ser Paratyphi-B, ph 2
m,t	ser Oranienburg	1,5	ser Thompson, ph 2
r	ser Rubislaw, ph 1	1,6	ser Anatum, ph 2
y	ser Madelia, ph 1	1,7	ser Bredeney, ph 2
z	ser Poona, ph 1	1,10	ser Cardiff, ph 2
z_4,z_{23},z_{25}	ser Cerro	1,11	ser Italiana, ph 2
z_4,z_{24}	ser Duesseldorf	1,12	ser 6,8:e,h:1,12,ph2
z_4,z_{32}	ser Tallahassee		(unnamed)

Ph, phase

* See text, Irreversible Variation and Multiple Phases

Use of H Antisera. Several of the H antisera listed in Table 103 may be used in the unabsorbed state for practical work in most laboratories, and for preliminary examinations in all laboratories. In addition, several of the absorbed antisera (Table 105) may be needed only in certain reference laboratories.

Unlike O antisera, H antisera are used in tube agglutination tests. The antigens employed are infusion broth cultures incubated for 18 to 24 hr (or 4 to 6 hr in a waterbath) and diluted with an equal volume of physiological saline solution containing 0.6% formalin. The bacteria must be actively motile to insure development of H antigens. The dilutions employed differ with particular antisera. *However, if H antisera are produced properly, the unabsorbed glycerolated antisera usually can be used in a dilution of 1:1,000* (0.02 ml of a 1:20 dilution of antiserum is placed in the bottom of a tube, e.g., 13 X 100 mm, and 1.0 ml of formalinized broth suspension is added). This dilution is high enough to avoid troublesome cross reactions yet sufficiently low that specific H agglutination appears very quickly. In all the sets of antisera produced by the author and colleagues it was possible to use dilutions of 1 to 1000 for H agglutination tests. The tests are incubated in a waterbath at 48-50 C for one hour and read. Using this method, the flocculation which characterizes H agglutination becomes apparent after a few minutes' incubation if actively motile cultures are employed as antigens. No O agglutination is apparent in the dilutions used after this short incubation period.

It is possible to do H agglutination tests by the slide technique and this method is recommended by Kauffmann (1941). If it is used, the H antisera either must be freed of O agglutinins by absorption or must be used in dilutions sufficiently high that O reactions do not occur. Unabsorbed H antisera usually should be diluted 1 to 50 or 1 to 100. The author has found it expedient to use tube agglutination tests for the determination of H antigens.

The reactions of H antisera produced and used by the author and colleagues, are given in Table 104. The antigens used in the titrations were formalinized broth cultures of bacteria that had reached maximum development of H antigens. If bacteria with a lesser development of H antigens were used, the titers of the antisera would be lower.

The reactions of the H antisera (Table 104) give evidence of antigenic relationships of varying extent. In those instances in which the relationships are extensive the antisera usually must be absorbed. In others the relationships are minor and unabsorbed antisera may be used at a dilution of 1 to 1,000 for identification. Many of the minor relationships are not expressed in the antigenic schema; others are peculiar to the particular lots of antisera employed. Since no two sets of antisera can be expected to react in exactly the same way, it is imperative that investigators determine the peculiarities of each of the antisera that they employ. To ascertain this each antiserum must be tested in a dilution of 1 to 100 with antigen suspensions prepared with each of the H antigens, not just with those that might be expected to react. Antigens that react in an antiserum then are tested in serial dilutions to determine the extent of the relationships. The results given in Table 104 were obtained by this procedure (final readings made after incubation for 1 hr). This work always proves to be valuable. Among other things it yields information as to which antisera must be absorbed and gives a rough estimate of the amount of absorbing suspension that must be used. Further, the procedure aids in selection of antisera that may be used in the unabsorbed state for identification. For example, if one has several 1 . . . antisera (l,v;l,w;l,z$_{13}$ etc), tests performed in the manner outlined above assist in the selection of one of the antisera that may be used in preliminary work (*prior* to the use of absorbed single factor antisera). Clearly, one should select the antiserum that reacts rapidly with all of the 1 . . . antigens *and* which exhibits the fewest cross agglutination reactions. Of the 1 . . . antisera listed in Table 104, the l,v or the l,w antiserum may be used in the unabsorbed state at a dilution of 1 to 1,000 for preliminary work. Examination of the data given in Table 104 reveals other cross reactions that would seem to be extensive enough to require absorption. However, many of the antigens that cause these reactions are not identified in practical work in most laboratories. For example, antigens z$_6$ and z$_{35}$ reacted to titers of 1 to 800 and 1:6,400, respectively, in antiserum for antigen a (Table 104). Nevertheless this antiserum may be used in the unabsorbed state for the identification of antigen a. Clearly such antisera must be employed judiciously, and if anomolous reactions occur, their cause must be determined. Other antisera that usually can be employed without absorption are those that represent antigens b, c, d, i, k, r, y, z, z$_{10}$, z$_{27}$, z$_{29}$, z$_{37}$, z$_{38}$, z$_{41}$, z$_{42}$, z$_{43}$, z$_{45}$, z$_{46}$, and z$_{47}$. These antisera should be used in a dilution of 1 to 1,000. The presence of certain H antigen factors is determined by agglutination in each of two unabsorbed antisera. Antigen e is detected by agglutination of an antigen suspension in e,h *and* e,n,x antisera; antigen l by agglutination in l,v *and* l,w antisera; and antigen 1 by reactions in antisera for 1,2 *and* 1,5. The remaining H antisera must be absorbed (v. inf.).

As mentioned above several of the characterized H antigens of salmonellae are not identified in practical work in most laboratories. These are H antigens

TABLE 104

Reactions of H Antisera

H Antisera	Homologous Titers	Titers with Heterologous Antigens
a	25,600	z_6-800; z_{35}-1600
b	102,400	1,2-800; 1,5-400; 1,6-400; 1,7-400
c	25,600	
d	51,200	
e,h	102,400	e,n,x-800; e,n,z_{15}-800
e,n,x	51,200	e,h-400; e,n,z_{15}-12800
e,n,z_{15}	25,600	e,h-400; e,n,x-12800
f,g	51,200	f,g,t-12800; g,m-12800; g,m,s-6400; g,m,t-1600; g,p-6400; g,p,u-6400; g,q-6400; g,s,t-6400; g,t-12800; g,z_{51}-800; m,t-800
g,m	51,200	f,g-6400; f,g,t-3200; g,m,s-25600; g,m,t-6400; g,p-25600; g,p,u-12800; g,q-25600; g,s,t-6400; g,t-25600; g,z_{51}-800; m,t-400; z_{42}-200
g,m,s	51,200	f,g-400; f,g,t-1600; g,m-51200; g,m,t-3200; g,p-12800; g,p,u-6400; g,q-12800; g,s,t-51200; g,z_{51}-400; m,t-400
g,p	102,400	f,g-6400; f,g,t-3200; g,m-25600; g,m,s-12800; g,m,t-6400; g,p,u-51200; g,q-12800; g,s,t-12800; g,t-12800; g,z_{51}-200
g,p,u	25,600	f,g-6400; f,g,t-3200; g,m-12800; g,m,s-3200; g,m,t-6400; g,p-12800; g,q-6400; g,s,t-3200; g,t-6400; g,z_{51}-400
g,q	12,800	f,g-800; f,g,t-400; g,m-6400; g,m,s-3200; g,m,t-3200; g,p-1600; g,p,u-1600; g,s,t-1600; g,t-6400
g,s,t	51,200	f,g-3200; f,g,t-3200; g,m-6400; g,m,s-12800; g,m,t-3200; g,p-6400; g,p,u-6400; g,q-3200; g,t-25600; m,t-200
g,z_{51}	12,800	f,g-1600; f,g,t-800; g,m-3200; g,m,t-1600; g,p-1600; g,p,u-1600; g,s,t-1600; g,t-800
i	51,200	r-400
k	25,600	1,v-400; 1,w-800; 1,z_{13}-400; 1,z_{28}-400; 1,z_{40}-800; z-100; z_{29}-200; z_{52}-200
l,v	25,600	1,w-12800; 1,z_{13}-6400; 1,z_{40}-6400; z_{52}-400; 1,2-400; 1,5-800; 1,6-400
l,w	25,600	k-200; 1,v-12800; 1,z_{13}-6400; 1,z_{28}-6400; 1,z_{40}-6400; z_{52}-400; 1,2-400; 1,5-400; 1,6-200
l,z_{13}	25,600	1,v-6400; 1,w-6400; 1,z_{28}-12800; 1,z_{40}-3200; z_{52}-200; 1,2-400; 1,5-6400; 1,6-400; 1,7-200
l,z_{28}	12,800	1,v-12800; 1,w-12800; 1,z_{13}-12800; 1,z_{40}-3200; 1,2-1600; 1,5-3200; 1,6-800; 1,7-200
l,z_{40}	12,800	k-200; 1,v-6400; 1,w-6400; 1,z_{13}-6400; 1,z_{28}-3200; z_{52}-400; 1,2-200; 1,7-200
m,t	25,600	f,g-200; f,g,t-3200; g,m-200; g,m,s-400; g,m,t-400; g,p-100; g,p,u-200; g,q-200; g,s,t-1600; g,t-1600
r	25,600	i-1600
y	12,800	
z	25,600	c-200; k-400; z_{35}-1600; z_{41}-800; z_{44}-1600
z_4,z_{23}	12,800	z_4,z_{24}-1600; z_4,z_{32}-6400
z_4,z_{24}	6,400	z_4,z_{23}-1600; z_4,z_{32}-3200

TABLE 104 (Continued)

Reactions of H Antisera

H Antisera	Homologous Titers	Titers with Heterologous Antigens
z_4,z_{32}	12,800	z_4,z_{23}-6400; z_4,z_{24}-3200
z_6	12,800	a-200; 1,5-200
z_{10}	51,200	
z_{27}	102,400	
z_{29}	51,200	
z_{35}	12,800	a-1600; b-400; z-3200; z_{44}-200; z_{45}-400
z_{36}	12,800	z_{38}-200; z_{36},z_{38}-1600
z_{36},z_{38}	25,600	z_{36}-3200; z_{38}-800
z_{37}	12,800	z_{43}-1600; z_{44}-200
z_{38}	25,600	z_{36},z_{38}-800; 1,5-200; 1,6-200
z_{39}	12,800	z_{41}-400; 1,2-200
z_{41}	25,600	z_{39}-800
z_{42}	12,800	f,g-400; g,m-100; g,p,u-200; g,q-100; g,s,t-100; g,t-200
z_{43}	25,600	z_{37}-400
z_{44}	6,400	z-400
z_{45}	51,200	f,g-200; f,g,t-200; g,m-400; g,m,s-800; g,p-400; g,p,u-400; g,q-200; g,s,t-3200; g,t-1600
z_{46}	25,600	
z_{47}	51,200	
z_{48}	12,800	z_{49}-1600
z_{49}	12,800	z_{48}-400
z_{50}	12,800	1,2-800; 1,5-400; 1,6-100; 1,7-800
z_{52}	12,800	k-800; 1,v-800; 1,w1600; 1,z_{13}-800; 1,z_{28}-1600; 1,z_{40}-800
z_{53}	12,800	1,2-400; 1,5-800; 1,6-400; 1,7-200
z_{59}	6,400	
1,2	25,600	b-200; 1,v-200; 1,w-200; 1,z_{13}-400; 1,z_{28}-200; 1,5-6400; 1,6-6400; 1,7-6400
1,5	102,400	1,v-100; 1,z_{13}-800; 1,z_{28}-400; 1,2-3200; 1,6-3200; 1,7-1600
1,6	204,800	b-400; e,n,x-200; e,n,z_{15}-200; 1,v-6400; 1,w-6400; 1,z_{13}-6400; 1,z_{28}-6400; 1,2-25600; 1,5-102400; 1,7-12800
1,7	25,600	b-200; 1,z_{13}-200; 1,2-12800; 1,5-3200; 1,6-3200

$0, z_3, z_5, z_7, z_8, z_9, z_{11}, z_{12}, z_{16}$ to $z_{19}, z_{25}, z_{26}, z_{33}, z_{34}$, and z_{40}. Since antisera produced specifically for these antigens are not required for serotyping they are not listed in Table 103. If desired antisera for these particular antigens may be produced with the serotypes listed by Kauffmann (1966, pp. 111-124). The occurrence of these antigens must be borne in mind and if anomolous reactions occur, they should be considered as a possible source of the difficulty.

Absorption of H Antisera. Cell suspensions (antigens) to be employed for absorption of H antisera are prepared in a manner similar to that suggested for O antigen absorbing suspensions (v. sup.) except that actively motile cultures in the desired phase[7] are employed to inoculate plates of infusion agar.

After an H antiserum has been absorbed it must be titrated with H antigen suspensions prepared from the absorbing culture(s) and with the homologous antigens. Titration is necessary in order to determine whether the absorption is complete and to ascertain the amount of antiserum to be used in subsequent work. This amount is not the same with all antisera (v. inf.). A convenient method for titrations follows. Using a 0.2 ml pipette, add 0.05 ml of antiserum to the first of a series of 13X 100 mm tubes, 0.02 ml to the second, 0.01 ml to the third, and 0.005 to the fourth. Then add 1.0 ml of formalinized broth antigen suspension to each tube and to a fifth tube, to which no antiserum has been added. Incubate at 48-50 C in a water bath. Read and record reactions after 15 min, 30 min, and 1 hour.

Often an absorbed antiserum is used to determine the presence of a single antigenic factor that occurs in a number of different combinations. For example, f,g antiserum is absorbed to obtain single factor f, and this antiserum is used to detect antigen f in the complex f,g,t as well as in f,g. The titer of the f antiserum must be determined for both f,g and f,g,t, since it will be different for these two antigens. Further, the titers must be determined each time a lot of antiserum is absorbed. One cannot rely upon results obtained in a previous absorption even when the same lot of antiserum is employed. In practice it is better to use absorbed antisera in double the amount the causes pronounced, specific agglutination. For instance, if a sample of absorbed single factor f antiserum agglutinates f,g antigen in the amount of 0.01 per ml of antigen and agglutinates f,g,t antigen in the amount of 0.02 per ml of antigen, this f antiserum should be used in the amount of 0.05 per ml of antigen. That is, the antiserum should be used in an amount that will agglutinate either f,g or f,g,t when these antigens are

[7]See *Selection and Reversal of Phases* and *Production of H Antisera.*

present in freshly isolated cultures that have not been passaged in semisolid agar medium to enhance motility and H antigen development. If absorbed antisera have high titers, the dilution may be adjusted to 1 to 5 or 1 to 10 and the amount to be used determined by retitration. It should be borne in mind that if the suggested method of absorption is used, the initial dilution of absorbed antisera will be 1 to 3 when the procedure is completed.

The manner in which H antisera may be absorbed is given in Table 105. Absorption of the antisera employed by the author and colleagues with suspensions prepared from the cultures listed in Table 103 resulted in satisfactory single factor H antisera. It is emphasized that cultures other than those listed in Tables 103 and 105 should not be used for antiserum production or for absorptions unless the *complete* antigenic structure of each of the proposed substitute strains is known with certainty. Kauffmann (1966, pp. 111-124) listed cultures that may be used in lieu of those given in the tables.

The factors marked with an asterisk in Table 105 usually may be prepared by single absorption of the indicated antisera while those not marked in this way should be prepared by double absorption (v. sup., Absorption of O Antisera). Generally, the growth from 5 to 10 plate cultures (90 mm petri dishes) of each absorbing strain is sufficient to remove the heterologous H agglutinins from 1.0 ml of antiserum. However, in some instances growth from additional plate cultures will be required, and in a few instances a lesser amount of absorbing suspension may suffice. In other words, the amount of suspension required to remove heterologous agglutinins from different lots of antisera must be determined by trial.

Selection of H Antisera in Practice. It is unnecessary to test each culture in all of the H antisera. After the O antigens have been identified, one may refer to the antigenic schema (Chapter 9) to ascertain which H antigens occur in combination with the O antigens identified in a particular culture. If a particular strain is agglutinated by antiserum for serogroup B (4,5,12), H antigen suspensions prepared from it should be tested first in the H antisera representative of the H antigens possessed by the serotypes that compose serogroup B. Very often, however, it is not necessary even to employ all of these H antisera since some serotypes occur commonly while others are rare. The sequences in which unabsorbed H antisera may be used are outlined in Table 106. The suggested sequences are based upon the frequency of occurrence of serotypes within the various O antigen groups. For example, if a culture of *Salmonella* belongs to serogroup B, and possesses O antigen 5 as determined by agglutination in

TABLE 105

Preparation of Absorbed H Antisera

Factor(s)	Antiserum	Absorbing Culture(s)
f	ser Derby:f,g,z_8	ser Dublin:g,p + ser Montevideo:g,m,s + ser Senftenberg:g,s,t,z_8,z_9
*h	ser Reading:e,h	ser Abortusequi:e,n,z_{16} + ser Chester:e,n,x,z_{17}
m	ser Enteritidis:g,m + ser Oranienburg:m,t	ser Senftenberg:g,s,t + ser Budapest:g,t + ser Berta:f,g,t + Dublin:g,p + Moscow: g,q
*n,z_{15},z_{17}	ser Salinatis (P230):e,n,z_{15},z_{17}	ser Reading: e,h
p	ser Dublin:g,p	ser Enteritidis:g,m + ser Senftenberg:g,s,t
q	ser Moscow:g,q	ser Enteritidis:g,m
s	ser Montevideo:g,m,s	ser Enteritidis:g,m + ser Oranienburg:m,t
*t	ser Oranienburg:m,t	ser Montevideo:g,m,s
u	ser Rostock:g,p,u	ser Dublin:g,p
v	ser London:l,v	ser Daressalaam,phl:l,w + ser Uganda,phl:l,z_{13}
w	ser Daressalaam:l,w	ser Bredeney,phl:l,v + ser Uganda,phl:l,z_{13}
x	ser Chester:e,n,x,z_{17}	ser Salinatis (P230):e,n,z_{15},z_{17} phase + ser Daressalaam,ph2:e,n,z_{16},z_{18}
*z_6	ser Kentucky:z_6	ser Thompson,ph2:1,5
z_{13}	ser Uganda:l,z_{13}	ser London,phl:l,v + ser Daressalaam,phl:l,w + ser Javiana,phl:l,z_{28}
z_{15}	ser Sandiego:e,n,z_{15},z_{17}	ser Chester,ph2:e,n,x,z_{17} + ser Daressalaam,ph2:e,n,z_{16},z_{18}
*z_{16}	ser Abortusequi:e,n,x,z_{16}	ser Chester,ph2:e,n,x,z_{17}
*z_{17}	ser Chester:e,n,x,z_{17}	ser Abortusequi,ph2:e,n,x,z_{16}
*z_{18}	ser Daressalaam:e,n,z_{16},z_{18}	ser Abortusequi,ph2:e,n,x,z_{16}
*z_{23}	ser Cerro:z_4,z_{23}	ser Dusseldorf:z_4,z_{24}
*z_{24}	ser Dusseldorf:z_4,z_{24}	ser Cerro:z_4:z_{32} + ser Tallahassee:z_4:z_{32}
z_{28}	ser Javiana:l,z_{28}	ser Uganda,phl:l,z_{13} + ser London,phl:l,v
z_{32}	ser Tallahassee:z_4,z_{32}	ser Cerro,z_4,z_{23} + ser Dusseldorf:z_4,z_{24} + ser Wangata:z_4,z_{23}
*z_{35}	ser Chittagong:z_{35}	bioser Paratyphi-A:a + ser Paratyphi-B,phl:b + ser Poona,phl:z
*z_{36}	ser Weslaco:z_{36}	ser Lille:z_{38}
*z_{37}	ser Wichita:z_{37}	ser Senftenberg (2170-58):z_{43} phase
*z_{38}	ser Lille:z_{38}	ser Weslaco:z_{36}

TABLE 105 (Continued)

Preparation of Absorbed H Antigen

Factor(s)	Antiserum	Absorbing Culture(s)
*z_{39}	ser Quimbama:z_{39}	ser Paratyphi-A:a
z_{40}	ser Rutgers:l,z_{40}·	ser Give,phl:l,v
*z_{44}	ser Quinhon:z_{44}	ser Poona,phl:z
*z_{48}	ser Cook:z_{48}	ser Infantis,z_{49} phase
z_{49}	ser Infantis:z_{49}	ser Thompson,ph2:1,5 + ser Cook:z_{48}
z_{50}	ser Mikawashima:z_{50}	ser Paratyphi-B,ph2:1,2 + ser Bredeney,ph2:1,7
*z_{51}	ser Maricopa:g,(p),z_{51}	ser Dublin:g,p + ser Budapest:g,t + ser Moscow:g,q
2	ser Paratyphi-B:1,2	ser Bredeney,ph2:1,7 + ser Thompson,ph2:1,5 + ser Panama,ph2:1,5
5	ser Thompson:1,5	ser Newport,ph2:1,2 + ser Anatum,ph2:1,6
6	ser Anatum:1,6	ser Muenster,ph2:1,5 + ser Thompson,ph2:1,5
7	ser Bredeney:1,7	ser Newport,ph2:1,2 + ser Muenster,ph2:1,5 + ser Panama,ph2:1,5
10	ser Cardiff:1,10	ser Berlin,ph2:1,5 + ser Bredeney,ph2:1,7 + Italiana,ph2:1,11 + ser 6,8:e,h:1,12 (unnamed),ph2:1,12
11	ser Italiana:1,11	ser Berlin,ph2:1,5 + ser Bredeney,ph2:1,7 + ser Cardiff,ph2:1,10 + ser 6,8:e,h:1,12 (unnamed), ph2:1,12 + ser Panama,ph2:1,5
12	ser 6,8:e,h:1,12 (unnamed),ph2	ser Berlin,ph2:1,5 + ser Cardiff,ph2:1,10 + ser Italiana,ph2:1,11

* Single absorptions. Factors not so marked should be prepared by double absorption of antisera (see text).

single factor 05 antiserum, it should be tested first in the H antisera representative of H antigens b;i;e,h;e,n;r; and 1,2. If a strain lacks O antigens 5 and 27, it should be tested first in H antisera i;e,h;e,n;f,g;r; and 1,2 (Table 106). If no reaction is obtained in these antisera, this culture (05-, 027-) then should be tested in H antisera d; l,w, and z. A serogroup B culture that possesses O antigen 27 is tested first in H antisera d;l,w; and 1,7. Rapid agglutination in any of the above-mentioned H antisera is followed by tests in indicated single factor H antisera and by reversal of the phase (v. inf.) if required. Reversal of the H antigen phase of a strain is not always required since both phases of a diphasic serotype may be present in the culture as

isolated or received for examination. For example, if an H antigen suspension prepared from a serogroup B serotype reacts in H antisera for both i and 1,2, both phases have been identified and all that remains to be done is to test in single factor H2 antiserum. However, if all of the above-mentioned tests (Table 106, 1 to 3 incl.) are negative, the suspension should be tested in H antisera for the remainder of the H antigens that are known to be associated with serotypes within the particular serogroup. To make this determination the antigenic schema (Chapter 9) should be consulted. If these tests are negative, the culture should be inoculated into semisolid agar motility medium.

Similar sequences may be followed with cultures

TABLE 106

Selection of H Antisera for use with Members
of Various O Antigen Groups of *Salmonella*

Serogroup		Single Factor O Antisera	Unabsorbed H Antisera
A	1,2,12	2	a
B	[1],4,5,12	5+	1.* b,i;e,h;e,n;r; and 1,2
	[1],4,12	5-,27-	2a.* i;e,h;e,n;f,g;r; and 1,2
			2b.* d;l,w; and z
	[1],4,12,27	27+	3.* d;l,w; and 1,7
			4.* Test for remainder of H antigens that occur in serogroup
C_1	6,7,(14)	7+,14-	1. c;g,m,s;m,t;r;y;k;z_{29}; and 1,5
			2. a;b;d;l,w;e,h; and e,n
		7+,14+	3. d;l,w
			4. See 4, above (serogroup B)
C_2	6,8,(20)	20-	1. d;e,h;k; and 1,2
			2. l,w;z_4,z_{24}; and z_4,z_{32}
		20+	3. i;z_6;z_4,z_{24}
		20- or +	4. See 4, above (serogroup B)
D	9,12 [Vi]	Vi+	1. d
		Vi-	2. d;g,m;l,w; and 1,5
			3. a; and f,g
			4. See 4, above (serogroup B)
E_1	3,10	10	1. e,h;l,w;y;1,6
			2. z_{10}; and g,s,t
E_2	3,15	15	(same as for E_1)
E_3	3,15,34	15,34	(same as for E_1)
			3. See 4, above (serogroup B)
E_4	1,3,19	19	1. g,s,t
			2. i;z_6; and z_{27}
			3. See 4, above (serogroup B)
F	011		1. r; and e,n,x
			2. i; and 1,2
			3. See 4, above (serogroup B)
G_1	13,22	22	1. z; and 1,6
			2. See 4, above (serogroup B)
G_2	13,23	23	1. z;l,w; and z_{29}
			2. b;d;f,g; and 1,7
			3. See 4, above (serogroup B)
	018		1. z_4,z_{23}
			2. See 4, above (serogroup B)
	030		1. b; and e,n,x
			2. See 4, above (serogroup B)

*1. Indicates antisera that should be tried first. If tests in these antisera are negative then the antisera listed opposite 2 (or 2a,2b), 3, and 4 should be employed, successively. However, if a serogroup B strain posses O Antigen 27 and H Antisera indicated (3) should be tried first. Similarly, if a serogroup C_2 culture reacts in single factor 020 antiserum, the H Antisera listed opposite 3 should be employed first.

NOTE: For selection of H Antisera for members of O Antigen groups not listed, the antigenic schema must be consulted.

that belong to serogroups A to G, 18 and 30. A very high percentage of salmonellae seen in daily practice belong to these O antigen serogroups.

The antigens of the g . . . complex are more difficult to determine, perhaps, than most others. The following notes may be helpful in their determination. If H antigen suspensions derived from actively motile cultures are tested in unabsorbed f,g;g,m,s; and m,t antisera at 1:1000 dilution, certain assumptions may be made based upon the reactions obtained. These are illustrated in the following examples:

Antisera

	f,g	g,m,s	m,t
1.	-	-	++++
2.	-	++++	++ or +++
3.	++++	++++	-
4.	++++	++++	++++

When an antigen reacts as illustrated in the first example it may be assumed that factor g is not present. Such an antigen then should be tested in single factor f,m,s,t,p,q, and u antisera. When a pattern similar to example two is obtained it may be assumed that factor f is not present and tests are made with single factor antisera m,s,t,p,q,u, and z_{51}. Reactions such as those in example three indicate that factor t is not present. In this event single factor f,m,s,p,q,u, and z_{51} antisera are used. When a pattern such as that indicated in the fourth example is seen tests must be made with single factor antisera for all members of the g . . . complex (f,m,p,q,s,t,u,z_{51}).

The speed and intensity with which antigen suspensions react in unabsorbed g . . . complex antisera react also is helpful in determination of g . . . complex antigens and in the selection of the proper single factor antisera. This is illustrated in Table 107. When a homologous antigen-antibody system exists (e.g., f,g antigen in f,g antiserum) agglutination reactions are rapid and complete (Table 107). Whether this occurs is dependent upon the dilution of antiserum used and upon the condition of the antigen suspension. The examples given in Table 107 resulted from tests with antisera diluted 1 to 1,000 and antigens prepared from actively motile cultures. Antigens that are related only partially to those represented by the antibody content

TABLE 107

Representative Reactions of Antigens of the g . . . Complex

Antigen suspensions	Antisera								
	f,g			g,m,s			m,t		
	15*	30	60	15	30	60	15	30	60
f,g	++++			-	-	-	-	-	-
f,g,t	++++			-	-	-	+	++	++
g,m	++++			++++			-	-	-
g,m,s	++	+++	+++	++++			-	-	-
g,p	++++			++++					
g,p,u	++++			+	+	+	-	-	-
g,q	+	++	+++	++	+++	+++	-	-	-
g,s,t	++++			++++			++++		
g,t	++++			tr (-)	-	-	++++		
g,z_{51}	+++	+++	+++	-	-	+	-	-	-
m,t	-	-	-	-	-	-	++++		
g,m,t	+	+	++	+++	+++	+++	tr	++	++
g,m,s,t	++++			++++			tr	+	+

* Minutes (15,30,60)

+ to ++++, degrees of agglutination. +++, complete. tr, trace.

of an antiserum either fail to react or react slowly and incompletely. The f,g antigen (Table 107) failed to react in g,m,s antiserum because the g antibody content of the g,m,s antiserum (diluted 1 to 1,000) was insufficient to cause agglutination. The f,g,t antigen reacted slowly and incompletely in m,t, antiserum, indicating a partial relationship between the antigens contained in the antigen suspension (f,g,t) and those reflected by the antibody content of the antiserum (m,t). Study of the reactions listed in Table 107 shows that the closer the relationships between the antigen factors and the agglutinin content of an antiserum, the more rapid and complete are the reactions. Cultures that possess g . . . complex antigens usually are monophasic, but there are enough exceptions to warrant an attempt to reverse the phases.

It is useless to test an antigen suspension with absorbed single factor antiserum unless it reacts with the unabsorbed antiserum containing the single factor. It is wasteful to test a suspension for the presence of v or w antigens unless it reacts in unabsorbed l,v or l,w antiserum. Similarly, a suspension should not be tested for antigens f,m,p,q,s,t,u, or z_{51} in single factor antisera unless it reacts with f,g;g,m;g,p;g,q;g,s,t;m,t; or g,z_{51} antiserum. If this principle is observed, much less absorbed antiserum will be used and much work necessary for its absorption will be saved.

Some investigators pool H antisera and the pools are used for preliminary tests. For example, all antisera that contain agglutinins for antigens f,g,m,p,q,s,t,u, and z_{51} may be included in the same pool. After it is determined with which pool an antigen suspension reacts, the suspension is tested in the component antisera for that pool. Spicer (1956) devised a more complex system of four pools or polyvalent mixtures. This system was based upon a simplified scheme for salmonellae (Kauffmann and Edwards, 1952). Readers who are interested in the Spicer pools are referred to the second edition of this book (pp. 123, 124).

Isolation of Phases, Phase Reversal. While a culture that contains H antigens of the g . . . complex, the z_4 . . . complex, z_{27}, z_{29}, z_{36}, z_{38}; $1,z_{40}$; or z_{44} is likely to be monophasic, there are exceptions, and an attempt should be made to isolate a second phase. However, if a strain possesses any of the other H antigens, it is likely to be diphasic. As noted above, it sometimes is possible to identify both phases of a diphasic culture when it is first isolated or received for examination. More often, however, only one phase can be detected initially in diphasic cultures, especially when they are recently isolated from single colonies on plating media used in the recovery of salmonellae.

A culture of ser Typhimurium, for example, just picked from a single colony on isolation media would be likely to react in i antiserum or in 1,2 antiserum but not in both. If the strain is in phase 1 (i), it is necessary to reverse the phases and isolate phase 2 (1,2). Likewise, if only phase 2 is apparent, isolation of phase 1 is required.

It is possible to isolate and identify the suppressed phase of a culture by plating it on infusion agar medium and examining single colonies. However, if a phase is present in such small amount that it is not detectable by agglutination it is clear that a large number of colonies may be examined before the desired phase is found. Such a process is time consuming and a much better method is the use of semisolid phase reversal medium to which H antiserum has been added.

If a culture suspected of being ser Typhimurium is in phase 1 (i), it should be inoculated into semisolid phase reversal medium to which i antiserum has been added. A simple way of doing this is to add a 3 mm loopful of sterile i antiserum to 3 ml of melted and cooled (40 to 45 C) semisolid medium in a 13 X 100 mm tube. After thorough mixing, the semisolid medium is allowed to gel and is inoculated by just puncturing the medium at the surface (3 or 4 mm stab). The H agglutinins in the antiserum immobilize the homologous flagella (i in this instance) and upon incubation growth of the phase that is not immobilized spreads throughout the medium. After overnight incubation the top portion of the medium is heated in a flame and the melted agar, including the point of inoculation, is poured from the tube. After the tube cools a loopful of the spreading growth is transferred to infusion broth which is incubated overnight (or for 4 to 6 hr in a waterbath) and formalinized. The desired phase may be obtained in this manner. Antisera to be added to phase reversal medium may be sterilized by addition of an excess of chloroform (add about 2 drops per ml, shake well, and refrigerate for 24 hr). Ordinarily, a 3 mm loopful of glycerolated H antiserum diluted 1 to 5 or 1 to 6 (sterile) added to 3 ml of semisolid medium yields the desired result. However, it may be necessary to employ greater or lesser amounts of antiserum on occasion. If sufficient O agglutinins are present to prevent rapid spread of a suppressed phase, or if an antiserum contains H agglutinins that might interfere with its spread, a smaller amount of antiserum may be used. If an antiserum has a titer of 1 to 10,000 to 1 to 20,000, 0.1 ml of a 1 to 100 dilution may be added to 3 ml of medium. When treated in the above-mentioned manner monophasic cultures do not spread through the medium. Since their flagella are immobilized, growth is confined to the point of inoculation. Useful modifications of this method have been devised by Hajna (1944), Levine (1945), Juenker (1946), Hinshaw and McNeil (personal communication, 1946), Kuhn (1947), and Costin (1967). Hinshaw and McNeil (1946) used a Craigie tube in which the surface of the medium in the small inside tube was moistened with a loopful of antiserum for the phase to be suppressed. Other methods for phase reversal, including use of the Craigie tube, are mentioned in the section entitled Production of H Antisera (v. inf.).

If a newly produced H antiserum does not permit phase reversal as it should, the amount of sterile antiserum employed may be too large or too small. This may be corrected easily by adjusting the amount. However, if the amount of antiserum used is not the cause, the antiserum may require absorption. For example, an unabsorbed antiserum for H antigen b may be satisfactory for use in agglutination tests, but may contain enough agglutinins for the second phase antigens to immobilize *both* phases when phase reversal is attempted. To correct this the b antiserum (produced with phase 1 of ser Paratyphi-B) should be absorbed with phase 2 of bioser Paratyphi-B, Java (1,2). If antisera are produced properly the number of absorptions similar to that mentioned above is minimized. Further, newly produced antisera should be tested thoroughly as mentioned elsewhere. If this is done, the need for absorption of certain antisera will be apparent.

Some monophasic serotypes of *Salmonella* possess H antigens that also are present in many diphasic serotypes. Notable examples of this are bioser Paratyphi-A (1,2,12:a:-) and *S. typhi* (9,12,Vi:d:-). By examination of the antigenic schema it is possible to determine the various combinations of O and H antigens of serotypes of salmonellae.

Certain species and bioserotypes of salmonellae are closely related antigenically, and for accurate characterization of these it is necessary to employ certain biochemical tests. For example, bioserotypes Paratyphi-C, Decatur, and Typhisuis and the species *S. cholerae-suis* all are related closely, and bioserotypes Pullorum and Gallinarum cannot be distinguished by serological methods alone. The distinctive biochemical reactions of these and other bioserotypes of salmonellae that must be differentiated by biochemical methods are listed in Tables 90 to 93 (v. sup.). The reactions given by the species *S. cholerae-suis* and *S. typhi* and of *S. enteritidis* bioser Paratyphi-A also are summarized in those tables.

STAGES IN SEROLOGICAL CLASSIFICATION OF *SALMONELLA*

The growing multiplicity of serotypes of *Salmonella* and the numerous antisera required for complete characterization of all of them make it impractical for the

personnel of most laboratories to attempt complete serological typing of all salmonellae. However, this situation should not deter the personnel of any laboratory, large or small, from making accurate identification of infections caused by salmonellae or from exact recognition of the serotypes that are of greatest importance in the epidemiology of salmonellosis. Exact recognition includes the best possible isolation procedures and adequate differential biochemical methods in addition to serological procedures, regardless of the extent to which the latter are pursued.

Various factors such as available space and personnel make it necessary for those in charge of each laboratory to decide to what extent work with salmonellae can be carried and how much time profitably can be devoted to it. As regards hospital laboratories, the census and its nature also are important factors to be considered in making this decision. Whatever the extent of the work may be, all work that is done should be performed with the greatest care and by the best methods.

Fortunately, there are a number of points at which more or less natural divisions can and should be made. It is at one of these points or stages that work with the majority of cultures may be stopped in a particular laboratory. If serologic identification of strains is not completed, they should be forwarded to a central or reference laboratory, together with the minimal required information[8] regarding each isolate. Of course, the biochemical reactions of cultures should be determined before they are forwarded to a reference laboratory. Also, strains should be inoculated into tubes of stock culture medium (see Chapter 18) for shipment. *Neither* TSI *nor* KI agar cultures should be used for this purpose.

The first of these stages is complete identification of certain important species and serotypes of salmonellae. These are as follows:

S. typhi: 9,12,Vi:d:-
S. enteritidis bioser Paratyphi-A:1,2,12:a:-
S. enteritidis ser Paratyphi-B:4,5,12:b:1,2
*S. cholerae - suis:*6,7:[c]:1,5
S. enteritidis ser Typhimurium 4,5,12:i:1,2.

The first three salmonellae listed above invade the blood stream and cause enteric fever. The fourth may produce enteric fever, particularly in children, but also may cause only primary gastroenteritis. These sero-

---

[8]Identifying information: Name, age, and sex of individual(s) involved.

Source of specimen: Feces, urine, blood, food, other (specify).

Clinical Diagnosis: Diarrheal disease, enteric fever, asymptomatic person, or other (specify).

Epidemiological relationships of cases, when known.

types and species should be identified and a presumptive report sent to the physician as soon as possible so that appropriate treatment can be instituted. Also hospital authorities including the hospital epidemiologist, if one is available, should be notified so that appropriate control measures may be taken. The fifth microorganism (ser Typhimurium) is listed because it occurs more frequently than any other serotype of *Salmonella* and should be identified. Parenthetically, almost any serotype of *Salmonella* may invade the blood stream occasionally and produce enteric fever and infrequently they may be isolated from other extraintestinal sources such as cerebrospinal fluid, bone marrow, etc. (See Report, Committee on Salmonella, 1969, or Ewing, 1969).

A culture that is anaerogenic, produces an alkaline slant and acid reaction in the butt, forms only a small amount of hydrogen sulfide in TSI or KI agar medium (K/A+)[9], and fails to hydrolyze urea rapidly (i.e., within 1 or 2 hr) should be considered as a possible strain of *S. typhi* and treated accordingly. A *dense* suspension prepared from such a culture first should be tested in the living (unheated) state in group D *(Salmonella)* antiserum and in Vi antiserum on a slide. After these tests are completed, the suspension should be heated in a beaker of boiling water for about 15 min, cooled, and retested in the same antisera (Table 102). Polyvalent antiserum, which contains Vi antibodies, also may be included in the above-mentioned tests. If the reactions of a culture are typical of *S. typhi* on TSI or KI agar medium and tests for urease are negative, and if a suspension reacts as shown in Table 102 (i.e., a reversal of the reactions in group D antiserum and Vi antiserum after heat treatment), a presumptive report can, and should, be made at this point. Confirmatory biochemical tests should be made, and the H (flagellar) antigens must be determined before a final report is issued.

Cultures that react in polyvalent antiserum for *Salmonella* but fail to agglutinate in Vi antiserum should be tested in O grouping antisera (Table 108).

S. enteritidis bioser Paratyphi-A is relatively uncommon in the United States but bacteriologists should be familiar with it since it sometimes is imported by tourists and others. The biochemical reactions of bioser Paratyphi-A, like those of *S. typhi,* are somewhat atypical as compared to salmonellae in general (see Tables 87 and 91). Hydrogen sulfide frequently is not formed in TSI or KI agar medium but gas is formed from glucose; hence the reactions in these media usually are K/Ag. Cultures that give this appearance in

one or the other of these media and are urease negative should be tested in O grouping antisera. If rapid and complete agglutination occurs in O group A (1,2,12) antiserum and confirmatory biochemical reactions obtained (20 to 24 hr) are consistent with those given by bioser Paratyphi-A, presumptive identification can be made and a preliminary report issued. The differential biochemical tests should be observed for a longer period and the H antigens must be determined before a final report is made.

The reactions of the three remaining salmonellae listed above in TSI or KI agar medium are like those given by the majority of salmonellae, i.e., K/Ag+++. This is true of *S. cholerae-suis* bioser Kunzendorf, which is the most commonly occurring form of *S. cholerae-suis*. Strains that yield this type of reaction in the above-mentioned primary differential media and are urease negative first should be tested in polyvalent antiserum for *Salmonella* and then in O grouping antisera for serogroups A through E (Table 108). Confirmatory biochemical tests should be made and the H antigens of such cultures must be determined before final reports can be made.

If a suspension reacts rapidly and completely in antiserum for *Salmonella* O group A, the H antigens of the culture should be tested in H a antiserum.

The H antigens of strains that react in serogroup B antiserum should be tested in H antisera b, i, and the 1 . . . complex (1,2; 1,5; 1,6; 1,7). Two of the microorganisms listed above may be expected to react in one or another of these three H antisera. Cultures of ser Paratyphi-B possess H antigen b in the first phase and H antigens 1,2 in the second. If, in the original tests, a reaction occurs in only one of H antisera mentioned, the phase must be reversed (v. sup., Isolation of Phases, Phase Reversal). Strains of ser Typhimurium possess H antigen i in the first phase and H antigen 1,2 in the second. Again, if agglutination occurs in antiserum for only one of these, the phase must be reversed.

Normally, the Kunzendorf bioserotype of *S. cholerae-suis* occurs as a monophasic form in the second phase (1,5). However, antiserum for H antigen c should be available for the identification of the diphasic form of *S. cholerae-suis* and for the identification of cultures of *S. enteritidis* ser Paratyphi C (6, 7, Vi: c: 1,5) should they be encountered.

The H antigens of cultures that are agglutinated by group D antiserum should be tested in H antisera a, d, and the 1 . . . complex. Phases should be reversed as required.

For exact characterization of serotypes Paratyphi-B and Typhimurium absorbed single factor H antiserum for H antigen factor 2 is required since both of these serotypes possess H antigens 1,2 in their second phase.

[9]See Chapter 2.

TABLE 108

Reactions of Representative Antigens with O Grouping Antisera

Antigen Suspensions	Antisera for Serogroups:					
	A 1,2,12	B 4,5,12	C₁ 6,7	C₂ 8	D 9,12	E 3,10,15
1,2,12	+ + + +	–	–	–	+	–
1,4,12	+ +	+ + + +	–	–	+	–
1,4,5,12	+ +	+ + + +	–	–	+	–
4,5,12	–	+ + + +	–	–	+	–
6,7	–	–	+ + + +	–	–	–
6,8	–	– or +	+ +	+ + + +	–	–
8	–	+	–	+ + + +	–	–
8,20	–	+	–	+ + + +	–	–
9,12	–(+)	+	–	–	+ + + +	–
1,9,12	+ +	+	–	–	+ + + +	–
3,10	–	–	–	–	–	+ + + +
3,15	–	–	–	–	–	+ + + +
1,3,19	+ +	–	–	–	–	+ + + +

+ to + + + + indicates different degrees of agglutination.

Adapted from Ewing, 1966

Similarly, absorbed single factor 5 antiserum is required for exact characterization of *S. cholerae-suis*.

The serogroup to which many additional salmonellae belong may be determined with O antisera for serogroups A to E incl (Table 108) although their H antigens cannot be determined with the limited number of H antisera referred to in the preceding paragraphs. If the biochemical reactions of such a strain are consistent with those given by salmonellae, it may be reported as a member of the serogroup to which it belongs. The strain then may be referred to another laboratory for complete serologic analysis.

Cultures that do not react in the above-mentioned O antisera (A to E) may be members of other serogroups (F, G etc.) and should be subjected to a sufficient number of biochemical tests to be certain that they are salmonellae and then forwarded to a reference laboratory along with the results obtained and the necessary information concerning each isolant. If a comprehensive polyvalent antiserum is available such cultures may be expected to react in it.

It is apparent that only a few antisera are required for the first stage. These are a polyvalent antiserum, O antisera for serogroups A to E incl. (the serogroup E antiserum referred to here is a mixed antiserum that agglutinates all members of E₁,E₂,E₃, and E₄), antisera for flagellar antigens a,b,c,d, and the 1 . . . complex (mixed 1,2; 1,5; 1,6; and 1,7 antisera) single factor H antisera for factors 2 and 5, and a Vi antiserum prepared with *Citrobacter* 029:Vi. These antisera permit complete characterization of several important salmonellae and partial characterization of many others. The procedures outlined in stage one are, with a few exceptions, similar to those mentioned by Kauffmann and Edwards (1947).

Stage 1 may be modified by deletion of the polyvalent and the two single factor H antisera if desired. Or, it may be augmented, for example, by the addition of single factor H antisera for factors 6 and 7. If this is done, a number of other serotypes may be characterized completely.

The second stage is intermediate and should be useful to investigators in many laboratories. In this the 50 or 60 commonly occurring serotypes, bioser

Paratyphi-A, and the species *S. cholerae-suis* and *S. typhi* are identified completely. These species and serotypes belong to serogroups A to H, K, L, and N and comprise a list that may be regarded as an abbreviated antigenic schema (Table 109). The 50 serotypes and variants listed in Table 109 accounted for about 96 per cent of salmonellae isolated from all sources in the United States (Edwards, 1962; Ewing, 1966, 1969; and Martin and Ewing, 1969). The same serotypes (Table 109) also were the most commonly occurring forms listed in other reports on incidence of serotypes (for references and additional data, see Ewing, 1966, 1969; van Oye, 1964; Kelterborn, 1967; and Martin and Ewing, 1969). (See also Development of the Schema, Chapter 9.)

Fifty-one unabsorbed and 32 absorbed antisera are required for complete serological identification of the above-mentioned commonly occurring serotypes. The antisera are listed in Table 110. Clearly, these antisera also permit complete serological identification of serotypes not listed in Table 109 and allow partial characterization of others.

Stage 2 may be modified by the addition of antisera for complete serotyping of particular strains that may become epidemic or endemic in an area or of serotypes that are of special interest. Further, the personnel of some laboratories may wish to limit their work to complete serological identification of about 25 serotypes and species which comprise the most frequently occurring salmonellae (e.g., see Ewing, 1969). For this 33 unabsorbed and 21 absorbed antisera are needed. The above-mentioned 25 serotypes and species comprised about 89 per cent of the isolates from human sources, and about 80 per cent of these from all other sources, reported during the period 1963 to 1967 incl. (loc. cit).

The third stage is complete serological identification of all serotypes of *Salmonella,* including characterization of new serotypes. This work should be restricted to a relatively small number of laboratories, which are equipped for the work and which may serve as reference centers. For comprehensive work of this sort the unabsorbed and absorbed antisera listed in Tables 99, 101, 103, and 105 are required.

It is the author's opinion that emphasis should be given to *complete* serological identification of the bacteria regardless of the stage that the personnel of a laboratory may adopt. If the first stage is adopted, then the few species and serotypes mentioned should be identified completely. There are at least two additional stages that should be mentioned. It is understood that there are many institutions in which the number of stool specimens submitted is so small that cultural examination is not warranted. In such

instances freshly collected specimens should be placed in transport media (Chapters 2 and 18) and sent to a central laboratory together with pertinent information (see stage one). This procedure also should be followed in institutions in which the volume of specimens of other kinds is so large that time and technical assistance are unavailable for adequate work with stool specimens. Another possible stage might be applicable in institutions in which adequate cultural work can be done, but time or assistance may not be available for serological identification. In this situation careful cultural examination of specimens should be done and isolates should be identified to the generic level by adequate biochemical tests (Chapters 2 and 3). With few exceptions, cultures of *S. typhi* should be identified, at least presumptively. Cultures of *Salmonella,* and other significant isolates, then should be sent to a central laboratory for serological identification. However, it is hoped that it will be feasible for the personnel of most laboratories to adopt stage one, two, or three, or some modification of one of them, if this has not been done.

So-called simplified methods for serological examination of cultures of *Salmonella* have been proposed by a number of investigators (e.g., see Edwards and Kauffmann, 1952; Kauffmann, 1957; Kauffmann, 1958; Edwards, 1962; and Kauffmann, 1966). Some of these were extensions of methods similar to those outlined in stage one (above). However, no absorbed O or H antisera were recommended and in some instances H antigens were grouped together into complexes labeled G, L, Z_4 etc. Very little has been reported concerning the epidemiological adequacy of these methods or their effectiveness when applied to large numbers of cultures. Since these methods do not permit exact identification of a large number of serotypes, including commonly occurring ones, the author does not subscribe to the use of so-called simplified methods.

PRODUCTION OF ANTISERA

Production of O Antisera. To produce antisera that contain only agglutinins for heat stable antigens, it is necessary to inactivate the H antigens. This may be accomplished by heating broth cultures of the organisms in a boiling water bath or an Arnold sterilizer for 2 hr. After the broth cultures are heated they may be preserved with 0.3% formalin. Rabbits are given five intravenous injections of the heated bacteria at intervals of 4 days. The amounts usually administered are 0.5 ml, 1.0 ml, 2.0 ml, 3.0 ml and 3.0 ml. Five injections usually result in the production of an antiserum of satisfactory titer but if, on test bleeding, the titer is low, a sixth injection may be given. If a satisfactory titer then has not been attained it is

TABLE 109

Abbreviated Antigenic Schema

Species and Serotype		Group	O Antigens	H Antigens Phase 1	H Antigens Phase 2
S. enteritidis,	bioser Paratyphi A	A	1,2,12	a	—
	ser Paratyphi B	B	1,4,5,12	b	1,2
	ser Paratyphi B, Odense		1,4,12	b	1,2
	bioser Java		1,4,5,12	b	[1,2]
	ser Stanley		4,5,12	d	1,2
	ser Schwarzengrund		1,4,12,27	d	1,7
	ser Saintpaul		1,4,5,12	e,h	1,2
	ser Reading		4,5,12	e,h	1,5
	ser Chester		4,5,12	e,h	e,n,x
	ser Sandiego		4,12	e,h	e,n,z_{15}
	ser Derby		1,4,5,12	f,g	—
	ser California		4,5,12	m,t	—
	ser Typhimurium		1,4,5,12	i	1,2
	ser Typhimurium, Copenhagen		1,4,12	i	1,2
	ser Bredeney		1,4,12,27	l,v	1,7
	ser Heidelberg		1,4,5,12	r	1,2
S. cholerae-suis		C_1	6,7	c	1,5
S. cholerae-suis,	bioser Kunzendorf		6,7	[c]	1,5
S. enteritidis,	ser Braenderup		6,7	eh	e,n,z_{15}
	ser Montevideo		6,7	g,m,s	—
	ser Oranienburg		6,7	m,t	—
	ser Thompson		6,7	k	1,5
	ser Infantis		6,7	r	1,5
	ser Bareilly		6,7	y	1,5
	ser Tennessee		6,7	z_{29}	—
	ser Muenchen	C_2	6,8	d	1,2
	ser Manhattan		6,8	d	1,5
	ser Newport		6,8	e,h	1,2
	ser Blockley		6,8	k	1,5
	ser Litchfield		6,8	l,v	1,2
	ser Tallahassee		6,8	z_4,z_{32}	—
	ser Kentucky		(8),20	i	z_6

TABLE 109 (Continued)
Abbreviated Antigenic Schema

Species and Serotype		Group	O Antigens	H Antigens Phase 1	Phase 2
S. typhi	bioser Miami	D	1,9,12	a	1,5
			9,12,Vi	d	—
S. enteritidis,	ser Enteritidis		1,9,12	g,m	—
	ser Berta		9,12	f,g,t	—
	ser Dublin		1,9,12	g,p	—
	ser Panama		1,9,12	l,v	1,5
	ser Javiana		1,9,12	l,z_{28}	1,5
	bioser Pullorum		9,12	—	—
	ser Anatum	E_1	3,10	e,h	1,6
	ser Meleagridis		3,10	e,h	l, w
	ser Give		3,10	l,v	1,7
	ser Newington	E_2	3,15	e,h	1,6
	ser Illinois	E_3	(3),(15),34	z_{10}	1,5
	ser Senftenberg	E_4	1,3,19	g,s,t	—
	ser Simsburgy		1,3,19	z_{27}	—
	ser Rubislaw	F	11	r	e,n,x
	ser Poona	G	13,22	z	1,6
	ser Worthington		1,13,23	z	l,w
	ser Cubana		1,13,23	z_{29}	—
	ser Florida	H	1,6,14,25	d	1,7
	ser Madelia		1,6,14,25	y	1,7
	ser Cerro	18 (K)	18	z_4,z_{23}	—
	ser Siegburg		6,14,18	z_4,z_{23}	—
	ser Minnesota	21 (L)	21	b	e,n,x
	ser Urbana	30 (N)	30	b	e,n,x

Adapted from Ewing, 1966.

TABLE 110

Antisera Needed for Typing the 50 Commonly Occurring Serotypes of *Salmonella*

O Antisera		H Antisera			
		Phase 1		Phase 2	
Unabsorbed	Absorbed	Unabsorbed	Absorbed	Unabsorbed	Absorbed
1,2,12	2	a	—	e,n,x	x
4,5,12	5	b	—	e,n,z_{15}	z_{15}
4,12,27	27	c	—	1,2	2
6,7	7	d	—	1,5	5
6,8	—	e,h	h	1,6	6
(8),20	20	f,g	f	1,7	7
9,12	—	g,m	m	z_6	z_6
3,10	10	g,m,s	m and s		
3,15	15	g,p	p		
(3),(15),34	34	g,s,t	s and t		
1,13,19	19	i	—		
11	—	k	—		
13,22	22	l,v	v		
1,13,23	23	l,w	w		
6,14,25	14 and 25	1,z_{13}	z_{13}		
18	—	1,z_{28}	z_{28}		
21	—	m,t	m and t		
30	—	r	—		
		y	—		
Vi	—	z	—		
		z_4,z_{23}	z_{23}		
		z_4,z_{32}	z_{32}		
		z_{10}	—		
		z_{27}	—		
		z_{29}	—		
Totals (18 + Vi)	13	25	12	7	7

Revised from Edwards, 1966.

advisable to discard the rabbit since further injections usually are without effect. The rabbits are bled on the sixth or seventh day after the last injection. It has been the author's practice to exsanguinate the rabbits at this time rather than to bleed them repeatedly and attempt to maintain the agglutinin titer by further injections. Prolonged immunization may result in an antiserum of such broad reactivity that it is not suitable for use. The antisera are preserved by the addition of an equal volume of glycerol.

A different method of O antiserum production in which bacteria extracted with alcohol and acetone were used was described by Roschka (1950). The bacteria are grown on infusion agar plates for 24 hr and removed with physiological saline solution so that a dense suspension results. The suspension is heated at 100 C for 2 hr to inactivate H antigens and centrifuged. The sediment is suspended in 95% alcohol and incubated 4 hr at 37 C. The suspension is again centrifuged, washed twice with acetone, dried overnight at 37 C, and ground to a fine powder. The dry powder is placed in a stoppered tube and kept at room temperature. If kept dry, the preparations are usable for several years.

The powder is suspended in sterile saline for injection. Five injections are given consisting respectively of 0.4, 0.8, 1.6, 3.2, and 6.4 mg of dried cells. These amounts of cells can be suspended conveniently in volumes of physiological saline solution varying from 0.5 to 3.0 ml. The author's experience indicates that high titered antisera may be obtained by this method with loss of fewer animals than by the usual method. Also, fewer animals fail to develop a satisfactory titer when the Roschka method is used. This method now is used exclusively by the author and colleagues for the production of O antisera for salmonellae and related bacteria.

Production of Vi Antiserum. To prepare Vi antiserum it is necessary that the culture be in the Vi (V) form. Any bacterium that possesses Vi antigen may be used to produce the antiserum. To insure a pure Vi culture the strain should be plated repeatedly and Vi colonies selected until no O (W) colonies appear upon the plates. Living broth cultures of the Vi form may be used to produce the antiserum, but when motile strains are used this procedure results in the production of an antiserum having a high H titer and it may be difficult to absorb the H agglutinins from such an antiserum. Since Peluffo (1941) showed that Vi antigen is not inactivated by treatment with absolute alcohol, the writer has produced antigen for Vi antiserum production by removing the growth of a pure Vi culture from agar plates, suspending it in absolute alcohol, and dehydrating the bacteria *in vacuo*. This treatment largely inactivates the H antigens. The dried cells are ground to a powder which is stored in a stoppered tube. Under these conditions the Vi antigen is stable and the powder can be stored without deterioration. It is suspended in physiological saline solution just before injection and is administered intravenously without further treatment. The rabbits usually withstand the injections well, and fairly large amounts can be given. Four or five injections of this material usually result in the production of a potent Vi antiserum.

Any antiserum produced by the above-mentioned method contains O agglutinins as well as Vi antibodies since the whole bacterial soma is injected. It also contains a low titer of H agglutinins since the agglutinogenic properties of H antigens are not annulled completely by alcohol treatment. The writer uses the Vi form of *Citrobacter* 029, to produce Vi antiserum. Since the O and H antigens of this culture are unrelated to those of salmonellae that are known to possess Vi antigen, the antiserum may be employed without absorption. If *S. typhi* is used for production of Vi antiserum, the O and H agglutinins must be removed by absorption with a motile O (W) form of this species.

Production of H antisera. The antigens used in the production of H antisera are actively motile broth cultures of the bacteria diluted with an equal amount of physiological saline solution containing 0.6% formalin. It is extremely important that only very actively motile forms be used for injection since H antigen is associated with the flagella of the bacteria and active motility is necessary to insure a maximum development of H antigens. Rarely one may encounter a nonmotile strain that is flagellated. These are flocculated by H antisera (Edwards, et al. 1946). As a rule the content of H antigen is directly proportional to the motility. If it is desired to use a given strain for antiserum production and the culture is not sufficiently motile, it may be passaged several times through semisolid agar to improve motility. The semisolid agar (phase reversal medium) used by the writer is a slight modification of that of Jordan et al. (1934).

If a culture of a motile bacterium is inoculated at the surface of a tube of semisolid agar (stab inoculation about 3 mm deep) and incubated overnight, the bacteria spread throughout the medium. The upper part of the medium may be heated and allowed to run out of the tube, leaving only the agar in the base. Inoculations made from the medium at the bottom of the tube yield actively motile cultures with maximum development of H antigen. Broth cultures inoculated with organisms passaged through this medium serve admirably as antigens for H antiserum production when diluted with formalized physiological saline solution. Craigie (1931) employed a tube of semisolid medium containing an inner tube which was open at each end with the upper end extending above the surface of the medium. Inoculation of the small inner tube was followed by migration of the organisms through its bore to the bottom and thence through the outer tube to the surface of the medium. Use of this method obviates the necessity of melting the upper portion of the medium and removing it from the tube. It is a very satisfactory method and is widely used. U tubes also are used for this purpose. When possible, monophasic cultures or stable monophasic variants of serotypes that normally are diphasic are employed for H antiserum production. A number of such cultures were listed by Edwards and Bruner (1946) and by Kauffmann (1965 or 1966).

An antigen such as that described above may be used in the production of H antiserum for any of the monophasic serotypes and monophasic variants (v. sup.). Unfortunately, it is not possible to produce all the H antisera from monophasic serotypes. Where this is not possible it is necessary to isolate one phase for use as an antigen. Each phase of a diphasic serotype contains traces of the other phase which are not detectable in agglutination tests but produce agglutinins for the temporarily suppressed phase when injected into rabbits. For this reason antiserum derived from one phase of a diphasic serotype is likely to contain a minor component of agglutinins for the other phase as well. *The amount of these minor agglutinins in the antiserum depends upon the cultures used and the care exercised in the isolation of the phase for which it is desired to produce agglutinins.*

One method of obtaining an isolated phase for antiserum production is by colony selection. If a diphasic culture is plated, both phase 1 and phase 2 colonies may be found. Use of one of these phase 1 colonies for antiserum production usually results in an antiserum, the phase 2 titer of which is from 5 to 20 per cent of its phase 1 titer. Continued plating and selection of phase 1 colonies until no phase 2 colonies can be found upon the plates reduces the tendency of the culture to produce phase 2 agglutinins. This is a long and laborious process and cannot be relied upon to produce antisera of desired specificity. A more rapid and reliable method of obtaining isolated phases is that of Gard (1938), in which agglutinating antiserum is added to semisolid agar to immobilize the phase to be suppressed. For example, to isolate phase 1 of ser Typhimurium (4,5,12:i:1,2), antiserum derived from phase 2 of ser Newport (1,2) may be used. If an antiserum having an H titer of 1 to 10,000 to 1 to 20,000 is available, 0.1 ml may be added to 25 ml of melted agar and the mixture poured into a petri dish. After the agar gels the medium is inoculated at a spot at one side of the dish. The dish is incubated in an upright position at 37 C. It is advisable to use a glazed porcelain top to prevent water of evaporation from dropping on the medium. It will be found that phase 1 of ser Typhimurium (4,5,12:i) will swarm across the plate, while phase 2 (4,5,12:1,2) is immobilized and held at the site of inoculation by the 1,2 agglutinins in the ser Newport antiserum. Bacteria taken from the side of the plate opposite the site of inoculation are planted in broth, incubated for 12 to 15 hours, and formalinized. Cultures prepared in this way are used to inject rabbits and the antisera derived from them

usually contains only traces of agglutinins for the suppressed phases. However, it is advisable to passage a culture on two or three Gard plates, each containing antiserum for the phase to be suppressed. The method of selection of phases given in the section entitled Isolation of Phases, Phase Reversal (v. sup.) is a modification of the Gard technique. It may be employed instead of the Gard technique for selection of phases for H antiserum production. The formalinized broth cultures[10] are used in the same dosage to produce H antisera as are the boiled cultures in the production of O antisera.

Production of Polyvalent Antisera. Since more than 95 per cent of salmonellae isolated from man and lower animals are members of the first five serogroups (A to E, incl.), it is possible to produce a simple polyvalent antiserum that contains O agglutinins for the O antigens possessed by members of these serogroups. A polyvalent O antiserum that contains agglutinins for O antigens 1 to 10, 12, 15, and 19 may be produced with smooth cultures of bioser Paratyphi-A, ser Paratyphi-B, ser Thompson, ser Newport, *S. typhi*, ser London, ser Newington, and ser Senftenberg. Antigen suspensions are prepared from each of the eight cultures by methods described for production of O antisera (v. sup.) and are mixed in equal amounts. Vi antigen may be added in the form of dry powder prepared from *Citrobacter* 029:Vi. Usually six injections of the mixture in amounts of 0.5, 1.0, 2.0, 4.0, 5.0, and 5.0 ml are given to rabbits at intervals of 4 days. After four injections a trial bleeding may be taken. If it is found that the agglutinin content for a particular component is low, the amount of that particular antigen in the mixture may be doubled. The rabbits are bled on the sixth or seventh day following the last injection and the antiserum preserved with an equal volume of glycerol. The antiserum then is standardized against all the antigens injected as described under standardization of O antisera. It is used in slide tests in the highest dilution in which it agglutinates all the injected antigens.

By selection of phases of the above-mentioned serotypes and adding a few more cultures, a polyvalent antiserum may be produced which will react with the H antigens of many common serotypes, as well as with their O antigens. For this purpose serotypes Typhimurium, Montevideo, and a member of serogroup E_3 should be added. A simple OH polyvalent

[10]Freshly prepared formalinized broth H antigen suspensions may be used throughout one course of immunization (i.e., for three or four weeks), but they *should not be used for immunization* after this period of time. However, formalinized H antigens may be stored at room temperature and used for test and control purposes.

such as this should be standardized for use in slide tests with O antigens and for use in tube agglutination tests with formalinized H antigen suspensions.

Kauffmann (1942, 1950, 1954) described polyvalent O and polyvalent H antisera which facilitated the recognition of salmonellae. Four O (OA, OB, OC, and OD) and four H antisera (HA, HB, HC, and HD) were devised. Additional polyvalent O antisera (OE and OF) and H antisera (HE and HF) subsequently were described (Kauffmann, 1966). The O antisera were prepared from agar cultures suspended in physiological saline solution, heated for 2.5 hr at 100 C, and treated with formalin. The antigens used in the preparation of the H antisera were organisms washed from Gard plates with formalinized saline. In preparing both the O and the H antisera, rabbits were injected 6 to 8 times with increasing amounts of vaccine. The six O antisera (OA to OF) were pooled in equal proportions to give a mixed O antiserum (OM). Likewise, the six H antisera were mixed to yield a mixed H antiserum (HM). In the examination of cultures, the OM and HM antisera first were used and if the organism reacted with these it was tested with the component O and H polyvalent antisera. Then it was tested with monovalent antisera for the antigens included in the polyvalent antisera with which reactions were obtained. Interested readers are

referred to Kauffmann (1966) for details regarding the strains used to produce these polyvalent antisera.

The author and colleagues have approached the subject of a comprehensive polyvalent antiserum in a somewhat different manner. Only one antiserum was produced and in this it was attempted to incorporate agglutinins for all the O and H antigens of salmonellae. The organisms to be injected first were passed repeatedly through tubes of semisolid agar to assure maximum motility, inoculated into infusion broth, and after overnight incubation killed by the addition of 0.5% formalin. Thirty-three antigens were required and these were pooled in three mixtures. Each mixture contained an equal amount of each of eleven antigens. Each mixture was injected into a separate lot of rabbits in increasing amounts until 8 to 10 injections had been given. After the second injection, dried organisms of the Vi form of *Citrobacter* 029:Vi were added to each vaccine mixture. The dried cells were suspended in physiological saline solution and added to the vaccines immediately before injection. The O and H titers were followed carefully through trial bleedings and if the agglutinin content for certain antigens was low, additional amounts of those antigens were added to the vaccines. The author and colleagues found the following mixtures of organisms suitable for the production of this kind of polyvalent antiserum:

Mixture 1

ser Paratyphi A	1,2,12:a	ser Newington, phase 1	3,15:e,h
ser Typhimurium, phase 1	4,5,12:i	ser Simsbury	$1,3,19:z_{27}$
ser Deversoir, phase 1	45:c	ser Rubislaw, phase 1	11:r
S. cholerae-suis, phase 2	6,7:1,5	ser Allandale, phase 2	1,40:1,6
ser Newport, phase 2	6,8:1,2	ser Milwaukee	43:f,g
S. typhi	9,12:d		

Mixture 2

ser California	4,12:g,m,t	ser Kentucky, phase 2	$8,20:z_6$
ser Essen	4,12:g,m	ser Minnesota, phase 1	21:b
ser Worthington, phase 1	1,13,23:l,w	ser Pomona, phase 2	28:1,7
ser Sundsvall, phase 1	1,6,14,25:z	ser Tallahassee	$6,8:z_4,z_{32}$
ser Hvittingfoss, phase 2	16:e,n,x	ser Niarembe, phase 2	44:l,w
ser Kirkee, phase 1	17:b		

Mixture 3

ser Tel-aviv, phase 1	28:y	ser Tennessee	$6,7:z_{29}$
ser Urbana, phase 2	30:e,n,x	ser Chittagong, phase 2	$1,3,10,19:z_{35}$
ser Adelaide	35:f,g	ser Waycross	$41:z_4z_{23}$
ser Inverness, phase 2	38:1,6	ser Weslaco	$42:z_{36}$
ser Champaign, phase 1	39:k	ser Deversoir, phase 2	45:e,n,x
ser Illinois, phase 1	$(3),(15),34:z_{10}$		

The three antisera are not used separately but are pooled in equal amounts and used as one polyvalent antiserum. It is necessary to determine in what dilution the complete antiserum may be used in slide tests to detect O antigens. This should be done with alcohol treated cells. The H titer in slide tests invariably exceeds the O titer. Also it is necessary to determine the highest dilution in which the antiserum agglutinates formalinized broth cultures of serotypes containing the various H antigen combinations. These tests should be done as tube tests in a water bath at 48 to 50 C and incubation should not exceed one hour.

The above-mentioned comprehensive polyvalent antiserum was produced in large amount and has been used by the author and colleagues for several years. Since the preparation of a satisfactory polyvalent antiserum of this kind is difficult and expensive, new polyvalent antisera have not been produced as new O and H antigens were described. Instead, pooled H antisera and pooled O antisera have been prepared to accommodate the additional antigens.

The antiserum mentioned above may be used in slide tests for the detection of both O and H antigens if living organisms are used, or for the detection of O antigens alone if alcohol treated cells are used as antigens. Also, it may be used in tube tests for the detection of H antigens. When used in slide tests, it will be found that this antiserum will react with many members of other genera, since there is a marked community of antigens among the Enterobacteriaceae. If it is used in the proper dilution in tube tests for the detection of H antigens, fewer extraneous reactions are encountered, but many members of the genus *Arizona* may be expected to react in it. (See Chapters 10 and 11).

From the foregoing description it may be seen that production of an antiserum that contains agglutinins for all the O and H antigens of salmonellae is not to be approached lightly. The production of such an anti-serum is, in fact, an arduous and time-consuming task and failures may result.

If used judiciously, polyvalent antisera often are useful in preliminary examination of cultures.

REFERENCES

Andrewes, F. W. 1922. J. Pathol. Bacteriol., **25**, 505.
_____. 1925. Ibid., **28**, 345.
Archer, G. T. L., and J. L. Whitby. 1957. J. Hyg., **55**, 513.
Braun, H. 1938. Schweiz. Z. f. allgem. Pathol. u. Bakt., **1**, 257.
_____. 1939. Ibid., **2**, 309.
Braun, H., and K. Guggenheim. 1932. Zentralbl. f. Bakt., I. Orig., **127**, 97.
Buelow, P. 1964. Acta Pathol. Microbiol. Scand., **60**, 376.
Buttiaux, R. 1952. Ann. Inst. Pasteur, **83**, 156.
Costin, I. D. 1967. Zentralbl. f. Bakt. I. Orig., **202**, 437.
Costin, I. D., L. Petrica, and M. Garoiu. 1964. Ibid., **194**, 342.
Craigie, J. 1931. J. Immunol., **21**, 417.
_____. 1936. J. Bacteriol., **31**, 56.
Craigie, J., and K. F. Brandon. 1936. J. Pathol. Bacteriol., **43**, 233.
Davis, B. R., and W. H. Ewing. 1966. The biochemical reactions of members of the genus *Citrobacter*. CDC Publ.*
Douglas, G. W., and P. R. Edwards. 1962. J. Gen. Microbiol., **29**, 367.
Eberth, C. J. 1880. Virchow's Arch., **81**, 58.
Edwards, P. R. 1945. Proc. Soc. Exp. Biol. Med., **59**, 49.
_____. 1950. Ibid., **74**, 471.
_____. 1962. Serologic examination of *Salmonella* cultures for epidemiologic purposes. CDC Publ.*
_____. 1963. Kauffmann-White Schema (modi-fied). CDC Publ.*
Edwards, P. R., and D. W. Bruner. 1938. J. Hyg., **38**, 716.
_____. 1942. J. Bacteriol., **44**, 289.
_____. 1946. Ibid., **52**, 493.
Edwards, P. R., A. B. Moran, and D. W. Bruner. 1946. Proc. Soc. Exp. Biol. Med., **62**, 296.
Edwards, P. R., and A. B. Moran. 1947. Ibid., **66**, 230.
Edwards, P. R., M. G. West, and D. W. Bruner. 1947. J. Infect. Dis., **81**, 24.
Edwards, P. R., and F. Kauffmann. 1952. Amer. J. Clin. Pathol., **22**, 292.
Edwards, P. R., A. C. McWhorter, and M. A. Fife. 1954. J. Bacteriol., **67**, 346.
Edwards, P. R., and M. A. Fife. 1956. Appl. Microbiol., **4**, 46.
Edwards, P. R., F. Kauffmann, and C. R. Huey. 1957. Acta Pathol. Microbiol. Scand., **41**, 517.
Edwards, P. R., A. C. McWhorter, and G. W. Douglas. 1962a. J. Bacteriol., **84**, 95.
Edwards, P. R., R. Sakazaki, and I. Kato. 1962b. Ibid., **84**, 99.
Edwards, P. R., M. A. Fife, and W. H. Ewing. 1965. Antigenic schema for the genus *Arizona*. CDC Publ.*

*CDC Publ. Publication from the Center for Disease Control (formerly the Communicable Disease Center), Atlanta, Ga. 30333.

Ellis, R. J., P. R. Edwards, and M. A. Fife. 1957. Pub. Health Lab.,**15, 89.

Ewing, W. H. 1963. Int. Bull. Bacteriol. Nomen. Tax., 13, 95.

_____. 1966a. Preliminary examination of *Salmonella* and *Shigella* cultures. CDC Publ*

1966b. Revised definitions for the family Enterobacteriaceae, its tribes and genera. CDC Publ.*

_____. 1969. Excerpts from an evaluation of the *Salmonella* problem. CDC Publ.*

Ewing, W. H., B. R. Davis, and R. W. Reavis. 1957. Pub. Health Lab.,**15, 153.

Ewing, W. H., B. R. Davis, and P. R. Edwards. 1960. Ibid., 18, 77.

Ewing, W. H., M. A. Fife, and B. R. Davis. 1965. The biochemical reations of *Arizona arizonae.* CDC Publ.*

Ewing, W. H., and M. A. Fife. 1966. Int. J. System. Bacteriol., 16, 427.

Ewing, W. H., and M. M. Ball. 1966. The biochemical reactions of members of the genus *Salmonella.* CDC Publ.*

Ewing, W. H., M. M. Ball, S. F. Bartes, and A. C. McWhorter. 1970. J. Infect. Dis., 121, 288.

Felix, A., and R. M. Pitt. 1934a. Lancet, ii, 186.

_____. 1934b. J. Pathol. Bacteriol., 38, 409.

Gaertner, A. 1888. Korres-Blatter allgem. ärzt. Vereins Thuringen, 17, 573.

Gaffky, G. 1884. Mitt. a.d. Kais. Ges., 2, 372.

Gard, S. 1938. Z. f. Hyg., 120, 615.

Giovanardi, A. 1938. Zentralbl. f. Bakt. I. Orig., 141, 341.

Hajna, A. A. 1944. J. Bacteriol., 48, 609.

Jordon, E. O., M. E. Caldwell, and D. Reiter. 1934. Ibid., 27, 165.

Juenker, A. P. 1946. Ibid., 52, 609.

Kauffmann, F. 1930. Zentralbl. f. Bakt. I. Orig., 119, 152.

_____. 1935. Ibid., 116, 617.

_____. 1936a. Ibid., 117, 778.

_____. 1936b. Ibid., 118, 318.

_____. 1937. Ibid., 120, 177.

_____. 1940. Acta Pathol. Microbiol. Scand., 17, 135.

_____. 1941a. Die Bakteriologie der Salmonella-Gruppe. Copenhagen: Munksgaard.

_____. 1941b. J. Bacteriol., 41, 127.

_____. 1942. Acta Pathol. Microbiol. Scand., 19, 248.

_____. 1944. Ibid., 21, 33.

**Public Health Laboratory: Journal of the Conference of Public Health Laboratory Directors.

_____. 1950. The diagnosis of *Salmonella* types. Springfield, Ill.: C. C. Thomas.

_____. 1954. Enterobacteriaceae. 2nd. Ed. Copenhagen: Munksgaard.

_____. 1956. Acta Pathol. Microbiol. Scand., 39, 299.

_____. 1957. Ibid., 40, 343.

_____. 1958. Zentralbl. f. Bakt. I. Ref., 166, 403.

_____. 1963. Acta Pathol. Microbiol. Scand., 58, 109.

_____. 1965. Ibid., 63, 261.

_____. 1966. The bacteriology of Enterobacteriaceae. Copenhagen: Munksgaard.

Kauffmann, F., and Mitsui, C. 1930. Z.f. Hyg., 111, 740.

Kauffmann, F., and E. Moeller. 1940. J. Hyg., 40, 246.

Kauffmann, F., and P. R. Edwards. 1947. J. Lab. Clin. Med., 32, 548.

Kauffmann, F., and A. Petersen. 1956. Acta Pathol. Microbiol. Scand., 38, 481.

_____. 1962. Ibid., 56, 343.

_____. 1963. Ibid., 58, 99.

Kauffmann, F., and R. Rhode. 1961. Ibid., 52, 211.

Kelterborn, E. 1967. *Salmonella* Species. The Hague: Dr. W. Junk.

Kuhn, L. R. 1947. Amer. J. Clin. Pathol., 17, 569.

Landy, M. 1952. Proc. Soc. Exp. Biol. & Med., 80, 55.

Landy, M., and M. E. Webster. 1952. J. Immunol., 69, 143.

Leifson, E. 1933. J. Bacteriol., 26, 329.

Le Minor, L., and P. R. Edwards. 1960. Ann. Inst. Pasteur, 99, 469.

Le Minor, L., and F. Ben Hamida. 1962. Ibid., 102, 267.

Levine, M. 1945. J. Lab. Clin. Med., 30, 716.

Levine, P., and A. W. Frisch. 1935. Proc. Soc. Exp. Biol. Med., 32, 883.

Lubin, A. H., and W. H. Ewing. 1964. Pub. Health Lab.,** 22, 83.

McWhorter, A. C., G. W. Douglas, and P. R. Edwards. 1962. Int. Bull. Bacteriol., Nomen. Tax., 12, 181.

McWhorter, A. C., and P. R. Edwards. 1963. J. Bacteriol., 85, 1440.

McWhorter, A. C., M. M. Ball, and B. O. Freeman. 1964a. Ibid., 87, 967.

McWhorter, A. C., and A. B. Moran. 1964b. Ibid., 87, 1248.

Martin, W. J., and W. H. Ewing. 1969. Appl. Microbiol., 17, 111.

Martin, W. J., W. H. Ewing, A. C. McWhorter, and M. M. Ball. 1969. Pub. Health Lab.,** 27, 61.

Moeller, V. 1954a. Acta Pathol. Microbiol. Scand., 34, 115.

_____. 1954b. Ibid., **35**, 259.

_____. 1955. Ibid., **36**, 158.

Moran, A. B., and P. R. Edwards. 1958. Cornell. Vet., **48**, 96.

Peluffo, C. A. 1941. Proc. Soc. Exp. Biol. Med., **48**, 340.

Roschka, R. 1950. Klin. Med., **5**, 88.

Salmon, D. E., and T. Smith. 1885. U. S. Bureau of Am. Indust., 2nd Ann. Rept., p. 184.

Salmonella Committee, Report. 1934. J. Hyg., **34**, 333.

_____. 1940. Proc. 3rd. Int. Congress Microbiol., New York., p. 832.

Salmonella, Committee on. Report. 1969. Publ. 1683, National Academy of Sciences. Washington, D. C.

Schaub, I. G. 1948. Bull. Johns Hopkins Hosp., **83**, 367.

Schütze, H. 1921. J. Hyg., **20**, 230.

Shaw, C. 1956. Int. Bull. Bacteriol. Nomen. Tax., **6**, 1.

Smith, T., and J. R. Stewart. 1897. J. Boston Soc. Med. Sci., **16**, 12.

Spaun, J. 1951. Acta Pathol. Microbiol. Scand., **29**, 416.

_____. 1952. Ibid., **31**, 462.

Spicer, C. C. 1956. J. Clin. Pathol., **9**, 378.

Stuart, C. A., and E. R. Kennedy. 1948. Proc. Soc. Exp. Biol. Med., **68**, 455.

Taylor, J., M. Lee, P. R. Edwards, and C. H. Ramsey, 1960. J. Gen. Microbiol., **23**, 583.

Trabulsi, L. R., and P. R. Edwards. 1962. Cornell Vet., **52**, 563.

van Oye, E. 1964. The world problem of Salmonellosis. The Hague: Dr. W. Junk.

Webster, M. E., M. Landy, and M. E. Freeman. 1951. J. Immunol., **69**, 135.

White, P. B. 1925. Med. Res. Council Gt. Brit., Spec. Rep. Ser. No. 91.

_____. 1926. Ibid., Spec. Rep. Ser. No. 103.

_____. 1929. Ibid., System of Bacteriol., **4**, 86.

Winslow, C. E. A., I. Kligler, and W. Rothberg. 1919. J. Bacteriol., **4**, 429.

Chapter 9

Antigenic Schema for *Salmonella*

Development of the Schema. As mentioned in Chapter 8, the work of White (1925, 1926) and of Kauffmann (e.g., 1930, 1941, 1966) made possible the rapid, accurate serological characterization of salmonellae. White recognized the importance of considering the then recent discoveries regarding bacterial variation such as those of Schuetze (1921) and Andrewes (1922), and Kauffmann reorganized, further systematized, and greatly extended the work and the schema. However, there were very many other investigators who also made important contributions to the development of the antigenic schema.

White (1925, 1926) classified the serotypes of *Salmonella* that he studied on the basis of their flagellar (H) antigens and devised a subdivision into monophasic and diphasic serotypes. Although White characterized and labeled a number of heat stable O antigen factors, he did not utilize these as a basis for primary grouping of serotypes. This was done by Kauffman. It is notable that the designations and symbols applied by White to the H antigens were different from those employed later by Kauffmann. Kauffmann (1930, 1941, 1966) reorganized the system and made the O antigens the basis for primary serological grouping.

Since serotypes such as Paratyphi-A, Senftenberg, Paratyphi-B, Typhimurium, etc., had been known and studied intensely for many years, their O antigens were characterized and labeled first. For example, when the O antigens of ser Paratyphi-A were compared to those of ser Senftenberg it was found that while the two serotypes shared an antigen, which was labeled 1, each serotype possessed another unrelated antigen and these were labeled 2 and 3, respectively. As other serotypes were studied, their O antigens were differentiated and labeled consecutively, more or less in the following manner:

ser Paratyphi-A	1,2	12
ser Senftenberg	1,3	
ser Paratyphi-B	1,4[5]	12
S. cholerae-suis	6,7	
ser Newport	6,8	
S. typhi	9,	12
ser London	3,10	

ser Aberdeen 11
etc.

As indicated above, antigen 12 was differentiated and labeled after a number of other antigens had been given their designations, hence it was labeled 12 and the designation was added to the complexes where it belonged. Ser Senftenberg subsequently was moved to serogroup E when additional O antigens were characterized in it.

Characterization, delineation, and labeling of the H antigens of salmonellae progressed in a manner analogous to that mentioned in the preceeding paragraph. The H antigens of ser Paratyphi-A were among the first to be characterized, hence they were labeled a. Those of the first phases of serotypes Paratyphi-B and Paratyphi-C were delineated and labeled b and c, respectively. The antigens of the so-called nonspecific phases of these two serotypes were related through a common factor, which was designated factor 1, but each serotype possessed other unrelated factors. These unrelated factors were labeled 2 in ser Paratyphi-B and 4,5 in ser Paratyphi-C. Factor 3 in this series was characterized in the nonspecific phase antigens of *S. cholerae-suis* (1,3,4,5). As additional serotypes were studied the H antigens of their specific (i.e., first) phases were labeled d,e etc., and those of their nonspecific (i.e., second) phases were designated by arabic numerals. However, two facts quickly became apparent. First, it was learned that the so-called specific phases seldom were specific to any particular serotype, and second, several of the specific phase antigen complexes originally labeled with letters actually were phase 2 antigens and had to be moved from phase 1 to phase 2. Parenthetically, the terms specific and nonspecific first were employed because, in the early developmental work, the antigens of the former were known to occur only in one serotype and therefore were considered to be specific for that serotype, while those of the nonspecific phases occurred in more than one serotype. As such the antigens themselves were quite specific, of course. The concept of specific phases was not held for long but the terminology was used for some time (e.g., see Report,

Salmonella Committee, 1934). The following examples may aid in clarification of the two points mentioned above.

	Phase 1	Phase 2
ser Paratyphi-A	a	-
ser Paratyphi-B	b	1,2
ser Paratyphi-C	c	1,5

The antigens of the first phases of these three serotypes each occurred only in a particular serotype so that the concept of specific and nonspecific phases was valid to this point. However, when *S. cholerae-suis* was characterized completely it was found that it possessed antigen c, the same as ser Paratyphi-C:

S. cholerae-suis   c   1,5.

Later serotypes such as Reading, Abortusequi, and London were described and their flagellar antigens were labeled:

ser Reading	e,h	1,5
ser Abortusequi	e,n	-
ser London	l,v	1,6

The phase 1 antigens of ser Reading and ser Abortusequi were related and the common antigen was labeled e, but each serotype possessed an unrelated factor, labeled h and n, respectively. Subsequently, serotypes that possessed antigens that were labeled l,e,n,w and e,h,l,v were characterized:

| ser Brandenburg | l,e,n,v |
| ser Daressalaam | l,e,n,w. |

At first these were characterized as specific (or phase 1 antigens) but phase variation was demonstrated in serotypes of this sort (e.g., Kauffmann and Mitsui, 1930), which indicated that these complexes actually were divisible into:

| ser Brandenburg | l,v | e,n . . . z_{15} |
| ser Daressalaam | l,w | e,n. |

Thus, the e,n complex was moved to phase 2. Although numerous examples could be given, those mentioned above should illustrate the fact that the so-called specific phases were not specific to particular serotypes and that antigens once labeled specific may occur as either phase 1 or phase 2 in certain serotypes. However, antigens labeled with arabic numerals always are phase 2. For these reasons the concept of specific and nonspecific phases was abandoned and the terms phase 1 and phase 2 were substituted.

An adaptation of one of the earliest antigenic schemata published by Kauffmann (1930) is given in Table 111 and the schema published by the Salmonella

Committee (see Report, 1934) is given in Table 112. These schemata are presented to indicate the small number of serotypes and biotypes recognized as late as 1934 and illustrate the fact that certain antigens usually are not expressed in antigenic schemata (e.g., O antigens 36 and 37 in serogroup G strains and flagellar antigen factors 3 and 4.

It will be noted that in 1934, 44 serotypes, variants, and biotypes were recognized. In 1941 Kauffmann listed 104 serotypes and biotypes. In 1951, 211 serotypes and biotypes were recorded; in 1954, 309; in 1961, 695; and in 1966, 962 serotypes were listed in the schema (Kauffmann, 1941, 1951, 1954, 1961 and 1966). Since then the number of delineated serotypes of *Salmonella* has continued to grow. However, the large number of delineated serotypes should not be a source of either dismay or alarm since it is well known (e.g., Kelterborn, 1967) that a very high percentage of cultures isolated in daily practice belong to a small number of serotypes (see also Table 109, Chapter 8). It should be noted that in his analysis of data concerning 547,386 cultures of *Salmonella* isolated in various parts of the world, Kelterborn (1967) reported that 47.1 per cent belonged to serogroup B, 13.3 percent to C_1, 7.1 per cent to C_2, 23.7 percent to D_1, and 4.4 per cent to serogroup E_1. That is, 95.7 percent of the cultures belonged to these serogroups. Further, this investigator reported that only 47 serotypes (and species), or 4.5 per cent of the 1043 serotypes listed in the schema at the time, were isolated more than 500 times from man, lower animals and all other sources. With a few exceptions the above-mentioned 47 serotypes and species all were among those given in the abbreviated schema (Table 109, Chapter 8).

Alphabetical List of Serotypes and an Antigenic Schema for Salmonella. These are given in the following pages. All serotypes characterized and described through the end of 1969 are given in the alphabetical list and in the schema. The O antigens of salmonellae are designated by arabic numerals and the H antigens are labeled with lower case roman letters or, as in the case of many phase 2 antigens, arabic numerals. If the designation for an antigen is enclosed in parentheses, e.g., (8), the indication is that the complete antigen is not present. Enclosure of an antigen designation in brackets indicates that the antigen or complex may be present or absent, e.g., [1], [1,2] etc.

Type facings sometimes may cause some difficulty because of the similarity of appearance of the roman letter l and the arabic numeral 1. However, this difficulty easily may be overcome if a simple rule is borne in mind. If the character l occurs with another

TABLE 111

Antigenic Schema, 1930*

Serotype	O Antigen	H Antigen specific	nonspecific
A	I.II	a	—
Senftenberg-Newcastle	I.III	g,s	—
(B) Schottmueller		b	1,2
Breslau-	IV.V	i	1,2,3
Binns		—	1,2,3
Stanley		d	1,2
Reading		e,h	1,4,5
Derby	IV	f,g	—
Abortus equi		e,n,x	—
Abortus ovis		—	1,4
Brandenburg		e,n,l,v	—
(C) Suipestifier American		c	1,3,4,5
Kunzendorf		—	1,3,4,5
Glaesser-		c	1,3,4,5
Voldagsen		—	1,3,4,5
Orient		c	1,4,5
Thompson-	VI.VII	k	1,3,4,5
Berlin		—	1,3,4,5
Virchow		r	1,2,3
Oranienburg		m,t	—
Potsdam		e,n,l,v	—
Newport	VI.VIII	e,h	1,3,4,5
Bovismorbificans		r	1,3,4,5
Gaertner Jena		g,o,m	—
Dublin-Kiel	IX	g,p	—
Rostock		g,p,u	—
Moskow		g,o,q	—
Typhus (*S. typhi*)		d	—
Sendai		a	1,4,5
Dar es Salaam		e,n,l,w	—
Pullorum		—	—
London	X.III	l,v	1,4

* Adapted from Kauffmann, 1930.

roman letter (e.g., 1,v; 1,w), it is the letter 1. If this character appears in conjunction with an arabic numeral (e.g., 1,2; 1,5), it is the arabic numeral one. Further, the arabic numerals 1,2; 1,5 etc. are employed to designate phase 2 antigens only.

It is emphasized that the formulas given in the antigenic schema are abbreviated. By no means do they include all of the antigens that may be present in a given serotype.

TABLE 112

Salmonella Subcommittee*
**The Kauffmann-White Schema Showing the Antigenic Structure
of the Various Species of *Salmonella* Recognized**

Species	O-Antigen	H-Antigen Specific	H-Antigen Non-specific
Group A:			
1 S. paratyphi-A	I, II	a	—
2 S. senftenberg	I, III	gs	—
3 S. senftenberg var. newcastle		gs	—
Group B:			
4 S. paratypi-B		b	1,2
5 S. typhi-murium		i	1, 2, 3
6 S. typhi-murium var. binns	IV, V	—	1, 2, 3
7 S. stanley		d	1, 2
8 S. heidelberg		r	1, 2, 3
9 S. reading		eh	1, 4, 5
10 S. derby		fg	—
11 S. abortus-equi	IV	enx	—
12 S. abortus-ovis		c	1,4,6
13 S. brandenburg		enlv	—
Group C:			
14 S. paratyphi-C		c	1, 4, 5
15 S. cholerae-suis		c	1, 3, 4, 5
16 S. cholerae-suis var. kunzendorf		—	1, 3, 4, 5
17 S. typhi-suis		c	1, 3, 4, 5
18 S. typhi-suis var. voldagsen		—	1, 3, 4, 5
19 S. thompson	VI, VII	k	1, 3, 4, 5
20 S. thompson var. berlin		—	1, 3, 4, 5
21 S. virchow		r	1, 2, 3

*From Report, Salmonella Committee, 1934. Reproduced from *Journal of Hygiene,* 1934, 34:333-350 by permission of Cambridge University Press.

TABLE 112 (Continued)

Salmonella Subcommittee
The Kauffmann-White Schema Showing the Antigenic Structure
of the Various Species of *Salmonella* Recognized

Species		O-Antigen	H-Antigen	
			Specific	Non-specific
22	S. oranienburg		mt	—
23	S. potsdam	VI, VII	enlv	—
24	S. bareilly		y	1, 3, 4, 5
25	S. newport		eh	1, 2, 3
26	S. newport var. puerto-rico		—	1, 2, 3
27	S. newport var. kottbus	VI, VIII	eh	1, 3, 4, 5
28	S. bovis-morbificans		r	1, 3, 4, 5
29	S. muenchen		d	1, 2
Group D:				
30	S. typhi		d	—
31	S. enteritidis		gom	—
32	S. enteritidis var. danysz		gom	—
33	S. enteritidis var. dublin		gp	—
34	S. enteritidis var. rostock		gpu	—
35	S. enteritidis var. moscow	IX	goq	—
36	S. sendai		a	1, 4, 5
37	S. dar-es-salaam		enlw	—
38	S. eastbourne		eh	1, 3, 4, 5
39	S. panama		lv	1, 3, 4, 5
40	S. gallinarum		—	—
41	S. pullorum		—	—
Group E:				
42	S. london		lv	1, 4, 6
43	S. anatum	X, III	eh	1, 4, 6
44	S. anatum var. muenster		eh	1, 4, 5

* From report, Salmonella Committee, 1934. Reproduced from *Journal of Hygiene*, 1934, 34: 333-350 by permission of Cambridge University Press.

ALPHABETICAL LIST OF THE SEROTYPES OF *SALMONELLA*

Salmonella enteritidis

ser Aba	$6,8:i:e,n,z_{15}$
ser Abadina	$28:g,m:[e,n,z_{15}]$
ser abaetetuba	$11:k:1,5$
ser Aberdeen	$11:i:1,2$
ser Abony	$1,4,5,12:b:e,n,x$
variant Haifa	$4,12:b:e,n,x$
ser Abortusbovis	$1,4,12,27:b:e,n,x$
ser Abortusequi	$4,12:-:e,n,x$
ser Abortusovis	$4,12:c:1,6$
ser Accra	$1,3,19:b:z_6$
*ser Acres	$1,13,23:b:z_{42}:[1,5]$
ser Adabraka	$3,10:z_4,z_{23}:[1,7]$
ser Adamstown	$28:k:1,6$
ser Adamstua	$11:e,h:1,6$
ser Adelaide	$35:f,g:-$
ser Adeoyo	$16:g,m:-$
ser Aderike	$28:z_{38}:-$
ser Adjame	$13,23:r:1,6$
ser Aequatoria	$6,7:z_4,z_{23}:e,n,z_{15}$
ser Aflao	$1,6,14,25:l,z_{28}:e,n,x$
ser Africana	$4,12:r,(i):l,w$
ser Agama	$4,12:i:1,6$
ser Agbeni	$13,23:g,m:-$
ser Agege	$3,10:c:e,n,z_{15}$
ser Ago	$30:z_{38}:-$
ser Agodi	$35:g,t:-$
ser Agona	$4,12:f,g,s:-$
ser Ahmadi	$1,3,19:d:1,5$
ser Ahuza	$43:k:1,5$
ser Ajiobo	$13,23:z_4,z_{23}:-$
ser Akanji	$6,8:r:1,7$
ser Akuafo	$16:y:1,6$
ser Alabama	$9,12:c:e,n,z_{15}$
ser Alachua	$35:z_4,z_{23}$
ser Alagbon	$6,8:y:1,7$
ser Alamo	$6,7:g,z_{51}:1,5$
ser Albany	$(8),20:z_4,z_{24}:$
ser Albert	$4,12:z_{10}:e,n,x$
ser Albuquerque	$6,14,24:d:z_6$
*ser Alexander	$3,10:z:1,5$
ser Alexanderplatz	$47:z_{38}:-$
ser Alexanderpolder	$(8):c:l,w$
ser Alger	$38:l,v:1,2$
ser Allandale	$1,40:k:1,6$
ser Allerton	$3,10:b:1,6$
*ser Alsterdorf	$1,40:g,m,t:-$
ser Altendorf	$4,12:c:1,7$

ser Altona	$(8),20:r(i):z_6$
ser Amager	$3,10:y:1,2$
ser Amba	$11:k:l,z_{13},z_{28}$
ser Amersfoort	$6,7:d:e,n,x$
ser Amherstiana	$(8):l,(v):1,6$
ser Amina	$16:i:1,5$
ser Aminatu	$3,10:a:1,2$
ser Amounderness	$3,10:i:1,5$
ser Amoutive	$28:d:1,5$
ser Amsterdam	$3,10:g,m,s$
ser Amunigun	$16:a:1,6$
ser Anatum	$3,10:e,h:1,6$
ser Anderlecht	$3,10:c:l,w$
ser Anfo	$39:y:1,2$
ser Angoda	$30:k:e,n,x$
*ser Angola	$1,9,12:z:z_6$
ser Ank	$28:k:e,n,z_{15}$
ser Annedal	$16:r(i):e,n,x$
ser Antsalova	$51:z:1,5$
ser Apapa	$45:m,t:-$
ser Aqua	$30:k:1,6$
ser Ardwick	$6,(7),(14):f,g:-$
ser Arechavaleta	$4,5,12:a:[1,7]$
*ser Argentina	$6,7:z_{36}:-$
ser Arkansas	$(3),(15),34:e,h:1,5$
*ser Artis	$56:b:-$
ser Aschersleben	$30:b:1,5$
ser Ashanti	$28:b:1,6$
*ser Askraal	$51:l,z_{28}:-$
ser Assen	$21:a:-$
ser Atlanta (Mississippi)	$[1],13,23:b:-$
*ser Atra	$50:m,t:z_6:z_{42}$
ser Augustenborg	$6,7:i:1,2$
ser Austin	$6,7:a:1,7$
ser Avonmouth	$1,3,19:i:e,n,z_{15}$
ser Ayinde	$4,12,27:d:z_6$
ser Ayton	$1,4,12,27:l,w:z_6$
ser Azteca	$4,5,12:l,v:1,5$
ser Babelsberg	$28:z_4,z_{23}:e,n,z_{15}$
*ser Bacongo	$6,7:z_{36}:z_{42}$
ser Baguirmi	$30:y:e,n,x$
ser Bahati	$13,22:b:e,n,z_{15}$
ser Bahrenfeld	$6,14,24:e,h:1,5$
ser Baibokoum	$6,7:k:1,7$
ser Baildon	$(9),46:a:e,n,x$
ser Ball	$1,4,12,27:y:e,n,x$
ser Bamboye	$(9),46:b:l,w$
ser Banalia	$6,8:b:z_6$
ser Bandia	$35:i:l,w$
*ser Baragwanath	$6,8:m,t:1,5$
ser Bardo	$8:e,h:1,2$

*biochemically aberrant

Note: ser., serotype; bioser., bioserotype. ( ), partial antigen. [ ], antigen present or absent.

ser Bareilly	$6,7,[14]:y:1,5$
ser Barmbek	$16:d:z_6$
ser Barranguilla	$16:d:e,n,x$
*ser Basel	$58:l,z_{13},z_{28}:1,5$
*ser Bechuana	$4,12,27:g,t:-$
ser Bedford	$1,3,19:l,z_{13}z_{28};e,n,z_{15}$
ser Belem	$6,8:c:e,n,x$
*ser Bellville	$16:e,n,x:1,7$
*ser Beloha	$18:z_{36}:-$
ser Benfica	$3,10:b:e,n,x$
variant T_1	$T_1:b:e,n,x$
ser Benguella	$40:b:z_6$
ser Bere	$47:z_4,z_{23}:z_6$
ser Bergedorf	$(9),46:e,h:1,2$
ser Bergen	$47:i:e,n,z_{15}$
ser Berkeley	$43:a:1,5$
ser Berlin	$17:d:1,5$
*ser Bern	$1,40:z_4,z_{32}:-$
ser Berta	$9,12:f,g,t:-$
*ser Betioky	$59:k:(z)$
ser Biafra	$3,10:z_{10}:z_6$
*ser Bilthoven	$47:a:[1,5]$
ser Bilu	$(1),3,10,(19):f,g,t:1,(2),7$
ser Binza	$3,15:y:1,5$
ser Birkenhead	$6,7:c:1,6$
ser Birmingham	$3,10:d:l,w$
ser Bispebjerg	$1,4,5,12:a:e,n,x$
*ser Blankenese	$1,9,12:b:z_6$
*ser Bleadon	$17:(f),g,t:[e,n,x,z_{15}]$
ser Bledgam	$9,12:g,m,q:-$
ser Blijdorp	$1,6,14,25:c:1,5$
ser Blockley	$6,8:k:1,5$
*ser Bloemfontein	$6,7:b:[e,n,x]:z_{42}$
ser Blukwa	$18:z_4,z_{24}:-$
ser Bochum	$4,5,12:r:l,w$
*ser Bockenheim	$1,53:z_{36},z_{38}:-$
ser Bodjonegoro	$30:z_4,z_{24}:-$
ser Boecker	$[1],6,14[25]:l,v:1,7$
ser Bokanjac	$28:b:1,7$
*ser Boksburg	$40:g,s:e,n,x,z_{15}$
ser Bolombo	$3,10:z_{38}:-$
ser Bolton	$3,10:y:e,n,z_{15}$
*ser Bonaire	$50:z_4,z_{32}:-$
ser Bonames	$17:a:1,2$
ser Bonariensis	$6,8:i:e,n,x$
ser Bongor	$48:z_{35}:-$
ser Bonn	$6,7:l,v:e,n,x$
ser Bootle	$47:k:1,5$
ser Borbeck	$13,22:l,v:1,6$
*ser Bornheim	$1,6,14,25:z_{10}:1,(2),7$
ser Bornum	$6,(7),(14):z_{38}:-$
*ser Boulders	$13,23:m,t:z_{42}$

*biochemically aberrant

ser Bournemouth	$9,12:e,h:1,2$
ser Bousso	$1,6,14,25:z_4,z_{23}:-$
ser Bovismorbificans	$6,8:r:1,5$
ser Bracknell	$13,23:b:1,6$
ser Bradford	$4,12,27:r:1,5$
ser Braenderup	$6,7:e,h:e,n,z_{15}$
ser Brancaster	$1,4,12,27:z_{29}:-$
ser Brandenburg	$4,12:l,v:e,n,z_{15}$
ser Brazil	$16:a:1,5$
ser Brazzaville	$6,7:b:1,2$
ser Bredeney	$1,4,12,27:l,v:1,7$
*ser Bremen	$45:g,m,s,t:e,n,x$
ser Breukelen	$6,8:l,z_{13}:e,n,z_{15}$
ser Brijbhumi	$11:i:1,5$
ser Brisbane	$28:z:e,n,z_{15}$
ser Bristol	$13,22:z:1,7$
ser Bron	$13,22:g,m:[e,n,z_{15}]$
ser Bronx	$6,8:c:1,6$
ser Broughton	$1,3,19:b:l,w$
ser Brunei	$(8),20:y:1,5$
ser Budapest	$1,4,12:g,t:-$
ser Bukavu	$1,40:l,z_{28}:1,5$
ser Bukuru	$6,8:b:l,w$
*ser Bulawayo	$1,40:z:1,5$
ser Bulbay	$11:l,v:e,n,z_{15}$
*ser Bunnik	$43:z_{42}:[1,5]$
ser Burgas	$16:l,v:e,n,z_{15}$
ser Bury	$4,12,27:c:z_6$
ser Businga	$6,7:z:e,n,z_{15}$
ser Butantan	$3,10:b:1,5$
ser Buzu	$1,6,14,25:i:1,7$
ser Cairina	$3,10:z_{35}:z_6$
ser Cairns	$45:k:e,n,z_{15}$
ser Cairo	$1,4,12,27:d:1,2$
ser Calabar	$1,3,19:e,h:l,w$
*ser Caledon	$4,12:g,m:e,n,x$
ser California	$4,5,12:m,t:-$
*ser Calvinia	$6,7:a:z_{42}$
ser Camberene	$35:z_{10}:1,5$
ser Cambridge	$3,15:e,h:l,w$
ser Canada	$4,12:b:1,6$
*ser Canastel	$9,12:z_{29}:1,5$
ser Canoga	$(3),(15),34:g,s,t:-$
ser Cannstatt	$1,3,19:m,t:-$
*ser Cape	$6,7:z_6:1,7$
ser Caracas	$1,6,14,25:g,m,s:-$
*ser Carletonville	$38:d:[1,5]$
ser Carmel	$17:l,v:e,n,x$
ser Carno	$1,3,19:z:l,w$
ser Carrau	$6,14,24:y:1,7$
ser Casablanca	$45:k:1,7$
*ser Ceres	$28:z:z_{39}$
ser Cerro	$18:z_4,z_{23}:[z_{45}]$

ser Ceyco	$(9),46:k:z_{35}$
ser Chagoua	$1,13,23:a:1,5$
ser Chailey	$6,8:z_4,z_{23}:[e,n,z_{15}]$
*ser Chameleon	$16:z_4,z_{32}:-$
ser Champaign	$39:k:1,5$
ser Chandans	$11:d:e,n,x$
ser Charity	$1,6,14,25:d:e,n,x$
*ser Chersina	$47:z:z_6$
ser Chester	$4,5,12:e,h:e,n,x$
ser Chicago	$28:r:1,5$
ser Chincol	$6,8:g,m,s:e,n,x$
ser Chingola	$11:e,h:1,2$
*ser Chinovum	$42:b:1,5$
ser Chittagong	$(1),3,10,(19):b:z_{35}$
Salmonella cholerae-suis	$6,7:c:1,5$
bioser Kunzendorf	$6,7:[c]:1,5$
S. enteritidis	
ser Christiansborg	$44:z_4,z_{24}:-$
* variant	$44:z_4,z_{24}:-$
*ser Chudleigh	$3,10:e,n,x:1,7$
ser Clackamas	$4,12:l,v,(z_{13}):1,6$
ser Claibornei	$1,9,12:k:,1,5$
ser Clerkenwell	$3,10:z:l,w$
ser Cleveland	$6,8:z_{10}:1,7$
*ser Clifton	$13,22:z_{29}:1,5$
*ser Clovelly	$1,44:z_{39}:[e,n,x,z_{15}]$
ser Cocody	$(8),20:r,(i):e,n,z_{15}$
ser Coeln	$4,5,12:y:1,2$
ser Coleypark	$6,7:a:l,w$
ser Colindale	$6,7:r:1,7$
ser Colombo	$38:y:1,6$
ser Colorado	$6,7:l,w:1,5$
ser Concord	$6,7:l,v:1,2$
ser Congo	$13,23:g,t:-$
*ser Constantia	$17:z:l,w:z_{42}$
ser Cook	$39:z_{48}:1,5$
ser Coquilhatville	$3,10:z_{10}:1,7$
ser Corvallis	$(8),20:z_4,z_{23}:-$
ser Cotham	$28:i:1,5$
ser Croft	$28:g,m,s:-$
ser Cubana	$1,13,23:z_{29}:-$
ser Curacao	$6,8:a:1,6$
ser Dahlem	$48:k:e,n,z_{15}$
ser Dakar	$28:a:1,6$
ser Dallgow	$1,3,19:z_{10}:e,n,z_{15}$
ser Dan	$51:k:e,n,z_{15}$
*ser Daressalaam	$1,9,12:l,w:e,n,x$
ser Daytona	$6,7:k:1,6$
bioser Decatur	$6,7:c:1,5$
*ser Degania	$40:z_4,z_{24}:-$
ser Dembe	$35:d:l,w$
ser Demerara	$13,23:z_{10}:l,w$
ser Denver	$6,7:a:e,n,z_{15}$
ser Derby	$1,4,5,12:f,g:[1,2]$
ser Dessau	$(1),3,15,(19):g,s,t:-$
*ser Detroit	$42:z:1,5$
ser Deversoir	$45:c:e,n,x$
ser Diguel	$1,13,22:d:e,n,z_{15}$
ser Diourbel	$21:i:1,2$
ser Djakarta	$48:z_4,z_{24}:-$
ser Djermaia	$28:z_{29}:-$
ser Djugu	$6,7:z_{10}:e,n,x$
ser Doncaster	$6,8:a:1,5$
ser Donna	$30:l,v:1,5$
ser Dougi	$50:y:1,6$
ser Dresden	$28:c:e,n,x$
ser Driffield	$1,40:d:1,5$
ser Drypool	$3,15:g,m,s:-$
ser Dublin	$1,9,12:g,p:-$
variant Vi +	$1,9,12:Vi:g,p:-$
*ser Dubrovnik	$41:z:1,5$
ser Duesseldorf	$6,8:z_4,z_{24}:-$
ser Dugbe	$45:d:1,6$
ser Duisburg	$[1],4,12,[27]:d:e,n,z_{15}$
*ser Duivenhoks	$(9),46:g,m,s,t:e,n,x$
ser Durban	$9,12:a:e,n,z_{15}$
*ser Durbanville	$[1],4,12,[27]:[z_{39}]:1,5,7$
ser Durham	$13,23:b:e,n,z_{15}$
ser Duval	$1,40:b:e,n,z_{15}$
ser Ealing	$35:g,m,s:-$
ser Eastbourne	$1,9,12:e,h:1,5$
ser Eberswalde	$28:c:1,6$
ser Ebrie	$35:g,m,t:-$
ser Echa	$38:k:1,2$
ser Edinburg	$6,7:b:1,5$
ser Edmonton	$6,8:l,v:e,n,z_{15}$
ser Egusi	$41:d:-$
ser Egusitoo	$1,42:b:z_6$
*ser Eilbek	$61:i:z$
ser Eimsbuettel	$6,(7),(14),:d:l,w$
*ser Ejeda	$45:a:z_{10}$
ser Ekotedo	$(9),46:z_4,z_{23}:-$
ser Elizabethville	$3,10:r:1,7$
ser Elomrane	$1,9,12:z_{38}:-$
*ser Elsiesrivier	$16:z_{42}:1,6$
ser Emek	$(8),20:g,m,s:-$
ser Emmastad	$38:r:1,6$
*ser Emmerich	$6,14:[m,t]:e,n,x$
ser Encino	$1,6,14,25:d:l,z_{13},z_{28}$
ser Enschede	$35:z_{10}:l,w$
ser Entebbe	$1,4,12,27:z:z_6$
ser Enteritidis	$1,9,12:g,m:-$

*biochemically aberrant

ser Enugu	$16:l,z_{13},z_{28}:-$
ser Epicrates	$3,10:b:l,w$
ser Eppendorf	$[1],4,12,[27]:d:1,5$
*ser Epping	$13,23:e,n,x:1,7$
*ser Erlangen	$48:g,m,t:-$
ser Escanaba	$6,7:k:e,n,z_{15}$
ser Eschersheim	$3,15:d:e,n,x$
ser Eschweiler	$6,7:z_{10}:1,6$
ser Essen	$4,12:g,m:-$
ser Etterbeek	$11:z_4,z_{23}:e,n,z_{15}$
ser Ezra	$28:z:1,7$
ser Faji	$1,42:a:e,n,z_{15}$
ser Falkensee	$3,10:i:e,n,z_{15}$
ser Fallowfield	$3,10:l,z_{13},z_{28}:e,n,z_{15}$
*ser Fandran	$1,40:z_{35}:e,n,x,z_{15}$
ser Fann	$11:l,v:e,n,x$
ser Fanti	$13,23:z_{38}:-$
ser Farcha	$43:y:1,2$
*ser Farmsen	$13,23:z:1,6$
*ser Faure	$50:z_{42}:1,7$
ser Fayed	$6,8:l,w:1,2$
ser Ferlac	$1,6,14,25:a:e,n,x$
*ser Finchley	$3,10:z:e,n,x$
ser Findorff	$11:d:z_6$
ser Finkenwerder	$1,6,14,25:d:1,5$
ser Fischerhuette	$16:a:e,n,z_{15}$
ser Fischerkietz	$1,6,14,25:y:e,n,x$
ser Fischerstrasse	$44:d:e,n,z_{15}$
ser Fitzroy	$48:e,h:1,5$
*ser Flint	$50:z_4,z_{23}:-$
ser Florida	$1,6,14,25:d:1,7$
ser Flottbek	$52:b:-$
ser Fluntern	$6,14,18:b:1,5$
ser Fortune	$4,12,27:z_{10}:z_6$
*ser Foulpointe	$38:g,t:-$
ser Frankfurt	$16:i:e,n,z_{15}$
ser Freetown	$38:y:1,5$
*ser Fremantle	$42:(f),g,t:-$
ser Fresno	$(9),46:z_{38}:-$
* variant	$(9),46:z_{38}:-$
ser Friedenau	$13,22:d:1,6$
ser Friedrichsfelde	$28:f,g:-$
ser Frintrop	$1,9,12:b:1,5$
*ser Fuhlsbuettel	$3,10:l,v:z_6$
ser Fulica	$4,5,12:a:1,5$
ser Gabon	$6,7:l,w:1,2$
ser Galiema	$6,7:k:1,2$
ser Galil	$3,10:a:e,n,z_{15}$
bioser Gallinarum	$1,9,12:-:-$
ser Gamaba	$44:g,m,s:-$
ser Gambaga	$21:z_{35}:e,n,z_{15}$
ser Gambia	$35:i:e,n,z_{15}$
ser Gaminara	$16:d:1,7$
ser Garba	$1,6,14,25:a:1,5$
ser Garoli	$6,7:i:1,6$
ser Gassi	$35:e,h:z_6$
ser Gateshead	$(9),46:g,s,t:-$
ser Gatow	$6,7:y:1,7$
ser Gatuni	$6,8:b:e,n,x$
ser Gdansk	$6,7:l,v:z_6$
ser Gege	$30:r:1,5$
ser Gelsenkirchen	$6,(7),(14):l,v:z_6$
ser Georgia	$6,7:b:e,n,z_{15}$
ser Gera	$1,42:z_4,z_{23}:1,6$
*ser Germiston	$6,8:m,t:e,n,x$
ser Ghana	$21:b:1,6$
ser Giessen	$30:g,m,s:-$
*ser Gilbert	$6,7:z_{39}:1,7$
ser Give	$3,10:l,v:1,7$
ser Glasgow	$16:b:1,6$
*ser Glencairn	$11:a:z_6:z_{42}$
ser Glostrup	$6,8:z_{10}:e,n,z_{15}$
ser Gloucester	$1,4,12,(27):i:l,w$
ser Gnesta	$1,3,19:b:1,5$
ser Godesberg	$30:g,m:-$
ser Goelzau	$3,10:a:1,5$
ser Goerlitz	$3,15:e,h:1,2$
ser Goeteborg	$9,12:c:1,5$
ser Goettingen	$9,12:l,v:e,n,z_{15}$
*ser Gojenberg	$1,13,23:g,t:1,5$
ser Gokul	$1,51:d:-$
ser Goldcoast	$6,8:r:l,w$
ser Gombe	$6,7:d:e,n,z_{15}$
ser Good	$21:f,g:e,n,x$
*ser Goodwood	$13,22:z_{29}:e,n,x$
ser Gori	$17:z:1,2$
ser Goulfy	$1,40:k:1,5(6)$
ser Goverdhan	$9,12:k:1,6$
*ser Grabouw	$11:g,m,s,t:[z_{39}]$
ser Graz	$43:a:1,2$
*ser Greenside	$50:z:e,n,x$
ser Greiz	$40:a:z_6$
ser Grumpensis	$13,23:d:1,7$
*ser Grunty	$1,40:z_{39}:1,6$
ser Guildford	$28:k:1,2$
ser Guinea	$44:z_{10}:[1,7]$
*ser Gwaai	$21:z_4,z_{24}:-$
ser Gwoza	$1,3,19:a:e,n,z_{15}$
ser Haardt	$(8):k:1,5$
*ser Haarlem	$(9),46:z:e,n,x$
ser Hadar	$6,8:z_{10}:e,n,x$
*ser Haddon	$16:z_4,z_{23}:-$
ser Haelsingborg	$6,7:m,p,t,[u]:-$

*biochemically aberrant

ser Haferbreite	42:k:[1,6]
*ser Hagenbeck	48:d:z_6
ser Haifa	1,4,5,12:z_{10}:1,2
variant afula	
01 & 05-	4,12:z_{10}:1,2
ser Halle	28a,28c:c:1,7
variant Vidin	28a,28b:c:1,7
ser Halmstad	3,15:g,s,t:-
*ser Hamburg	1,9,12:g,t:-
ser Hamilton	3,15:z_{27}:-(=ser. Goerlitz)
*ser Hammonia	48:e,n,x,z_{15}:z_6
ser Hannover	16:a:1,2
ser Haouaria	13,22:c:e,n,x,z_{15}
ser Harburg	1,6,14,25:k:1,5
*ser Harmelen	51:z_4,z_{23}:-
ser Harrisonburg	(3),(15),34:z_{10}:1,6
ser Hartford	6,7:y:e,n,x
ser Harvestehude	1,42:y:z_6
ser Hato	4,5,12:g,m,s:-
ser Havana	1,13,23:f,g,[s]:-
ser Heerlen	11:i:1,6
ser Heidelberg	[1],4,[5],12:r:1,2
*ser Heilbron	6,7:l,z_{28}:1,5:[z_{42}]
*ser Helsinki	1,4,12:z_{29}:[e,n,x]
*ser Hennepin	41:d:z_6
ser Hermannswerder	28:c:1,5
ser Heron	16:a:z_6
ser Herston	6,8:d:e,n,z_{15}
ser Herzliya	11:y:e,n,x
ser Hessarek	4,12,[27]:a:1,5
ser Heves	6,14,24:d:1,5
ser Hidalgo	6,8:r:e,n,z_{15}
*ser Hillbrow	17:b:e,n,x,z_{15}
ser Hillegersberg	(9),46:z_{35}:1,5
ser Hillsborough	6,7:z_{41}:l,w
ser Hilversum	30:k:1,2
ser Hindmarsh	(8),r:1,5
ser Hisingen	48:a:1,5,7
ser Hofit	39:i:1,5
ser Holcomb	6,8:l,v:e,n,x
ser Homosassa	1,6,14,25:z:1,5
ser Honelis	28:a:e,n,z_{15}
*ser Hooggraven	50:z_{10}:z_6:z_{42}
ser Horsham	1,6,14,25:l,v:e,n,x
*ser Houten	43:z_4,z_{23}:-
*ser Hueningen	9,12:z:z_{39}
*ser Huila	11:l,z_{28}:e,n,x
ser Hull	16:b:1,2
*ser Humber	53:z_4,z_{24}:-
ser Huvudsta	3,10:b:1,7
ser Hvittingfoss	16:b:e,n,x

*biochemically aberrant

ser Ibadan	13:22:b:1,5
ser Idikan	13,23:i:1,5
ser Ilala	28:k:1,5
ser Illinois	(3),(15),34:z_{10}:1,5
ser Ilugun	(1),3,10,(19):z_4,z_{23}:z_6
ser Indiana	1,4,12:z:1,7
ser Infantis	6,7,[14]:r:1,5
ser Inganda	6,7:z_{10}:1,5
ser Inglis	(9),46:z_{10}:e,n,x
ser Inpraw	41:z_{10}:e,n,x
ser Inverness	38:k:1,6
ser Ipeko	9,12:c:1,6
ser Ipswich	41:z_4,z_{24}:-
ser Irenea	17:k:1,5
ser Irigny	43:z_{38}:-
ser Irumu	6,7:l,v:1,5
*ser Islington	3,10:g,t:-
ser Israel	9,12:e,h:e,n,z_{15}
ser Isuge	13,23:d:z_6
ser Ituri	1,4,12:z_{10}:1,5
ser Itutaba	(9),46:c:z_6
*ser Jacksonville	16:z_{29}:-
ser Jaffna	1,9,12:d:z_{35}
ser Jaja	4,12,27:z_4,z_{23}:-
ser Jamaica	9,12:r:1,5
ser Jangwani	17:a:1,5
bioser Java	1,4,5,12:b:[1,2]
ser Javiana	1,9,12:l,z_{28}:1,5
ser Jedburgh	3,10:z_{29}:-
ser Jericho	1,4,12,27:c:e,n,z_{15}
ser Jerusalem	6,(7),[14]:z_{10}:l,w
ser Jodhpur	45:z_{29}:-
ser Joenkoeping	4,5,12:g,s,t:-
ser Johannesburg	1,40:b:e,n,x
ser Jos	1,4,12,27:y:e,n,z_{15}
ser Jukestown	13,23:i:e,n,z_{15}
ser Kaapstad	4,12:e,h:1,7
ser Kaduna	6,(7),(14):c:e,n,z_{15}
ser Kahla	1,42:z_{35}:1,6
ser Kaitaan	1,6,14,25:m,t:-
ser Kalamu	1,4,12:z_4,z_{24}:[1,5]
ser Kalina	3,10:b:1,2
*ser Kaltenhausen	28:b:z_6
ser Kamoru	4,12,27:y:z_6
ser Kampala	1,42:c:z_6
ser Kandla	17:z_{29}:-
ser Kaneshie	1,42:i:l,w
ser Kaolack	47:z:1,6
ser Kapemba	9,12:l,v:1,7
ser Karachi	45:d:e,n,x
ser Karamoja	40:z_{41}:1,2
ser Kasenyi	38:e,h:1,5

ser Kassberg	$1,6,14,25:c:1,6$	ser Langensalza	$3,10:y:l,w$
*ser Katesgrove	$1,13,23:m,t:1,5$	ser Langford	$28:b:e,n,z_{15}$
ser Kentucky	$(8),20:i:z_6$	ser Lanka	$3,15:r:z_6$
variant Jerusalem	$(8):i:z_6$	ser Lansing	$38:i:1,5$
ser Kenya	$6,7:l,z_{13}:e,n,x$	ser Larochelle	$6,7:e,h:1,2$
*ser Khami	$47:b:e,n,x,z_{15}$	ser Lattenkamp	$45:z_{35}:1,5$
ser Khartoum	$(3),(15),34:a:1,7$	ser Lawndale	$1,9,12:z:1,5$
ser Kiambu	$4,12:z:1,5$	ser Lawra	$44:k:e,n,z_{15}$
ser Kibi	$16:z_4,z_{23}:-$	ser Leeuwarden	$11:b:1,5$
ser Kibusi	$28:r:e,n,x$	ser Legon	$[1],4,12,[27]:c:1,5$
ser Kidderminster	$38:c:1,6$	ser Leiden	$13,22:z_{38}:-$
ser Kiel	$1,2,12:g,p:-$	ser Leipzig	$41:z_{10}:1,5$
ser Kikoma	$16:y:e,n,x$	ser Leith	$6,8:a:e,n,z_{15}$
*ser Kilwa	$4,12:l,w:e,n,x$	ser Leoben	$28:l,v:1,5$
ser Kimberly	$38:l,v:1,5$	ser Leopoldville	$6,7:b:z_6$
ser Kimuenza	$1,4,12,27:l,v:e,n,x$	*ser Lethe	$41:g,t:-$
ser Kingabwa	$43:y:1,5$	ser Lexington	$3,10:z_{10}:1,5$
ser Kingston	$1,4,12,27:g,s,t:-$	ser Lezennes	$6,8:z_4,z_{23}:1,7$
variant Copenhagen	$4,12:g,s,t:-$	*ser Lichtenberg	$41:z_{10}:[z_6]$
ser Kinondoni	$17:a:e,n,x$	ser Ligeo	$30:l,v:1,2$
ser Kinshasa	$3,15:l,z_{13}:1,5$	ser Ligna	$35:z_{10}:z_6$
ser Kintambo	$13,23:m,t:-$	ser Lille	$6,7:z_{38}:-$
ser Kirkee	$17:b:1,2$	*ser Limbe	$1,13,22:g,m,t:[1,5]$
ser Kisangani	$1,4,5,12:a:1,2$	ser Limete	$1,4,12,27:b:1,5$
ser Kisarawe	$11:k:e,n,x$	*ser Lincoln	$11:m,t:e,n,x$
ser Kitenge	$28:y:e,n,x$	ser Lindenburg	$6,8:i:1,2$
ser Kivu	$6,7:d:1,6$	ser Lindern	$6,14,24:d:e,n,x$
*ser Klapmuts	$45:z:z_{39}$	ser Lindi	$38:r:1,5$
*ser Kluetjenfelde	$4,12:d:e,n,x$	*ser Lindrick	$9,12:e,n,x:1,[5],7$
ser Kokemlemle	$39:l,v:e,n,x$	ser Lingivala	$16:z:1,7$
*ser Kommetje	$43:b:z_{42}$	ser Linton	$13,23:r:e,n,z_{15}$
ser Korbol	$(8),20:b:1,5(6)$	ser Lisboa	$16:z_{10}:1,6$
ser Korlebu	$1,3,19:z:1,5$	ser Lishabi	$(9),46:z_{10}:1,7$
ser Korovi	$38:g,m,s:-$	ser Litchfield	$6,8:l,v:1,2$
ser Kottbus	$6,8:e,h:1,5$	ser Liverpool	$1,3,19:d:e,n,z_{15}$
ser Kotte	$6,7:b:z_{35}$	ser Livingstone	$6,7:d:l,w$
ser Kralendyk	$6,7:z_4,z_{24}$	ser Ljubljana	$4,12,27:k:e,n,x$
*ser Kraaifontein	$1,13,23:g,(m),t:[e,n,x]$	ser Llandoff	$1,3,19:z_{29}:-$
ser Kralingen	$(8),20:y:z_6$	*ser Llandudno	$28:g,s,t:1,5$
ser Krefeld	$1,3,19:y:l,w$	ser Loanda	$6,8:l,v:1,5$
ser Kristianstad	$3,10:z_{10}:e,n,z_{15}$	*ser Lobatsi	$52:-:1,5,7$
*ser Krugersdorp	$50:e,n,x:1,7$	*ser Locarno	$57:z_{29}:z_{42}$
ser Kuessel	$28:i:e,n,z_{15}$	ser Loenga	$1,42:z_{10}:z_6$
*ser Kuilsrivier	$1,9,12:g,m,s,t:e,n,x$	ser Logone	$39:d:1,5$
ser Kumasi	$30:z_{10}:e,n,z_{15}$	*ser Lohbruegge	$44:z_4,z_{32}:-$
ser Kunduchi	$1,4,[5],12,27:l,z_{28}:1,2$	ser Lokstedt	$1,3,19:l,z_{13},z_{28}2$
ser Kuru	$6,8:z:l,w$	ser Lomalinda	$9,12:a:e,n,x$
		ser Lome	$9,12:r:z_6$
ser Labadi	$6,8:d:z_6$	ser Lomita	$6,7:e,h:1,5$
ser Lagos	$1,4,12:i:1,5$	ser London	$3,10:l,v:1,6$
ser Landala	$41:z_{10}:1,6$	ser Losangeles	$16:l,v:z_6$
ser Landau	$30:i:1,2$	ser Louga	$30:b:1,2$
ser Langenhorn	$18:m,t:-$	*ser Louwbester	$16:z:e,n,x$
		ser Lovelace	$13,22:l,v:1,5$
*biochemically aberrant		*ser Luanshya	$13,23:g,s,(t):-$

ser Luciana	$11:a:e,n,z_{15}$
ser Luckenwalde	$28:z_{10}:e,n,z_{15}$
ser Luke	$1,47:g,m:-$
*ser Lundby	$(9),46:b:e,n,x$
*ser Lurup	$41:z_{10}:e,n,x,z_{15}$
*ser Luton	$60:z:e,n,x$
ser Lyon	$47:k:e,n,z_{15}$
*ser Maarssen	$(9),46:z_4,z_{24}:z_{39}:z_{42}$
ser Maastricht	$11:z_{41}:1,2$
ser Macallen	$3,10:z_{36}:-$
ser Machaga	$1,3,19:i:e,n,x$
ser Madelia	$1,6,14,25:y:1,7$
ser Madiago	$1,3,19:c:1,7$
ser Madigan	$44:c:1,5$
ser Madjorio	$3,10:d:e,n,z_{15}$
ser Magumeri	$1,6,14,25:e,h:1,6$
ser Magwa	$21:d:e,n,x$
ser Maiduguri	$1,3,19:f,g,t:e,n,z_{15}$
ser Makiso	$6,7:l,z_{13},z_{28}:z_6$
*ser Makoma	$\cdot4,12:a:-$
*ser Makumira	$[1],4,12,[27]:e,n,x:1,7$
ser Malakal	$16:e,h:1,2$
ser Malstatt	$16:b:z_6$
ser Mampeza	$1,6,14,25:i:1,5$
ser Mampong	$13,22:z_{35}:1,6$
ser Manchester	$6,8:l,v:1,7$
ser Mandera	$16:l,z_{13}:e,n,z_{15}$
ser Manhattan	$6,8:d:1,5$
*ser Manica	$1,9,12:g,m,s,t:z_{42}$
ser Manila	$3,15:z_{10}:1,5$
*ser Manombo	$57:z_{39}:e,n,x,z_{15}$
ser Mapo	$6,8:z_{10}:1,5$
ser Mara	$39:e,h:[1,5]$
ser Maracaibo	$11:l,v:1,5$
ser Maricopa	$1,42:g,z_{51}:1,5$
ser Marienthal	$3,10:k:e,n,z_{15}$
*ser Marina	$48:g,z_{51}:-$
ser Maron	$3,10:d:z_{35}$
ser Marseille	$11:a:1,5$
ser Marylebone	$(9),46:k:1,2$
ser Massakory	$35:r:l,w$
ser Massenya	$1,4,12,27:k:1,5$
ser Matadi	$17:k:e,n,x$
ser Mathura	$(9),46:i:e,n,z_{15}$
ser Matopeni	$30:y:1,2$
*ser Matroosfontein	$3,10:a:e,n,x$
ser Mayday	$(9),46:y:z_6$
ser Mbandaka	$[1],6,7,[25]:z_{10}:e,n,z_{15}$
ser Mbao	$43:i:1,2$
ser Meleagridis	$3,10:e,h:l,w$
ser Memphis	$18:k:1,5$
ser Menden	$6,7:z_{10}:1,2$

ser Mendoza	$9,12:l,v:1,2$
ser Menhaden	$(3),(15),34:l,v:1,7$
ser Menston	$6,7:g,s,t:-$
*ser Merseyside	$16:g,t:1,5$
ser Mesbit	$47:m,t:e,n,z_{15}$
ser Meskin	$51:e,h:1,2$
ser Messina	$30:d:1,5$
ser Mgulani	$38:i:1,2$
bioser Miami	$1,9,12:a:1,5$
ser Michigan	$17:l,v:1,5$
ser Middlesbrough	$1,42:i:z_6$
*ser Midhurst	$53:l,z_{28}:z_{39}$
ser Mikawasima	$6,7:y:e,n,z_{15}$
ser Millesi	$1,40:l,v:1,2$
ser Milwaukee	$43:f,g:-$
ser Mim	$13,22:a:1,6$
ser Minneapolis	$(3),(15),34:e,h:1,6$
ser Minnesota	$21:b:e,n,x$
ser Mishmarhaemek	$1,13,23:d:1,5$
ser Mission	$6,7:d:1,5$
ser Mississippi (Atlanta)	$1,13,23:b:1,5$
ser Missouri	$11:g,s,t:-$
ser Miyazaki	$9,12:l,z_{13}:1,7$
*ser Mjimwema	$1,9,12:b:e,n,x$
*ser Mobeni	$16:g,m,s,t:-$
ser Mocamedes	$28:d:e,n,x$
ser Moero	$28:b:1,5$
ser Mokola	$3,10:y:1,7$
ser Molade	$(8),20:z_{10}:z_6$
*ser Mondeor	$39:l,z_{28}:e,n,x$
ser Mons	$1,4,12,[27]:d:l,w$
ser Monschaui	$35:m,t:-$
ser Montevideo	$6,7:g,m,s:-$
*ser Montgomery	$11:d,a:d,e,n,z_{15}$
ser Morehead	$30:i:1,5$
ser Morocco	$30:l,z_{13},z_{28}:e,n,z_{15}$
ser Morotai	$17:l,v:1,2$
ser Moroto	$28:z_{10}:l,w$
ser Moscow	$9,12:g,q:-$
*ser Mosselbay	$43:g,s,(t):z_{42}$
ser Moualine	$47:y:1,6$
ser Mountpleasant	$47:z:1,5$
ser Mowanjum	$6,8:z:1,5$
*ser Mpila	$3,10:z_{38}:z_{42}$
ser Muenchen	$6,8:d:1,2$
ser Muenster	$3,10:e,h:1,5$
ser Muguga	$44:m,t:-$
*ser Muizenberg	$9,12:g,m,s,t:1,5$
ser Mundonobo	$28:d:1,7$
*ser Mundsburg	$11:g,z_{51}:-$
ser Mura	$1,4,12:z_{10}:l,w$
*ser Nachshonim	$1,13,23:z:1,5$
ser Maestved	$1,9,12:g,p,s:-$

*biochemically aberrant

ser Nagoya	$6,8:b:1,5$
*ser Nairobi	$42:r:-$
ser Nakura	$1,4,12,27:a:z_6$
*ser Namib	$50:g,m,s,t:1,5$
ser Napoli	$1,9,12:l,z_{13}:e,n,x$
ser Narashino	$6,8:a:e,n,x$
ser Nashua	$28:l,v:e,n,z_{15}$
ser Nchanga	$3,10:l,v:1,2$
ser Ndolo	$[1],9,12:d:1,5$
*ser Neasden	$9,12:g,s,t:e,n,x$
*ser Negev	$41:z_{10}:1,2$
ser Nessa	$1,6,14,25:z_{10}:1,5$
ser Nessziona	$6,7:l,z_{13}:1,5$
ser Neukoelln	$6,7:l,z_{13},z_{28}:e,n,z_{15}$
ser Neumuenster	$1,4,12,27:k:1,6$
* variant	$1,4,12,27:k:1,6$
ser Newbrunswick	$3,15:l,v:1,7$
ser Newhaw	$3,15:e,h:1,5$
ser Newington	$3,15:e,h:1,6$
ser Newlands	$3,10:e,h:e,n,x$
ser Newmexico	$9,12:g,z_{51}:1,5$
ser Newport	$6,8:e,h:1,2$
ser New rochelle	$3,10:k:l,w$
ser Ngili	$6,7:z_{10}:1,7$
ser Ngor	$1,3,19:l,v:1,5$
*ser Ngozi	$48:z_{10}:[1,5]$
ser Niamey	$17:d:l,w$
ser Niarembe	$44:a:l,w$
ser Nienstedten	$6,(7),(14):b:[l,w]$
ser Nieukerk	$6,(7),(14):d:z_6$
variant	
zollenspicker	$6,7:d:z_6$
ser Nigeria	$6,7:r:1,6$
ser Nikolaifleet	$16:g,m,s:-$
ser Niloese	$1,3,19:d:z_6$
ser Nima	$28:y:1,5$
ser Nitra	$2,12:g,m:-$
*ser Noordhoek	$16:l,w:z_6$
*ser Nordenham	$1,4,12,27:z:e,n,x$
ser Nordufer	$6,8:a:1,7$
ser Norton	$6,7:i:l,w$
ser Norwich	$6,7:e,h:1,6$
ser Nottingham	$16:d:e,n,z_{15}$
ser Nowawes	$40:z:z_6$
ser Nuatja	$16:k:e,n,x$
*ser Nuernberg	$42:z:z_6$
ser Nyanza	$11:z:z_6$
ser Nyborg	$3,10:e,h:1,7$
ser Oakland	$6,7:z:1,6,(7)$
ser Obogu	$6,7:z_4,z_{23}:1,5$
ser Ochsenwerder	$54:k:1,5$

*biochemically aberrant

*ser Ochsenzoll	$16:z_4,z_{23}:-$
*ser Odijk	$30:a:z_{39}$
ser Odozi	$30:k:e,n,x,z_{15}$
* variant	$30:k:e,n,x,z_{15}$
*ser Oevelgoenne	$28:r:e,n,z_{15}$
ser Offa	$41:z_{38}:-$
ser Ohio	$6,7:b:l,w$
ser Ohlstedt	$3,10:y:e,n,x$
ser Okatie	$13,23:g,s,t:-$
ser Okefoko	$3,10:c:z_6$
ser Okerara	$3,10:z_{10}:1,2$
ser Oldenburg	$16:d:1,2$
ser Omderman	$6,(7),(14):d:e,n,x$
ser Omifisan	$40:z_{29}:-$
ser Ona	$28:g,s,t:-$
ser Onarimon	$1,9,12:b:1,2$
ser Onderstepoort	$1,6,14,25:e,h:1,5$
ser Onireke	$3,10:d:1,7$
ser Oranienburg	$6,7:m,t:-$
ser Ordonez	$1,13,23,37:y:l,w$
ser Orientalis	$16:k:e,n,z_{15}$
ser Orion	$3,10:y:1,5$
ser Oritamerin	$6,7:i:1,5$
ser Os	$9,12:a:1,6$
ser Oskarshamn	$28:y:1,2$
ser Oslo	$6,7:a:e,n,x$
ser Osnabrueck	$11:1,z_{13},z_{28}:e,n,x$
ser Othmarschen	$6,7:g,m,[t]:-$
*ser Ottershaw	$40:d:-$
ser Ouakam	$(9),[12],[34]:46:z_{29}:-$
ser Overschie	$51:l,v:1,5$
ser Overvecht	$30:a:1,2$
ser Oxford	$3,10:a:1,7$
*ser Oysterbeds	$6,7:z:z_{42}$
ser Pakistan	$(8):l,v:1,2$
ser Panama	$1,9,12:l,v:1,5$
ser Pankow	$3,15:d:1,5$
ser Papuana	$6,7:r:e,n,z_{15}$
bioser Paratyphi A	$1,2,12:a:-$
variant Durazzo	$2,12:a:-$
ser Paratyphi B	$1,4,5,12:b:1,2$
variant Odense	$1,4,12:b:1,2$
bioser Paratyphi C	$6,7,[Vi]:c:1,5$
*ser Parera	$11:z_4,z_{23}:-$
*ser Parow	$3,15:g,m,s,t:-$
ser Paris	$(8):20:z_{10}:1,5$
ser Patience	$28:d:e,n,z_{15}$
ser Penarth	$9,12:z_{35}:z_6$
ser Pensacola	$9,12:m,t:-$
*ser Perinet	$45:m,t:e,n,x,z_{15}$
ser Perth	$38:y:e,n,x$
ser Pharr	$11:b:e,n,z_{15}$
*ser Phoenix	$47:b:1,5$

ser Pikine	$(8),20:r:z_6$
ser Plymouth	$(9),46:d:z_6$
ser Poano	$1,6,14,25:z:l,z_{13},z_{28}$
ser Poeseldorf	$54:i:z_6$
ser Pomona	$28:y:1,7$
ser Poona	$[1],13,22,[37]:z,1,6$
*ser Portbech	$42:l,v:e,n,x,z_{15}$
ser Portland	$9,12:z_{10}:1,5$
ser Portsmouth	$3,15:l,v:1,6$
ser Potsdam	$6,7:l,v:e,n,z_{15}$
ser Potto	$(9),12,46:i:z_6$
ser Praha	$6,8:y:e,n,z_{15}$
ser Pramiso	$3,10:c:1,7$
ser Presov	$6,8:b:e,n,z_{15}$
ser Preston	$1,4,12:z:l,w$
ser Pretoria	$11:k:1,2$
bioser Pullorum	$9,12:-:-$
ser Putten	$13,23:d:l,w$
*ser Quimbamba	$47:d:z_{39}$
ser Quinhon	$47:z_{44}:-$
ser Quiniela	$6,8:c:e,n,z_{15}$
ser Ramatgen	$30:k:1,5$
*ser Rand	$42:z:e,n,x,z_{15}$
ser Raus	$13,22:f,g:e,n,x$
ser Reading	$4,[5],12:e,h:1,5$
ser Rechovot	$(8),20:e,h:z_6$
ser Redhill	$11:e,h:l,z_{13},z_{28}$
ser Redlands	$16:z_{10}:e,n,z_{15}$
ser Regent	$3,10:f,g:-$
ser Remo	$1,4,12,27:r:1,7$
*ser Rhodesiense	$9,12:d:e,n,x$
ser Rhone	$21:c:e,n,x$
ser Richmond	$6,7:y:1,2$
ser Rideau	$1,3,19:f,g:-$
ser Ridge	$9,12:c:z_6$
ser Riggil	$6,7:g,t:-$
ser Riogrande	$40:b:1,5$
ser Rissen	$6,7:f,g:-$
ser Riverside	$45:b:1,5$
ser Roan	$38:l,v:e,n,x$
*ser Rochdale	$50:b:e,n,x$
*ser Roggeveld	$51:-:1,7$
ser Rogy	$28:z_{10}:1,2$
ser Romanby	$13,23:z_4,z_{24}:-$
ser Roodepoort	$[1],13,22,[37]:z_{10}:1,5$
*ser Rooikrantz	$1,6,14:m,t:1,5$
ser Rosenthal	$3,15:b:1,5$
ser Rossleben	$54:e,h:1,6$
ser Rostock	$1,9,12:g,p,u:-$
*ser Roterberg	$6,7:z_4,z_{23}:-$

*ser Rotterdam	$1,13,22:g,t:1,5$
*ser Rowbarton	$16:m,t:-$
ser Rubislaw	$11:[d],r:[d],e,n,x$
ser Ruiru	$21:y:e,n,x$
ser Ruki	$4,5,12:y:e,n,x$
ser Rutgers	$3,10:l,z_{40}:1,7$
ser Ruzizi	$3,10:l,v:e,n,z_{15}$
ser Saarbruecken	$[1],9,12:a:1,7$
*ser Sachsenwald	$1,40:z_4,z_{23}:-$
ser Saintmarie	$52:g,t:-$
ser Saintpaul	$1,4,[5],12:e,h:1,2$
ser Saka	$47:b:-$
*ser Sakaraha	$48:[k]:z_{39}$
ser Salford	$16:l,v:e,n,x$
ser Salinatis	$4,12:d,e,h:d,e,n,z_{15}$
ser Sandiego	$4,12:e,h:e,n,z_{15}$
ser Sandow	$6,8:f,g:e,n,z_{15}$
ser Sanga	$(8):b:1,7$
ser Sanjuan	$6,7:a:1,5$
ser Sanktgeorg	$28:r,(i):e,n,z_{15}$
ser Sanktmarx	$1,3,19:e,h:1,7$
ser Santhiaba	$40:l,z_{28}:1,6$
ser Sao	$1,3,19:e,h:e,n,z_{15}$
ser Saphra	$16:y:1,5$
ser Sara	$1,6,14,25:z_{38}:[e,n,x]$
ser Sarajane	$4,12,27:d:e,n,x$
*ser Sarepta	$16:l,z_{28}:z_{42}$
ser Schalkwijk	$6,14,(24):i:e,n....$
ser Schleissheim	$4,12,27:b,z_{12}:-$
ser Schoeneberg	$1,3,19:z:e,n,z_{15}$
ser Schwarzengrund	$1,4,12,27:d:1,7$
ser Schwerin	$6,8:k:e,n,x$
*ser Seaforth	$50:k:z_6$
ser Seattle	$28:a:e,n,x$
ser Sedgwick	$44:b:e,n,z_{15}$
ser Seegefeld	$3,10:r(i):1,2$
ser Sekondi	$3,10:e,h:z_6$
ser Selandia	$3,15:e,h:1,7$
*ser Seminole	$1,40a,40b:g:z_{51}$
bioser Sendai	$1,9,12:a:1,5$
ser Senegal	$11:r:1,5$
ser Senftenberg	$1,3,19:g,s,t:-$
ser Seremban	$9,12:i:1,5$
*ser Setubal	$60:g,m,t:z_6$
ser Shamba	$16:c:e,n,x$
ser Shangani	$3,10:d:1,5$
ser Shanghai	$16:l,v:1,6$
ser Sharon	$11:k:1,6$
ser Sheffield	$38:c:1,5$
ser Shikmonah	$40:a:1,5$
ser Shipley	$(8),20:b:e,n,z_{15}$
ser Shomolu	$28:y:l,w$
*ser Shomron	$18:z_4,z_{32}:-$

ser Shoreditch	$(9),46:r:e,n,z_{15}$
ser Shubra	$4,5,12:z:1,2$
ser Siegburg	$6,14,18:z_4,z_{23}:[1,5]$
ser Simi	$3,10:r:e,n,z_{15}$
*ser Simonstown	$1,6,14:z_{10}:1,5$
ser Simsbury	$1,3,19:z_{27}:-$
ser Singapore	$6,7:k:e,n,x$
ser Sinstorf	$3,10:l,v:1,5$
ser Sinthia	$18:z_{38}:-$
ser Sladun	$1,4,12,27:b:e,n,x$
*ser Slangkop	$1,6,14:z_{10}:z_6z_{42}$
*ser Slatograd	$30:f,g,(p),t:-$
ser Sljeme	$1,47:f,g:-$
ser Sloterdijk	$1,4,12,27:z_{35}:z_6$
ser Soahamina	$6,14,24:z:e,n,x$
ser Soerenga	$30:i:l,w$
*ser Soesterberg	$21:z_4,z_{23}:-$
*ser Sofia	$4,12,[27]:b:[e,n,x]$
ser Solna	$28:a:1,5$
ser Solt	$11:y:1,5$
ser Southbank	$3,10:m,t:-$
*ser Soutpan	$11:z:z_{39}$
ser Souza	$3,10:d:e,n,x$
ser Spartel	$21:d:1,5$
*ser Springs	$40:a:z_{39}$
*ser Srinagar	$11:b:e,n,x$
ser Stanley	$4,5,12:d:1,2$
ser Stanleyville	$1,4,5,12:z_4,z_{23}:[1,2]$
ser Steinplatz	$30:y:1,6$
*ser Stellenbosch	$1,9,12:z:1,7$
ser Stellingen	$47:d:e,n,x$
ser Stendal	$11:l,v:1,2$
ser Sternchanze	$30:g,s,t:-$
ser Sterrenbos	$6,8:d:e,n,x$
*ser Stevenage	$1,13,23:[z_{42}]:1,7$
*ser Stikland	$3,10:m,t:e,n,x$
ser Stockholm	$3,10:y:z_6$
ser Stormont	$3,10:d:1,2$
ser Stourbridge	$6,8:b:1,6$
ser Straengnaes	$11:z_{10}:1,5$
ser Strasbourg	$(9),46:d:1,7$
ser Stratford	$1,3,19:i:1,2$
*ser Suarez	$1,40:c:e,n,x,z_{15}$
ser Suberu	$3,10:g,m:-$
*ser Suederelbe	$1,9,12:b:z_{39}$
ser Suelldorf	$45:f,g:-$
*ser Sullivan	$6,7:z_{42}:1,7$
ser Sundsvall	$1,6,14,25:z:e,n,x$
ser Sunnycove	$(8):y:e,n,x$
*ser Sunnydale	$1,40:k:e,n,x,z_{15}$
ser Surat	$1,6,14,25:r(i):e,n,z_{15}$
variant Hr-	$1,6,14,25:i:e,n,z_{15}$
*ser Sydney	$48:i:z$
ser Szentes	$16:k:1,2$
*ser Tafelbaai	$3,10:z:z_{39}$
ser Tafo	$1,4,12,27:z_{35}:1,7$
ser Takoradi	$6,8:i:1,5$
ser Taksony	$1,3,19:i:z_6$
ser Tallahassee	$6,8:z_4,z_{32}:-$
ser Tamale	$(8),20:z_{29}:-$
ser Tananarive	$6,8:y:1,5$
ser Tanger	$1,13,22:y:1,6$
ser Tarshyne	$9,12:d:1,6$
ser Taunton	$28:k:e,n,x$
ser Tchad	$35:b:-$
ser Techimani	$28:c:z_6$
ser Teddington	$4,12,27:y:1,7$
ser Tees	$16:f,g:-$
ser Tejas	$4,12:z_{36}:-$
ser Teko	$1,6,14,25:d:e,n,z_{15}$
ser Telaviv	$28:y:e,n,z_{15}$
ser Telelkebir	$13,23:d:e,n,z_{15}$
ser Telhashomer	$11:z_{10}:e,n,x$
ser Teltow	$28:z_4,z_{23}:1,6$
ser Tennessee	$6,7:z_{29}:-$
ser Teshie	$1,47:l,z_{13},z_{28}:e,n,z_{15}$
ser Texas	$4,5,12:k:e,n,z_{15}$
ser Thiaroye	$38:e,h:1,2$
ser Thielallee	$6,(7),(14):m,t:-$
ser Thomasville	$(3),(15),34:y:1,5$
ser Thompson	$6,7,[14]:k:1,5$
ser Tilburg	$1,3,19:d:l,w$
ser Tilene	$1,40:e,h:1,2$
ser Tinda	$1,4,12,27:a:e,n,z_{15}$
ser Tione	$51:a:e,n,x$
ser Togo	$4,12:l,w:1,6$
*ser Tokai	$57:z_{42}:1,6:z_{53}$
ser Tokoin	$4,12:z_{10}:e,n,z_{15}$
ser Tonev	$54:b:e,n,x$
ser Tornow	$45:g,m:-$
*ser Tosamanga	$6,7:z:1,5$
ser Tournai	$3,15:y:z_6$
ser Trachau	$4,12,27:y:1,5$
*ser Tranoroa	$55:k:z_{39}$
ser Travis	$4,5,12:g,z_{51}:1,7$
ser Treforest	$1,51:z:1,6$
ser Trotha	$40:z_{10}:z_6$
ser Tshiongwe	$6,8:e,h:e,n,z_{15}$
ser Tucson	$1,6,14,25:b:[1,7]$
ser Tudu	$4,12:z_{10}:1,6$
ser Tuebingen	$3,15:y:1,2$
*ser Tuindorp	$43:z_4,z_{32}:-$
*ser Tulear	$6,8:a:z_{52}$
ser Tunis	$1,13,23:y:z_6$
*ser Tygerberg	$1,13,23:a:z_{42}$

*biochemically aberrant

Salmonella typhi	9,12,[Vi]:d:-	ser	1,40:-:1,7
		ser	41:k:-
S. enteritidis		ser	42:m,t:e,n,x,z_{15}
ser Typhimurium	1,4,5,12:i:1,2	ser	42:-:1,6
variant		ser	43:e,n,x,z_{15}:1,(5),7
Copenhagen	1,4,12:i:1,2	ser	43:e,n,z_{15}:1,6
bioser Typhisuis	6,7:[c]:1,5	ser	43:z:1,5
		ser	44:g,z_{51}:-
ser Uccle	54:g,s,t:-	ser	44:z_4,z_{23}:-
ser Uganda	3,10:l,z_{13}:1,5	ser	44:z_{36},z_{38}:-
ser Ughelli	3,10:r:1,5	ser	45:g,z_{51}:-
ser Uhlenhorst	44:z:l,w	ser	47:z_6:1,6
ser Ullevi	1,13,23,27:b:e,n,x	ser	48:a:z_6
ser Umbilo	28:z_{10}:e,n,x	ser	48:z_4,z_{32}:-
ser Umhlali	6,7:a:1,6	ser	50:l,w:e,n,x;z_{15}:z_{42}
ser Umhlatazana	35:a:e,n,z_{15}	ser	50:l,z_{28}:z_{42}
		ser	50:z_4,z_{24}:-
*Unnamed serotypes		ser	52:d:e,n,x,z_{15}
ser	4,12:(f),g:-	ser	53:z:z_6
ser	4,12:-:1,6	ser	56:e,n,x:1,7
ser	6,7:a:z_6	ser	57:g,m,s,t:z_{42}
ser	6,7:g,t:e,n,x:z_{42}	ser	58:a:1,5
ser	6,7:k:[z_6]	ser	64:k:e,n,x,z_{15}
ser	6,7:z:z_6	ser	64:z_{29}:-
ser	6,7:z_{10}:z_{35}	ser Uno	6,8:z_{29}:-
ser	6,7:z_{29}:-	*ser Uphill	42:b:e,n,x,z_{15}
ser	6,7:z_{42}:e,n,x:1,6	ser Uppsala	4,12,27:b:1,7
ser	6,8:g,(m),t:e,n,x	ser Urbana	30:b:e,n,x
ser	9,12:e,n,x:1,6	ser Ursenbach	1,42:z:1,6
ser	(9),46:z_{10}:z_6	ser Usumbura	18:d:1,7
ser	1,9,12,(46),27:y:z_{39}	ser Utah	6,8:c:1,5
ser	3,10:l,z_{28},z_{39}	*ser Utbremen	35:z_{29}:e,n,x
ser	11:b:1,7	ser Utrecht	52:d:1,5
ser	11:z_4,z_{23}:-	ser Uzaramo	1,6,14,25:z_4,z_{24}:-
ser	13,23:l,z_{28}:z_6	126,l119,,	
ser	(6),14:k:[e,n,x]	ser Vaertan	13,22:b:e,n,x
ser	1,(6),14:k:z_6:z_{42}	ser Vancouver	16:c:1,5
ser	1,(6),14:z_{42}:1,6	*ser Veddel	43:g,t:-
ser	16:b:z_{42}	ser Vejle	3,10:e,h:1,2
ser	16:l,z_{40}:-	ser Vellore	1,4,12,27:$z_{10}$$z_{35}$
ser	17:k:-	ser Veneziana	11:i:e,n,x
ser	18:b:1,5	*ser Verity	17:e,n,x,z_{15}:1,6
ser	18:m,t:1,5	ser Victoria	1,9,12:l,w:1,5
ser	18:y:e,n,x,z_{15}	ser Victoriaborg	17:c:1,6
ser	28:e,n,x:1,7	ser Vietnam	41:b:-
ser	30:z_{39}:1,(7)	* variant	41:b:-
ser	35:g,m,s,t:-	ser Vinohrady	28:m,t:-
ser	35:l,z_{28}:-	* variant	28:m,t:-
ser	40:b:-	ser Virchow	6,7:r:1,2
ser	1,40a,40c:g,z_{51}:-	ser Virginia	(8):d:[1,2]
ser	1,40:m,t:z_{42}	ser Visby	1,3,19:b:1,6
ser	1,40:z_6:1,5	ser Vitkin	28:l,v:e,n,x
		ser Vleuten	44:f,g:-
*biochemically aberrant		ser Volkmarsdorf	28:i:1,6

*ser Volksdorf	$43:z_{36},z_{38}:-$
ser Volta	$11:r:l,z_{13},z_{28}$
ser Vom	$4,12,27:l,z_{13},z_{28}:e,n,z_{15}$
*ser Vredelust	$1,13,23:l,z_{28}:z_{42}$
*ser Vrindaban	$45:a:e,n,x$
ser Wa	$16:b:1,5$
ser Wagenia	$1,4,12,27:b:e,n,z_{15}$
*ser Wandsbek	$21:z_{10}:z_6$
ser Wandsworth	$39:b:1,2$
ser Wangata	$9,12:z_4,z_{23}:[1,7]$
ser Warnow	$6,8:i:1,6$
ser Warragul	$1,6,14,25:g,m:-$
*ser Wassenaar	$50:g,z_{51}:-$
ser Waycross	$41:z_4,z_{23}:-$
* variant	$41:z_4,z_{23}:-$
ser Wayne	$30:g,z_{51}:-$
ser Wedding	$28:c:e,n,z_{15}$
ser Welikade	$16:l,v:1,7$
ser Weltevreden	$3,10:r:z_6$
ser Wentworth	$11:z_{10}:1,2$
ser Wernigerode	$(9),46:f,g:-$
ser Weslaco	$42:z_{36}:-$
ser Westerstede	$1,3,19:l,z_{13}:-$
ser Westhampton	$3,10:g,s,t:-$
ser Weston	$16:e,h:z_6$
*ser Westpark	$3,10:l,z_{28}:e,n,x$
ser Westphalia	$35:z_4,z_{24}:-$
ser Weybridge	$3,10:d:z_6$
ser Wichita	$1,13,23:d:z_{37}$
ser Wien	$1,4,12,[27]:b;l,w$
ser Wil	$6,7:d:l,z_{13},z_{28}$
ser Wildwood	$(3),(15),34:e,h:l,w$

ser Wilhelmsburg	$4,[5],12,[27]:z_{38}:-$
variant	
Teufelsbrueck	$1,4,12:z_{38}:-$
*ser Wilhelmstrasse	$52:z_{44}:1,5,7$
ser Willemstad	$1,13,22:e,h:1,6$
ser Wimborne	$3,10:k:1,2$
*ser Winchester	$3,10:z_{39}:1,7$
ser Windermere	$39:y:1,5$
*ser Windhoek	$45:g,t:1,5$
ser Wingrove	$6,8:c:1,2$
ser Wippra	$6,8:z_{10}:z_6$
*ser Woerden	$17:c:z_{39}$
ser Womba	$4,12,27:c:1,7$
*ser Woodstock	$16:z_{42}:1,(5),7$
*ser Worcester	$1,13,23:m,t:e,n,x$
ser Worthington	$1,13,23:z:l,w$
*ser Wynberg	$1,9,12:z_{39}:1,7$
ser Yaba	$3,10:b:e,n,z_{15}$
ser Yalding	$1,3,19:r:e,n,z_{15}$
ser Yarm	$6,8:z_{35}:1,2$
ser Yarrabah	$13,23:y:1,7$
ser Yeerongpilly	$3,10:i,z_6$
ser Yoff	$38:z_4,z_{23}:1,2$
ser Yolo	$35:c:-$
ser Zadar	$(9),46:b:1,6$
ser Zanzibar	$3,10:k:1,5$
ser Zega	$9,12:d:z_6$
ser Zehlendorf	$30:a:1,5$
*ser Zeist	$18:z_{10}:z_6$
*ser Zuerich	$1,9,12:c:z_{39}$
ser Zuilen	$1,3,19:i:l,w$
ser Zwickau	$16:r(i):e,n,z_{15}$

*biochemically aberrant

ANTIGENIC SCHEMA FOR *SALMONELLA*

Serogroup A

Salmonella enteritidis			
bioser Paratyphi-A	1,2,12	a	—
variant Durazzo	2,12	a	—
ser Nitra	2,12	g,m	—
ser Kiel	1,2,12	g,p	—

Serogroup B

S. enteritidis			
*ser Makoma	4,12	a	—
ser Kisangani	1,4,5,12	a	1,2
ser Hessarek	4,12,[27]	a	1,5
ser Fulica	4,5,12	a	1,5
ser Arechaveleta	4,5,12	a	[1,7]
ser Bispebjerg	1,4,5,12	a	e,n,x
ser Abortusequi	4,12	—	e,n,x
ser Tinda	1,4,12,27	a	e,n,z_{15}
ser Nakura	1,4,12,27	a	z_6
ser Paratyphi-B	1,4,5,12	b	1,2
variant Odense	1,4,12	b	1,2
bioser Java	1,4,5,12	b	[1,2]
*ser Sofia	4,12,[27]	b	[e,n,x]
ser Limete	1,4,12,27	b	1,5
ser Canada	4,12	b	1,6
ser Uppsala	4,12,27	b	1,7
ser Schleissheim	4,12,27	b,z_{12}	—
ser Abony	1,4,5,12	b	e,n,x
variant Haifa	4,12	b	e,n,x
ser Abortusbovis	1,4,12,27	b	e,n,x
ser Slädun	1,4,12,27	b	e,n,x
ser Wagenia	1,4,12,27	b	e,n,z_{15}
ser Wien	1,4,12,[27]	b	l,w
ser Legon	[1],4,12,[27]	c	1,5
ser Abortusovis	4,12	c	1,6
ser Altendorf	4,12	c	1,7
ser Womba	4,12,27	c	1,7
ser Jericho	1,4,12,27	c	e,n,z_{15}
ser Bury	4,12,27	c	z_6
ser Stanley	4,5,12	d	1,2
ser Cairo	1,4,12,27	d	1,2

*biochemically aberrant

Note: ser., serotype; bioser., bioserotype. ( ) partial antigen. [ ], antigen present or absent.

ser Eppendorf	[1],4,12,[27]	d	1,5
ser Schwarzengrund	1,4,12,27	d	1,7
*ser Kluetjenfelde	4,12	d	e,n,x
ser Sarajane	4,12,27	d	e,n,x
ser Duisburg	[1],4,12,[27]	d	e,n,z_{15}
ser Mons	1,4,12,[27]	d	l,w
ser Ayinde	4,12,27	d	z_6
ser Salinatis	4,12	d, e, h	d,e,n,z_{15}
ser Saintpaul	1,4,[5],12	e,h	1,2
ser Reading	4,[5],12	e,h	1,5
ser Kaapstad	4,12	e,h	1,7
ser Chester	4,5,12	e,h	e,n,x
ser San diego	4,5,12	e,h	e,n,z_{15}
*ser Makumira	[1],4,12,[27]	e,n,x	1,7
*ser _____	4,12	(f),g	–
ser Derby	1,4,5,12	f,g	[1,2]
ser Agona	4,12	f,g,s	–
ser Essen	4,12	g,m	–
*ser Caledon	4,12	g,m	e,n,x
ser Hato	4,5,12	g,m,s	–
ser Joenkoeping	4,5,12	g,s,t	–
ser Kingston	[1],4,12,[27]	g,s,t	–
variant Copenhagen	4,12	g,s,t	–
ser Budapest	1,4,12	g,t	–
*ser Bechuana	4,12,27	g,t	–
ser Travis	4,5,12	g,z_{51}	1,7
ser California	4,5,12	m,t	–
ser Typhimurium	1,4,5,12	i	1,2
variant Copenhagen	1,4,12	i	1,2
ser Lagos	1,4,12	i	1,5
ser Agama	4,12	i	1,6
ser Gloucester	1,4,12,(27)	i	l,w
ser Massenya	1,4,12,27	k	1,5
ser Neumuenster	1,4,12,27	k	1,6
* variant	1,4,12,27	k	1,6
ser Ljubljana	4,12,27	k	e,n,x
ser Texas	4,5,12	k	e,n,z_{15}
ser Azteca	4,5,12	l,v	1,5
ser Bredeney	1,4,12,27	l,v	1,7
ser Kimuenza	1,4,12,27	l,v	e,n,x
ser Brandenburg	4,12	l,v	e,n,z_{15}
ser Clackamas	4,12	l,v,(z_{13})	1,6
*ser Kilwa	4,12	l,w	e,n,x
ser Ayton	1,4,12,27	l,w	z_6
ser Vom	4,12,27	l,z_{13},z_{28}	e,n,z_{15}
ser Kunduchi	1,4,[5],12,27	l,z_{28}	1,2

*biochemically aberrant

ser Heidelberg	[1],4,5,[12]	r	1,2
ser Bradford	4,12,27	r	1,5
ser Remo	1,4,12,27	r	1,7
ser Bochum	4,5,12	r	l,w
ser Africana	4,12	r(i)	l,w
ser Coeln	4,5,12	y	1,2
ser Trachau	4,12,27	y	1,5
ser Teddington	4,12,27	y	1,7
ser Ruki	4,5,12	y	e,n,x
ser Ball	1,4,12,27	y	e,n,x
ser Jos	1,4,12,27	y	e,n,z_{15}
ser Kamoru	4,12,27	y	z_6
ser Shubra	4,5,12	z	1,2
ser Kiambu	4,12	z	1,5
ser Indiana	1,4,12	z	1,7
*ser Nordenham	1,4,12,27	z	e,n,x
ser Preston	1,4,12	z	l,w
ser Entebbe	1,4,12,27	z	z_6
ser Stanleyville	1,4,5,12	z_4,z_{23}	[1,2]
ser Jaja	4,12,27	z_4,z_{23}	—
ser Kalamu	(1),4,12	z_4,z_{24}	[1,5]
ser Haifa	1,4,5,12	z_{10}	1,2
variant Afula, 01 & 05—	4,12	z_{10}	1,2
ser Ituri	1,4,12	z_{10}	1,5
ser Tudu	4,12	z_{10}	1,6
ser Albert	4,12	z_{10}	e,n,x
ser Tokoin	4,12	z_{10}	e,n,z_{15}
ser Mura	1,4,12	z_{10}	l,w
ser Fortune	4,12,27	z_{10}	z_6
ser Vellore	1,4,12,27	z_{10}	z_{35}
ser Brancaster	1,4,12,27	z_{29}	—
*ser Helsinki	1,4,12	z_{29}	[e,n,x]
ser Tafo	1,4,12,27	z_{35}	1,7
ser Sloterdijk	1,4,12,27	z_{35}	z_6
ser Tejas	4,12	z_{36}	—
ser Wilhelmsburg	4,12,27	z_{38}	—
variant Teufelsbrueck	1,4,12	z_{38}	—
*ser Durbanville	[1],4,12,[27]	$[z_{39}]$	1,[5],7
*ser _____	4,12	—	1,6

*biochemically aberrant

	Serogroup C_1		
S. enteritidis			
ser Sanjuan	6,7	a	1,5
ser Umhlali	6,7	a	1,6
ser Austin	6,7	a	1,7
ser Oslo	6,7	a	e,n,x
ser Denver	6,7	a	e,n,z_{15}
ser Coleypark	6,7	a	l,w
*ser _____	6,7	a	z_6
*ser Calvinia	6,7	a	z_{42}
ser Brazzaville	6,7	b	1,2
ser Edinburg	6,7	b	1,5
ser Georgia	6,7	b	e,n,z_{15}
ser Ohio	6,7	b	l,w
ser Leopoldville	6,7	b	z_6
ser Kotte	6,7	b	z_{35}
*ser Bloemfontein	6,7	b	[e,n,x] :z_{42}
Salmonella cholerae-suis	6,7	c	1,5
bioser Kunzendorf	6,7	[c]	1,5
S. enteritidis			
bioser Paratyphi-C	6,7[Vi]	c	1,5
bioser Decatur	6,7	c	1,5
bioser Typhisuis	6,7	[c]	1,5
ser Birkenhead	6,7	c	1,6
ser Mission	6,7	d	1,5
ser Kivu	6,7	d	1,6
ser Amersfoort	6,7	d	e,n,x
ser Gombe	6,7	d	e,n,z_{15}
ser Livingstone	6,7	d	l,w
ser Wil	6,7	d	l,z_{13},z_{28}
ser Nieukerk variant Zollenspicker	6,7	d	z_6
ser Larochelle	6,7	e,h	1,2
ser Lomita	6,7	e,h	1,5
ser Norwich	6,7	e,h	1,6
ser Braenderup	6,7	e,h	e,n,z_{15}
ser Rissen	6,7	f,g	—
ser Montevideo	6,7	g,m,s	—
ser Othmarschen	6,7	g,m,[t]	—
ser Menston	6,7	g,s,t	—
*ser _____	6,7	g,t	e,n,x:z_{42}
ser Riggil	6,7	g,t	—
ser Alamo	6,7	g,z_{51}	1,5
ser Haelsingborg	6,7	m,p,t,[u]	—
ser Oranienburg	6,7	m,t	—

*biochemically aberrant

ser Augustenborg	6,7	i	1,2
ser Oritamerin	6,7	i	1,5
ser Garoli	6,7	i	1,6
ser Norton	6,7	i	1,w
ser Galiema	6,7	k	1,2
ser Thompson	6,7	k	1,5
ser Daytona	6,7	k	1,6
ser Baiboukoum	6,7	k	1,7
ser Singapore	6,7	k	e,n,x
ser Escanaba	6,7	k	e,n,z_{15}
*ser _____	6,7	k	[z_6]
ser Concord	6,7	l,v	1,2
ser Irumu	6,7	l,v	1,5
ser Bonn	6,7	l,v	e,n,x
ser Potsdam	6,7	l,v	e,n,z_{15}
ser Gdansk	6,7	l,v	z_6
ser Gabon	6,7	l,w	1,2
ser Colorado	6,7	l,w	1,5
ser Nessziona	6,7	l,z_{13}	1,5
ser Kenya	6,7	l,z_{13}	e,n,x
ser Neukoelln	6,7	l,z_{13},z_{28}	e,n,z_{15}
ser Makiso	6,7	l,z_{13},z_{28}	z_6
*ser Heilbron	6,7	l,z_{28}	1,5:[z_{42}]
ser Virchow	6,7	r	1,2
ser Infantis	6,7,[14]	r	1,5
ser Nigeria	6,7	r	1,6
ser Colindale	6,7	r	1,7
ser Papuana	6,7	r	e,n,z_{15}
ser Richmond	6,7	y	1,2
ser Bareilly	6,7	y	1,5
ser Gatow	6,7	y	1,7
ser Hartford	6,7	y	e,n,x
ser Mikawasima	6,7	y	e,n,z_{15}
*ser Tosamanga	6,7	z	1,5
ser Oakland	6,7	z	1,6,(7)
ser Businga	6,7	z	e,n,z_{15}
*ser _____	6,7	z	z_6
*ser Oysterbeds	6,7	z	z_{42}
*ser Roterberg	6,7	z_4,z_{23}	—
ser Obogu	6,7	z_4,z_{23}	1,5
ser Aequatoria	6,7	z_4,z_{23}	e,n,z_{15}
*ser Kralendyk	6,7	z_4,z_{24}	—
*ser Cape	6,7	z_6	1,7
ser Menden	6,7	z_{10}	1,2

*biochemically aberrant

ser Inganda	6,7	z_{10}	1,5
ser Eschweiler	6,7	z_{10}	1,6
ser Ngili	6,7	z_{10}	1,7
ser Djugu	6,7	z_{10}	e,n,x
ser Mbandaka	[1],6,7,[25]	z_{10}	e,n,z_{15}
ser Jerusalem variant 014−	6,7	z_{10}	l,w
*ser _____	6,7	z_{10}	z_{35}
ser Tennessee	6,7	z_{29}	−
*ser _____	6,7	z_{29}	−
*ser Argentina	6,7	z_{36}	−
*ser Bacongo	6,7	z_{36}	z_{42}
ser Lille	6,7	z_{38}	−
*ser Gilbert	6,7	z_{39}	1,7
ser Hillsborough	6,7	z_{41}	l,w
*ser _____	6,7	z_{42}	e,n,x:1,6
*ser Sullivan	6,7	z_{42}	1,7

ser Nienstedten	6,(7),(14)	b	l,w
ser Kaduna	6,(7),(14)	c	e,n,z_{15}
ser Omderman	6,(7),(14)	d	e,n,x
ser Eimsbuettel	6,(7),(14)	d	l,w
ser Nieukerk	6,(7),(14)	d	z_6
ser Ardwick	6,(7),(14)	f,g	−
ser Thielallee	6,(7),(14)	m,t	−
ser Thompson variant 014+	6,(7),(14)	k	1,5
ser Gelsenkirchen	6,(7),(14)	l,v	z_6
ser Bareilly variant 014+	6,(7),(14)	y	1,5
ser Jerusalem	6,(7),(14)	z_{10}	l,w
ser Bornum	6,(7),(14)	z_{38}	−

Serogroup C₂

S. enteritidis

ser Doncaster	6,8	a	1,5
ser Curacao	6,8	a	1,6
ser Nordufer	6,8	a	1,7
ser Narashino	6,8	a	e,n,x

*biochemically aberrant

ser Leith	6,8	a	e,n,z_{15}
*ser Tulear	6,8	a	z_{52}
ser Nagoya	6,8	b	1,5
ser Stourbridge	6,8	b	1,6
ser Gatuni	6,8	b	e,n,x
ser Presov	6,8	b	e,n,z_{15}
ser Bukuru	6,8	b	1,w
ser Banalia	6,8	b	z_6
ser Wingrove	6,8	c	1,2
ser Utah	6,8	c	1,5
ser Bronx	6,8	c	1,6
ser Belem	6,8	c	e,n,x
ser Quiniela	6,8	c	e,n,z_{15}
ser Muenchen	6,8	d	1,2
ser Manhattan	6,8	d	1,5
ser Sterrenbos	6,8	d	e,n,x
ser Herston	6,8	d	e,n,z_{15}
ser Labadi	6,8	d	z_6
ser Newport	6,8	e,h	1,2
ser Kottbus	6,8	e,h	1,5
ser Tshiongwe	6,8	e,h	e,n,z_{15}
ser Sandow	6,8	f,g	e,n,z_{15}
ser Chincol	6,8	g,m,s	e,n,x
*ser _____	6,8	g,(m),t	e,n,x
*ser Baragwanath	6,8	m,t	1,5
*ser Germiston	6,8	m,t	e,n,x
ser Lindenburg	6,8	i	1,2
ser Takoradi	6,8	i	1,5
ser Warnow	6,8	i	1,6
ser Bonariensis	6,8	i	e,n,x
ser Aba	6,8	i	e,n,z_{15}
ser Blockley	6,8	k	1,5
ser Schwerin	6,8	k	e,n,x
ser Litchfield	6,8	l,v	1,2
ser Loanda	6,8	l,v	1,5
ser Manchester	6,8	l,v	1,7
ser Holcomb	6,8	l,v	e,n,x
ser Edmonton	6,8	l,v	e,n,z_{15}
ser Fayed	6,8	l,w	1,2
ser Breukelen	6,8	l,z_{13}	e,n,z_{15}
ser Bovismorbificans	6,8	r	1,5
ser Akanji	6,8	r	1,7
ser Hidalgo	6,8	r	e,n,z_{15}

*biochemically aberrant

ser Goldcoast	6,8	r	l,w
ser Tananarive	6,8	y	1,5
ser Alagbon	6,8	y	1,7
ser Praha	6,8	y	e,n,z_{15}
ser Mowanjum	6,8	z	1,5
ser Kuru	6,8	z	l,w
ser Lezennes	6,8	z_4,z_{23}	1,7
ser Chailey	6,8	z_4,z_{23}	$[e,n,z_{15}]$
ser Duesseldorf	6,8	z_4,z_{24}	—
ser Tallahassee	6,8	z_4,z_{32}	—
ser Mapo	6,8	z_{10}	1,5
ser Cleveland	6,8	z_{10}	1,7
ser Hadar	6,8	z_{10}	e,n,x
ser Glostrup	6,8	z_{10}	e,n,z_{15}
ser Wippra	6,8	z_{10}	z_6
ser Uno	6,8	z_{29}	—
ser Yarm	6,8	z_{35}	1,2

---

ser Korbol	(8),20	b	1,5,(6)
ser Sanga	(8)	b	1,7
ser Shipley	(8),20	b	e,n,z_{15}
ser Alexanderpolder	(8)	c	l,w
ser Virginia	(8)	d	[1,2]
ser Bardo	(8)	e,h	1,2
ser Rechovot	(8),20	e,h	z_6
ser Emek	(8),20	g,m,s	—
ser Kentucky	(8),20	i	z_6
variant Jerusalem	(8)	i	z_6
ser Haardt	(8)	k	1,5
ser Pakistan	(8)	l,v	1,2
ser Amherstiana	(8)	l,(v)	1,6
ser Hindmarsh	(8)	r	1,5
ser Pikine	(8),20	r	z_6
ser Cocody	(8),20	r(i)	e,n,z_{15}
ser Altona	(8),20	r(i)	z_6
ser Brunei	(8),20	y	1,5
ser Sunnycove	(8)	y	e,n,x

ser Kralingen	(8),20	y	z_6
ser Corvallis	(8),20	z_4,z_{23}	—
ser Albany	(8),20	z_4,z_{24}	—
ser Paris	(8),20	z_{10}	1,5
ser Molade	(8),20	z_{10}	z_6
ser Tamale	(8),20	z_{29}	—

Serogroup D_1

S. enteritidis

ser Miami	1,9,12	a	1,5
bioser Sendai	1,9,12	a	1,5
ser Os	9,12	a	1,6
ser Saarbruecken	[1],9,12	a	1,7
ser Lomalinda	9,12	a	ė,n,x
ser Durban	9,12	a	e,n,z_{15}
ser Onarimon	1,9,12	b	1,2
ser Frintrop	1,9,12	b	1,5
*ser Mjimwema	1,9,12	b	e,n,x
*ser Blankenese	1,9,12	b	z_6
*ser Suederelbe	1,9,12	b	z_{39}
ser Goeteborg	9,12	c	1,5
ser Ipeko	9,12	c	1,6
ser Alabama	9,12	c	e,n,z_{15}
ser Ridge	9,12	c	z_6
*ser Zuerich	1,9,12	c	z_{39}
Salmonella typhi	9,12,Vi	d	—

S. enteritidis

ser Ndolo	[1],9,12	d	1,5
ser Tarshyne	9,12	d	1,6
*ser Rhodesiense	9,12	d	e,n,x
ser Zega	9,12	d	z_6
ser Jaffna	1,9,12	d	z_{35}
ser Bournemouth	9,12	e,h	1,2
ser Eastbourne	1,9,12	e,h	1,5
ser Israel	9,12	e,h	e,n,z_{15}
*ser Lindrick	9,12	e,n,x	1,[5],7
*ser _____	9,12	e,n,x	1,6
ser Berta	9,12	f,g,t	—
ser Enteritidis	1,9,12	g,m	—
ser Blegdam	9,12	g,m,q	—
*ser Muizenberg	9,12	g,m,s,t	1,5

*biochemically aberrant

*ser Kuilsrivier	1,9,12	g,m,s,t	e,n,x
*ser Manica	1,9,12	g,m,s,t	z_{42}
ser Dublin	1,9,12	g,p	—
ser Naestved	1,9,12	g,p,s	—
ser Rostock	1,9,12	g,p,u	—
ser Moscow	9,12	g,q	—
*ser Neasden	9,12	g,s,t	e,n,x
*ser Hamburg	1,9,12	g,t	—
ser Newmexico	9,12	g,z_{51}	1,5
ser Pensacola	9,12	m,t	—
ser Seremban	9,12	i	1,5
ser Claibornei	1,9,12	k	1,5
ser Goverdhan	9,12	k	1,6
ser Mendoza	9,12	l,v	1,2
ser Panama	1,9,12	l,v	1,5
ser Kapemba	9,12	l,v	1,7
ser Goettingen	9,12	l,v	e,n,z_{15}
ser Victoria	1,9,12	l,w	1,5
*ser Daressalaam	1,9,12	l,w	e,n,x
ser Miyazaki	9,12	l,z_{13}	1,7
ser Napoli	1,9,12	l,z_{13}	e,n,x
ser Javiana	1,9,12	l,z_{28}	1,5
ser Jamaica	9,12	r	1,5
ser Lome	9,12	r	z_6
ser Lawndale	1,9,12	z	1,5
*ser Stellenbosch	1,9,12	z	1,7
*ser Angola	1,9,12	z	z_6
*ser Hueningen	9,12	z	z_{39}
ser Wangata	9,12	z_4,z_{23}	[1,7]
ser Portland	9,12	z_{10}	1,5
*ser Canastel	9,12	z_{29}	1,5
ser Penarth	9,12	z_{35}	z_6
ser Elomrane	1,9,12	z_{38}	—
*ser Wynberg	1,9,12	z_{39}	1,7
bioser Gallinarum	1,9,12	—	—
bioser Pullorum	9,12	—	—

Serogroup D_2

ser Baildon	(9),46	a	e,n,x

*biochemically aberrant

ser Zadar	(9),46	b	1,6
*ser Lundby	(9),46	b	e,n,x
ser Bamboye	(9),46	b	l,w
ser Itutaba	(9),46	c	z_6
ser Strasbourg	(9),46	d	1,7
ser Plymouth	(9),46	d	z_6
ser Bergedorf	(9),46	e,h	1,2
ser Wernigerode	(9),46	f,g	—
*ser Duivenhoks	(9),46	g,m,s,t	e,n,x
ser Gateshead	(9),46	g,s,t	—
ser Mathura	(9),46	i	e,n,z_{15}
ser Marylebone	(9),46	k	1,2
ser Ceyco	(9),46	k	z_{35}
ser India	(9),46	l,v	1,5
ser Shoreditch	(9),46	r	e,n,z_{15}
ser Mayday	(9),46	y	z_6
*ser Haarlem	(9),46	z	e,n,x
ser Ekotedo	(9),46	z_4,z_{23}	—
*ser Maarssen	(9),46	z_4,z_{24}	$z_{39}:z_{42}$
ser Lishabi	(9),46	z_{10}	1,7
ser Inglis	(9),46	z_{10}	e,n,x
*ser _____	(9),46	z_{10}	z_6
ser Ouakam	(9),[12],[34],46	z_{29}	—
ser Hillegersberg	(9),46	z_{35}	1,5
ser Fresno	(9),46	z_{38}	—
* variant	(9),46	z_{38}	—
ser Potto	(9),12,46	i	z_6
*ser _____	1,9,12,(46),27	y	z_{39}

Serogroup E_1

S. enteritidis
ser Aminatu	3,10	a	1,2
ser Goelzau	3,10	a	1,5
ser Oxford	3,10	a	1,7

*biochemically aberrant

236   Antigenic Schema for *Salmonella*

*ser Matroosfontein	3,10	a	e,n,x
ser Galil	3,10	a	e,n,z$_{15}$
ser Kalina	3,10	b	1,2
ser Butantan	3,10	b	1,5
ser Allerton	3,10	b	1,6
ser Huvudsta	3,10	b	1,7
ser Benfica	3,10	b	e,n,x
ser Yaba	3,10	b	e,n,z$_{15}$
ser Epicrates	3,10	b	l,w
ser Pramiso	3,10	c	1,7
ser Agege	3,10	c	e,n,z$_{15}$
ser Anderlecht	3,10	c	l,w
ser Okefoko	3,10	c	z$_6$
ser Stormont	3,10	d	1,2
ser Shangani	3,10	d	1,5
ser Onireke	3,10	d	1,7
ser Souza	3,10	d	e,n,x
ser Madjorio	3,10	d	e,n,z$_{15}$
ser Birmingham	3,10	d	l,w
ser Weybridge	3,10	d	z$_6$
ser Maron	3,10	d	z$_{35}$
ser Vejle	3,10	e,h	1,2
ser Muenster	3,10	e,h	1,5
ser Anatum	3,10	e,h	1,6
ser Nyborg	3,10	e,h	1,7
ser Newlands	3,10	e,h	e,n,x
ser Meleagridis	3,10	e,h	l,w
ser Sekondi	3,10	e,h	z$_6$
*ser Chudleigh	3,10	e,n,x	1,7
ser Regent	3,10	f,g	—
ser Suberu	3,10	g,m	—
ser Amsterdam	3,10	g,m,s	—
ser Westhampton	3,10	g,s,t	—
*ser Islington	3,10	g,t	—
ser Southbank	3,10	m,t	—
*ser Stikland	3,10	m,t	e,n,x
ser Amounderness	3,10	i	1,5
ser Falkensee	3,10	i	e,n,z$_{15}$
ser Yeerongpilly	3,10	i	z$_6$
ser Wimborne	3,10	k	1,2
ser Zanzibar	3,10	k	1,5
ser Marienthal	3,10	k	e,n,z$_{15}$
ser Newrochelle	3,10	k	l,w
ser Nchanga	3,10	l,v	1,2

*biochemically aberrant

ser Sinstorf	3,10	l,v	1,5
ser London	3,10	l,v	1,6
ser Give	3,10	l,v	1,7
ser Ruzizi	3,10	l,v	e,n,z_{15}
*ser Fuhlsbuettel	3,10	l,v	z_6
ser Uganda	3,10	$1,z_{13}$	1,5
ser Fallowfield	3,10	$1,z_{13},z_{28}$	e,n,z_{15}
*ser Westpark	3,10	$1,z_{28}$	e,n,x
*ser _____	3,10	$1,z_{28}$	z_{39}
ser Rutgers	3,10	$1,z_{40}$	1,7
ser Seegefeld	3,10	r(i)	1,2
ser Ughelli	3,10	r	1,5
ser Elisabethville	3,10	r	1,7
ser Simi	3,10	r	e,n,z_{15}
ser Weltevreden	3,10	r	z_6
ser Amager	3,10	y	1,2
ser Orion	3,10	y	1,5
ser Mokola	3,10	y	1,7
ser Ohlstedt	3,10	y	e,n,x
ser Bolton	3,10	y	e,n,z_{15}
ser Langensalza	3,10	y	l,w
ser Stockholm	3,10	y	z_6
*ser Alexander	3,10	z	1,5
*ser Finchley	3,10	z	e,n,x
ser Clerkenwell	3,10	z	l,w
*ser Tafelbaai	3,10	z	z_{39}
ser Adabraka	3,10	z_4,z_{23}	[1,7]
ser Okerara	3,10	z_{10}	1,2
ser Lexington	3,10	z_{10}	1,5
ser Coquilhatville	3,10	z_{10}	1,7
ser Kristianstad	3,10	z_{10}	e,n,z_{15}
ser Biafra	3,10	z_{10}	z_6
ser Jedburgh	3,10	z_{29}	—
ser Cairina	3,10	z_{35}	z_6
ser Macallen	3,10	z_{36}	—
ser Bolombo	3,10	z_{38}	—
*ser Mpila	3,10	z_{38}	z_{42}
*ser Winchester	3,10	z_{39}	1,7

Serogroup E_2

S. enteritidis

ser Rosenthal	3,15	b	1,5

*biochemically aberrant

ser Pankow	3,15	d	1,5
ser Eschersheim	3,15	d	e,n,x
ser Goerlitz	3,15	e,h	1,2
ser Newhaw	3,15	e,h	1,5
ser Newington	3,15	e,h	1,6
ser Selandia	3,15	e,h	1,7
ser Cambridge	3,15	e,h	l,w
ser Drypool	3,15	g,m,s	—
*ser Parow	3,15	g,m,s,t	—
ser Halmstad	3,15	g,s,t	—
ser Portsmouth	3,15	l,v	1,6
ser New brunswick	3,15	l,v	1,7
ser Kinshasa	3,15	l,z_{13}	1,5
ser Lanka	3,15	r	z_6
ser Tuebingen	3,15	y	1,2
ser Binza	3,15	y	1,5
ser Tournai	3,15	y	z_6
ser Manila	3,15	z_{10}	1,5
ser Hamilton	3,15	—	z_{27}

Serogroup E_3

S. enteritidis

ser Khartoum	(3),(15),34	a	1,7
ser Arkansas	(3),(15),34	e,h	1,5
ser Minneapolis	(3),(15),34	e,h	1,6
ser Wildwood	(3),(15),34	e,h	l,w
ser Canoga	(3),(15),34	g,s,t	—
ser Menhaden	(3),(15),34	l,v	1,7
ser Thomasville	(3),(15),34	y	1,5
ser Illinois	(3),(15),34	z_{10}	1,5
ser Harrisonburg	(3),(15),34	z_{10}	1,6

Serogroup E_4

S. enteritidis

ser Gwoza	1,3,19	a	e,n,z_{15}

*biochemically aberrant

ser Gnesta	1,3,19	b	1,5
ser Visby	1,3,19	b	1,6
ser Broughton	1,3,19	b	l,w
ser Accra	1,3,19	b	z_6
ser Madiago	1,3,19	c	1,7
ser Ahamdi	1,3,19	d	1,5
ser Liverpool	1,3,19	d	e,n,z_{15}
ser Tilburg	1,3,19	d	l,w
ser Niloese	1,3,19	d	z_6
ser Sanktmarx	1,3,19	e,h	1,7
ser Sao	1,3,19	e,h	e,n,z_{15}
ser Calabar	1,3,19	e,h	l,w
ser Rideau	1,3,19	f,g	—
ser Maiduguri	1,3,19	f,g,t	e,n,z_{15}
ser Senftenberg	1,3,19	g,s,t	—
ser Cannstatt	1,3,19	m,t	—
ser Stratford	1,3,19	i	1,2
ser Machaga	1,3,19	i	e,n,x
ser Avonmouth	1,3,19	i	e,n,z_{15}
ser Zuilen	1,3,19	i	l,w
ser Taksony	1,3,19	i	z_6
ser Ngor	1,3,19	l,v	1,5
ser Westerstede	1,3,19	l,z_{13}	—
ser Lokstedt	1,3,19	l,z_{13},z_{28}	1,2
ser Bedford	1,3,19	l,z_{13},z_{28}	e,n,z_{15}
ser Yalding	1,3,19	r	e,n,z_{15}
ser Krefeld	1,3,19	y	l,w
ser Korlebu	1,3,19	z	1,5
ser Schoeneberg	1,3,19	z	e,n,z_{15}
ser Carno	1,3,19	z	l,w
ser Dallgow	1,3,19	z_{10}	e,n,z_{15}
ser Simsbury	1,3,19	—	z_{27}
ser Llandoff	1,3,19	z_{29}	—
ser Chittagong	(1),3,10,(19)	b	z_{35}
ser Bilu	(1),3,10,(19)	f,g,t	1,(2),7
ser Ilugun	(1),3,10,(19)	$z_4,z23$	z_6
ser Dessau	(1),3,15,(19)	g,s,t	—

	Serogroup F		

S. enteritidis

ser Marseille	11	a	1,5
ser Luciana	11	a	e,n,z_{15}
*ser Glençairn	11	a	$z_6 : z_{42}$
ser Leeuwarden	11	b	1,5
*ser _____	11	b	1,7
*ser Srinagar	11	b	e,n,x
ser Pharr	11	b	e,n,z_{15}
ser Chandans	11	d	e,n,x
*ser Montgomery	11	d,a	d,e,n,z_{15}
ser Findorff	11	d	z_6
ser Chingola	11	e,h	1,2
ser Adamstua	11	e,h	1,6
ser Redhill	11	e,h	$1,z_{13},z_{28}$
*ser Grabouw	11	g,(m),s,t	$[z_{39}]$
ser Missouri	11	g,s,t	—
*ser Mundsburg	11	g,z_{51}	—
*ser Lincoln	11	m,t	e,n,x
ser Aberdeen	11	i	1,2
ser Brijbhumi	11	i	1,5
ser Heerlen	11	i	1,6
ser Veneziana	11	i	e,n,x
ser Pretoria	11	k	1,2
ser Abaetetuba	11	k	1,5
ser Sharon	11	k	1,6
ser Kisarawe	11	k	e,n,x
ser Amba	11	k	$1,z_{13},z_{28}$
ser Stendal	11	l,v	1,2
ser Maracaibo	11	l,v	1,5
ser Fann	11	l,v	e,n,x
ser Bulbay	11	l,v	e,n,z_{15}
ser Osnabrueck	11	l,z_{13},z_{28}	e,n,x
ser Huila	11	l,z_{28}	e,n,x
ser Senegal	11	r	1,5
ser Rubislaw	11	[d] ,r	[d] ,e,n,x
ser Volta	11	r	$1,z_{13},z_{28}$
ser Solt	11	y	1,5
ser Herzliya	11	y	e,n,x
ser Nyanza	11	z	z_6

*biochemically aberrant

*ser Soutpan	11	z	z_{39}
*ser Parera	11	z_4,z_{23}	—
ser Etterbeek	11	z_4,z_{23}	e,n,z_{15}
*ser	11	z_4,z_{23}	—
ser Wentworth	11	z_{10}	1,2
ser Straengnaes	11	z_{10}	1,5
ser Tel hashomer	11	z_{10}	e,n,x
ser Maastricht	11	z_{41}	1,2

Serogroup G_1

S. enteritidis			
ser Mim	13,22	a	1,6
ser Ibadan	13,22	b	1,5
ser Vaertan	13,22	b	e,n,x
ser Bahati	13,22	b	e,n,z_{15}
ser Haouaria	13,22	c	e,n,x,z_{15}
ser Friedenau	13,22	d	1,6
ser Diguel	1,13,22	d	e,n,z_{15}
ser Willemstad	1,13,22	e,h	1,6
ser Raus	13,22	f,g	e,n,x
ser Bron	13,22	g,m	$[e,n,z_{15}]$
*ser Limbe	1,13,22	g,m,t	[1,5]
*ser Rotterdam	1,13,22	g,t	[1,5]
ser Lovelace	13,22	l,v	1,5
ser Borbeck	13,22	l,v	1,6
ser Tanger	1,13,22	y	1,6
ser Poona	[1],13,22	z	1,6
ser Bristol	13,22	z	1,7
ser Roodepoort	[1],13,22	z_{10}	1,5
*ser Clifton	13,22	z_{29}	1,5
*ser Goodwood	13,22	z_{29}	e,n,x
ser Mampong	13,22	z_{35}	1,6
ser Leiden	13,22	z_{38}	—

Serogroup G_2

S. enteritidis			
ser Chagoua	1,13,23	a	1,5
*ser Tygerberg	1,13,23	a	z_{42}
ser Mississippi: (Atlanta)	1,13,23	b	1,5

*biochemically aberrant

ser Bracknell	13,23	b	1,6
ser Ullevi	13,23	b	e,n,x
ser Durham	13,23	b	e,n,z_{15}
*ser Acres	1,13,23	b	$z_{42}:[1,5]$
ser Mishmar haemek	1,13,23	d	1,5
ser Grumpensis	13,23	d	1,7
ser Telelkebir	13,23	d	e,n,z_{15}
ser Putten	13,23	d	l,w
ser Isuge	13,23	d	z_6
ser Wichita	1,13,23	d	z_{37}
*ser Epping	1,13,23	e,n,x	1,7
ser Havana	1,13,23	f,g,[s]	—
ser Agbeni	13,23	g,m	—
*ser Kraaifontein	1,13,23	g,(m),t	[e,n,x]
*ser Luanshya	13,23	g,s,(t)	—
ser Okatie	13,23	g,s,t	—
ser Congo	13,23	g,t	—
*ser Gojenberg	1,13,23	g,t	1,5
ser Kintambo	13,23	m,t	—
*ser Katesgrove	1,13,23	m,t	1,5
*ser Worcester	1,13,23	m,t	e,n,x
*ser Boulders	13,23	m,t	z_{42}
ser Idikan	13,23	i	1,5
ser Jukestown	13,23	i	e,n,z_{15}
*ser _____	13,23	$1,z_{28}$	z_6
*ser Vredelust	1,13,23	$1,z_{28}$	z_{42}
ser Adjame	13,23	r	1,6
ser Linton	13,23	r	e,n,z_{15}
ser Yarrabah	13,23	y	1,7
ser Ordonez	1,13,23	y	l,w
ser Tunis	1,13,23	y	z_6
*ser Nachshonim	1,13,23	z	1,5
*ser Farmsen	13,23	z	1,6
ser Worthington	1,13,23	z	l,w
ser Ajiobo	13,23	z_4,z_{23}	—
ser Romanby	13,23	z_4,z_{24}	—
ser Demerara	13,23	z_{10}	l,w
ser Cubana	1,13,23	z_{29}	—
ser Fanti	13,23	z_{38}	—
ser Stevenage	1,13,23	$[z_{42}]$	1,7

*biochemically aberrant

	Serogroup H		
S. enteritidis			
ser Garba	1,6,14,25	a	1,5
ser Ferlac	1,6,14,25	a	e,n,x
ser Tucson	1,6,14,25	b	[1,7]
ser Blijdorp	1,6,14,25	c	1,5
ser Kassberg	1,6,14,25	c	1,6
ser Heves	6,14,24	d	1,5
ser Finkenwerder	1,6,14,25	d	1,5
ser Florida	1,6,14,25	d	1,7
ser Lindern	6,14,24	d	e,n,x
ser Charity	1,6,14,25	d	e,n,x
ser Teko	1,6,14,25	d	e,n,z_{15}
ser Encino	1,6,14,25	d	l,z_{13},z_{28}
ser Albuquerque	6,14,24	d	z_6
ser Bahrenfeld	6,14,24	e,h	1,5
ser Onderstepoort	1,6,14,25	e,h	1,5
ser Magumeri	1,6,14,25	e,h	1,6
ser Warragul	1,6,14,25	g,m	—
ser Caracas	1,6,14,25	g,m,s	—
ser Kaitaan	1,6,14,25	m,t	—
*ser Rooikrantz	1,6,14	m,t	1,5
*ser Emmerich	6,14	[m,t]	e,n,x
ser Mampeza	1,6,14,25	i	1,5
ser Buzu	1,6,14,25	i	1,7
ser Schalkwijk	6,14,(24)	i	e,n.....
*ser _____	(6),14	k	[e,n,x]
ser Harburg	1,6,14,25	k	1,5
*ser _____	1,(6),14	k	$z_6:z_{42}$
ser Boecker	[1],6,14,[25]	l,v	1,7
ser Horsham	1,6,14,25	l,v	e,n,x
ser Surat	1,6,14,25	r(i)	e,n,z_{15}
variant Hr-	1,6,14,25	i	e,n,z_{15}
ser Carrau	6,14,24	y	1,7
ser Madelia	1,6,14,25	y	1,7
ser Fischerkietz	1,6,14,25	y	e,n,x
ser Homosassa	1,6,14,25	z	1,5
ser Soahamina	6,14,24	z	e,n,x
ser Sundsvall	1,6,14,25	z	e,n,x
ser Poano	1,6,14,25	z	l,z_{13},z_{28}

*biochemically aberrant

ser Bousso	1,6,14,25	z_4,z_{23}	—
ser Uzaramo	1,6,14,25	z_4,z_{24}	—
ser Nessa	1,6,14,25	z_{10}	1,2
*ser Bornheim	1,6,14,25	z_{10}	1,(2),7
*ser Simonstown	1,6,14	z_{10}	1,5
*ser Slangkop	1,6,14	z_{10}	$z_6:z_{42}$
ser Sara	1,6,14,25	z_{38}	[e,n,x]
ser _____	1,(6),14	z_{42}	1,6

Serogroup I

S. enteritidis

ser Hannover	16	a	1,2
ser Brazil	16	a	1,5
ser Amunigun	16	a	1,6
ser Fischerhuette	16	a	e,n,z_{15}
ser Heron	16	a	z_6
ser Hull	16	b	1,2
ser Wa	16	b	1,5
ser Glasgow	16	b	1,6
ser Hvittingfoss	16	b	e,n,x
ser Malstatt	16	b	z_6
*ser _____	16	b	z_{42}
ser Vancouver	16	c	1,5
ser Shamba	16	c	e,n,x
ser Oldenburg	16	d	1,2
ser Gaminara	16	d	1,7
ser Barranguilla	16	d	e,n,x
ser Nottingham	16	d	e,n,z_{15}
ser Barmbek	16	d	z_6
ser Malakal	16	e,h	1,2
ser Weston	16	e,h	z_6
*ser Bellville	16	e,n,x	1,7
ser Tees	16	f,g	—
ser Adeoyo	16	g,m	—
ser Nikolaifleet	16	g,m,s	—
*ser Mobeni	16	g,m,s,t	—
*ser Merseyside	16	g,t	1,5
*ser Rowbarton	16	m,t	—
ser Amina	16	i	1,5
ser Frankfurt	16	i	e,n,z_{15}
ser Szentes	16	k	1,2

*biochemically aberrant

ser Nuatja	16	k	e,n,x
ser Orientalis	16	k	e,n,z_{15}
ser Shanghai	16	l,v	1,6
ser Welikade	16	l,v	1,7
ser Salford	16	l,v	e,n,x
ser Burgas	16	l,v	e,n,z_{15}
ser Losangeles	16	l,v	z_6
*ser Noordhoek	16	l,w	z_6
ser Mandera	16	l,z_{13}	e,n,z_{15}
ser Enugu	16	l,z_{13},z_{28}	—
*ser Sarepta	16	l,z_{28}	z_{42}
ser _____	16	l,z_{40}	—
ser Annedal	16	r(i)	e,n,x
ser Zwickau	16	r(i)	e,n,z_{15}
ser Saphra	16	y	1,5
ser Akuafo	16	y	1,6
ser Kikoma	16	y	e,n,x
ser Lingivala	16	z	1,7
*ser Louwbester	16	z	e,n,x
ser Kibi	16	z_4,z_{23}	—
*ser Haddon	16	z_4,z_{23}	—
*ser Ochsenzoll	16	z_4,z_{23}	—
*ser Chameleon	16	z_4,z_{32}	—
ser Lisboa	16	z_{10}	1,6
ser Redlands	16	z_{10}	e,n,z_{15}
*ser Jacksonville	16	z_{29}	—
*ser Woodstock	16	z_{42}	1,(5),7
*ser Elsiesrivier	16	z_{42}	1,6

Further Serogroups

S. enteritidis

ser Bonames	17	a	1,2
ser Jangwani	17	a	1,5
ser Kinondoni	17	a	e,n,x
ser Kirkee	17	b	1,2
*ser Hillbrow	17	b	e,n,x,z_{15}
ser Victoriaborg	17	c	1,6
*ser Woerden	17	c	z_{39}
ser Berlin	17	d	1,5
ser Niamey	17	d	l,w

*biochemically aberrant

*ser Verity	17	e,n,x,z_{15}	1,6
*ser Bleadon	17	$(f),g,t,$	$[e,n,x,z_{15}]$
*ser _____	17	k	—
ser Irenea	17	k	1,5
ser Matadi	17	k	e,n,x
ser Morotai	17	l,v	1,2
ser Michigan	17	l,v	1,5
ser Carmel	17	l,v	e,n,x
ser Gori	17	z	1,2
*ser Constantia	17	z	$l,w:z_{42}$
ser Kandla	17	z_{29}	—
ser Fluntern	6,14,18	b	1,5
ser _____	18	b	1,5
ser Usumbura	18	d	1,7
ser Langenhorn	18	m,t	—
*ser _____	18	m,t	1,5
ser Memphis	18	k	1,5
*ser _____	18	y	e,n,x,z_{15}
ser Siegburg	6,14,18	z_4,z_{23}	[1,5]
ser Cerro	18	z_4,z_{23}	$[z_{45}]$
ser Blukwa	18	z_4,z_{24}	—
*ser Shomron	18	z_4,z_{32}	—
*ser Zeist	18	z_{10}	z_6
*ser Beloha	18	z_{36}	—
ser Sinthia	18	z_{38}	—
ser Assen	21	a	—
ser Ghana	21	b	1,6
ser Minnesota	21	b	e,n,x
ser Rhône	21	c	e,n,x
ser Spartel	21	d	1,5
ser Magwa	21	d	e,n,x
ser Good	21	f,g	e,n,x

*biochemically aberrant

ser Diourbel	21	i	1,2
ser Ruiru	21	y	e,n,x
*ser Soesterberg	21	z_4,z_{23}	—
*ser Gwaai	21	z_4,z_{24}	—
*ser Wandsbek	21	z_{10}	z_6
ser Gambaga	21	z_{35}	e,n,z_{15}

ser Solna	28	a	1,5
ser Dakar	28	a	1,6
ser Seattle	28	a	e,n,x
ser Honelis	28	a	e,n,z_{15}
ser Möëro	28	b	1,5
ser Ashanti	28	b	1,6
ser Bokanjac	28	b	1,7
ser Langford	28	b	e,n,z_{15}
*ser Kaltenhausen	28	b	z_6
ser Hermannswerder	28	c	1,5
ser Eberswalde	28	c	1,6
ser Halle	28a,28c	c	1,7
variant vidin	28a,28b	c	1,7
ser Dresden	28	c	e,n,x
ser Wedding	28	c	e,n,z_{15}
ser Techimani	28	c	z_6
ser Mundonobo	28	d	1,7
ser Mocamedes	28	d	e,n,x
ser Patience	28	d	e,n,z_{15}
*ser _____	28	e,n,x	1,7
ser Friedrichsfelde	28	f,g	—
ser Abadina	28	g,m	[e,n,z_{15}]
ser Croft	28	g,m,s	—
ser Ona	28	g,s,t	—
*ser Llandudno	28	g,s,t	1,5
ser Vinohrady	28	mt	—
*ser variant	28	m,t	—
ser Cotham	28	i	1,5
ser Volkmarsdorf	28	i	1,6
ser Kuessel	28	i	e,n,z_{15}
ser Guildford	28	k	1,2
ser Ilala	28	k	1,5
ser Adamstown	28	k	1,6
ser Taunton	28	k	e,n,x

*biochemically aberrant

ser Ank	28	k	e,n,z_{15}
ser Leoben	28	l,v	1,5
ser Vitkin	28	l,v	e,n,x
ser Nashua	28	l,v	e,n,z_{15}
ser Chicago	28	r	1,5
ser Kibusi	28	r	e,n,x
*ser Oevelgoenne	28	r	e,n,z_{15}
ser Sankt georg	28	r(i)	e,n,z_{15}
ser Oskarshamn	28	y	1,2
ser Nima	28	y	1,5
ser Pomona	28	y	1,7
ser Kitenge	28	y	e,n,x
ser Telaviv	28	y	e,n,z_{15}
ser Shomolu	28	y	1,w
ser Ezra	28	z	1,7
ser Brisbane	28	z	e,n,z_{15}
*ser Ceres	28	z	z_{39}
ser Babelsberg	28	z_4,z_{23}	e,n,z_{15}
ser Teltow	28	z_4,z_{23}	1,6
ser Rogy	28	z_{10}	1,2
ser Umbilo	28	z_{10}	e,n,x
ser Luckenwalde	28	z_{10}	e,n,z_{15}
ser Moroto	28	z_{10}	1,w
ser Djermaia	28	z_{29}	—
ser Aderike	28	z_{38}	—
ser Overvecht	30	a	1,2
ser Zehlendorf	30	a	1,5
*ser Odijk	30	a	z_{39}
ser Louga	30	b	1,2
ser Aschersleben	30	b	1,5
ser Urbana	30	b	e,n,x
ser Messina	30	d	1,5
*ser Slatograd	30	f,g,(p),t	—
ser Godesberg	30	g,m	—
ser Giessen	30	g,m,s	—
ser Sternchanze	30	g,s,t	—
ser Wayne	30	g,z_{51}	—
ser Landau	30	i	1,2
ser Morehead	30	i	1,5
ser Soerenga	30	i	1,w

*biochemically aberrant

ser Hilversum	30	k	1,2
ser Ramat gan	30	k	1,5
ser Aqua	30	k	1,6
ser Angoda	30	k	e,n,x
ser Odozi	30	k	e,n,x,z_{15}
* variant	30	k	e,n,x,z_{15}
ser Ligeo	30	l,v	1,2
ser Donna	30	l,v	1,5
ser Morocco	30	l,z_{13},z_{28}	e,n,z_{15}
ser Gege	30	r	1,5
ser Matopeni	30	y	1,2
ser Steinplatz	30	y	1,6
ser Baguirmi	30	y	e,n,x
ser Bodjonegoro	30	z_4,z_{24}	—
ser Kumasi	30	z_{10}	e,n,z_{15}
ser Ago	30	z_{38}	—
*ser _____	30	z_{39}	1,(7)
ser Umhlatazana	35	a	e,n,z_{15}
ser Tchad	35	b	—
ser Yolo	35	c	—
ser Dembe´	35	d	1,w
ser Gassi	35	e,h	z_6
ser Adelaide	35	f,g	—
ser Ealing	35	g,m,s	—
ser Ebrie	35	g,m,t	—
*ser _____	35	g,m,s,t	—
ser Agodi	35	g,t	—
ser Monschaui	35	m,t	—
ser Gambia	35	i	e,n,z_{15}
ser Bandia	35	i	1,w
ser _____	35	l,z_{28}	—
ser Massakory	35	r	1,w
ser Alachua	35	z_4,z_{23}	—
ser Westphalia	35	z_4,z_{24}	—
ser Camberene	35	z_{10}	1,5

*biochemically aberrant

ser Enschede	35	z_{10}	l,w
ser Ligna	35	z_{10}	z_6
*ser Utbremen	35	z_{29}	e,n,x
ser Sheffield	38	c	1,5
ser Kidderminster	38	c	1,6
*ser Carletonville	38	d	[1,5]
ser Thiaroye	38	e,h	1,2
ser Kasenyi	38	e,h	1,5
ser Korovi	38	g,m,s	—
*ser Foulpointe	38	g,t	—
ser Mgulani	38	i	1,2
ser Lansing	38	i	1,5
ser Echa	38	k	1,2
ser Inverness	38	k	1,6
ser Alger	38	l,v	1,2
ser Kimberly	38	l,v	1,5
ser Roan	38	l,v	e,n,x
ser Lindi	38	r	1,5
ser Emmastad	38	r	1,6
ser Freetown	38	y	1,5
ser Colombo	38	y	1,6
ser Perth	38	y	e,n,x
ser Yoff	38	z_4, z_{23}	1,2
ser Wandsworth	39	b	1,2
ser Logone	39	d	1,5
ser Mara	39	e,h	[1,5]
ser Hofit	39	i	1,5
ser Champaign	39	k	1,5
ser Kokomlemle	39	l,v	e,n,x
*ser Mondeor	39	l,z_{28}	e,n,x
ser Anfo	39	y	1,2
ser Windermere	39	y	1,5
ser Cook	39	z_{48}	1,5

*biochemically aberrant

ser Shikmonah	40	a	1,5
ser Greiz	40	a	z_6
*ser Springs	40	a	z_{39}
*ser _____	40	b	—
ser Riogrande	40	b	1,5
ser Johannesburg	1,40	b	e,n,x
ser Duval	1,40	b	e,n,z_{15}
ser Benguella	40	b	z_6
*ser Suarez	1,40	c	e,n,x,z_{15}
*ser Ottershaw	40	d	—
ser Driffield	1,40	d	1,5
ser Tilene	1,40	e,h	1,2
*ser Alsterdorf	1,40	g,m,t	—
*ser Boksburg	40	g,s	e,n,x,z_{15}
*ser Seminole	1,40a,40b	g,z_{51}	—
*ser _____	1,40a,40c	g,z_{51}	—
*ser _____	1,40	m,t	z_{42}
ser Goulfy	1,40	k	1,5,(6)
ser Allandale	1,40	k	1,6
*ser Sunnydale	1,40	k	e,n,x,z_{15}
ser Millesi	1,40	l,v	1,2
ser Bukavu	1,40	l,z_{28}	1,5
ser Santhiaba	40	l,z_{28}	1,6
*ser Bulawayo	(1),40	z	1,5
ser Nowawes	40	z	z_6
*ser Sachsenwald	1,40	z_4,z_{23}	—
*ser Degania	40	z_4,z_{24}	—
*ser Bern	1,40	z_4,z_{32}	—
*ser _____	1,40	z_6	1,5
ser Trotha	40	z_{10}	z_6
ser Omifisan	40	z_{29}	—
*ser Fandran	1,40	z_{35}	e,n,x,z_{15}
*ser Grunty	1,40	z_{39}	1,6
ser Karamoja	40	z_{41}	1,2
*ser _____	40	—	1,7

*biochemically aberrant

ser Vietnam	41	b	—
* variant	41	b	—
ser Egusi	41	d	—
*ser Hennepin	41	d	z_6
*ser Lethe	41	g,t	—
*ser _____	41	k	—
*ser Dubrovnik	41	z	1,5
ser Waycross	41	z_4,z_{23}	—
* variant	41	z_4,z_{23}	—
ser Ipswich	41	z_4,z_{24}	—
*ser Negev	41	z_{10}	1,2
ser Leipzig	41	z_{10}	1,5
ser Landala	41	z_{10}	1,6
ser Inpraw	41	z_{10}	e,n,x
*ser Lurup	41	z_{10}	e,n,x,z_{15}
*ser Lichtenberg	41	z_{10}	[z_6]
ser Offa	41	z_{38}	—

ser Faji	1,42	a	e,n,z_{15}
*ser Chinovum	42	b	1,5
*ser Uphill	42	b	e,n,x,z_{15}
ser Egusitoo	1,42	b	z_6
ser Kampala	1,42	c	z_6
*ser Fremantle	42	(f),g,t	—
ser Maricopa	1,42	g,z_{51}	1,5
*ser _____	42	m,t	e,n,x,z_{15}
ser Kaneshie	1,42	i	l,w
ser Middlesbrough	1,42	i	z_6
ser Haferbreite	42	k	[1,6]
*ser Portbech	42	l,v	e,n,x,z_{15}
*ser Nairobi	42	r	—
ser Harvestehude	1,42	y	z_6
*ser Detroit	42	z	1,5
ser Ursenbach	1,42	z	1,6
*ser Rand	42	z	e,n,x,z_{15}

*biochemically aberrant

*ser Nuernberg	42	z	z_6
ser Gera	1,42	z_4, z_{23}	1,6
ser Loenga	1,42	z_{10}	z_6
ser Kahla	1,42	z_{35}	1,6
ser Weslaco	42	z_{36}	—
*ser _____	42	—	1,6
ser Graz	43	a	1,2
ser Berkeley	43	a	1,5
*ser Kommetje	43	b	z_{42}
*ser _____	43	e,n,x,z_{15}	1,(5),7
*ser _____	43	e,n,z_{15}	1,6
ser Milwaukee	43	f,g	—
*ser Mosselbay	43	g,s,(t)	z_{42}
*ser Veddel	43	g,t	—
ser Mbao	43	i	1,2
ser Ahuza	43	k	1,5
ser Farcha	43	y	1,2
ser Kingabwa	43	y	1,5
*ser _____	43	z	1,5
*ser Houten	43	z_4, z_{23}	—
*ser Tuindorp	43	z_4, z_{32}	—
*ser Volksdorf	43	z_{36}, z_{38}	—
ser Irigny	43	z_{38}	—
*ser Bunnik	43	z_{42}	[1,5]

ser Niarembe	44	a	l,w
ser Sedgwick	44	b	e,n,z_{15}
ser Madigan	44	c	1,5
ser Bobo	44	d	1,5
ser Fischerstrasse	44	d	e,n,z_{15}
ser Vleuten	44	f,g	—

*biochemically aberrant

ser Gamaba	44	g,m,s	—
*ser _____	44	g,z_{51}	—
ser Muguga	44	m,t	—
ser Lawra	44	k	e,n,z_{15}
ser Uhlenhorst	44	z	l,w
*ser _____	44	z_4,z_{23}	—
*ser Christiansborg	44	z_4,z_{24}	—
* variant	44	z_4,z_{24}	—
*ser Lohbruegge	44	z_4,z_{32}	—
ser Guinea	44	z_{10}	[1,7]
*ser _____	44	z_{36},z_{38}	—
*ser Clovelly	1,44	z_{39}	$[e,n,x,z_{15}]$
*ser Vrindaban	45	a	e,n,x
*ser Ejeda	45	a	z_{10}
ser Riverside	45	b	1,5
ser Deversoir	45	c	e,n,x
ser Dugbe	45	d	1,6
ser Karachi	45	d	e,n,x
ser Suelldorf	45	f,g	—
ser Tornow	45	g,m	—
*ser Bremen	45	g,m,s,t	e,n,x
*ser Windhoek	45	g,t	1,5
*ser	45	g,z_{51}	—
ser Apapa	45	m,t	—
*ser Perinet	45	m,t	e,n,x,z_{15}
ser Casablanca	45	k	1,7
ser Cairns	45	k	e,n,z_{15}
*ser Klapmuts	45	z	z_{39}
ser Jodhpur	45	z_{29}	—
ser Lattenkamp	45	z_{35}	1,5
*ser Bilthoven	47	a	[1,5]
ser Saka	47	b	—
*ser Phoenix	47	b	1,5
*ser Khami	47	b	e,n,x,z_{15}
ser Stellingen	47	d	e,n,x

*biochemically aberrant

*ser Quimbamba	47	d	z_{39}
ser Sljeme	1,47	f,g	—
ser Luke	1,47	g,m	—
ser Mesbit	47	m,t	e,n,z_{15}
ser Bergen	47	i	e,n,z_{15}
ser Bootle	47	k	1,5
ser Lyon	47	k	e,n,z_{15}
ser Teshie	1,47	l,z_{13},z_{28}	e,n,z_{15}
ser Moualine	47	y	1,6
ser Mountpleasant	47	z	1,5
ser Kaolack	47	z	1,6
*ser Chersina	47	z	z_6
ser Bere	47	z_4,z_{23}	z_6
*ser _____	47	z_6	1,6
ser Alexanderplatz	47	z_{38}	—
ser Quinhon	47	z_{44}	—
ser Hisingen	48	a	1,5,7
*ser _____	48	a	z_6
*ser Hagenbeck	48	d	z_6
ser Fitzroy	48	e,h	1,5
*ser Hammonia	48	e,n,x,z_{15}	z_6
*ser Erlangen	48	g,m,t	—
*ser Marina	48	g,z_{51}	—
*ser Sydney	48	i	z
ser Dahlem	48	k	e,n,z_{15}
*ser Sakaraha	48	[k]	z_{39}
ser Djakarta	48	z_4,z_{24}	—
*ser _____	48	z_4,z_{32}	—
*ser Ngozi	48	z_{10}	[1,5]
ser Bongor	48	z_{35}	—
ser Rochdale	50	b	e,n,x

*biochemically aberrant

*ser Krugersdorp	50	e,n,x	1,7
*ser Namib	50	g,m,s,t	1,5
*ser Wassenaar	50	g,z_{51}	—
*ser Atra	50	m,t	$z_6:z_{42}$
*ser Seaforth	50	k	z_6
*ser _____	50	l,w	$e,n,x,z_{15}:z_{42}$
*ser _____	50	l,z_{28}	z_{42}
ser Dougi	50	y	1,6
*ser Greenside	50	z	e,n,x
*ser Flint	50	z_4,z_{23}	—
*ser _____	50	z_4,z_{24}	—
*ser Bonaire	50	z_4,z_{32}	—
*ser Hooggraven	50	z_{10}	$z_6:z_{42}$
*ser Faure	50	z_{42}	1,7

ser Tione	51	a	e,n,x
ser Gokul	1,51	d	—
ser Meskin	51	e,h	1,2
ser Dan	51	k	e,n,z_{15}
ser Overschie	51	l,v	1,5
*ser Askraal	51	l,z_{28}	—
ser Antsalova	51	z	1,5
ser Treforest	1,51	z	1,6
*ser Harmelen	51	z_4,z_{23}	—
*ser Roggeveld	51	—	1,7

ser Flottbek	52	b	—
ser Utrecht	52	d	1,5
*ser _____	52	d	e,n,x,z_{15}
ser Sainte marie	52	g,t	—
*ser Wilhelmstrasse	52	z_{44}	1,5,7
*ser Lobatsi	52	—	1,5,7

*ser Midhurst	53	l,z_{28}	z_{39}

*biochemically aberrant

*ser _____	53	z	z_6
*ser Humber	53	z_4,z_{24}	—
*ser Bockenheim	1,53	z_{36},z_{38}	—
ser Tonev	54	b	e,n,x
ser Rossleben	54	e,h	1,6
ser Uccle	54	g,s,t	—
ser Poeseldorf	54	i	z_6
ser Ochsenwerder	54	k	1,5
*ser Tranoroa	55	k	z_{39}
*ser Artis	56	b	—
*ser _____	56	e,n,x	1,7
*ser _____	57	g,m,s,t	z_{42}
*ser Locarno	57	z_{29}	z_{42}
*ser Manombo	57	z_{39}	e,n,x,z_{15}
*ser Tokai	57	z_{42}	$1,6:z_{53}$
*ser _____	58	a	1,5
*ser Basel	58	$1,z_{13},z_{28}$	1,5
*ser Betioky	59	k	(z)
*ser Setubal	60	g,m,t	z_6
*ser Luton	60	z	e,n,x
*ser Eilbek	61	i	z
*ser _____	64	k	e,n,x,z_{15}
*ser _____	64	z_{29}	—

*biochemically aberrant

REFERENCES

Andrewes, F. W. 1922. J. Pathol. Bacteriol., **25**, 505.

Kauffmann, F. 1930. Zentralbl. f. Bakt. I. Orig., **119**, 152.

————. 1941. Die Bakteriologie der Salmonella-Gruppe. Copenhagen: Munksgaard.

————. 1951. Enterobacteriaceae. Copenhagen: Munksgaard.

————. 1954. Enterobacteriaceae, 2nd Ed. Copenhagen: Munksgaard.

————. 1961. Die Bakteriologie der Salmonella-Species. Copenhagen: Munksgaard.

————. 1966. The Bacteriology of Enterobacteriaceae. Copenhagen: Munksgaard.

Kauffmann, F., and Mitsui, C. 1930. Z.f. Hyg., **111**, 740.

Kelterborn, E. 1967. Salmonella Species. The Hague: Dr. W. Junk.

Salmonella Committee. Report. 1934. J. Hyg., **34**, 333.

Schütze, H. 1921. Ibid., **20**, 230.

White, P. B. 1925. Med. Res. Council Gt. Brit., Spec. Rep. Ser. No. 91.

————. Ibid., Spec. Rep. Ser. No. 103.

Chapter 10
The Genus *Arizona*

Members of the genus *Arizona* originally were recovered from cold-blooded animals (Caldwell and Reyerson, 1939), but since have been isolated from a wide variety of animals including man. Serotypes of *Arizona* may cause gastroenteritis in man, and quite often are involved in localized lesions in man and lower animals (Edwards et al., 1956, 1959; Martin et al., 1967). In some areas serotypes of *Arizona* also are very important because of their etiological role in disease processes in turkey poults in particular.

The genus *Arizona* was defined as follows by Ewing et al. (1965), Ewing and Fife (1966), and Ewing (1967):

The Genus *Arizona* is composed of motile bacteria that conform to the definitions of the family ENTEROBACTERIACEAE and the tribe SALMONELLEAE. Urease is not produced and growth does not occur in medium containing potassium cyanide. Lysine, arginine, and ornithine are decarboxylated, sodium malonate is utilized, gelatin is liquefied slowly in nutrient medium, and lactose is fermented by the majority of cultures. With few exceptions acid is not produced in Jordan's tartrate medium. Dulcitol and inositol are not fermented and salicin is utilized infrequently. The type species is *Arizona hinshawii* (Ewing and Fife) Ewing.

This definition should be used in conjunction with the definition of the tribe Salmonelleae (Chapter 1).

The first description of a bacterium now included in the genus *Arizona* was given by Caldwell and Ryerson (1939). These investigators isolated several similar cultures from reptiles (horned lizards, *Phrynosoma solare;* Gila monster, *Heloderma suspectum;* and chuckawalla, *Sauromalus ater*), in all three of which the bacteria apparently produced fatal infections. The isolants were designated *Salmonella* sp. (Dar-es-Salaam type, variety from Arizona) because the isolants liquefied gelatin in the same manner as *S. enteritidis* ser Daressallam. Kauffmann (1941) studied one of the cultures of Caldwell and Ryerson and designated it as *Salmonella arizona* although it was recognized that the organism fermented lactose and liquefied gelatin. Its classification in the genus *Salmonella* was based upon the relationship of its H antigens to those of *S.*

enteritidis ser Cerro. Later, studies by Peluffo et al. (1942), Edwards et al. (1943), Edwards and West (1945), and Edwards et al. (1947) indicated that the strains described by Caldwell and Ryerson (1939) actually were representatives of a large group of biochemically and serologically related bacteria. The name "Arizona group" was applied to these particular bacteria by Edwards et al. (1947).

The generic term *Arizona* and the specific name *Arizona arizonae* were employed by Kauffmann and Edwards (1952), who listed means by which these bacteria could be differentiated from salmonellae. Acting on a proposal (Ewing, 1966a) for validation of the species name *Arizona arizonae,* the Judicial Commission of the International Committee on Nomenclature of Bacteria ruled (file C68/20) that the generic term *Arizona* was not validly published by Kauffmann and Edwards (1952). This generic term subsequently was validly published and characterized by Ewing (1963), Ewing et al. (1965), Ewing and Fife (1966), and Ewing (1967, 1969a, 1970). The specific epithet *arizonae* also had been found to be illegitimate under Rule 25 of the revised (1966) International Code of Nomenclature of Bacteria. The specific epithet *hinshawii* was proposed (Ewing, 1969b) in honor of Dr. William R. Hinshaw, who was responsible for much of the pioneer work with members of the genus *Arizona.* Thus, the correct name for these bacteria is *Arizona hinshawii.*

BIOCHEMICAL REACTIONS OF CULTURES OF *ARIZONA*

The results of investigations of the biochemical reactions given by cultures of *Arizona* reported by several investigators were reviewed by Ewing et al. (1965). Portions of this review are given in Tables 77-81 (Chapter 8). The results obtained by Ewing et al. (1965) are summarized in Table 113. These results are based upon complete examination of 150 cultures: 100 different serotype strains and 50 taken at random from materials submitted for serologic characterization during a period of a few months. These isolants were representative of more than 4000 cultures of *Arizona* received during a 17-year period.

TABLE 113

Summary of the Biochemical Reactions of *A. hinshawii*

Test or substrate	Sign	%+	(%+)*	Test or substrate	Sign	%+	(%+)*
Hydrogen sulfide	+	98.7		Rhamnose	+	93	(5)
Urease	–	0		Malonate	+	92.6	(0.7)
Indol	–	2		Mucate	d	56.6	(2.7)
Methyl red (37 C)	+	100		Jordan's tartrate	–	5.3	
Voges-Proskauer (37 C)	–	0		Stern's glycerol	+ or –	86.5	
Citrate (Simmons')	+	98.7	(1.3)	Sodium acetate	+ or (+)	76.9	(17.3)
KCN	–	8.7		Sodium alginate	–	0	
Motility	+	100		Lipase: Corn oil	–	0	
Gelatin (22 C)	(+)	92		Maltose	+	96	
Lysine decarboxylase	+	100		Xylose	+	100	
Arginine dihydrolase	(+) or +	12.7	(84.6)	Trehalose	+	100	
Ornithine decarboxylase	+	100		Cellobiose	d	1	(72)
Phenylalanine deaminase	–	0		Glycerol	– or (+w)		(37.8)
Glucose acid	+	100		Alpha methyl glucoside	–	0.9	
gas	+	99.3		Erythritol	–	0	
Lactose	d	61.3	(16.7)	Esculin	–	0.9	
Sucrose	–	4.7		Beta galactosidase	+	92.8[a]	
Mannitol	+	100		Nitrate to nitrite	+	100	
Dulcitol	–	0		Oxidation – fermentation	F	100	
Salicin	–	4.7	(3.3)	Oxidase	–	0	
Adonitol	–	0					
Inositol	–	0		Organic acids**			
Sorbitol	+	97	(2)				
Arabinose	+	98	(1)	citrate	(+)	4.7	(93.3)
Raffinose	–	5	(1)	D-tartrate	(+) or –		(83.3)

*Figures in parentheses indicate percentages of delayed reactions (3 or more days).
**Method of Kauffmann and Petersen, 1956.
[a]Lubin and Ewing, 1964.

F Glucose utilized fermentatively.
w Weakly positive reactions.
+ 90% or more positive within one or two days' incubation.
(+) Positive reaction after 3 or more days.
– No reaction (90% or more).
+ or – Majority of strains positive, some cultures negative.
– or + Majority of cultures negative, some strains positive.
(+) or + Majority of reactions delayed, some occur within 1 or 2 days.
d Different reactions: +, (+), –.

All of the above-mentioned 150 cultures gave positive results in tests for beta-D-galactosidase activity (ONPG test). However, in a study of 446 cultures of *Arizona*, Lubin and Ewing (1964) reported that 92.8 percent yielded positive results in the ONPG test, as follows:

No. of cultures tested	ONPG+ Lactose+	ONPG+ Lactose-	ONPG- Lactose-
446	349	65	32
		92.8%	7.2%

It is notable that Edwards and Fife (1956) reported that 39 (7.8 percent) of 501 strains of *Arizona* were able to grow in the presence of KCN. However, of the 39 positive cultures, 22 belonged to O antigen group 21. Only one strain of O group 21 tested failed to grow in KCN medium.

The reactions given by members of the genus *Arizona* in commonly employed tests are summarized in Table 6 (Chapter 3), and the results of tests in some of the organic acid media of Kauffmann and Petersen (1956) are given in Table 14 (Chapter 3). As mentioned above, the results obtained by various investigators with certain tests and substrates are listed in Tables 77-81 (Chapter 8).

Biochemical tests that are of particular value in the differentiation of members of the genera *Salmonella*, *Arizona*, and *Citrobacter* are given in Table 114. As regards Table 114, it should be noted that the reactions given by cultures of *Citrobacter* on urea medium are in no way comparable to the rapid, strong reactions given

TABLE 114

Differentiation of *Salmonella*, *Arizona*, and *Citrobacter*

Test or substrate	*Salmonella enteritidis* Sign	%+	(%+)*	*Arizona hinshawii* Sign	%+	(%+)*	*Citrobacter freundii* Sign	%+	(%+)*
Urease	–	0		–	0		d	69.4	(6.9)
KCN	–	0.1	(0.1)	–	8.7		+	96.2	(0.9)
Gelatin (22 C)	–		(0.4)	(+)		(92)	–		(0.9)
Lysine decarboxylase	+	94.4	(0.1)	+	100		–	0	
Ornithine decarboxylase	+	100		+	100		d	17.2	(0.2)
Lactose	–	0.3		d	61.3	(16.7)	(+) or +	39.3	(50.9)
Sucrose	–	0.2		–	4.7		d	15.3	(9.4)
Dulcitol	+	97.7		–	0		d	59.8	(0.7)
Inositol	d	43.8	(2)	–	0		–	3.3	(1.9)
Malonate	–	1	(0.1)	+	92.6	(0.7)	d	21.8	(0.7)
Jordan's tartrate	d	84.2	(1)	–	5.3		+	100	
Beta galactosidase	–	2.1		+	92.8		+ or –	74.4	
D-tartrate	+	92.3	(4)	(+) or –		(83.3)	(+)		(90.9)

Figures in parentheses indicate percentages of delayed reactions (3 or more days).

  + 90% or more positive within one or two days' incubation.

(+) Positive reaction after 3 or more days.

  – No reaction (90% or more negative).

+ or –   Majority of strains positive, some cultures negative.

– or +   Majority of cultures negative, some strains positive.

(+) or +   Majority of reactions delayed, some occur within 1 or 2 days.

  d Different reactions: +, (+), –.

by strains of the genus *Proteus*. On the contrary, urease activity in cultures of *Citrobacter* is weak and 18 to 20 or more hours of incubation are required before evidence of urease activity appears. Nonetheless, when positive, even weakly so, the test for urease is of considerable differential value.

From the data given in the tables mentioned above it is apparent that the genus *Arizona* (and the species *A. hinshawii*) is a distinct entity, which may be characterized and which may be differentiated from the genus *Salmonella* and the genus *Citrobacter*, as well as from other Enterobacteriaceae.

For additional information on the biochemical reactions given by cultures of *A. hinshawii,* the reader is referred to the reports of Kauffmann (1954, 1963, 1966), Moeller (1954), Kauffmann and Moeller (1955), and Kauffmann and Petersen (1956).

SEROLOGICAL CHARACTERIZATION OF CULTURES OF *ARIZONA*

The variational phenomena that affect serotypes of *Arizona* are mentioned in the general discussion given in Chapter 4. With a few possible exceptions, the phenomena that affect the serotyping of salmonellae also affect serotyping of members of the genus *Arizona*. Reversible phase variation in the flagellar (H) antigens occurs in serotypes of *Arizona* in the same manner that it occurs in salmonellae (Edwards and West 1945).

Peluffo et al. (1942) discussed seven cultures including the original strain isolated by Caldwell and Ryerson (1939). All were serologically related and all fermented lactose and liquefied gelatin. The strains were divisible into five serologic types. Edwards et al. (1943) studied 44 cultures from reptiles, fowls, mammals, and man which possessed biochemical properties similar to the original culture from Caldwell and Ryerson and which were related to it serologically (directly or indirectly). The 44 strains composed 15 serological types that were significant epidemiologically. Thus it became apparent that these microorganisms made up a separate group of bacteria which later became known as the genus *Arizona*. Knowledge of the bacteria was expanded by Hinshaw and McNeil (1944, 1946) and by Edwards et al. (1947). The latter workers investigated 382 cultures and established 25 O antigen groups and 61 serotypes among which both monophasic and diphasic serotypes were delineated. Since 1947 members of the genus *Arizona* have been studied by Edwards and West (1945, 1950), Le Minor et al. (1953), Edwards et al. (1953a, 1953b, 1956, 1959, 1965), Martin et al. (1967), Fife and Martin (1967, 1969), and others. As a result of continued investigations 34 O antigen groups and 43 H antigens have been delineated and about 300

serotypes have been characterized (see Antigenic Schema). The bacteria have continued to appear in fowls, mammals, and man, in all of which they may produce severe and fatal infections. The role of the organisms in the production of disease in some reptiles is not clear, since Le Minor et al. (1958) found the organisms in the feces of 44.8 per cent of 310 apparently normal snakes. As in the genus *Salmonella,* delineated serotypes are epidemiologically significant.

The methods described in Chapter 8 are recommended for use with members of the genus *Arizona*. This applies to methods for production of antisera, O and H agglutination tests, absorption of agglutinins, determination of single factor antigenic components (both O and H), and reversal of phases.

Since arabic numerals are employed for designation of both the O and the H antigens of serotypes of *Arizona,* the antigenic formulas are written in a form that is slightly different from that used for salmonellae. Commas are used to separate O antigen factors, a colon should be employed to distinguish the O and H antigens, commas should be used to indicate H antigen factors within a single phase, and a hyphen or a dash is employed to separate designations for the first phase from those of the second phase flagellar antigens, the second phase from the third, etc., (e.g., 9:26-21). When a monophasic variety of a normally diphasic form occurs, this fact is noted in writing the formula, e.g., 1,4:26 monophasic.

Determination of O Antigens. The antisera required for determination of the O antigens of cultures of *Arizona* are listed in Table 115. A number of the antisera usually may be used in the unabsorbed state (Table 115). However, many cultures possess related but distinct O antigens which must be distinguished by means of absorbed single factor antisera. All lots of newly produced antisera must be tested to determine their reactivity with homologous and heterologous cultures, to ascertain whether absorption is necessary, and to determine the dilution at which they can be used in daily work.

The agglutinin absorptions that are necessary for preparation of single factor antisera for exact determination of the O antigens are indicated in Table 115.

The O antigens of cultures of *Arizona* are determined by slide agglutination tests, first in unabsorbed antisera and then in single factor O antisera if required. The cross reactions that may be expected in properly diluted unabsorbed antisera are shown in Table 116. Thirty-four O antigens are listed in this table, which is a simplification since O antigens 7, 9, and 10 are subdivided. These sub-O groups are distinct and can be differentiated without difficulty.

The results of tube agglutination tests, in which

TABLE 115

The O Antigens of Serotypes of *Arizona*
Test Strains and Preparation of Single Factor O Antisera

O Factor	Culture		O Antigens Used in Absorptions
	No.	Formula	
2	1	1,2:1,2,5	1,3 + 1,33
3	699-52	1,3:1,2,5	1,2
4	292-53	1,4:1,2,5	1,2 + 1,3
5	1840-54	5:17,20	14 + 29
6	16	6:13,14	*
7b	2432-53	7a,7b:1,2,6	7a,7c
7c	143-57	7a,7c:27-31	7a,7b + 32
8	19	8:1,7,8	*
9b	25	9a,9b:13,14	9a,9c
9c	129	9a,9c:33-31	9a,9b + 18 + 20
10b	3970-53	10a,10b:1,2,5	10a,10c
10c	184	10a,10c:1,10	10a,10b + 32
11	34	11:16,17,18	*
12	124-57	12:27-28	*
13	1715-50	13:1,2,5	*
14	44	14:1,6,7,9	*
15	79	15:1,3,11	*
16	46	16:13,14	*
17	M 98	17:29-25	1,3 + 13
18	48	18:13,14	*
19	49	19:1,2,5	*
20	50	20:1,2,6	*
21	1450-53	21:1,2,6	*
22	162	22:13,14	*
23	89	23:33-25	1,2 + 28
24	196	24:26-25	32
25	63	25:27-28	30
26	4859-52	26:23-30	*
27	68	27:23-25	13 + 32
28	128	28:23-25	1,3 + 23 + 29
29	557-52	29:33-31	5 + 28
30	111	30:23-31	26
31	112	31:23-25	*
32	108	32:1,2,6	24
33	152	1,33:23-21	1,2 + 1,3
34	195	34:33-28	*

* Unabsorbed antiserum may be used.

TABLE 116
Reactions of the O Antigens of *Arizona* (Slide Tests)

O Antigens Suspensions	Unabsorbed O Antisera																																	
	1,2	1,3	1,4	5	6	7	8	9	10	11	12	13	14	15	16	17	18	19	20	21	22	23	24	25	26	27	28	29	30	31	32	1,33	34
1,2	++++	+	–	–	–	–	–	–	–	–	–	–	–	–	–	–	–	–	–	–	–	–	–	–	–	–	–	–	–	–	–	–	–
1,3	++++	++++	+	–	–	–	–	–	–	–	–	–	–	–	–	–	–	–	–	–	–	–	–	–	–	–	–	–	–	–	–	–	–
1,4	–	+	++++	–	–	–	–	–	–	–	–	–	–	–	–	–	–	–	–	–	–	–	–	–	–	–	–	–	–	–	–	–	–
5	–	–	–	++++	–	–	–	–	–	–	–	–	–	–	–	–	–	–	–	–	–	–	–	–	–	–	–	++++	–	–	–	–	–
6	–	–	–	+	++++	–	–	–	–	–	–	–	–	–	–	–	–	–	–	–	–	–	–	–	–	–	–	–	–	–	–	–	–
7	–	–	–	–	–	++++	–	–	–	–	–	–	–	–	–	–	–	–	–	–	–	–	–	–	–	–	–	–	–	–	–	–	–
8	–	–	–	–	–	–	++++	–	–	–	–	–	–	–	–	–	–	–	–	–	–	–	–	–	–	–	–	–	–	–	–	–	–
9	–	+	–	–	–	–	–	++++	+	–	–	–	–	–	–	–	–	–	–	–	–	–	–	–	–	–	–	–	–	–	–	–	–
10	–	–	–	–	–	–	–	+	++++	–	–	–	–	–	–	–	–	–	–	–	–	–	–	–	–	–	–	–	–	–	+	–	–
11	–	–	–	–	–	–	–	–	–	++++	+	–	–	–	–	–	–	–	–	–	–	–	–	–	–	–	–	–	–	–	–	–	–
12	–	–	–	–	–	–	–	–	–	+	++++	–	–	–	–	–	–	–	–	–	–	–	–	–	–	–	–	–	–	–	–	–	–
13	–	–	–	–	–	–	–	–	–	–	–	++++	–	–	–	–	–	–	–	–	–	–	–	–	–	–	–	–	–	–	–	–	–
14	–	–	–	–	–	–	–	–	–	–	–	–	++++	+	–	–	–	–	–	–	–	–	–	–	–	–	–	–	–	–	–	–	–
15	–	–	–	+	–	–	–	+	–	–	–	–	+	++++	+	–	–	–	–	–	–	–	–	–	–	–	–	–	–	–	–	–	–
16	–	–	–	–	–	–	–	–	–	–	–	–	–	+	++++	–	–	–	–	–	–	–	–	–	–	–	–	–	–	–	–	–	–
17	–	–	–	–	–	–	–	–	–	–	–	–	–	–	–	++++	+	–	–	–	–	–	–	–	–	–	–	–	–	–	–	–	–
18	–	+	–	–	–	–	–	+	–	–	–	–	–	–	–	+	++++	+	–	–	–	–	–	–	–	–	–	–	–	–	–	–	–
19	–	–	–	–	–	+	–	–	–	–	–	–	–	–	–	–	+	++++	–	–	–	–	–	–	–	–	–	–	–	–	–	–	–
20	–	–	+	–	–	–	–	–	–	–	–	–	–	–	–	–	–	–	++++	+	–	–	–	–	–	–	–	–	–	–	–	–	–
21	–	–	–	–	–	–	–	+	–	–	–	–	–	–	–	–	–	–	+	++++	+	–	–	–	–	–	–	–	–	–	–	–	–
22	–	–	–	–	–	–	–	–	–	–	–	–	–	–	–	–	–	–	–	+	++++	–	–	–	–	–	–	–	–	–	–	–	–
23	–	–	–	–	–	–	–	–	–	–	–	–	–	–	–	–	–	–	–	–	–	++++	–	–	–	–	+	–	–	–	–	–	–
24	–	–	–	–	–	+	–	+	–	–	–	–	–	–	–	–	–	–	–	–	–	+	+++	+	–	–	–	–	–	–	–	–	–
25	–	–	–	–	–	–	–	–	–	–	–	–	–	–	–	–	–	–	–	–	–	–	–	++++	–	–	–	–	–	–	–	–	–
26	–	–	–	–	–	–	–	–	–	–	–	–	–	–	–	–	–	–	–	–	–	–	–	–	++++	–	–	–	–	–	–	–	–
27	–	–	–	–	–	+	–	–	+	–	–	–	–	–	–	+	–	–	–	–	–	–	–	–	+	++++	+	+	–	–	–	–	–
28	–	–	–	–	–	–	–	–	–	–	–	–	–	–	–	–	–	–	–	–	–	+	+	–	–	+	++++	–	–	–	–	–	–
29	–	–	–	–	–	–	–	–	–	–	–	–	–	–	–	–	–	–	–	–	–	–	–	–	–	–	–	++++	+	–	–	–	–
30	–	–	–	–	–	–	–	–	–	–	–	–	–	–	–	–	–	–	–	–	–	–	–	–	–	–	–	+	++++	+	–	–	–
31	–	–	–	–	–	–	–	–	+	–	–	–	–	–	–	–	–	–	–	–	–	–	–	–	–	–	–	–	–	++++	–	–	–
32	++++	–	–	–	–	–	–	–	–	–	–	–	–	–	–	–	–	–	–	–	–	–	–	–	–	–	–	–	–	–	++++	+	–
1,33	++++	+	–	–	–	–	–	–	–	–	–	–	–	–	–	–	–	–	–	–	–	–	–	–	–	–	–	–	–	–	–	++++	++++
34	–	–	–	–	–	+	–	–	–	–	–	–	–	–	–	–	–	–	–	–	–	–	–	–	–	–	–	–	–	–	–	–	++++

Antisera used at highest dilution that gave strong, rapid agglutination with homologous antigens.
—, no reaction. + + + +, complete agglutination. +, partial reaction.

suspensions of all of the O antigens were tested in all of the O antisera, are recorded in Table 117. These reactions are included because all of the O antigenic relationships detected by tube agglutination may not be apparent in slide tests.

Occasionally a culture of *Arizona* that is inagglutinable in O antisera is encountered. Such forms generally possess an M (mucoid) antigen. The M antigen of serotypes of *Arizona* may be identified by agglutination in MO antiserum produced for the purpose or by agglutination in *Klebsiella* capsule type 21 antiserum. In order to determine the O antigen group of a serotype that possesses M antigen a suspension may be heated at 100 C for 15 to 30 min, cooled, and retested in O antisera. If this procedure fails to yield satisfactory results, the culture may be plated on infusion agar (incubation at 35 to 37 C) and a search made for normal (N) colonies. Or, the strain may be inoculated into semisolid motility medium as described for the isolation of N forms of salmonellae (Chapter 8). As far as is known, the M antigen of all cultures of *Arizona* is the same, regardless of the serotype in which it occurs.

Determination of H Antigens. The cultures used by the writer and colleagues for the production of antisera for the H antigens of serotypes of *Arizona* are listed in Table 118. Clearly, the desired phase of diphasic or triphasic cultures must be isolated for production of H antisera and for absorption of antisera. Since the H antigens are complex, the majority of the antisera must be absorbed in order to identify the delineated H antigen factors of the serotypes. The agglutinin absorptions required for preparation of factor antisera also are indicated in Table 118. As in the genus *Salmonella,* it is necessary to determine the titer of a factor antiserum each time it is prepared. Further, it is necessary to determine the titer of the absorbed antiserum for each antigenic complex in which the factor occurs. For example, each time that H antiserum 1, 2, 6 is absorbed for preparation of H factor 6, it is necessary to titrate the absorbed antiserum not only with antigen 1,2,6, but with 1,6,7 and 1,6,7,9 antigens as well.

The homologous titers and cross agglutination reactions exhibited by the H antisera produced by the author and colleagues are shown in Table 119. It is notable that cross reactions among the H antigens that contain 1 are particularly strong, numerous, and complex. Antigen 1 of the genus *Arizona* is analogous to antigen z_4 of the genus *Salmonella* and it occurs in combination with a variety of antigens so that a number of absorbed antisera are necessary to distinguish them. Antigenic relationships among the antigens that occur ordinarily in diphasic cultures are less complex and many of the cross reactions among

these are caused by incomplete suppression of phase 1 of a diphasic culture used to produce antiserum. Occasionally strains that possess three or more reversible phases are encountered but, as in the genus *Salmonella,* the occurrence of multiple phases was recognized comparatively recently and has not been studied extensively. Such cultures have been described by Edwards et al. (1960) and Fife et al. (1960).

When phases have been reversed as necessary and the O and H antigens of a culture have been determined, the antigenic formula of the serotype may be written. Any investigator who has had experience with absorption of antisera for salmonellae and in serotyping strains of *Salmonella* should have no difficulty with serological typing of cultures of *Arizona.*

ANTIGENIC SCHEMA FOR THE GENUS *ARIZONA*

All known serotypes of *Arizona* characterized prior to the end of 1969 are included in the antigenic schema (Table 120). This schema is a revision of that published by Edwards et al. (1965) and of supplements thereto (Fife and Martin, 1967, 1969). Five monophasic serotypes listed in the 1965 schema (Edwards et al., 1965) have been deleted (Table 120). These serotypes (12:1,2,6–; 12:1,7,8–; 13:13,15–; 15:1,2,6–; and 27:17,20–) were characterized by the late Dr. P. R. Edwards, but the author and colleagues have been unable to locate either the stock cultures of these serotypes or data regarding their characterization. Under these circumstances the author thought it best not to include the above-mentioned serotypes in the schema. In addition, several serotypes have been deleted from the schema because of their biochemical reactions. The majority of these were bacteria that were classified as members of the genus *Arizona* before the advent of a number of differential tests that are of considerable value in the differentiation of salmonellae and arizonae.

The antigenic properties of members of the genus *Arizona* cannot be expressed adequately by means of the antigenic formulas of serotypes of *Salmonella* (v. inf., Extrageneric Relationships). Further, the so-called simplified formulas proposed by Kauffmann and Rohde (1962a, 1962b) fail to distinguish epidemiologically significant differences between and within each of the two genera. Therefore a separate antigenic schema for members of the genus *Arizona* is necessary.

EXTRAGENERIC RELATIONSHIPS

The relationships of the O and H antigens of arizonae and salmonellae are listed in Tables 121 and 122. The O antigens of serotypes of *Arizona* are closely related to those of salmonellae in many instances. However, the O antigens of members of each of these genera also

TABLE 117

Relationships of the O Antigens of *Arizona*

O Antigens	Homologous Titer	Titers with Heterologous Antisera
1,2	5120	1,3-320; 1,33-20; 9a,9b-320
1,3	2560	1,2-5,120; 1,4-40; 7a,7b-160; 9a,9b-160; 12-80; 17-160; 27-20; 28-160; 1,33-20
1,4	2560	12-40; 17-40
1,33	5120	1,2-640; 1,3-20; 7a,7b-20; 9a,9b-20; 12-20; 14-20; 16-20; 25-40; 26-20
5	5120	26-20; 29-640
6	5120	7a,7b-20; 7a,7c-80; 8-20
7a, 7b	2560	7a,7c-640; 8-40; 25-20; 27-40; 32-40
7a,7c	2560	7a,7b-160; 8-80; 27-20; 32-40
8	1280	
9a,9b	2560	9a,9c-10,240; 18-20
9a,9c	5120	9a,9b-640; 18-20
10a,10b	1280	1,2-80; 10a,10c-320; 24-20
10a,10c	640	6-40; 10a,10b-40; 14-20
11	1280	9a,9c-40
12	2560	1,4-40; 28-40; 30-20
13	1280	9a,9b-80; 17-320; 18-20; 27-320; 28-40; 30-40
14	1280	1,2-40; 5-320; 10a,10c-20; 17-20; 30-20; 32-20
15	2560	1,2-20; 9a,9b-160; 9a,9c-20; 17-40; 28-20
16	1280	1,2-160; 22-20
17	5120	1,4-160; 10a,10c-40; 12-80; 25-20; 28-80
18	2560	9a,9c-160; 21-40
19	1280	5-80; 18-20; 28-20
20	1280	9a,9c-160; 10a,10c-20
21	1280	8-20; 10a,10c-20; 18-40
22	2560	10a,10c-160; 31-20
23	1280	1,2-160; 1,3-40; 28-160
24	1280	8-40
25	2560	8-20; 17-20
26	2560	5-40; 9a, 9c-40; 27-40; 30-160
27	2560	7a,7b-80; 7a,7c-80; 17-20; 24-40
28	640	5-160; 23-20; 29-320
29	640	5-160
30	2560	5-20; 8-40; 15-20; 23-80; 25-160; 28-40; 31-40
31	640	
32	2560	1,4-20; 7a,7c-320; 10a,10c-320; 24-160; 31-20
1,33	5120	1,2-640; 1,3-20; 7a,7b-20; 9a,9b-20; 12-20; 14-20; 16-20; 25-40; 26-20
34	2560	16-80

TABLE 118

H Antigens of *Arizona*
Cultures Used for Antiserum Production and
Preparation of Factor Antisera

| H Factors | Culture | | H Antigens Used for Absorption |
	No.	Formula	
2	1	1,2:1,2,5	1,10 + 1,3,11
3,11	11	5:1,3,11	1,2,5 + 1,10
5	1	1,2:1,2,5	1,2,6
6	4	1,3:1,2,6	1,2,5 + 1,7,8
7	17	7:1,7,8	1,2,6 + 1,10
8	17	7:1,7,8	1,6,7
9	7	1,3:1,6,7,9	1,2,6 + 1,6,7
(10)	5	1,3:1,10	1,2,6 + 1,3,11 + 1,2,36
14	40	13:13,14	13,15
15	26	9,13,15	13,14
16,18	34	11:16,17,18	17,20
20	14	5:17,20	16,17,18
21	56	12:23-21	35
22	64	27:22-31	29 + 31
23	100	28:23-28	26 + 30
24	217	24:24 – 28	*
25	60	23:24 – 25	24
26	90	9:26 – 21	*
27	63	25:27 – 28	*
28	63	25:27 – 28	27
29	66	26:29 – 30	22 + 30
30	93	10:33 – 30	22 + 23 + 33
31	103	16:27 – 31	23 + 29
32b	70	28:32a,32b – 28	28 + 31 + 32a,32c
32c	259	22:32a,32c – 28	28 + 31 + 30a,30b
33	85	1,4:33 – 31	24
34	97	16:23 – 24	23 + 30
35	123	11:35 – 28	21 + 25 + 27
36	121	17:1,2,36	1,10 + 1,3,11
37	126	16:22 – 37	40
38	145	7:27 – 38	27 + 40
39	161	16:39 – 25	*
40b	168	29:33 – 40a,40b	33 + 37 + 38 + 40a,40c
40c	248	19:27 – 40a,40c	27 + 37 + 38 + 40a,40b
41	221	20:24 – 41	24
42	3046 – 61	28: 23 – 30 – 48	*
43	269	7a,7c:43 – 28	*

* These antisera may be used without absorption if produced with cultures in proper phase.

TABLE 119

Relationships of H Antigens of *Arizona*
(Unabsorbed Antisera)

H Antigens	Homologous Titer	Titers with Heterologous Antisera
1,2,5	12,800	1,2,6-3,200; 1,10-1,600; 1,2,36-12,800; 1,3,11-1,600; 1,6,7-1,600; 1,6,7,9-1,600; 1,7,8-200
1,2,6	12,800	1,2,5-3,200; 1,10-6,400; 1,2,36-25,600; 1,3,11-1,600; 1,6,7-25,600; 1,6,7,9-6,400; 1,7,8-400
1,10	25,600	1,2,5-400; 1,2,6-1,600; 1,2,36-12,800; 1,3,11-3,200; 1,6-7-6,400; 1,6,7,9-3,200; 1,7,8-800
1,2,36	25,600	1,2,5-25,600; 1,2,6-3,200; 1,10-12,800; 1,3,11-1,600; 1,6,7-1,600; 1,6,7,9-3,200; 1,7,8-200; 17,20-200
1,3,11	12,800	1,2,5-400; 1,2,6-800; 1,10-6,400; 1,2,36-6,400; 1,6,7-400; 1,6,7,9-800; 1,7,8-400
1,6,7	25,600	1,2,5-200; 1,10-6,400; 1,2,36-1,600; 1,3,11-1,600; 1,6,7,9-6,400; 1,7,8-1,600
1,6,7,9	6,400	1,2,5-800; 1,2,6-3,200; 1,10-3,200; 1,2,36-3,200; 1,3,11-800; 1,6,7-25,600; 1,7,8-800
1,7,8	6,400	1,2,5-800; 1,2,6-6,400; 1,10-6,400; 1,2,36-3,200; 1,3,11-3,200; 1,6,7-12,800; 1,6,7,9-6,400
13,14	25,600	13,15-25,600
13,15	25,600	13,14-3,200
16,17,18	25,600	17,20-3,200
17,20	25,600	1,2,36-400; 16,17,18-400; 35-200
21	6,400	35-1,600
22	25,600	29-6,400; 30-800; 37-200
23	25,600	21-200; 29-1,600; 31-200; 34-3,200
24	25,600	25-400; 33-800
25	6,400	31-200; 35-800
26	12,800	23-1,600
27	25,600	28-1,600; 35-800; 37-1,600
28	25,600	31-200
29	25,600	22-400; 28-200; 31-400
30	25,600	21-400; 23-12,800; 29-800; 34-1,600
31	25,600	22-400; 32-6,400
32a,32b	25,600	27-100; 28-200; 31-100; 32a,32c-3,200
32a,32c	12,800	28-400; 29-200; 31-800; 32a,32b-1,600
33	25,600	40-400
34	12,800	38-100
35	25,600	25-100; 27-200
37	3,200	17,20-100; 31-200; 35-100; 38-200; 40-400
38	25,600	40-800
39	25,600	
40a,40b	25,600	37-200; 38-1,600; 40a,40c-1,600
40a,40c	12,800	38-800; 40a,40b-6,400; 42-800
41	3,200	
42	12,800	
43	12,800	

TABLE 120

Antigenic Sehema for the Genus *Arizona*

O Antigens	H Antigens		O Antigens	H Antigens	
	Phase 1	Phase 2		Phase 1	Phase 2
1,2	1,2,5	—	5	27	28
1,2	1,3,11	—	5	29	21
1,2	13,14	—	5	29	25
1,2	23	31	5	29	28
			5	29	30
1,3	1,2,5	—	(5)	29	30
1,3	1,2,6	—	5	33	25
1,3	1,10	—			
1,3	1,3,11	—	6	1,2,5	—
1,3	1,6,7,9	—	6	1,7,8	—
1,3	1,7,8	—	6	13,14	—
1,4	1,2,5	—	7a,7b	1,2,5	—
1,4	1,2,6	—	7a,7b	1,2,6	—
1,4	1,3,11	—	7a,7b	1,7,8	—
1,4	1,6,7	—	7a,7b	13,14	—
1,4	1,6,7,9	—	7a,7b	22	34
1,4	16,17,18	—	7a,7b	24	31
1,4	22	21			
1,4	23	28	7a,7c	27	31
1,4	24	21	7a,7c	27	38
1,4	24	31	7a,7c	27	31 – 38*
1,4	24	38	7a,7c	29	25
1,4	26	21	7a,7c	29	31
1,4	26	25	7a,7c	43	28
1,4	27	21			
1,4	29	28	8	1,7,8	—
1,4	29	31	8	13,14	—
1,4	30	31	8	17,20	—
1,4	33	31			
			9a,9b	1,2,5	—
5	1,2,5	—	9a,9b	1,3,11	—
5	1,3,11	—	9a,9b	1,6,7	—
5	1,6	—	9a,9b	1,7,8	—
5	1,6,7	—	9a,9b	13,14	—
5	1,7,8	—	9a,9b	16,17,18	—
5	13,14	—	9a,9b	17,20	—
5	16,17,18	—	9a,9b	22	21
5	17,20	—	9a,9b	22	31
5	22	25	9a,9b	23	28
5	23	30	9a,9b	24	21
5	23	39 – 30*	9a,9b	24	25
5	24	28	9a,9b	24	30
5	26	31	9a,9b	24	31

—, monophasic   *Phase 3   ( ), partial antigen
**Fourth phase

TABLE 120 Continued)

Antigenic Sehema for the Genus *Arizona*

O Antigens	H Antigens Phase 1	H Antigens Phase 2	O Antigens	H Antigens Phase 1	H Antigens Phase 2
9a,9b	26	21	12	23	21
9a,9b	26	25	12	23	28
9a,9b	26	30	12	24	21
9a,9b+	26	31	12	27	28
9a,9b	29	21	12	27	31
9a,9b	29	31			
			13	1,2,5	—
9a,9c	23	21	13	1,2,6	—
9a,9c	23	31	13	1,3,11	—
9a,9c	26	21	13	1,6,7	—
9a,9c	26	25	13	1,7,8	—
9a,9c	26	31	13	13,14	—
9a,9c	27	31	13	16,17,18	—
9a,9c	29	25	13	17,20	—
9a,9c	29	28	13	22	—
9a,9c	29	30			
9a,9c	29	31	14	1,2,5	—
9a,9c	33	28	14	1,2,6	—
9a,9c	33	30	14	1,6,7,9	—
9a,9c	33	31			
			15	1,3,11	—
10a,10b	1,2,5	—	15	13,14	—
10a,10b	1,2,6	—	15	22	21
10a,10b	1,3,11	—	15	23	30
10a,10b	1,7,8	—	15	24	25
10a;10b	13,14	—	15	24	31
10a,10b	13, (14)	28	15	24	42
10a,10b	16,17,18	—	15	26	31
10a,10b	17,20	—	15	27	28
10a,10b	23	25	15	27	31
10a,10b	29	31 – 40a,40c*	15	29	—
10a,10b	33	30			
			16	13,14	—
10a,10c	1,10	—	16	22	21
10a,10c	27	21	16	22	30
			16	22	31
11	13,14	—	16	22	34
11	16,17,18	—	16	22	37
			16	23	21
12	1,2,5	—	16	23	25
12	1,3,11	—	16	23	31
12	1,6,7	—	16	23	34
12	1,6 ...	—	16	24	21
12	1,6,7,9	—	16	24	31
12	17,20	—			

TABLE 120 (Continued)

Antigenic Sehema for the Genus *Arizona*

O Antigens	H Antigens		O Antigens	H Antigens	
	Phase 1	Phase 2		Phase 1	Phase 2
16	24	31 – 40a,40b*	20	27	21
16	27	25	20	29	25
16	27	31	20	29	31
16	29	25	20	33	28
16	29	31	20	33	31
16	33	25			
16	39	25	21	1,2,5	–
			21	1,2,6	–
17	1,2,5	–	21	1,3,11	–
17	29	25	21	17,20	–
			21	23	25
18	1,3,11	–	21	24	28
18	1,6,7	–	21	29	31
18	13,14	–			
			22	1,3,6	–
19	1,2,5	–	22	13,14	–
19	1,2,6	–	22	16,17,18	–
19	16,17,18	–	22	23	31
19	17,20	–	22	27	28
19	22	21	22	27	31
19	22	28	22	29	28
19	22	31	22	32	28
19	23	31	22	33	28
19	26	–			
19	27	25	23	23	30 – 42*
19	27	40a,40c	23	24	21
19	29	25	23	24	25
19	32	28	23	24	31
19	33	21	23	32	28
19	33	31	23	33	21
			23	33	25
20	1,2,6	–	23	33	28
20	13,14	–	23	33	28 – 42*
20	16,17,18	–			
20	22	21	24	22	25
20	22	31	24	23	31
20	23	21	24	24	25
20	23	30	24	24	28
20	24	21	24	24	31
20	24	28	24	25	26
20	24	41	24	26	21
20	26	21	24	27	21
20	26	28	24	27	31
20	26	30	24	29	21

TABLE 120 (Continued)

Antigenic Sehema for the Genus *Arizona*

O Antigens	H Antigens			O Antigens	H Antigens	
	Phase 1	Phase 2			Phase 1	Phase 2
24	29	31		28	32	31
24	33	21		28	33	21
				28	33	25
25	22	21		28	33	25 – 40a,40c*
25	23	21		28	33	31
25	23	25		28	43	–
25	23	30				
25	23	31		29	23	31
25	27	28		29	24	31
25	29	25		29	26	28
25	29	31		29	27	31
25	33	21		29	29	25
				29	31	32
26	22	25		29	33	21
26	23	22		29	33	31
26	23	30		29	33	40a,40b
26	23	(30)				
26	23	30 – 40a,40c*		5,29	17,20	–
26	23	31		5,29	24	31
26	24	25		5,29	33	21 – 40a,40b*
26	24	30				
26	26	25		30	22	21
26	29	30		30	22	25
26	32	21		30	22	31
26	32	30		30	23	25
26	33	25		30	23	31
26	33	28		30	26	21
26	33	31		30	26	25
				30	26	31
27	22	31		30	27	28
27	23	25		30	32	25
28	16,17,18	–		31	23	25
28	23	21		31	29	21
28	23	25				
28	23	28		32	1,2,6	–
28	26	21				
28	26	28		1,33	23	21
28	26	30		1,33	23	28
28	26	31		1,33	24	25
28	27	30		1,33	24	25 – 39* – 40a,40c**
28	27	31		1,33	24	28
28	29	28		1,33	26	21
28	29	31		1,33	26	31
28	32	28		34	33	28
28	32	28 – 40a,40c*		34	33	31

–, monophasic   ( ), Partial antigen   * Phase 3   **Fourth Phase

are related to those of other genera such as *Citrobacter* (Chapter 11), *E. coli* and *Shigella* (Frantzen, 1950; Kampelmacher, 1959) and *Providencia* (Chapter 17), as might be expected.

The H antigens of serotypes of *Arizona* are related to some of those of salmonellae, but it will be noted (Table 122) that there are only a few instances in which H antigen factors of the former are identical with factors of the latter. The antigenic relationships of the H antigens of members of the above-mentioned genera also were investigated by Guinée et al. (1962). The flagellar antigens of some strains of *Citrobacter* are related to those of some arizonae and salmonellae (see Chapter 11).

ANTISERUM PRODUCTION

As mentioned above, O and H antisera for serotypes of *Arizona* are produced in the same manner as for salmonellae (Chapter 8).

Polyvalent Antisera. The author and colleagues employ two polyvalent antisera for preliminary work with cultures suspected of being arizonae. One of these contains agglutinins for the H antigens of all monophasic serotypes while the other contains agglutinins for the H antigens of diphasic serotypes. The following cultures are employed for production of the first (monophasic) polyvalent antiserum:

Strain No.	H Antigen	Strain No.	H Antigen
DC 5	1,2,5	143	1,6,7,9
So 89	1,2,6	CDAI 184	1,7,8
N 178	1,10	NJ 4	13,14
1189-56	1,2,36	N 99	13,15
S 39	1,3,11	Ore 181	16,17,18
Sp 21 X 34	1,6,7	107	17,20

H antigen suspensions are prepared from actively motile forms of each culture and equal quantities of each suspension are pooled. Since a relatively small

TABLE 121

Relationships of the O Antigens of *Arizona* and *Salmonella*

Arizona O Antigen	Related *Salmonella* O Antigen	*Arizona* O Antigen	Related *Salmonella* O Antigen
1,2	51	19	59
1,3	44	20	35
1,4	53 and 59	21	43
5	48a,48b,48c	22	21
6	*	23	47a,47b
7a,7b	18	24	60
8	*	25	16
9a,9b	50a,50b,50d	26a,26b	61
10a,10b	40a,40b	26a,26c	61
10a,10c	40a,40c	27	6,7
11	45a,45c	28	47a,47c
12	17	29	64
13	41	30	*
14	56	31	52
15	42	32	6,14,24
16	38	1,33	58
17	11	34	57
18	1,13,23		

* No known relationship.

NOTE: Underscores indicate O antigen identity.

TABLE 122

H Antigenic Relationships of *Arizona* and *Salmonella*

Arizona H Antigen	Related *Salmonella* H Antigen	*Arizona* H Antigen	Related *Salmonella* H Antigen
1	z_4	26	z_{52}
1,10	z_4,z_{32}	27	z_{10b}
1,36	z_4,z_{23}	28	$e,n \ldots$
2	z_{23}	29	k_1,k_2
3,11	z_{23}	30	1,5
13,14	$g,(p),z_{51}$	31	z_1,z_2
16	z_{29}	32a,32b;32a,32c	c_1,c_2
20	z_{36}	33a,33b	i_2
21	z	33a,33c	i_1
22	k_1,k_2	35	a
23	l,z_{13}	39	z_{47}
24	r_1,r_2	42	z_{50}
25	z_{53}	43	b

NOTE: Underscores indicate identity.

number of cultures are employed for production of this polyvalent antiserum, the H antigen suspensions may be made into a single pool. This pooled antigen suspension then is inoculated into a lot of several rabbits using a schedule of inoculation dosage sizes, etc., similar to that described for the preparation of a comprehensive polyvalent antiserum for salmonellae (Chapter 8).

For production of the second polyvalent antiserum mentioned above, i.e., that for diphasic (and multiphasic) serotypes, the following cultures are used:

Strain No.	H Antigen	Strain No.	H Antigen
166,ph[1]1	21	1971-51,ph1	32a,32b
Cal 1141,ph2	22	1048-61,ph1	32a,32c
3209-54,ph1	23	466-52,ph2	33
217,ph1	24	594-54,ph2	34
155,ph2	25	142-56,ph1	35
196,ph1	26	2224-56,ph2	37
110,ph1	27	1995-57,ph2	38
110,ph2	28a,28b	1158-58,ph1	39
211,ph2	28a,28c	2907-58,ph2	40a,40b
142-56,ph2	28a,28d	3061-60,ph2	40a,40c
K16019, monoph	29	740-60,ph2	41
456-53,ph1	30	3064-61,ph3	42
Cal 1141,ph1	31	1196-62,ph1	43

[1]ph., phase

After motility of the cultures has been enhanced by passage through tubes of semisolid agar medium H antigen suspensions are prepared from each and the antigens are pooled for production of polyvalent antiserum for the diphasic and multiphasic serotypes of *Arizona* it is advisable to prepare two pooled antigen suspensions, each containing half of the original H antigens. Each of the pooled antigens then is injected into separate lots of rabbits. The two lots of antiserum should be pooled when production is complete.

The above-mentioned polyvalent antisera are used in tube agglutination tests at a dilution of 1 to 1,000. Formalinized broth cultures are used as H antigens. Tests are incubated at 35 to 37 C in a waterbath and read at 15, 30, and 60 min. When employed judicially, polyvalent antisera for their H antigens are helpful in preliminary work with cultures that may be arizonae.

REFERENCES

Caldwell, M. E., and D. L. Ryerson. 1939. J. Infect. Dis., **65**, 242.

Edwards, P. R., W. B. Cherry, and D. W. Bruner. 1943. J. Infect. Dis., **73**, 229.

Edwards, P. R., and M. G. West. 1945. Ibid., **77**, 185.

Edwards, P. R., M. G. West, and D. W. Bruner. 1947. Ky. Agric. Exp. Sta. Bull., No. 499, Lexington, Ky.

Edwards, P. R., and M. G. West. 1950. J. Infect. Dis., **87**, 184.

Edwards, P. R., F. Kauffmann, and E. van Oye. 1953a. Acta Pathol. Microbiol. Scand., **31**, 5.

Edwards, P. R., F. Kauffmann, and A. Fain. 1953b. Ibid., **33**, 191.

Edwards, P. R., A. C. McWhorter, and M. A. Fife. 1956. Bull. Wld. Health Org., **14**, 511.

Edwards, P. R., and M. A. Fife. 1956. Appl. Microbiol., **4**, 46.

Edwards, P. R., M. A. Fife, and C. H. Ramsey. 1959. Bacteriol. Rev., **23**, 155.

Edwards, P. R., M. A. Fife, and L. LeMinor. 1960. J. Bacteriol., **80**, 259.

Edwards, P. R., M. A. Fife, and W. H. Ewing. 1965. Antigenic schema for the genus *Arizona* CDC Publ.*

Ewing, W. H. 1963. Int. Bull. Bacteriol. Nomen. Tax., **13**, 95.

———. 1966. Int. J. System. Bacteriol., **16**, 423.

———. 1967. Revised definitions for the family Enterobacteriaceae, its tribes, and genera. CDC Publ.*

———. 1969a. Biochemical reactions given by Enterobacteriaceae in commonly used tests. CDC Publ.*

———. 1969b. Int. J. System. Bacteriol., **19**, 1.

———. 1970. Differentiation of Enterobacteriaceae by biochemical reactions. CDC Publ.*

Ewing, W. H., M. A. Fife, and B. R. Davis. 1965. The biochemical reactions of *Arizona arizonae.* CDC Publ.*

Ewing, E. H., and M. A. Fife. 1966. Int. J. System. Bacteriol., **16**, 427.

Fife, M. A., P. R. Edwards, R. Sakazaki, M. Nozawa, and M. Murata. 1960. Jap. J. Med. Sci. Biol., **13**, 173.

Fife, M. A., and W. J. Martin. 1967. Antigenic schema for the genus *Arizona.* Suppl. I. CDC Publ.*

———. 1969. Antigenic schema for the genus *Arizona.* Supl. II. CDC Publ.*

Frantzen, A. 1950. Acta Pathol. Microbiol. Scand., **27**, 647.

Guinee, P. A. M., E. H. Kampelmacher, and H. M. C. C. Willems. 1962. Antonie v. Leeuwenhoek, **28**, 17.

Hinshaw, W. R., and E. McNeil. 1944. Cornell Vet., **34**, 248.

———. 1946. J. Bacteriol., **51**, 281.

International Code of Nomenclature of Bacteria. 1966. Inter. J. System. Bacteriol., **16**, 459.

Kampelmacher, E. H. 1959. Antonie v. Leeuwenhoek, **25**, 289.

Kauffmann, F. 1941. Acta Pathol. Microbiol. Scand., **18**, 351.

———. 1954. Enterobacteriaceae. 2nd Ed. Copenhagen: Munksgaard.

———. 1963. Acta Pathol. Microbiol. Scand., **58**, 109.

———. 1966. The bacteriology of Enterobacteriaceae. Copenhagen: Munksgaard.

Kauffmann, F., and P. R. Edwards. 1952. Int. Bull. Bacteriol. Nomen. Tax., **2**, 2.

Kauffmann, F., and V. Moeller. 1955. Acta Pathol. Microbiol. Scand., **36**, 173.

Kauffmann, F., and A. Petersen. 1956. Ibid., **38**, 481.

Kauffmann, F., and R. Rohde. 1962a. Ibid., **54**, 474.

———. 1962b. Ibid., **56**, 341.

LeMinor, L., P. R. Edwards, and H. Fossaert. 1953. Ann. Inst. Pasteur, **84**, 1056.

LeMinor, L., M. A. Fife, and P. R. Edwards. 1958. Ibid., **95**, 326.

Lubin, A. H., and W. H. Ewing. 1964. Publ. Health Lab., ** **22**, 83.

Martin, W. J., M. A. Fife, and W. H. Ewing. 1967. The occurrence and distribution of serotypes of *Arizona.* CDC Publ.*

Moeller, V. 1954. Acta Pathol. Microbiol. Scand., **34**, 115.

Peluffo, C. A., P. R. Edwards, and D. W. Bruner. 1942. J. Infect. Dis., **70**, 185.

*CDC Publ., Publication from the Center for Disease Control (formerly the Communicable Disease Center), Atlanta, Ga. 30333.

**Public Health Laboratory: Journal of the Conference of Public Health Laboratory Directors.

Chapter 11

The Genus *Citrobacter*

The genus *Citrobacter,* as presently defined, is composed of cultures previously classified as *Escherichia freundii* (v. inf.). The biochemical reactions of these microorganisms more closely resemble those of members of the genera *Salmonella* and *Arizona* than those of any other Enterobacteriaceae. Therefore the genus *Citrobacter* is placed in the tribe Salmonellae (Chapter 1).

DEFINITION

The results of an investigation (Davis and Ewing, 1966) of the biochemical reactions of relatively large numbers of cultures of *Citrobacter* made possible the framing (Ewing, 1967) of the following definition:

The genus *Citrobacter* is composed of motile bacteria that conform to the definitions of the family ENTEROBACTERIACEAE and the tribe SALMONELLEAE. Lysine is not decarboxylated and less than 20 per cent of strains possess ornithine decarboxylase. Urease is produced slowly by the majority of cultures, but the reactions are weak. Growth occurs in medium that contains potassium cyanide and acid is produced in Jordan's tartrate medium. Gelatin is not liquefied in nutrient medium. Dulcitol and cellobiose are fermented rapidly by the majority of cultures. Lactose is utilized but the reactions frequently are delayed. The type species is *Citrobacter freundii* (Braak) Werkman and Gillen.

This generic definition should be used in conjunction with the definition of the tribe Salmonelleae (Chapter 1).

Werkman and Gillen (1932) proposed the generic term *Citrobacter* for the "citrate-positive, coliaerogenes intermediates," and from a group of fifteen cultures these investigators described seven species and designated *Citrobacter freundii* Braak as the type species. Tittsler and Sandholzer (1935) and Carpenter and Fulton (1937) suggested that *"Escherichia-Aerobacter* intermediate" cultures, which did not produce acetylmethycarbinol, but utilized sodium citrate as a sole source of carbon, should be classified as *Escherichia.* Such strains were classified as *Escherichia freundii* in the 1939 and 1946 editions of *Bergey's Manual.* Borman et al. (1944) in their suggested reclassification of the family Enterobacteriaceae proposed that strains that fermented lactose rapidly, failed to produce acetylmethylcarbinol, and utilized citrate should be classified as *Colobactrum freundii.*

As presently constituted the genus *Citrobacter* is composed of similar bacteria that have been described under a variety of designations. In 1940 Kauffmann and Moeller described a microorganism under the name *Salmonella ballerup.* Later this type was removed from the genus *Salmonella* because of its biochemical reactions (Harhoff, 1949; Bruner et al., 1949). In their studies, Stuart et al. (1943) characterized five strains under the designation 14011. These cultures subsequently were found to be members of the Bethesda group, characterized by Barnes and Cherry (1946), and so named by Edwards et al. (1948) and Moran and Bruner (1949). West and Edwards (1954) investigated 585 cultures of Ballerup, Bethesda, and *E. freundii.* They combined the first two groups into the Bethesda-Ballerup group and called attention to the close biochemical relationships of members of the combined group and strains of *E. freundii (C. freundii).* Also, if the genus *Paracolobactrum (Bergey's Manual,* 7th ed.) existed, certain strains of *Citrobacter* would have to be included in it along with a variety of other quite different bacteria.

BIOCHEMICAL REACTIONS

As noted above, Werkman and Gillen (1932) described seven species of *Citrobacter.* However, the strain designated as the type species, *C. freundii,* was judged to be atypical. This made it necessary for the International Enterobacteriaceae Subcommittee to designate a neotype culture (see Appendix 3 to Report, 1958). Tittsler and Sandholzer (1935) studied the biochemical reactions of 29 strains of so-called intermediates. Examination of the data recorded by these workers indicated that 26 of the cultures probably were members of the genus *Citrobacter.* Vaughn and Levine (1942) examined a large number of intermediate coliforms, 196 of which may have belonged to the genus *Citrobacter.* In 1937 Carpenter and Fulton recorded the biochemical reactions of 125 isolants of intermediates. These probably were strains of *Citrobacter*. West and Edwards (1954) studied 585 cultures of the Bethesda-Ballerup group and of *E. freundii*, and Moeller (1954a) included 72 cultures of these microorganisms in his work on the usefulness of KCN medium. Also, Kauffmann (1956a, 1956b) included cultures of these bacteria in his studies on the biochemical reactions of Enterobacteriaceae, and Sedlak and Slajsova (1962) reported the reactions given by 3586 isolants of *Citrobacter* in a limited number of tests. Of the data given in the publication cited only that recorded by Tittsler and Sandholzer (1935), Carpenter and Fulton (1937), and by West and

Edwards (1954) is presented in a way that lends itself to summarization in tabular form. The data obtained with the 736 cultures reported upon by these investigators are summarized in Table 123.

The reactions given by cultures of *Citrobacter* in KCN medium, the decarboxylase, malonate, and beta galactosidase (ONPG) tests, and in organic acid media as reported by several investigators are summarized in Tables 77 to 81 (Chapter 8). In addition, Ewing et al. (1957) reviewed the literature on the production of phenylalanine deaminase and utilization of malonate by Enterobacteriaceae. These workers reported that 166 cultures of *Citrobacter* failed to deaminate

TABLE 123

Summary of the Biochemical Reactions of *Citrobacter* as Reported by Several Investigators[1]

Test or substrate	Sign	%+	(%+)*	Test or substrate	Sign	%+	(%+)*
Hydrogen sulfide	+	92.4		Inositol	–	3.8	
Urease	–	0		Sorbitol	+	96.2	
Indol	–	1.2		Arabinose	+	100	
Methyl red	+	100		Raffinose	– or +	46.2	
Voges-Proskauer	–	0		Rhamnose	+	100	
Citrate (Simmons')	+	99.3		Maltose	+	100	
Motility	+	99.2		Xylose	+	100	
Gelatin	–	0		Trehalose	+	100	
Jordan's tartrate	+	100		Cellobiose	+	99.3	
Glucose	+	100		Glycerol	+	100	
Lactose	d	63.9	(0.5)	Alpha methyl glucoside	+ or –	63.6	
Sucrose	d	20.9	(0.8)	Erythritol	–	0	
Mannitol	+	100		Esculin	– or +	19.2	
Dulcitol	– or +	42.1		Galactose	+	100	
Salicin	d	31.2	(7.1)	Nitrate to nitrite	+	100	
Adonitol	–	0					

*Figures in parentheses indicate percentages of delayed reactions (3 or more days).

[1] Results obtained with 736 cultures (see text). All substrates were not used with all strains. For details see Davis and Ewing (1966).

+ 90% or more positive within one or two days' incubation.

(+) Postive reaction after 3 or more days.

– No reaction (90% or more).

+ or – Majority of strains positive, some cultures negative.

– or + Majority of cultures negative, some strains positive.

(+) or + Majority of reactions delayed, some occur within 1 or 2 days.

d Different reactions: +, (+), –.

phenylalanine. Ewing and Johnson (1960) reported that Enterobacteriaceae, including 108 strains of *Citrobacter,* gave negative reactions in the indophenol oxidase test. Davis and Ewing (1964) studied the pectinolytic, alginolytic, and lipolytic capacities of Enterobacteriaceae and states that none of the cultures of *Citrobacter* examined was pectinolytic, alginolytic, or was able to utilize aglinate as a sole source of carbon, and none was lipolytic. Martin and Ewing (1967) failed to detect DNase activity in a small number of isolates of *Citrobacter.*

The biochemical reactions given by 582 cultures of *Citrobacter* are summarized in Table 124. These data are adapted from the work of Davis and Ewing (1966). Although the data presented in Table 124 largely are self-explanatory, a few comments are in order. Many strains produced weakly positive reactions on Christensen's urea agar after one or two days of incubation. Usually the weak color changes in the indicator occurred in the upper quarter of the medium column, and were in no way comparable to the rapid, strong reactions given by members of the genus *Proteus.* Three strains (0.5 per cent) yielded negative results in the methyl red test. However, this type of anomalous reaction was noted occasionally in cultures of many genera of Enterobacteriaceae and was reported to have occurred in a small percentage of isolants of *Citrobacter* by Moeller (1954b). Seventeen isolates failed to grow in medium containing KCN. In addition, five cultures grew in KCN medium after 3 to 7 days of incubation. These also should be regarded as negative since 48 hours is the recommended time for final reading of this test. Five strains liquefied gelatin, but the reactions were weak and delayed (22 or more days). None of the strains decarboxylated lysine in Moeller's medium during the recommended period (4 days) of incubation. Of the results of other investigators reviewed, only Moeller (1954b, 1955) reported positive reactions in lysine medium, and these were weak, erratic, and occurred only after prolonged incubation. Hence they may be disregarded as alluded to by Moeller. Analysis of the remaining decarboxylase reactions given by members of the genus *Citrobacter* indicated that 50 (8.6 per cent) of the strains examined failed to decarboxylate either arginine or ornithine. Obviously, these 50 cultures gave negative results in all three of the decarboxylase tests, since lysine was not decarboxylated by any of the 582 isolates.

Ninety-three (16 per cent) of the cultures (Table 124) failed to produce hydrogen sulfide in TSI agar medium. Thirty of these strains produced indol and 39 decarboxylated ornithine, which would yield higher percentages of positive reactions for these tests than those usually seen (Table 124). Otherwise, the

reactions given by the 93 isolates were similar to those given by typical cultures (see Davis and Ewing, 1966 for details).

The complete biochemical reactions of biotypes of *C. freundii* that formed indol or failed to grow on Simmons' citrate agar medium are given in Table 125. It should be noted (Table 125) that none of the 39 isolants that produced indol failed to grow on Simmons' citrate agar, and conversely none of the citrate negative strains (31) produced indol. In other words, the two sets of reactions should be considered separately. Washington et al. (1969, 1970) reported the isolation of a number of cultures of *C. freundii* that were similar to the hydrogen sulfide negative, indol positive biotype mentioned by Davis and Ewing, 1966 (see also Table 125).

Adonitol was not fermented by any of the biotypes of *C. freundii* studied by Davis and Ewing, 1966 (Table 124). With the exception of Werkman and Gillen (1932), who described two species that fermented adonitol (v. inf.), investigators have excluded bacteria that produced acid from this substrate from *Citrobacter* (e.g., Table 123).

With very few exceptions the cultures reported upon by Davis and Ewing (1966) were taken at random from cultures submitted for identification over a period of years. Twelve of the microorganisms designated as type 4485-65 by Slopek and Dabrowski (1960) were received in 1960. Since the biochemical reactions of these twelve were similar only four were included in the above-mentioned investigations. As far as the author and colleagues were able to determine, the type 4485-65 strains constituted a biotype of *C. freundii.* These cultures were anaerogenic, failed to produce hydrogen sulfide in TSI agar medium, did not ferment sorbitol, failed to decarboxylate either arginine or ornithine, and fermented lactose slowly, but in all other respects were similar to *C. freundii.* In 1961 five isolates of *Kluyvera* (Asai et al. 1957) were received from Dr. Mortimer Starr of the University of California at Davis. One of these (84c) proved to be a strain of *C. freundii* (indol produced, hydrogen sulfide not produced).

Although there were some differences between the results reported by other investigators (Table 123) and those recorded by the author and coworkers (Table 124), these were minor and usually resulted in slight differences in percentages. It was the author's belief that the 582 cultures of *C. freundii* represented the genus *Citrobacter* as a whole much better than did the materials studied by most of the other investigators. For example, most of the 736 strains reported upon by others constituted a selected group, since the majority of the 585 isolants studied by West and Edwards

TABLE 124

Summary of the Biochemical Reactions Given by 582 Cultures of *C. freundii*

Test or substrate	Sign	%+	(%+)*	Test or substrate	Sign	%+	(%+)*
Hydrogen sulfide TSI	d	81.6	(2.4)	Arabinose	+	100	
PIA	d	87.3	(1.5)	Raffinose	d	14.2	(0.8)
LIA	d	85.9	(2.3)	Rhamnose	+	99.4	
Urease	d	69.4	(6.9)	Malonate	d	21.8	(0.7)
Indol	–	6.7		Mucate	+	95.2	(1)
Methyl red (37 C)	+	99.5		Jordan's tartrate	+	99.1	(0.9)
Voges-Proskauer (37 C)	–	0		Sodium alginate	–	0	
Citrate (Simmons')	+	90.4	(4.3)	Sodium acetate	+ or (+)	77.5	(13.5)
KCN	+	96.2	(0.9)	Lipase, Corn oil	–	0	
Motility	+	95.7		Maltose	+	98.5	(1.5)
Gelatin (22 C)	–	0	(0.9)	Xylose	+	99.8	(0.2)
Lysine decarboxylase	–	0		Trehalose	+	100	
Arginine dihydrolase	d	43.6	(44.4)	Cellobiose acid	+ or (+)	60.8	(38)
Ornithine decarboxylase	d	17.2	(0.2)	gas	d	11.7	(3.1)
Phenylalanine deaminase	–	0		Glycerol acid	+	97.9	(0.9)
Glucose acid	+	100		gas	+ or –	83.5	
gas	+	90.9		Alpha methyl glucoside	d	4.3	(7.9)
Lactose	(+) or +	39.4	(50.8)	Erythritol	–	0	
Sucrose	d	15.3	(9.4)	Esculin	–	0.9	(3.0)
Mannitol acid	+	99.8	(0.2)	Nitrate to nitrite	+	98.6	
gas	+	92.1		Oxidase	–	0	
Dulcitol	d	59.8	(0.7)				
Salicin	d	4.1	(23)	Organic acids **			
Adonitol	–	0					
Inositol	–	3.3	(1.9)	Citrate	(+) or +	48.7	(50)
Sorbitol	+	98	(1)	D-tartrate	(+)		(90.9)

*Figures in parentheses indicate percentages of delayed reactions (3 or more days).
**Method of Kauffmann and Petersen (Chapter 18).

  + Postive within one or two days' incubation.
 (+) 90% or more positive reactions after 3 or more days.
  – No reaction (90% or more).
 + or – Majority of strains positive, some cultures negative.
 – or + Majority of cultures negative, some strains positive.
(+) or + Majority of reactions delayed, some occur within 1 or 2 days.
  d Different reactions: +, (+), –.

TABLE 125

The Biochemical Reactions of Cultures of *Citrobacter* that Gave Positive Indol Reactions or Failed to Grow on Simmons' Citrate Medium

Test or substrate	Indol positive (39)					Citrate negative (31)				
	+ 1-2*	(+) 3-7	(+) 8-14	(+) >14	–	+ 1-2*	(+) 3-7	(+) 8-14	(+) >14	–
Hydrogen sulfide (TSI)	8	1			30	17	3			11
(PI)	8				31	27	2			2
Urease	32				7	12	5			14
Indol	39				0	0				31
Methyl red (37 C)	39				0	31				0
Voges-Proskauer (37 C)	0				39	0				31
Citrate (Simmons')	38	1			0	0				31
KCN	37				2	30	1			0
Motility	39				0	27				4
Gelatin (22 C)	0			1	38	0				31
Lysine decarboxylase	0				39	0				31
Arginine dihydrolase	25	10			4	3	17			11
Ornithine decarboxylase	35				4	2	1			28
Phenylalanine deaminase	0				39	0				31
Glucose acid	39				0	31				0
gas	37				2	25				6
Lactose	35	4			0	13	13		1	4
Sucrose	4	1			34	11		1	1	18
Mannitol	39				0	31				0
Dulcitol	5				34	17				14
Salicin	12	23	2		2	1	2	3	2	23
Adonitol	0				39	0				31
Inositol	0				39	1				30
Sorbitol	37				2	29	2			0
Arabinose	35				0	31				0
Raffinose	4	1			31	10	1			20
Rhamnose	39				0	31				0
Malonate	4				35	4	1			26
Mucate	39				0	28				3
Christensen's citrate	38				0	27	4			0
Sodium acetate	25	4			2	1	2			26
Lipase, corn oil	0				28	0				18
Maltose	39				0	31				0
Xylose	39				0	31				0
Trehalose	39				0	31				0
Cellobiose acid	38	1			0	19	10	2		0
gas	33				6	1	2	1		27
Glycerol acid	34	1			4	31				0
gas	13				26	23				8
Alpha methyl glucoside	2	2			28	3	1			25
Erythritol	0				32	0				29
Esculin	1	4	4	2	21	0		1	1	27
Nitrate to nitrite	39				0	30				1
Oxidase	0				39	0				31

Organic acids **	1-2*	7	14	–	1-2*	7	14	–
Citrate	8	18	1	1	9	9		1
D-tartrate	0	15	7	6	0	8	8	3

* Days of incubation

** Media of Kauffmann and Petersen. Tested with lead acetate after 1 or 2, 7, 14 days.

NOTE: All cultures that were indol + grew on Simmons' citrate medium, and all that were citrate – failed to form indol (see text).

(1954) were members of the so-called Bethesda-Ballerup group, i.e., they were slow fermenters of lactose. Such selection was believed to account for the differences in percentages positive for hydrogen sulfide production, growth on Simmons' citrate medium, and indol production.

Data available in the literature indicate that the most commonly occurring bacteria of the genus *Citrobacter* are typical strains of *C. freundii* although biotypes of this species that are somewhat atypical occasionally are encountered as already mentioned. It is understandable that attention of workers should have been drawn to indol negative strains of *Citrobacter* that failed to ferment lactose or did so slowly, since colonies of these bacteria resemble salmonellae on most plating media used for isolation. Further, the reactions of such cultures on TSI agar or Kligler's medium resemble those of salmonellae. Since bacteria of the sort just mentioned frequently do not ferment lactose, sucrose, or salicin promptly, they often are confused with salmonellae. However, cultures of *Citro-*

bacter, irrespective of their avidity toward lactose, can be differentiated from members of the genera *Salmonella* and *Arizona* by means of the tests and substrates listed in Table 125. With few exceptions cultures of *C. freundii* can be differentiated without difficulty from members of other genera of Enterobacteriaceae (see tabular data, Chapter 3).

Davis and Ewing (1966) directed attention to the fact that no difficulty was encountered in the differentiation of the usual Voges-Proskauer positive culture of *Enterobacter cloacae* and a strain of *Citrobacter* even though their decarboxylase reactions may be the same. However, these investigators mentioned that they had studied cultures that resembled *E. cloacae* except that they gave negative results in the Voges-Proskauer test. It was stated that these cultures apparently occurred relatively uncommonly and that usually they could be differentiated from hydrogen sulfide negative cultures of *Citrobacter* when all of their biochemical reactions including utilization of adonitol were considered collectively and compared

TABLE 126

Differentiation of *Citrobacter, Salmonella,* and *Arizona*

Test or substrate	Citrobacter freundii			Salmonella enteritidis			Arizona hinshawii		
	Sign	%+	(%+)*	Sign	%+	(%+)*	Sign	%+	(%+)*
Urease	d	69.4	(6.9)	–	0		–	0	
KCN	+	96.2	(0.9)	–	0.1	(0.1)	–	8.7	
Gelatin (22 C)	–		(0.9)	–	0.4		(+)		(92)
Lysine decarboxylase	–	0		+	94.4	(0.1)	+	100	
Ornithine decarboxylase	d	17.2	(0.2)	+	100		+	100	
Lactose	(+) or +	39.3	(50.9)	–	0.3		d	61.3	(16.7)
Sucrose	d	15.3	(9.4)	–	0.2		–	4.7	
Dulcitol	d	59.8	(0.7)	+	97.7		–	0	
Inositol	–	3.3	(1.9)	d	43.8	(2)	–	0	
Malonate	d	21.8	(0.7)	–	1	(0.1)	+	92.6	(0.7)
Jordan's tartrate	+	100		d	84.2	(1)	–	5.3	
Beta galactosidase	+ or –	74.4		–	2.1		+	92.8	
D-tartrate**	(+)		(90.9)	+	92.3	(4)	(+) or –		(83.3)

+ 90% or more positive within 1 or 2 days.
– 90% or more, no reaction.
(+) Delayed positive 3 or more days.
d Different biochemical reactions, +, (+), –.
+ or –   Majority of strains positive, some cultures negative.
– or +   Majority of cultures negative, some strains positive.

*Figures in parentheses indicate percentages of delayed reactions (3 or more days).
**Method of Kauffmann and Petersen (see Chapter 18).

with those given by members of the genus *Citrobacter*. However, it was mentioned that in some instances this differentiation was difficult, particularly when the above-mentioned aberrant cultures failed to ferment adonitol. The author and colleagues continued to collect and study cultures of this sort and for a short time they were referred to as aberrant or atypical *E. cloacae* for want of a better designation, but this tentative designation soon was discarded.[2] The collection of these cultures numbers in excess of 200 and analysis of data regarding them indicates that several different bacterial types are represented (unpublished data). A few of these may be atypical strains of *Enterobacter*. However, the majority appear to compose two or three groups that could be placed in the genus *Citrobacter* but probably should not be included in the species *C. freundii*. If the above-mentioned groups of bacteria eventually are added to the genus *Citrobacter* as species other than *C. freundii*, it is probable that specific epithets employed by Werkman and Gillen (1932) could be used. Certainly epithets already extant in the literature should be employed if possible.

The author has been unable to obtain cultures of the bacteria described by Macierewitz (1966) under the name *Padewskia*, but judging from the data presented in the publication, the bacteria are similar to members of one of the above-mentioned groups characterized by the author and colleagues, and therefore could be added to the genus *Citrobacter*.

SEROLOGICAL REACTIONS

Intensive serological investigations have been made with only a portion of the genus *Citrobacter*, i.e., with the so-called Bethesda-Ballerup bacteria, which ferment lactose slowly or fail to utilize it. These bacteria resemble the more strictly characterized biotype 14011 of Stuart et al. (1943). The first detailed description of the serologic properties of an organism of this group was the report of Kauffmann and Moeller (1940) on the organism which they called *Salmonella ballerup*. This organism resembled members of the genus *Salmonella* in its fermentative properties and possessed an antigen related to the Vi antigen of *S. typhi*. The O and H antigens of the culture were not related to known salmonellae. Monteverde (1944) and Monteverde and Leiguarda (1944, 1945) studied cultures with similar O antigens which also contained a Vi-related antigen. These were called *S. hormaechei*. Harhoff (1949) investigated a group of strains that had related O antigens, and several of which fermented lactose.

Bruner et al. (1949) extended the work of Harhoff in a study of 45 cultures, the majority of which fermented lactose and all but two of which fermented salicin. Thus microorganisms of this sort became known as the Ballerup bacteria. The earliest investigation of the serological properties of the microorganisms that became known as the Bethesda bacteria was that of Edwards et al. (1948) who worked with cultures isolated by Barnes and Cherry (1946) in Bethesda, Maryland. Study of these strains, which were biochemically similar to the Ballerup bacteria, was continued by Moran and Bruner (1949). Finally, West and Edwards (1954), in the study of 506 cultures, demonstrated that many antigens were shared by the Bethesda and Ballerup bacteria and that the microorganisms actually comprised one group, the Bethesda-Ballerup bacteria.

In the study of 506 cultures, West and Edwards (1954) established 32 O antigen groups and 167 serologic types, which contained 75 distinct combinations of H antigens. An antigenic schema was established and the results correlated with the reports of other workers. The antigenic relationships of the Bethesda-Ballerup bacteria to salmonellae, and certain other Enterobacteriaceae also were investigated (v. inf.).

During their investigations West and Edwards (1954) examined the serological characteristics of 79 typical cultures of *C. freundii* as determined by their reactions in antisera for the O and H antigens they had delineated in the Bethesda-Ballerup bacteria (v. sup.). Lactose was fermented within 24 hr by 68 of the cultures, within 48 hr by seven strains, and two isolates required 72 hr to utilize the substrate. The results of serological examinations were as follows:

Serotype determined (O and H)	33	(41.8%)
O antigen (only) determined	14	(17.7%)
H antigen (only) determined	4	(5.1%)
Development of H antigen insufficient	4	(5.1%)
Neither O nor H determined	22	(27.8%)
Rough	2	(2.5%)
	79	(100%)

Seventeen of the typable cultures were serotypes that had not been encountered in the examination of the 506 cultures that fermented lactose slowly or failed to utilize it. One strain, a member of O group 29 possessed the antigen related to Vi antigen. These results led West and Edwards (1954) to conclude that there seemed to be no correlation between rapidity of lactose utilization and serological characteristics among the cultures examined.

Using the antisera employed by West and Edwards (1954), Davis (unpublished data, 1964) obtained the following results with 368 cultures of *Citrobacter*

---

[2] The name *Citrobacter diversus* has been proposed for this bacteria. (See Ewing and Davis, 1971.)

taken at random (i.e. without respect to reactions in lactose medium);

51 serotypes determined (O and H)	146	(39.7%)
O antigen determined, H undetermined	89	(24.2%)
O antigen undetermined, H determined	32	(8.7%)
O undetermined, H undetermined	101	(27.4%)
Total	368	(100%)

It may be noted that it was possible to determine the O antigen group of 235 or 63.9 per cent of the above-mentioned cultures. Sakazaki et al. (1960) employed the antisera described by West and Edwards (1954) and determined the O antigens of 445 (66.5 per cent) and the H antigens of 493 (73.7 per cent), respectively, of 669 strains of *Citrobacter*. They were able to serotype (O and H antigens determined) 382 or 57.1 per cent of the 669 cultures. Seventy-three serotypes occurred among the above-mentioned 382 strains.

Sedlak and Slajsova (1962) reported that they were able to determine the O antigen group of about 70 percent of 3,586 cultures of *Citrobacter* with 35 antisera. These investigators (1966) added ten provisional O antigen groups and with the 42 O antisera they were able to determine the O group of more than 90 per cent of 4,378 strains of *Citrobacter* isolated over a period of 15 years. The standard or test strains of five of the ten O antigens groups added to the schema by Sedlak and Slajsova (1966), were cultures that had been characterized as members of other genera but which had been removed from those genera. Strain number 2624-36 (*S. coli* 1) originally was classified in the antigenic schema for *Salmonella* because of its antigenic relationship to members of that genus; this became the standard culture for *Citrobacter* O group 37. Four of the cultures originally had been designated as standard strains for *E. coli* O groups 67, 72, 94, and 122, but were deleted (see Chapter 5). These strains were added to the antigenic schema for *Citrobacter* as O antigen groups 33, 34, 35, and 36, respectively. The remaining five O antigen groups were characterized among other cultures of *Citrobacter* examined by Sedlak and coworkers.

Since it is possible to determine the serotype of only a relatively small percentage of all strains of *Citrobacter* with the O and H antisera presently available, it does not seem appropriate to present the antigenic schema of West and Edwards (1954). Interested readers are referred to the original publication of West and Edwards for the schema and for methods of preparation of the absorbed O and H antisera required for serotyping.

In the examination of cultures on TSI agar medium in polyvalent antiserum for salmonellae, it should be borne in mind that certain serotypes of *Citrobacter* that belong to O antigen groups 5, 12, and 29 possess antigen related to the Vi antigen of *S. typhi*. Living (unheated) suspensions prepared from strains of *Citrobacter* that contain this antigen may be expected to react in most polyvalent antisera for *Salmonella* because agglutinins for the Vi-related antigen of *Citrobacter* are represented. Also O antigen suspensions prepared from some cultures of *Citrobacter* react in polyvalent antisera for salmonellae (v. inf.).

INTERGENERIC ANTIGENIC RELATIONSHIPS

O Antigenic Relationships. The O antigens of many serotypes of *Citrobacter* are related to those of many *Salmonella, Arizona, Escherichia,* etc. Mustakallio (1944) reported upon the relationship of the O antigens of strains that now would be called O group 29 of *Citrobacter* and O antigens 1,3,19 of salmonellae, and West and Edwards (1954) examined the relationships of the characterized O antigens of *Citrobacter* and those of *Salmonella, Arizona,* and *E. coli*. Although strong reciprocal, as well as less extensive minor, relationships were detected, reciprocal agglutinin absorption tests were not done. The interrelationship of the O antigens of *Citrobacter, Arizona,* and *Salmonella* mentioned by West and Edwards (1954) are depicted in Tables 127 and 128 in terms of reactions that may be expected in slide agglutination tests.

Edwards et al. (1954) described the O and H antigenic relationships of a culture (4598-53) of *C. freundii*. The O antigens of this culture were very closely related to those of O antigens 3,15 of salmonellae. Sakazaki and Namioka (1957) described eleven cultures of *C. freundii* which were reported to have the complete O antigen complex 3,10 of salmonellae (see also Table 128). Sedlak and Slajsova (1966) reported that the O antigens of several of their newly established provisional O antigen groups were related to those of salmonellae, as follows:

Citrobacter	Related Antigen of *Salmonella*
O37	O48
O39	O3,10
O40	O57
O41	O55
O42	O54

In a preliminary investigation Bartes et al. (unpublished data, 1969) detected reciprocal relationships between the O antigens of *Citrobacter* cultures submitted for identification and the following O antigens of salmonellae: O1,6,7,9,10,11,15,20,27,30,35,37,47b, 50,51,55, and 64. Some of these relationships extend to the O antigens of *Arizona,* e.g., O1,2,9,20,23, and 29.

TABLE 127

The Reactions of O Antigens of *Citrobacter* in Antisera for *Salmonella* and *Arizona** (Slide Tests)

O Antigens	O Antisera											
	Salmonella											
Citrobacter	**Arizona**				12	7				10		8
	4,5,12	6,7	6,14,24	(1),6,14,25	17	18	(8),20	30	38	40	1,2	8
9a,9b	–	–	–	–	–	–	–	++++	–	–	–	–
11	–	–	–	–	–	–	–	–	–	++	–	–
12a,12c	–	–	–	–	–	–	–	–	–	–	++	–
14	–	–	–	–	–	–	+	–	++++	–	–	–
20	–	–	–	–	++++	–	–	–	–	–	–	–
21a,21b	–	–	+	–	–	–	–	–	–	–	–	–
22	++++	–	–	–	–	–	–	–	–	–	–	–
23	–	–	–	–	–	++	–	–	–	–	–	–
28,1c	–	++++	–	++++	–	–	++++	–	–	–	–	–
32	–	–	–	–	–	–	–	–	–	–	–	++

– No agglutination.

+ to ++++, degree of agglutination

* From West and Edwards (1954).

TABLE 128

The Reactions of O Antigens of *Salmonella* and *Arizona* in Antisera for *Citrobacter** (Slide Tests)

O Antigens		O Antisera (*Citrobacter*)													
Salmonella	*Arizona*	3a,3b,1c	9a,9b	11	12a,12b	12a,12c	19	20	21a,21b	22	23	24	26	31	32
1,2,12	–	–	–	++	–	–	–	–	++	–	–	–	–	–	–
4,5,12	–	–	–	–	–	–	–	–	–	++++	–	–	–	–	–
1,4,27,12	–	–	–	–	–	–	–	–	–	++++	–	–	–	–	–
6,7	27	–	–	–	–	–	–	–	+	–	–	–	–	–	–
6,8	–	–	–	–	–	–	–	–	+	–	–	–	–	–	–
9,12	–	–	–	–	–	–	+	–	–	–	–	–	–	–	–
3,10	–	++	–	++	–	–	–	–	–	–	–	–	–	–	–
6,14,24	32	–	–	–	–	–	–	–	++++	–	–	–	–	++++	–
(1),6,14,25	–	–	–	–	–	–	++++	–	–	–	–	–	–	–	–
17	12	–	–	–	–	–	–	++++	–	–	–	–	–	–	–
18	7	–	–	–	–	–	–	–	–	–	++	–	–	–	–
21	22	–	–	–	–	–	–	–	–	–	–	–	++++	–	–
28	–	–	–	–	++	+	++	–	–	–	–	–	–	–	–
30	–	–	++++	–	–	–	–	–	–	–	–	–	–	–	–
40	10	–	–	++++	–	–	–	–	–	–	–	–	–	–	–
41	13	++++	–	–	–	–	–	–	–	–	–	–	–	–	–
	1,2	–	–	–	–	–	–	–	–	–	–	–	–	–	–
	1,3	–	–	–	–	–	–	–	–	–	–	–	–	–	+
	8	–	–	–	–	–	–	–	–	–	–	++	–	–	–
	11	–	–	–	–	–	–	–	–	–	–	–	–	–	–

– No agglutination.

+ to + + + + , degree of agglutination.

* From West and Edwards (1954).

The antigenic relationships between cultures of *Citrobacter* and *E. coli* as described by West and Edwards (1954) and by Davis and Ewing (1963) are shown in Table 129. Since the O antigens of some strains of *Citrobacter* (e.g., 1431-68) are related to O50 of salmonellae and 09 of arizonae, it is probable that they also are related to *E. coli* 055 and *Providencia* 06 (Chapter 17).

H Antigen Relationships. West and Edwards (1954) detected no relationships between the H antigens of the strains of *Citrobacter* that they studied and those of salmonellae. However, Edwards et al. (1954) reported that the H antigens of culture 4598-53 (v. sup.) were related to antigen z_{36} of serotypes Weslaco

and Mcallen. These relationships are given in Table 130. The data indicate that the H antigens of culture 4598-53, ser Weslaco, and ser Mcallen all are related reciprocally, but none is identical with either of the others. Further, it is clear that the z_{36} antigen of ser Weslaco is not identical with that of ser Mcallen (Table 130).

During the above-mentioned preliminary studies Bartes et al. (unpublished data, 1969) found a number of cultures of *Citrobacter* that possessed H antigens related to H z_{36} of salmonellae (antiserum z_{36} derived from ser Weslaco). In addition these investigators encountered strains of *Citrobacter* the H antigens of which were related to H antigens z_{27} and z_{46} of salmonellae and H antigens 17,20, and 29 of *Arizona*.

TABLE 129

Relationship of the O Antigens
of *Citrobacter* and *E. coli*

Citrobacter O Antigen	*E. coli* O Antigen
1a,1b,1c	9*
7,3b,1c	99*
8a,8b	93*
9a,9b	7*
10a,9b	71
12a,12c	38*
15	57*,65
17	101*
18	1,15*,100
21	73*
26	76*
28,1c	9*,44,73
29	53,62,78
31	52
Citrobacter culture No.	
777	26*
126	125a,125b*
6604	127a*
4473	128a,128b*

* Reciprocal relationships

Adapted from West and Edwards (1954) and Davis and Ewing (1963).

TABLE 130

H Antigens of *C. freundii*, 4598-53*

H Antisera	H Antigens		
	4598-53	Weslaco (z_{36})	Mcallen (z_{36})
4598-53			
Unabsorbed	16,000	1,000	4,000
Absorbed by ser			
Weslaco (z_{36})	8,000	0	500
Mcallen (z_{36})	4,000	0	0
ser Weslaco			
Unabsorbed	2,000	4,000	2,000
Absorbed by			
4598-53	0	2,000	1,000
Mcallen	0	1,000	0
ser Mcallen			
Unabsorbed	8,000	4,000	8,000
Absorbed by			
4598-53	0	2,000	4,000
Weslaco	500	0	4,000

*Adapted from Edwards et al. (1954).

With the exception of the relationship to H antigen z_{27}, all of these were extensive reciprocal relationships.

PRODUCTION OF ANTISERA

The methods employed for production of O and H antisera for salmonellae or for *E. coli* may be used with members of the genus *Citrobacter*. Also, the technique described in Chapter 8 may be employed for preparation of single factor O and H antisera. True phase variation of Andrewes is not known to occur in serotypes of *Citrobacter*. However, it is known that a variation takes place in the H antigens of some serotypes which results in a segregation of the antigens (see Chapter 4). Some serotypes of *Citrobacter* possess an alpha antigen related to that described by Stamp and Stone (1944). This fact should be borne in mind when selecting strains for antiserum production, particularly H antisera.

Polyvalent Antisera. Galton and Stuart (1949) prepared a pooled polyvalent antiserum with antisera produced with six serotypes which occurred commonly among materials examined. This pooled antiserum was found to be useful for preliminary examination of cultures although it was reported that many salmonellae were agglutinated by it.

West and Edwards (1954) produced a more comprehensive polyvalent antiserum in which all of the O and H antigens recognized in the 1954 schema were represented. This antiserum was difficult to produce and had distinct disadvantages, since many cultures of commonly occurring serotypes of *Salmonella* as well as strains of the genera *Arizona* and *Escherichia* were agglutinated by it. It was clear that members of *Citrobacter* O antigen groups 22 and 28 possessed strong antigenic relationships to commonly occurring salmonellae of serogroups B and C_1 (Tables 127 to 129, incl.), and that the majority of the intergeneric relationships involved higher O groups in the 1954 schema for *Citrobacter* (12 to 32, incl.). Further, it was estimated that more than 90 per cent of the strains examined by West and Edwards (1954) were members of *Citrobacter* O antigen groups 1 to 9, inclusive, and 29, which is not known to be shared by salmonellae.

Therefore, careful selection of members of these serogroups for production of polyvalent antiserum should result in an antiserum of practical value, particularly since it is possible to select cultures that possess H antigens that occur in serotypes of many O groups.

Edwards and colleagues obtained good results with a polyvalent antiserum produced with the following cultures:

Culture	O antigens	H antigens
Na 1a	1a,1b,1c	1,2
Na 4	2a,1b	7,(8),10
Md 10	2a,1b	21,22
Na 11	3a,3b,1c	8,9
Mich 7	3a,3b,1c	(13),17
Mich 1	4a,4b	(9),29,31
Md 2	7,3b,1c	(9),13,14
Md 1	7,3b,1c	(21),25,27
5131/51	8a,8c	5,6
Mich 11	8a,8c	32,33
3038	9a,9b	4,5
Br. Univ.	9a,9b	39
LC 54	29	74,75

The cultures should be passed repeatedly through semisolid medium until maximum motility is attained. Infusion broth cultures of the motile forms are incubated overnight and killed by addition of 0.6% formalin. The broth cultures are mixed in equal amount and used to inject rabbits in the usual manner. Six to eight injections should be administered. It probably will be found helpful to add alcohol-acetone treated cells (Roschka, 1950; Edwards, 1951) to the vaccine in order to assure adequate titres of O agglutinins. These cells should be added to the formalinized broth cultures immediately before injection. Vaccine may be administered in amounts of 0.3, 0.5, 1.0, 2.0, 3.0 and 3.5 ml at four-day intervals. Trial bleedings should be taken during the latter part of the course of immunization and the vaccine should be reinforced as indicated. Rabbits should be bled on the sixth or eighth day after the final injection. The serum may be preserved with an equal volume of glycerol or with another preservative. The serum should be carefully titrated with the various O and H components to determine the maximum dilution in which it can be used to detect all the injected antigens.

In practice the above-mentioned polyvalent antiserum may be used as an O antiserum in slide tests and in tube agglutination tests at a suitable dilution for detection of H antigens. For preliminary examination of cultures it is advisable to employ this antiserum in conjunction with polyvalent antisera for salmonellae and arizonae. The author and colleagues use all of the above-mentioned polyvalent antisera in tube agglutination tests for detection of H antigens only. Because of the community of antigens among Enterobacteriaceae it may be expected that some cross agglutination reactions will occur. Further, it is clear that by no means all cultures of *Citrobacter* are agglutinated by the above-mentioned polyvalent antiserum for *Citrobacter*.

REFERENCES

Asai, T. 1957. J. Gen. Appl. Microbiol. (Japan), **3**, 13.

Barnes, L. A., and W. B. Cherry. 1946. Amer. J. Pub. Health, **36**, 481.

Borman, E. K., C. A. Stuart, and K. M. Wheeler. 1944. J. Bacteriol., **48**, 351.

Bruner, D. W., P. R. Edwards, and A. S. Hopson. 1949. J. Infect. Dis., **85**, 290.

Carpenter, P. L., and M. Fulton. 1937. Amer. J. Pub. Health, **27**, 822.

Davis, B. R., and W. H. Ewing. 1963. Amer. J. Clin. Pathol., **39**, 198.

————. 1964. J. Bacteriol., **88**, 16.

————. 1966. The biochemical reactions of *Citrobacter freundii*. CDC Publ.*

Edwards, P. R. 1951. Pub. Health Repts., **66**, 837.

Edwards, P. R., M. G. West, and D. W. Brunner. 1948. J. Bacteriol., **55**, 711.

Ewing, W. H. 1967. Revised definitions for the family Enterobacteriaceae, its tribes and genera. CDC Publ.*

Ewing, W. H., and B. R. Davis. 1971. Biochemical characterization of *Citrobacter freundii* and *Citrobacter diversus*. CDC Publ.*

Ewing, W. H., B. R. Davis, and R. W. Reavis. 1957. Pub. Health Lab.,**15**, 151.

Ewing, W. H., and J. G. Johnson. 1960. Int. Bull. Bacteriol. Nomen. Tax., **10**, 223.

Galton, M. M., and C. A. Stuart. 1949. J. Bacteriol., **58**, 545.

Harhoff, N. 1949. Acta Pathol. Microbiol. Scand., **26**, 167.

International Enterobacteriaceae Subcommittee Report, Appendix 3. 1958. Int. Bull. Bacteriol. Nomen. Tax., **8**, 179.

Kauffmann, F. 1956a. Acta Pathol. Microbiol. Scand., **39**, 85.

————. 1956b. Ibid., **39**, 103.

Kauffmann, F., and E. Moeller. 1940. J. Hyg., **40**, 246.

Macierewitz, M. 1966. Med. Dosw. I. Microbiol., **18**, 337.

*CDC Publ., Publication from the Center for Disease Control (formerly the Communicable Disease Center), Atlanta, Ga. 30333.

**Public Health Laboratory: Journal of the Conference of Public Health Laboratory Directors.

Martin, W. J., and W. H. Ewing. 1967. Can. J. Microbiol., **13**, 616.

Moeller, V. 1954a. Acta Pathol. Microbiol. Scand., **34**, 115.

————. 1954b. Ibid., **35**, 259.

————. 1955. Ibid., **36**, 158.

Monteverde, J. J. 1944. Rev. Facultat Agron. Vet., Buenos Aires, **2**, 1.

Monteverde, J. J., and R. H. Leiguarda. 1944. Bol. Admin. Nac. de Agua, Buenos Aires, **87**, 168.

————. 1945. Ibid., **91**, 24.

Moran, A. B., and D. W. Bruner. 1949. J. Bacteriol., **58**, 695.

Mustakallio, E. 1944. Acta. Soc. Med. Fennicae, Duodecim. Ser. A., **24**, 131.

Roschka, R. 1950. Klin. Med., **5**, 88.

Sakazaki, R., and S. Namioka. 1957. Bull. No. 32, Nat. Inst. Animal Health, Tokyo.

Sedlak, J, and M. Slajsova. 1962. Wiss. Z. Karl-Marx Univ. Leipzig., **11**, 748.

————. 1966. J. Gen. Microbiol., **43**, 151.

Slopec, S., and L. Dabrowski. 1952. Arch. Immunol. Terakii Dosw., **6**, 103.

Stamp, L., and D. M. Stone. 1944. H. Hyg., **43**, 266.

Stuart, C. A., K. M. Wheeler, R. Rustigian, and A. Zimmerman. 1943. J. Bacteriol., **45**, 101.

Tittsler, R. P., and L. A. Sandholzer. 1935. J. Bacteriol., **29**, 349.

Vaughn, R. H., and M. Levine. 1942. J. Bacteriol., **44**, 487.

Washington, J. A., P. Yu, and W. J. Martin. 1969. Appl. Microbiol., **17**, 843.

————. 1970. Ibid., **20**, 587.

West, M. G., and P. R. Edwards. 1954. The Bethesda-Ballerup group of paracolon bacteria. U. S. Pub. Health Serv., Monograph No. 22, Govt. Print. Off., Washington, D. C.

Werkman, C. H., and G. F. Gillen. 1932. J. Bacteriol., **23**, 167.

Chapter 12
The Genus *Klebsiella*

The genus *Klebsiella* is composed of nonmotile bacteria that typically do not produce hydrogen sulfide, yield negative methyl red and positive Voges-Proskauer reactions, and do not possess a phenyl-alanine deaminase. When they occur urease reactions are slower and less intense than those exhibited by members of the genus *Proteus*. Some cultures of *Klebsiella* are methyl red positive and give negative reactions in the Voges-Proskauer test (v. inf.). However, strains of this sort do not occur as commonly as the more typical forms. Moreover, they generally belong to certain capsule types, which traditionally have been included in the genus *Klebsiella,* and rightly so, if their entire biochemical behavior is considered.

The following definition of the genus *Klebsiella* has been adopted for use herein (Ewing, 1967):

The genus *Klebsiella* is composed of nonmotile bacteria that conform to the definitions of the family ENTEROBACTERIACEAE and the tribe KLEBSIELLEAE. The Voges-Proskauer test is positive, gelatin is not liquefied. Lysine decarboxylase is produced, but arginine dihydrolase and ornithine decarboxylase are not. The majority of cultures utilize sodium alginate as a sole source of carbon and esculin is hydrolyzed. Gas is formed from inositol and glycerol, and by the majority of strains from adonitol. Acid is produced from sorbitol, rhamnose, arabinose, and raffinose. The type species is *Klebsiella pneumoniae* (Schroeter) Trevisan.

This definition should be used in conjunction with the definition for the tribe Klebsielleae given in Chapter 1.

The literature on the subjects of nomenclature and taxonomy of the genus *Klebsiella* has been reviewed by a number of investigators and need not be discussed in detail here. Interested readers are referred to Chapter 1 and to Winslow et al. (1919), Weldin (1927), Sedlak and Slajsova (1959), Cowan et al. (1960), Grimes (1961), Ewing et al. (1962), and Ewing (1963). In addition, this subject was discussed at some length in the second edition of this book, but the need for that sort of review no longer seems to exist.

Three species of *Klebsiella* are recognized: *K. pneumoniae, K. ozaenae,* and *K. rhinoschleromatis,* and isolants are identified and classified by means of their biochemical reactions regardless of the sources from which they are isolated. Parenthetically, typing of the capsular antigens of members of all three species is accomplished with one set of antisera.

BIOCHEMICAL REACTIONS

The biochemical reactions given by members of the three species of *Klebsiella* are recorded in some detail in Table 131. These results are based upon data obtained from the examination of 844 cultures of *Klebsiella* (Fife et al., 1965). Further, tests that are of particular value for differentiation of the three species are listed in Table 132.

K. pneumoniae. It is notable that 42 (6 per cent) of 705 cultures of *K. pneumoniae* produced indol and that 23 (3.3 per cent) liquefied gelatin (Table 131). Closer examination of results obtained in these two tests revealed the following relationships:

	Indol+	Indol-	Total
Gelatin +	22	1	23
Gelatin -	20	662	682
Total	42	663	705

The 22 cultures that liquefied gelatin and produced indol are regarded as a biotype of *K. pneumoniae*. If desired such strains could be referred to as the oxytoca biotype. The only essential difference between the oxytoca cultures and other indol-positive klebsiellae is gelatin liquefaction. Therefore it is believed that the oxytoca cultures are not sufficiently different to warrant status as a separate species as suggested by some writers. For additional information regarding this subject readers are referred to Lautrop (1956), Ørskov (1955a), Hugh (1959), Cowan et al. (1960), and Ewing (1963).

The decarboxylase reactions given by the above-

TABLE 131

Summary of the Biochemical Reactions Given by *Klebsiella*

est or substrate	K. pneumoniae			K. ozaenae			K. rhinoschleromatis		
	Sign	%+	(%+)*	Sign	%+	(%+)*	Sign	%+	(%+)*
ydrogen sulfide	–		(4)	–	0		–	0	
rease	+	94.5		d	9.5	(10.3)	–	0	
dol	–	6		–	0		–	0	
ethyl red 37 C	– or +	13.3		+	99.1		+	100	
oges-Proskauer 37 C	+	91.1		–	0		–	0	
trate (Simmons')	+	97.7		d	31.9	(31)	–	0	
CN (growth)	+	97.9		+ or –	88		+	100	
otility	–	0		–	0		–	0	
elatin 22 C	–	3.3		–	0		–	0	
sine decarboxylase	+	97.2	(2.8)	– or +	48		–	0	
rginine dihydrolase	–	0.9		–	6		–	0	
rnithine decarboxylase	–	0		–	4		–	0	
enylalanine deaminase	–	0		–	0		–	0	
ucose:									
acid	+	100		+	100		+	100	
gas	+	96.5		d	64	(2)	–	0	
ctose	+	98.2	(1.4)	d	24.1	(70.7)	(+) or –		(72.8)
crose	+	98.9		d	16.3	(17.3)	+ or (+)	68.2	(31.8)
annitol	+	100		+	100		+	100	
lcitol	– or +	31.5		–	0		–	0	
licin	+	99.7	(.3)	+	97.4	(2.6)	+	100	
lonitol:									
acid	+ or –	87.7		+	98.3	(1.7)	+	100	
gas	d	83.4	(.3)	d	60	(2)	–	0	
ositol:									
acid	+	97.9	(.8)	d	58.6	(21.6)	+	95.5	(4.5)
gas	+	91.9	(2)	d	28	(8)	–	0	
rbitol	+	99.4	(.3)	d	78	(10)	+	100	
abinose	+	99.9		+	100		+	100	
ffinose	+	99.7		+	90		+ or (+)	68.2	(31.8)
amnose	+	99.3	(.4)	d	60	(8)	+	95.5	(4.5)
lonate	+	92.5		–	4		+	95.5	
cate	+	92.8		– or +	24		–	0	

TABLE 131 (Continued)

Summary of the Biochemical Reactions Given by *Klebsiella*

Test or substrate	K. pneumoniae			K. ozaenae			K. rhinoschleromatis		
	Sign	%+	(%+)*	Sign	%+	(%+)*	Sign	%+	(%+)*
Nitrite from nitrate	+	99.9		+	92		+	100	
Organic acids**									
Citrate	+ or –	64.4		– or +	18		–	0	
D-tartrate	+ or –	67.1		– or +	36		–	0	
Jordan's tartrate	+	94.4		+ or –	56.9		– or +	36.4	
Sodium alginate, synthetic	+ or –	88.5	(9.2)	– or (+)		(37.5)	–	0	
nutrient	–	0		–	0		–	0	
Lipases:									
corn oil	–	0		–	0		–	0	
Triacetin	–	1.1		–	0		–	0	
Tribuyrin	–	0		–	0		–	0	
Pectate	–	0		–	0		–	0	
Starch	+ or (+)	71.6	(28.1)	d	34.3	(59.3)	+w	100	
Xylose	+	99.5	(.4)	+	92	(6)	+	100	
Cellobiose:									
acid	+	99.9	(.1)	+	98		+	100	
gas	+	95.7	(.3)	+ or –	70		–	0	
Glycerol:									
acid	+	97.2	(2.7)	d	68.2	(21.5)	+	100	
gas	+	92.5	(3.7)	d	30	(24)	–	0	
Alpha methyl glucoside	+ or (+)	86.3	(13.7)	+ or (+)	75	(25)	–	0	
Erythritol	–	0		–	0		–	0	
Esculin	+	98.9	(1.1)	+ or –	75		– or +	15.4	

+ Positive within 1 or 2 days' incubation.

(+) Positive reaction after 3 or more days.

– No reaction.

+ or –   Majority of strains positive, some cultures negative.

– or +   Majority of cultures negative, some strains positive.

(+) or +   Majority of reactions delayed, some occur within 1 or 2 days.

d Different reactions: +, (+), –.

*Numerals in parentheses indicate percentage of delayed reactions (3 or more days).

**Method of Kauffmann and Petersen, 1956.

TABLE 132

Tests of Value in Differentiation of the Three Species of *Klebsiella*

Test or substrate	K. pneumoniae			K. ozaenae			K. rhinoschleromatis		
	Sign	%+	(%+)*	Sign	%+	(%+)*	Sign	%+	(%+)*
Urease	+	94.5		d	9.5	(10.3)	–	0	
Methyl red	– or +	13.3		+	99.1		+	100	
Voges-Proskauer	+	91.1		–	0		–	0	
Citrate (Simmons')	+	97.7		d	31.9	(31)	–	0	
Organic acids:**									
citrate	+ or –	64.4		– or +	18		–	0	
D-tartrate	+ or –	67.1		– or +	36		–	0	
Malonate	+	92.5		–	4		+	95.5	
Mucate	+	92.8		– or +	24		–	0	
Lysine decarboxylase	+	97.2		– or +	48		–	0	
Gas from glucose	+	96.5		d	64	(2)	–	0	
Lactose	+	98.2	(1.4)	d	24.1	(70.7)	(+) or –		(72.8)
Dulcitol	– or +	31.5		–	0		–	0	

   + Positive (90% or more) within 1 or 2 days' incubation.
(+) Positive reaction after 3 or more days.
  – No reaction (90% or more).
+ or – Majority of strains positive, some cultures negative.
– or + Majority of cultures negative, some strains positive.
(+) or – Majority of reactions delayed, some occur within 1 or 2 days.
  d Different reactions: +, (+), –.

*Numerals in parentheses indicate percentage of delayed reactions (3 or more days)

mentioned 705 cultures of *K. pneumoniae* may be summarized as follows:

No.	Lysine	Arginine	Ornithine
683	+	–	–
2	+	+	–
4	–	+	–
16	–	–	–

The patterns of reactions given by 705 strains of *K. pneumoniae* in certain differential tests are given in Table 133. The data given in this table should be useful when used in conjunction with that given in Tables 7, 18, and 22 (Chapter 3). It may be added that the majority of aberrant cultures of *K. pneumoniae* belong to capsule types 1 and 2 (e.g., see Epstein, 1959).

K. ozaenae. The results obtained by Fife et al. (1965) in a study of the biochemical reactions of 117 strains

of *K. ozaenae* are recorded in Table 131 and means by which this species may be differentiated from the other species of *Klebsiella* are listed in Table 132. It must be said that cultures of *K. ozaenae* are more likely to yield aberrant biochemical reactions than strains of *K. pneumoniae*. Nevertheless cultures of *K. ozaenae* generally are recognized easily and can be differentiated from other klebsiellae and other Enterobacteriaceae by means of the biochemical reactions listed in the aforementioned tables. Of some assistance in this regard is the fact that almost all strains of *K. ozaenae* are members of capsule type 4. However, a few cultures belong to capsule types 5 and 6.

K. rhinoschleromatis. This microorganism occurs rarely in the United States in the experience of the author and colleagues. The majority of the 22 cultures examined by Fife et al. (1965) originated in other countries. However, the biochemical reactions given by

TABLE 133

Patterns of Reactions Given by Cultures of *K. pneumoniae* in Certain Tests

No. of cultures in each group	Decarboxylases			Gas from:			Gelatin
	Lysine	Arginine	Ornithine	Inositol	Adonitol	Glycerol	
530	+	−	−	+	+	+	−
88	+	−	−	+	−	+	−
3	+	−	−	+	+	−	−
14	+	−	−	−	+	+	−
2	+	−	−	−	−	+	−
22	+	−	−	−	−	−	−
22	+	−	−	+	+	+	+
1	+	−	−	−	−	−	+
4	−	+	−	+	+	+	−
2	+	+	−	+	+	+	−
12	−	−	−	+	+	+	−
1	−	−	−	+	−	+	−
1	−	−	−	−	−	+	−
2	−	−	−	−	+	+	−
1	−	−	−	−	−	−	−

these cultures (Tables 131, 132) are quite uniform and the reactions recorded are similar to those reported by others (e.g., Ørskov, 1955a). Cultures of *K. rhinoschleromatis* usually are members of capsule type 3, but all type three strains are not *K. rhinoschleromatis*.

Since cultures of *K. ozaenae* and *K. rhinoschleromatis* yielded negative results in the Voges-Proskauer test, they must be differentiated from other members of the tribe Klebsielleae that may yield similar results in this test (e.g., *Enterobacter hafniae, Enterobacter liquefaciens, Pectobacterium,* and *Serratia marcescens* subspecies *kiliensis*). These differentiations generally may be accomplished without difficulty if investigators compare the data given in Table 131 with that recorded for the above-mentioned bacteria (see Chapters 13, 14, and 15).

In the experience of the author and colleagues cultures of *K. pneumoniae* occur much more commonly in the United States than members of the other two species of klebsiellae (Fife et al., 1965). For this reason the reactions of *K. pneumoniae* only are given in several tables of differential reactions (e.g., Tables 4, 7, and 18, Chapter 3). However, the fact that strains of *K. ozaenae* are not rare should be borne in mind.

For additional data concerning the biochemical behavior of cultures of *Klebsiella* and for further references, the reader is referred to Moeller (1954), Edwards and Fife (1955), Ørskov (1955a), Kauffmann (1954, 1956a, 1956b), Lautrop (1956), Hormaeche and Munilla (1957), Sakazaki and Namioka (1957), Hormaeche and Edwards (1958), Hugh (1959), Sedlak and Slajsova (1959), Cowan et al. (1960), Grimes (1961), and Fife et al. (1965).

SEROLOGICAL REACTIONS

A notable characteristic of members of the genus *Klebsiella* is the production of large, moist colonies which often are very mucoid. Examination of the microorganisms in moist India ink mounts usually reveals that prominent capsules are present.

The writer and colleagues are indebted to Dr. J. Howard Brown for pointing out the eminent suitability of Pelikan India Ink, manufactured by Gunther Wagner, Hannover, in the preparation of moist mounts for the determination of capsules. The carbon particles of this ink are very finely divided and it is free of bacteria. In the examination, a drop of young (4-6-hr) broth culture or a drop of dilute saline suspension from a colony should be placed on a slide alongside a droplet of ink. The two should be joined with a loop and, without mixing, covered with a coverslip. Different degrees of density of the ink will result and some portion of the slide will be of the proper density for examination. This method has been used for many years with excellent results.

Toenniessen (1914, 1921) was first to demonstrate that two distinct antigens were present in capsulated strains of *Klebsiella,* one in the capsule and the other in the soma. Further, the capsular antigen was found by Toenniessen to be carbohydrate in nature. Julianelle (1926) established three serological (capsule) types among klebsiellae, which he labeled A, B, and C. In addition there was a heterologous group labeled X. Goslings and Snijders (1936) extended the work of Julianelle and established three additional types, D, E, and F, among cultures from ozaena. The types of Julianelle and of Goslings and Snijiders were based entirely upon capsular antigens. There was disagreement among these workers and others who investigated the somatic antigens of the bacteria. In general, it may be said that the somatic antigens did not exhibit the specificity displayed by the capsular antigens and were not as useful in typing as were the latter. Viewed in retrospect, it is not clear whether Julianelle, Goslings and Snijders, and Elbert and Gerkess (1929) were dealing with smooth acapsular forms or with rough variants, and no orderly arrangement of the O antigens of the group was achieved until they were investigated by Kauffmann (1949) and Ørskov (1954). For many years the capsular antigens were the sole basis for typing the bacteria and capsule typing still remains the only method in general use.

Capsular Antigens. As stated above, the capsular and somatic antigens of klebsiellae long have been known but it remained for Kauffmann (1949, 1966) to set forth clearly the antigenic and cultural forms which occurred in the bacteria. He depicted these forms as follows:

I Smooth Forms

   1. MKO form — mucoid, capsulated with O antigen
   2. KO form — nonmucoid, capsulated with O antigen
   3. MO form — mucoid, noncapsulated with O antigen
   4. O form — nonmucoid, noncapsulated with O antigen

II Rough Forms

   1. MKR form — mucoid, capsulated, without O antigen
   2. KR form — nonmucoid, capsulated, without O antigen
   3. MR form — mucoid, noncapsulated, without O antigen
   4. R form — nonmucoid, noncapsulated without O antigen

This tabulation requires some explanation. It is more than an exposition of the principal antigens of the bacteria since it expresses their form of growth as well. For instance, Edwards and Fife (1952) found that the M antigen and the K antigen apparently were the same. This conclusion was confirmed by Wilkinson et al. (1954). It is extremely difficult to draw a definite line between distinct capsules (K of Kauffmann) and bacterial slime (M of Kauffmann). In a given culture the two substances are serologically related if not identical. Some cultures produce distinct capsules which have little tendency to disintegrate and become dissipated into the surrounding medium. Other strains have little tendency to produce capsules but do produce large amounts of loose slime which is distributed throughout the menstruum. These two extremes of behavior may be found in variants of the same culture. Between these two extremes one finds the usual culture of *Klebsiella* that produces a distinct capsular or slime layer which is slowly liberated into the suspending medium. The conditions that influence capsule and slime production in *Klebsiella* have been studied by Duguid (1948, 1951) and Duguid and Wilkinson (1953). In general, it may be said that the tendency to produce capsules or slime is a characteristic of the individual strain or variant thereof and, further, that this tendency is influenced by the environment. A high carbohydrate to nitrogen ratio in the medium encourages the production of the polysaccharide substances.

Capsulated forms (MKO, KO, MKR, and KR of Kauffmann) are not acted upon by somatic (O or R) agglutinins, just as Vi cultures of *S. typhi* are resistant to O agglutination. On the contrary the noncapsulated forms (MO, O, MR, and R of Kauffmann) are susceptible to the action of their corresponding somatic agglutinin. Rough forms occur with surprising frequency in cultures of *Klebsiella,* even among recently isolated strains. The occurrence of rough forms is not easily detectable in capsulated or mucoid, noncapsulated cultures since colonies of these forms have a smooth appearance. The true nature of a strain (i.e., rough or smooth) can be seen with ease only when nonmucoid, noncapsulated forms are present. In these forms roughness can be detected by colony appearance and by the usual criteria of roughness (Chapter 4).

O ANTIGENS. The O antigens of cultures of *Kleb-*

siella are difficult to determine because treatment to rid cells of capsular material frequently renders them rough. Passage once or twice a week on 50% bile broth sometimes yields O forms which may be used for production of antisera. Kauffmann (1949) characterized O antigens 1, 2, and 3 in *Klebsiella* and Ørskov (1954) added O groups 4 and 5. O antigen group 2 may be subdivided into a number of fractions (2a, 2b, 2c etc.), but some of these are not always distinct. Most of the known O antigens of *Klebsiella* are related to those of *E. coli*, as follows:

Klebsiella	*E. coli*
01	019 (019a, 19b, identical)
02	No relationship known
03	09, identical
04	02, close relationship
05	08, identical

In addition to the above-mentioned antigens, Pickett and Cabelli (1953) postulated a group specific antigen (Sm) which they felt underlaid the capsular portion of the cell. The relationship of this antigen to the cell constituents is not clear.

THE QUELLUNG REACTIONS. Since noncapsulated cells are agglutinated by antisera that contain O or R agglutinins, and since these antibodies occur in antisera produced with capsulated forms, it is imperative in capsule typing to distinguish clearly between reactions caused by O or R agglutinins and those caused by capsular relationships. This can be done most surely and simply by means of the quellung (Neufeld) reaction. While this has been referred to as the capsule swelling reaction, it is the opinion of the writer and colleagues that actual swelling of the capsule does not occur. Rather it appears to be a precipitin reaction which occurs at the surface of the capsule or slime layer when appropriate antiserum is added to a suspension of capsulated bacteria. This precipitation of antibody on the surface of the capsule or slime layer causes it to become highly refractile, and thus easily visible in unstained moist mounts. The size of the layer thus outlined does not differ perceptibly from the size of capsules or slime layers seen in moist India ink mounts. Not only does the capsule or slime layer take part in the above-mentioned precipitin reactions, but the loose slime (so-called soluble specific substance) reacts as well. In suspension that contain large amounts of loose slime the resulting precipitate may be seen throughout the microscopic field, often with cells entangled within it. If loose slime is present in sufficient amount it often seems to unite with the antibody at the expense of the capsule or slime layer so that quellung may fail. For this reason, particularly in the case of mucoid strains which form loose slime, it always is best to use a fresh suspension so that the polysaccharide which has become detached from the cells is reduced to a minimum. It also is advisable to use a very dilute suspension for the same reason. In the performance of quellung reactions two drops of a freshly prepared dilute suspension of the bacterium are placed upon a slide. To one, India ink is added to make a moist mount, to the other a loopful of antiserum is added and mixed well. Cover slips are placed over both drops. The ink mount is used to ascertain the size of the capsule or slime layer. If the size of the capsule is known it is much easier to judge whether quellung has occurred than if the worker has no concept of its size. It is difficult to judge whether quellung has occurred when the capsule is very small, since the reaction takes place in such close apposition to the cell wall that it is not easily discernible. In such instances it is advisable to isolate forms which produce larger capsules. Cultivation on the Worfel-Ferguson medium is very useful in such instances. In 1951, Worfel and Ferguson devised a liquid medium that is excellent for promotion of capsule production by cultures of *Klebsiella*. Unfortunately, the formula of the medium never was published and the writer and colleagues are indebted to Dr. W. W. Ferguson for providing it (see Chapter 18).

DETERMINATION OF CAPSULE ANTIGENS

Capsule antigens may be determined by a variety of methods including slide and tube agglutination, precipitin techniques, etc. Klebsiellae produce large amounts of soluble specific substance and the supernatant fluid from centrifuged broth cultures or suspensions in physiological saline solution produce specific precipitation with appropriate antisera. Likewise, capsulated strains agglutinate with antisera produced with capsulated klebsiellae. In tube agglutination tests, capsular agglutination is characterized by a disclike precipitate which resists disintegration when the tube is shaken. Slide agglutination tests have been found very useful for preliminary work (v. inf.).

Through the work of Julianelle (1926), Goslings and Snijders (1936), Kauffmann (1949), Brooke (1951a, 1951b), Worfel and Ferguson (1951), Edwards and Fife (1952, 1955), Edmunds (1954), and Ørskov (1955b), 72 capsule types have been delineated within the genus *Klebsiella*. For preliminary examination of cultures, antisera are pooled. Since cross reactions occur between certain capsule types, antisera for related types are placed in the same pool when possible. The antisera employed by the author and colleagues (Table 124) are pooled in the following manner:

Pool No.	Antisera for capsule types
1	1,2,3,24
2	4,5,6,7
3	8,9,10,25
4	13,14,15,16
5	17,18,19,20
6	11,21,22,23
7	26,27,28,30
8	12,29,40,41
9	31,32,43,44
10	33,34,35,36
11	37,38,39,42
12	45,46,47
13	48,49,50,51
14	52,53,54,55
15	56,57,58,59
16	60,61,62,63
17	64,65,66,67,68,69
18	70,71,72.

These pooled antisera are employed without further dilution. It should be clear that it might be necessary for investigators who produce other lots of antisera to prepare pools in a different way.

A dense suspension of the microorganisms in phenolized or formalized physiological saline solution is prepared and droplets (one 3 mm loopful or equivalent) are distributed on a glass slide or plate. To each droplet is added a loopful of each of the undiluted pooled antisera. Agglutination usually appears immediately in one or more pools. It then is necessary to test the suspension in the constituent antisera of each of the pools in which agglutination is observed. If agglutination occurs in only one individual antiserum, a dilute suspension of the organism is made and tests for a quellung reaction performed with that antiserum. A positive quellung reaction assures that the agglutination observed was a capsular reaction and not somatic agglutination. Thus, it becomes apparent that the culture under examination possesses capsular antigens related to those of one of the type strains. If the extent of this relationship is to be determined it is necessary to perform quellung tests with serial dilutions of the antiserum in order to determine whether the culture under examination reacts to the homologous titer of the antiserum. In order to establish identity it is necessary to do absorption tests to determine whether the culture will remove all capsular antibody from the antiserum.

ABSORPTION TESTS

From the results of Kauffmann (1949), Brooke (1951), Edwards and Fife (1952), and Edmunds (1954) it is

evident that many relationships exist among the different capsule types. While the cross reactions will vary to some extent with each set of antisera produced, on the whole, they are quite similar. The reactions obtained with antisera produced by the writer and colleagues are given in Table 134. It is absolutely necessary that each antiserum produced be tested with all known capsular types since a fresh antiserum produced as a replacement for a previous lot may exhibit slightly different cross reactions. Many of the cross reactions are so weak and so low in titer that they are not apparent in the typing procedure and these may be disregarded. Others are so strong that they cause confusion and in these instances it is necessary to use absorbed antisera. The preparation of specific absorbed capsular antisera for members of the genus *Klebsiella* is not difficult. The organisms produce large amounts of capsular substance and hence a relatively small amount of growth serves to remove antibodies common to two or more capsule types. Usually the growth from one slant of Worfel-Ferguson medium in a 18 X 150 mm tube is sufficient to remove heterologous antibodies from 1.0 ml of antiserum.

A convenient method of absorbing common capsular antibodies is to suspend the growth from a slant of the absorbing organism in 1.0 ml of physiological saline solution and to add this drop by drop to 1.0 ml of the antiserum to be absorbed. During the addition the tube containing the antiserum should be shaken or rocked continuously. As the first drops are added it will be noted that agglutination takes place almost instantly. As more suspension is added the reaction occurs more slowly. When the point is reached where the supernatant fluid remains slightly cloudy no more suspension should be added since it is difficult to remove a large excess of slimy, mucoid bacteria. The antiserum should then be centrifuged and the supernatant fluid used in agglutination and quellung tests with the absorbing and the homologous cultures to assure that all cross reactions have been removed and that specific capsular antibody remains. Absorbed antisera may be used both in slide agglutination and in quellung tests. In some instances it is necessary to absorb an antiserum with two or more heterologous types. In such cases the absorptions should be done successively, absorbing first with one heterologous type and then with a second. No table listing the absorptions necessary to prepare specific capsular fractions is included, since these will vary with each lot of antisera produced.

IRREGULARLY CAPSULATED CULTURES

Occasionally cultures are encountered which, in spite

TABLE 134

Capsular (Quellung) Reactions of Klebsiellae

Antiserum for Capsule Type	Homologous Titer (Quellung)	Quellung Titers with Heterologous Antigens	Antiserum for Capsule Type	Homologous Titer (Quellung)	Quellung Titers with Heterologous Antigens
1	32	6(8) *	37	16	
2	32	13(2), 55(4), 69(2)	38	32	
3	64	68(16)	39	64	65(16), 71(8)
4	16		40	64	
5	16		41	64	12(32)
6	32	4(2), 5(4), 7(4)	42	16	12(8)
7	64	5(2), 6(8)	43	16	41(2)
8	64	9(4), 25(4), 71(2)	44	32	18(2)
9	64		45	16	71(8), 72(2)
10	32	7(4), 61(8)	46	16	27(8)
11	32	21(2)	47	32	
12	64	29(8), 41(8)	48	128	49(4)
13	32	26(8), 30(16)	49	16	
14	16	64(16)	50	32	
15	64		51	64	
16	64		52	16	53(4), 70(2)
17	16		53	32	52(2)
18	16		54	16	
19	32		55	32	
20	16	8(4), 15(4), 23(8), 45(4)	56	32	
21	64	11(16), 26(8), 33(8), 69(2)	57	64	
22	32	23(16), 37(16), 41(4)	58	32	
23	16	22(8)	59	64	
24	32		60	64	31(2)
25	32	8(2)	61	128	7(4), 10(4), 41(16)
26	32		62	128	
27	32		63	32	49(4)
28	32		64	128	14(32), 31(2), 36(2), 56(2)
29	128	12(8)	65	32	28(2), 29(2), 40(2)
30	16	2(2)	66	64	
31	32	60(2)	67	64	
32	16		68	32	3(4), 14(2), 36(2)
33	16		69	32	2(2), 21(2)
34	64		70	64	34(4), 72(4)
35	32	33(4)	71	128	28(4), 38(2), 45(2), 46(2), 72(8)
36	64	25(2)	72	64	6(4), 12(2), 29(4), 44(4) 45(4), 70(4), 71(2)

* Numerals in parentheses indicate quellung titers.

NOTE: Weak reactions that may occur at dilution of 1:1 are not listed.

of repeated selection, persist in producing a majority of noncapsulated cells. A few cells may develop prominent capsules but most of the cells are not visibly capsulated when examined in moist India ink mounts. Such cultures are very likely to be susceptible to somatic (O or R) agglutination and may agglutinate in most of the pooled antisera. It is impossible to group such a culture by slide agglutination and it is necessary to resort to primary grouping in pooled antisera by quellung reactions. A dilute suspension of the organism is tested for quellung with all the pooled antisera and often it is possible to detect a capsular reaction with the few cells which are well capsulated. If this can be done, one may continue the tests with the constituent antisera of the positive pool and thus determine the capsular type. Occasionally, otherwise typical cultures of *Klebsiella* are encountered which, in spite of all efforts to isolate capsulated forms, fail to produce capsules. Such cultures must be reported as "*Klebsiella* species, insufficient capsules for typing."

In conclusion, it may be added that if cultures are selected carefully for production of capsular antibodies, if the resulting antisera are standardized and absorbed as necessary, and if typing procedures are executed with care, little difficulty should be encountered in the capsule typing of klebsiellae. As pointed out by Henriksen (1954), cultures will be encountered that possess capsular relationships to two or more type cultures which are not in themselves related. In such instances one can only report the culture as a strain of *Klebsiella* related to the types in question. It is always necessary to examine the biochemical characteristics of the organisms to determine that they are members of the genus *Klebsiella* and to ascertain the species.

The author and colleagues have observed biochemically typical cultures of *Enterobacter,* particularly *E. cloacae* and *E. aerogenes,* which possessed prominent capsules. The majority of such strains yield positive agglutination and quellung reactions in antisera produced with klebsiellae, including antisera for capsule types 8, 11, 21, 26, and 69. Only rarely do such cultures react in a single antiserum.

PRODUCTION OF ANTISERA

In the production of antisera for the detection of capsular antigens it obviously is necessary to use capsulated forms. Many cultures of *Klebsiella* produce large, prominent capsules so that it is no problem to obtain suitable antigens. However, other strains may produce extremely small capsules or the majority of cells may be noncapsulated. Such cultures should be plated and suitable colonies selected for antigen production. The author and colleagues have found Worfel-Ferguson medium most useful for this purpose.

Colonies of organisms that produce capsules usually can be distinguished by their greater size, more moist appearance, and greater opacity. For antiserum production, it has been found that cultures that are regularly capsulated but which produce capsules of moderate size are more useful than strains that produce very large capsules. Apparently an excess of capsular substance inhibits production of capsular antibodies. It should be emphasized that if a capsulated, nonmucoid form can be obtained, it is preferable to a highly mucoid, slimy culture. A suitable colony is selected and inoculated into 50 ml of infusion broth to which 0.2 per cent of glucose has been added. After 4 or 5 hr of incubation the culture is killed by addition of 0.5% formalin. The broth culture is examined for capsules and if these are present in sufficient number and amount the killed suspension is used to inject rabbits in amounts of 0.5, 1.0, 2.0, 3.0, 3.0, and 3.0 ml at intervals of 4 days. The rabbits should be test bled 4 days after the last injection and the antiserum tested by quellung reactions. If a sufficient titer (1 to 16 or greater) has not been reached, the animals should be given two additional injections and retested. Rabbits should be bled 5 to 7 days following the last injection. Some animals never produce a satisfactory titer. In such instances the animals must be discarded and replacements must be injected with a newly prepared antigen suspension. Alternatively, dried antigens may be prepared from suitably capsulated cultures. These are prepared by a modification of the Roschka technique in which heat treatment is omitted. Methods for preparation of dried antigens are given in connection with *E. coli* K antiserum production (Chapter 5) and production of Vi antiserum (Chapter 8).

REFERENCES

Brooke, M. S. 1951a. Acta Pathol. Microbiol., Scand., **28**, 313.

_____. 1951b. Ibid., **28** 328.

Cowan, S. T., K. J. Steel, C. Shaw, and J. P. Duguid. 1960. J. Gen. Microbiol., **23**, 601.

Duguid, J. P. 1948. J. Pathol. Bacteriol., **60**, 265.

_____. 1951. Ibid., **63**, 673.

Duguid, J. P., and J. F. Wilkinson. J. Gen. Microbiol., **9**, 174.

Edmunds, P. N. 1954. J. Infect. Dis., **94**, 65.

Edwards, P. R., and M. A. Fife. 1952. J. Infect. Dis., **91**, 92.

_____. 1955. J. Bact., **70**, 382.

Elbert, B. J. and W. M. Gerkess. 1929. Zentralbl. f. Bakt. I. Orig., **112**, 116.

Epstein, S. S. 1959. J. Clin. Path., **12**, 52.

Ewing, W. H. 1963. Int. Bull. Bacteriol. Nomen. Tax., **13**, 95.

Ewing, W. H. 1967. Revised definitions for the family

Enterobacteriaceae, its tribes, and genera. CDC Publ.*

Ewing, W. H., B. R. Davis, and J. G. Johnson. 1962. Int. Bull. Bacteriol. Nomen. Tax., **12**, 47.

Fife, M. A., W. H. Ewing, and B. R. Davis. 1965. The biochemical reactions of the tribe Klebsielleae. CDC Publ.*

Goslings, W. R. O. and Snijders, E. P. 1936. Zentralbl. f. Bakt. I. Orig., **136**

Grimes, M. 1961. Int. Bull. Bacteriol. Nomen. Tax., **11**, 111.

Henriksen, S. D. 1954. Acta Pathol. Microbiol., Scand., **34**, 249.

Hormaeche, E., and M. Munilla 1957. Int. Bull. Bacteriol., Nomen. Tax., **7**, 1.

Hormaeche, E., and P. R. Edwards 1958. Ibid., **8**, 111.

Hugh, R. 1959. Can J. Microbiol., **5**, 251.

Julianelle, L. A. 1926. J. Exp. Med., **44**, 113.

Kauffmann, F. 1949. Acta Pathol. Microbiol. Scand., **26**, 381.

_____. 1956a. Ibid., **39**, 85.

_____. 1956b. Ibid., **39**, 103.

_____. 1966. The bacteriology of Enterobacteriaceae. Copenhagen: Munksgaard.

Kauffmann, F., and A. Peterson. 1956. Acta Pathol. Microbiol. Scand., **38**, 481.

Lautrop, H. 1956. Acta Pathol. Microbiol., Scand., **39**, 375.

Moeller, V. 1954. Ibid., **35**, 259.

Ørskov, I. 1954. Ibid., **34**, 145.

_____. 1955a. Ibid., **37**, 353.

_____. 1955b. Ibid., **36**, 449.

Pickett, M. J., and V. J. Cabelli. 1953. J. Gen. Microbiol., **9**, 249.

Sakazaki, R., and S. Namioka. 1957. Jap. J. Exp. Med., **27**, 273.

Sedlak, J., and M. Slajsova. 1959. Folia Microbiol., **4**, 229.

Toenniessen, E. 1914. Zentralbl. f. Bakt. I. Orig., **75**, 329.

_____. 1921. Ibid., **85**, 225.

Weldin, J. C. 1927. Iowa State Coll., J. Sci., **1**, 120.

Wilkinson, J. F., J. P. Duguid, and P. N. Edmunds. 1954. J. Gen. Microbiol., **11**, 59.

Winslow, C. E. A., J. Kligler, and W. Rothberg. 1919. J. Bacteriol., **4**, 429.

Worfel, M. T., and W. W. Ferguson. 1951. Amer. J. Clin. Path., **21**, 1096.

*CDC Publ., Publication from the Center for Disease Control (formerly the Communicable Disease Center), Atlanta, Ga. 30333.

Chapter 13
The Genus *Enterobacter*

When used in conjunction with the definition for the tribe Klebsielleae (Chapter 1) the following definition (Ewing, 1967) of the genus *Enterobacter* provides a convenient starting point for a discussion of the characteristics of members of the genus.

The genus *Enterobacter* is composed of motile bacteria that conform to the definitions of the family ENTEROBACTERIACEAE and the tribe KLEBSIELLEAE. The Voges-Proskauer reaction is positive, gelatin is liquefied slowly by the most commonly occurring forms *(Enterobacter cloacae).* Lysine decarboxylase is not produced by *E. cloacae,* but other species of the genus possess this enzyme system. Ornithine decarboxylase is produced. Sodium alginate is not utilized as a sole source of carbon. Gas is not formed from inositol and glycerol by cultures of *E. cloacae.* Acid is produced from sorbitol, rhamnose, arabinose, and raffinose by the majority of species. One species *(Enterobacter hafniae)* does not ferment sorbitol or raffinose. Only one species *(Enterobacter liquefaciens)* is lipolytic. The type species is *Enterobacter cloacae* (Jordan) Hormaeche and Edwards.

The history of the taxonomy and nomenclature of the bacteria classified in the genus *Enterobacter* is complicated. A thorough review of it would be long and involved and would serve no useful purpose as far as this book is concerned. However, it is necessary to include a brief discussion of the synonymy involved, since the older names occasionally are seen in the literature.

In the past many nonmotile cultures that actually were klebsiellae were classified as members of the genus *Aerobacter* (now *Enterobacter*). This fact was proven by numerous investigators (e.g., Edwards, 1929). When these particular nonmotile bacteria were classified correctly in the genus *Klebsiella* on the basis of their biochemical reactions as well as their lack of motility, there remained a large group of motile bacteria (and occasional nonmotile variants thereof) which required classification elsewhere. Kauffmann (1953) attempted to solve this problem by reintroducing the genus *Cloaca* of Castellani and Chalmers (1920) and by referring to the bacteria as *Cloaca cloacae.* Hormaeche

and Munilla (1957) distinguished two groups of bacteria among cultures classified as *Cloaca,* called Cloaca A and Cloaca B. Hormaeche and Edwards (1958) proposed a system in which the Cloaca A bacteria were labeled *Aerobacter cloacae* and *Aerobacter aerogenes* was redefined to correspond to the bacteria of Cloaca group B. In an attempt to avoid confusion that might result from reclassification in the genus *Klebsiella* of many nonmotile bacteria previously classified as *A. aerogenes,* Hormaeche and Edwards (1960a, 1960b) proposed the substitution of the generic term *Enterobacter* for *Aerobacter* and defined two species, *Enterobacter cloacae* and *Enterobacter aerogenes.* The Judicial Commission of the International Committee on Nomenclature of Bacteria acted upon the proposal of Hormaeche and Edwards (1960a, 1960b) and conserved the generic name *Enterobacter* (see Opinion 28, 1963; or Ewing, 1963). Thus, the most important synonyms are as follows:

Current Usage	Synonyms
E. cloacae	*A. cloacae,* Aerobacter A, Cloaca A
E. aerogenes	*A. aerogenes,* Aerobacter B, Cloaca B

The bacteria described by Grimes and Hennerty (1931) and Grimes (1961) under the designation *Aerobacter liquefaciens* were referred to for a time as Aerobacter C, but became *Enterobacter liquefaciens* in the revised nomenclatural system, as given in Chapter 1 for example. Parenthetically, some investigators have suggested that this species should be transferred to the genus *Serratia.* This subject is discussed in Chapter 14.

The remaining species presently recognized in the genus *Enterobacter* is *Enterobacter hafniae.* As mentioned in the previous editions of this book and elsewhere (e.g., Ewing and Fife, 1968), these microorganisms have been known by several designations e.g., the 32011 bacteria, the Hafnia group, *B. cadaveris, B. asiaticus,* and *Enterobacter alvei.*

For additional information regarding the nomenclature and taxonomy of bacteria of the genus *Enterobacter* interested readers are referred to the citations in Chapter 1 and at the beginning of Chapter 12, as well as to the papers cited above.

BIOCHEMICAL REACTIONS

The biochemical reactions given by members of the several species of *Enterobacter* in commonly employed tests are listed in Tables 7, 8, and 22 (Chapter 3). Of the three species of klebsiellae only the reactions given by strains of *K. pneumoniae* are listed in Tables 7 and 22 for comparative and differential purposes. In the experience of the author and colleagues (Fife et al., 1965 and unpublished data), *K. pneumoniae, E. cloacae,* and *E. hafniae* occur more commonly than the other species in these genera.

The biochemical reactions given by 444 cultures that belonged to the four species of *Enterobacter* are summarized in Table 135. The data given in this table are adapted from the work of Fife et al. (1965). Although numerous strains belonging to the several species of *Enterobacter* have been received and studied since that publication appeared, comparison of the data obtained indicates that there has been no appreciable change in the percentages of positive, positive delayed, and negative reactions. The patterns of reactions obtained with respect to the decarboxylases, gas production from inositol, adonitol, and glycerol, gelatin liquefaction, and motility are given in Tables 136, 137 138, and 139 for *E. cloacae, E. aerogenes, E. hafniae,* and *E. liquefaciens,* respectively. The information in these tables should be of value to investigators who wish to identify as closely as possible cultures that may be somewhat aberrant with respect to the above-mentioned reactions as well as typical strains.

It is notable that Fife et al. (1965) reported that high percentages of the cultures examined yielded typical patterns of reaction in the three decarboxylase tests (lysine, arginine, and ornithine). For example, 95 per cent of the strains of *E. cloacae* gave reactions typical of the species (-++) and 97.3 per cent of the isolants of *E. aerogenes* were typical (+-+) in this regard (see also Tables 136-139).

Tests and substrates that are of particular value in the differentiation of *K. pneumoniae* and *E. cloacae* are given in Table 18; for differentiation of *E. cloacae* and *E. aerogenes,* in Table 19; for differentiation of *E. aerogenes* and *E. hafniae,* in Table 20; and for differentiation of *E. liquefaciens* and *Serratia marcescens, marcescens* in Table 21. Tables 18 to 21 are located in Chapter 3.

The data presented in the above-mentioned tables largely are self-explanatory and little comment is necessary. Regarding cultures of *E. hafniae,* the failure of typical isolates to ferment adonitol, inositol, sorbitol, and raffinose is a notable characteristic. Further, the effect of the temperature of incubation of cultures upon the results of methyl red and Voges-Proskauer tests is worthy of mention. At 37 C the methyl red test may be positive or negative, whereas at 22 C to 25 C it is almost always negative. The results of the Voges-Proskauer test at 37 C also may be positive or negative, but cultures incubated at 22 to 25 C are positive (Table 135). Utilization of sodium citrate also is increased when cultures are incubated at 22 to 25 C.

Similarly, a number of the physiological reactions of *E. liquefaciens* are increased when strains are incubated at 22 C to 25 C. Among these are motility, the Voges-Proskauer reaction, lactose utilization, and decarboxylation of lysine. For example, 82 per cent of cultures tested decarboxylated lysine at 37 C whereas all strains produced positive reactions in lysine medium at 22 C to 25 C (Fife et al., 1965). The original isolations of these microorganisms (Grimes and Hennerty, 1931) were from dairy products and the strains were recovered after prolonged incubation at about 7 C. However, the author and colleagues have examined a number of cultures from specimens from humans and lower animals, a fact that occasionally has been reported by others (e.g., Richard, 1970).

Finally, it should be noted that very occasionally one may encounter an isolant of *E. cloacae* that produces a definite yellow pigment. The author and colleagues have received six or eight such cultures during the past 15 years. Without exception these have proven to be typical strains of *E. cloacae* in every respect except for pigment production. Nissen et al. (1965) also reported the isolation of a similar culture from the spinal fluid of an infant.

With very few exceptions the results of investigations of the biochemical reactions given by members of the genus *Enterobacter* reported by others were similar to those mentioned in the preceding pages and listed in the tables. For additional information regarding this subject, the reader is referred to the publications of Sakazaki and Namioka (1957), Hormaeche and Munilla (1957), Sedlak and Slajsova (1959) Fife et al. (1965), and Richard (1970).

SEROLOGICAL INVESTIGATIONS

An antigenic schema for *E. cloacae* was proposed by Sakazaki and Namioka (1960). These investigators characterized 53 O antigen groups, 57 H antigens, and 79 serotypes among 170 cultures. The occurrence of K antigens and alpha antigen was reported in some strains. Sedlak et al. (1958, 1962) reported the characterization of eight O antigen groups of *E. cloacae* with which they were able to determine the O group of about 46 per cent of 983 cultures examined. These investigators also delineated eight H antigens.

Stuart and Rustigian (1943) reported that they were able to classify 127 (85 percent) of 149 cultures of 32011 bacteria *(E. hafniae)* as one or another of eight serotypes. Fifty-seven of the strains belonged to one

TABLE 135

The Biochemical Reactions of *Enterobacter* Species, Summarized

Test or substrate	E. cloacae			E. aerogenes			E. hafniae			E. liquefaciens		
	Sign	%+	(%+)*	Sign	%+	(%+)*	Sign	%+	(%+)*	Sign	%+	(%+)*
Hydrogen sulfide (TSI)	-	0		-	0		-	0		-	0	
Urease	+ or -	64.7		-	2.7		-	3		d	4.4	(19.1)
Indol	-	0.5		-	0		-	0		-	0	
Methyl red:												
(37 C)	-	3		-	0		+ or -	54		+ or -	75	
(22 C)							-	1		- or +	33.3	
Voges-Proskauer:												
(37 C)	+	99.5		+	100		+ or -	65		- or +	30.9	
(22 C)	+	99.5		+	100		+	99		+ or -	79.4	
Citrate (Simmons')												
(37 C)	+	99.5		+	93.7		(+) or -		(58)	+	91.2	(7.3)
(22 C)							d	3	(79)			
KCN (growth)	+	98		+	98.7		+	96	(1)	+	98.5	
Motility:												
(37 C)	+	94.5		+	97.3		+	93		d	79.4	(17.7)
(22 C)										+	100	
Gelatin (22 C)	(+)	1	(96)	(+) or -		(77.3)	-	0		+	98.5	(1.5)
Lysine decarboxylase	-	0.5		+	98.7		+	100		+ or -	82.4	
Arginine dihydrolase	+	96.5		-	0		-	9		-	4.4	
Ornithine decarboxylase	+	96		+	98.7		+	100		+	98.5	
Phenylalanine deaminase	-	0		-	0		-	0		-	1.5	
Glucose:												
Acid	+	100		+	100		+	100		+	100	
Gas	+	100		+	100		+	100		+	94.1	(1.5)
Lactose	+	93.5	(5.5)	+	92.1	(5.3)	- or (+)		(23)	d	1.5	(29.4)

TABLE 135 (Continued)

The Biochemical Reactions of *Enterobacter* Species, Summarized

Test or substrate	E. cloacae			E. aerogenes			E. hafniae			E. liquefaciens		
	Sign	%+	(%+)*	Sign	%+	(%+)*	Sign	%+	(%+)*	Sign	%+	(%+)*
Sucrose	+	96.5	(0.5)	+	100		d	12	(61)	+	100	
Mannitol	+	100		+	100		+	100		+	100	
Dulcitol	- or +	12.9		-	4		-	1		-	0	
Salicin	d	75.6	(18.4)	+	98.7	(1.3)	d	13	(8)	+	100	
Adonitol:												
Acid	- or +	28.4		+	98.7		-	0		d	8.8	(2.9)
Gas	- or +	28.4		+	98.7		-	0		-	1.5	(1.5)
Inositol:												
Acid	d	21.9	(12.4)	+	100		-	0		+	97	(1.5)
Gas	-	4.5		+	100		-	0		d	1.5	(22)
Sorbitol	+	94.5	(0.5)	+	100		-	0		+	97	
Arabinose	+	99.5		+	100		+	96		+	92.6	
Raffinose	+	97		+	96		-	0		d	86.8	(2.9)
Rhamnose	+	92	(1.5)	+	98.7		+	93	(7)	-	0	
Malonate	+ or -	80.6		+ or -	74.7		+ or -	74		-	7.4	
Mucate	+ or -	75.6		+	94.7		-	0		-	0	
Nitrite from nitrate	+	100		+	100		+	100		+	100	
Organic acids**												
Citrate	+ or -	86.1		+	97.3		+	99.1		+ or -	64.7	
D-tartrate	- or +	16.9		- or +	40		-	0		- or +	22.1	
Jordan's tartrate	- or +	27.8		+ or -	89.3		+ or -	75.4		+ or -	75	
Sodium alginate, synthetic	-	0		-	0		-	0		-	0	
nutrient	-	0		-	0		-	0		-	0	

TABLE 135 (Continued)

The Biochemical Reactions of *Enterobacter* Species, Summarized

Test or substrate	E. cloacae			E. aerogenes			E. hafniae			E. liquefaciens		
	Sign	+%	(%+)*	Sign	%+	(%+)*	Sign	%+	(%+)*	Sign	%+	(%+)*
Lipases:												
corn oil	-	0.8	(1.5)	-	0		-	0		+ or (+)	86	(8.8)
triacetin	-	0		-	2		-	0		d	19.3	(73.7)
tributyrin	-		(.8)	-	0		-	0		+ or (+)	82.5	(17.5)
Pectate	-	0		-	0		-	0		-	0	
Starch	d	15.9	(33.4)	d	38.7	(20)	d	50	(16)	d	25	(67.7)
Xylose	+	98	(.5)	+	100		+	97		+	92.6	(1.5)
Cellobiose:												
acid	+	100		+	98.7		d	76	(16)	d	26.5	(44.1)
gas	+	100		+	98.7		d	67	(22)	d	5.9	(33.8)
Glycerol:												
acid	d	43.3	(44.8)	+	100		+	100		+	98.5	(1.5)
gas	d	5.5	(15.9)	+	98.7	(1.3)	+	95	(5)	d	45.6	(38.2)
Alpha methyl glucoside	+ or (+)	84.2	(12.8)	+	96	(2)	-	0		- or +	21.7	
Erythritol	-	0		-	0		-	0		-	0	
Esculin	- or +	29.3		+	98		-	6	(2)	d	73.3	(1.7)

*Numerals in parentheses indicate percentage of delayed reactions (3 or more days)

**Method of Kauffmann and Petersen, 1956

+ Positive within one or two days' incubation.

(+) Positive reaction after 3 or more days.

− No reaction.

+ or − Majority of strains positive, some cultures negative.

− or + Majority of cultures negative, some strains positive.

(+) or + Majority of reactions delayed, some occur within 1 or 2 days.

d Different reactions: +, (+), −.

Patterns of Reactions Given by Cultures of *Enterobacter* Species in Selected Tests

TABLE 136

Patterns of Reactions Given by Cultures of *E. cloacae* in Selected Tests

No. of cultures in each group	Decarboxylases			Gas from:			Gelatin	Motility
	Lysine	Arginine	Ornithine	Inositol	Adonitol	Glycerol		
96	-	+	+	-	-	-	(+)	+
33	-	+	+	-	+	-	(+)	+
5	-	+	+	+	-	-	(+)	+
22	-	+	+	-	-	+	(+)	+
1	-	+	+	+	+	+	(+)	+
1	-	+	+	+	-	+	(+)	+
16	-	+	+	-	+	+	(+)	+
2	-	+	+	-	-	-	-	+
3	-	+	+	-	-	-	-	+
1	-	+	+	+	-	+	-	+
6	-	+	+	-	+	-	(+)	-
2	-	+	+	+	-	+	(+)	-
1	-	+	+	-	+	+	(+)	-
2	-	+	+	-	-	-	(+)	-
1	-	+	+	-	+	-	(+)	+
1	-	-	+	-	-	+	(+)	+
1	-	+	-	-	-	-	(+)	+
3	-	-	-	-	+	-	(+)	+
1	-	-	-	-	+	+	(+)	+
2	-	-	-	-	+	-	(+)	+

TABLE 137

Patterns of Reactions Given by Cultures of *E. aerogenes* in Selected Tests

No. of cultures in each group	Decarboxylases			Gas from:			Gelatin	Motility
	Lysine	Arginine	Ornithine	Inositol	Adonitol	Glycerol		
54	+	-	+	+	+	+	(+)	+
1	+	-	+	+	-	+	(+)	+
17	+	-	+	+	+	+	-	+
1	+	-	-	+	+	+	(+)	+
1	-	-	+	+	+	+	(+)	-
1	+	-	+	+	+	+	(+)	-

TABLE 138

Patterns of Reactions Given by Cultures of *E. hafniae* in Selected Tests

No. of cultures in each group	Decarboxylases			Gas from:			Motility
	Lysine	Arginine	Ornithine	Inositol	Adonitol	Glycerol	
84	+	-	+	-	-	+	+
6	+	-	+	-	-	+	-
9	+	(+w)	+	-	-	+	+
1	+	-	+	-	-	+	+

TABLE 139

Patterns of Reactions Given by Cultures of *E. liquefaciens* in Selected Tests

No. of cultures in each group	Decarboxylases			Gas from:			Gelatin	Motility
	Lysine	Arginine	Ornithine	Inositol	Adonitol	Glycerol		
33	+	-	+	-	-	+	+	+
13	+	-	+	+	-	+	+	+
6	+	-	+	+	-	-	+	+
1	+	-	+	+	-	+	+	+
1	+	-	+	+	-	-	+	+
1	+	+	+	-	-	+	+	+
1	-	+	+	-	+	+	+	+
1	-	+	-	-	-	+	+	+
9	+	-	+	-	-	+	+	+
1	+	-	-	-	-	-	+	+

w = weakly positive reaction. All strains, gelatin negative.

serotype. Antigenic schemata for *E. hafniae* have been proposed by several investigators, as follows:

No. of cultures	No. of O Groups	No. of H antigens	No. of serotypes	References
18	12	6	14	Deacon, 1952
58	21	22		Eveland and Faber, 1953
294	29	23	51	Sakazaki, 1961, extended by
327	32	9	130	Matsumoto, 1963
30	7	2	11	Matsumoto, 1964

When the data reported by Sakazaki (1961) and Matsumoto (1963, 1964) were combined it was evident that 68 O antigen groups, 34 H antigens, and 192 serotypes were characterized among 651 strains. Unfortunately the O groups and H antigens characterized by Deacon (1962) and by Eveland and Faber have not been compared with those delineated by Sakazaki and Matsumoto.

The above-mentioned authors demonstrated the presence of K antigens and alpha antigen in certain cultures of *E. hafniae*. Deacon (1952) made note of variation in the flagellar antigens of the strains studied but Eveland and Faber (1953) and Sakazaki (1961) failed to record such variation.

Sakazaki (1961) and Matsumoto (1963) demonstrated interrelationships between several of the O antigens of *E. hafniae* and those of certain *E. cloacae* and *E. coli.*

The author has no knowledge of investigations leading to the characterization of the O and H antigens of cultures of *E. liquefaciens* and *E. aerogenes,* although some of the strains examined by Sedlak and Matejovska (1958) may have been members of the latter species. Further, it appears that there has been no attempt to correlate or integrate the schemata that have been proposed for *E. cloacae* and *E. hafniae.*

REFERENCES

Castellani, A., and A. J. Chalmers. 1920. Ann. Inst. Pasteur., **34**, 600.

Deacon, W. E. 1952. Proc. Soc. Exper. Biol. Med., **81**, 165.

Edwards, P. R. 1929. J. Bacteriol., **17**, 339.

Eveland, W. C., and J. E. Faber. 1953. J. Infect. Dis., **93**, 226.

Ewing, W. H. 1967. Revised definitions for the family Enterobacteriaceae, its tribes, and genera. CDC Publ.*

Ewing, W. H., and M. A. Fife. 1968. Int. J. System. Bacteriol., **18**, 263.

Fife, M. A., W. H. Ewing, and B. R. Davis. 1965. The biochemical reactions of the tribe Klebsielleae. CDC Publ.*

Grimes, M. 1961. Int. Bull. Bacteriol. Nomen. Tax., **11**, 111.

Grimes, M., and A. J. Hennerty. 1931. Sci. Proc. Roy. Dublin Soc. (Ireland), 20(NS), 89.

Hormaeche, E., and M. Munilla. 1957. Int. Bull. Bacteriol., Nomen. Tax., **7**, 1.

Hormaeche, E., and P. R. Edwards. 1958. Ibid., **8**, 111.

_____. 1960a. Ibid., **10**, 71.

_____. 1960b. Ibid., **10**, 75.

Kauffmann, F. 1953. Riv. Inst. Sieoterap., Ital., 28, 485.

Matsumoto, H. 1963. Jap. J. Microbiol., **7**, 105.

_____. 1964. Ibid., **8**, 139.

Nissen, R., T. Nørholm, and K. E. Siboni. 1965. Danish Med. Bull., **12**, 128.

Opinion 28. 1963. International Nomenclature Committee. Int. Bull. Bacteriol., Nomen. Tax., **13**, 38.

Richard, C. 1970. Ann. Biol. Clin. (France), **28**, 185.

Sakazaki, R. 1961. Jap. J. Med. Sci. Biol., **14**, 223.

Sakazaki, R., and S. Namioka. 1957. Jap. J. Exp. Med., **27**, 273.

_____. 1960. Jap. J. Med. Sci. Biol., **13**, 1.

Sedlak, J., and D. Matejovska. 1958. J. Hyg. Epidemiol. Microbiol. and Immunol., **2**, 258.

Sedlak, J., and M. Slajsova. 1959. Folia Microbiol., **4**, 229.

_____. 1962. Wiss. Karl-Marx-Univ. (Leipzig), 11, 748.

Stuart, C. A., and R. Rustigian. 1943. Amer. J. Pub. Health., **33**, 1323.

*CDC Publ., Publication from the Center for Disease Control (formerly the Communicable Disease Center), Atlanta, Ga. 30333.

Chapter 14
The Genus *Serratia*

Investigations of the biochemical reactions of members of the genus *Serratia* during the past 15 years have led to a clearer understanding of the constitution of the genus and its relationships to other genera within the family Enterobacteriaceae. In various editions of *Bergey's Manual* (e.g., 6th Edition, 1948; 7th Edition, 1957) the genus *Serratia* was classified in the tribe Serrateae as the sole genus while others, e.g., Krassilnikov (1949, quoted by Martinec and Kocur, 1961a) placed the serratiae in the genus *Chromobacterium*. On the basis of their studies, Davis and Ewing (1957), Ewing et al. (1959, 1962a), Ewing (1963), Fife et al. (1965) placed the genus *Serratia* in the same principal division (tribe) with *Klebsiella* and *Aerobacter* (now *Enterobacter*) because of the obvious relationships of serratiae to members of those genera. These investigators stated that there was no need for a separate tribe Serrateae erected as it was on the basis of pigment formation and rapid liquefaction of gelatin, since pigment formation was not a cardinal characteristic of *Serratia* and members of other genera rapidly liquefied gelatin. Incorporation of the serratiae into the genus *Chromobacterium* is regarded as untenable.

Davis and Ewing (1957) suggested that only one species of *Serratia (Serratia marcescens)* was needed and Ewing et al. (1959) concluded that the single species with one subspecies or variety *(Serratia marcescens* var. *kiliensis)* was sufficient. The subspecific or varietal name was applied to strains of *Serratia* that differed from *S. marcescens* in their negative Voges-Proskauer reactions. Cultures labeled with various epithets such as *Serratia indica, Serratia anolium, Serratia pyoseptica,* etc., could not be distinguished from *S. marcescens* or *S. marcescens* var. *kiliensis* by means at the disposal of the abovementioned investigators. Martinec and Kocur (1960, 1961a) studied strains of *S. indica* and *S. plymuthica,* concluded that they were indistinguishable from *S. marcescens,* and agreed with the conclusions of Davis and Ewing (1957) regarding a single species in the genus. Also, Martinec and Kocur (1961b) studied a number of cultures that had been described as species of *Serratia* by other authors and reported that some of

these (e.g., *Serratia piscatorum* and *Serratia fuchsina,* were *S. marcescens* while others, such as those described as *Serratia saponaria* I, II, were not *Serratia* and classified them provisionally in the genus *Alcaligenes.* These investigators agreed (1961c) with the conclusions of Ewing et al. (1959) regarding the usage of the designations *S. marcescens* and *S. marcescens* var. *kiliensis.* Parenthetically, it should be noted that the late Professor R. S. Breed previously had concluded that several of the above-mentioned bacteria could not be distinguished from *S. marcescens* (see *Bergey's Manual,* 5th Edition, 1939; and 6th Edition, 1948). For example, he considered *Serratia essyana, S. fuchsina,* and *S. pyoseptica* all to be *S. marcescens.* Also, *S. anolium* was omitted from the 6th Ed. (1948), although it had appeared as a separate species in the 5th Ed. (1939) of *Bergey's Manual.* Two of the strains described by Stevenson (1959) as *Aeromonas margarita* were found by Ewing (1961) to be typical cultures of *S. marcescens.*

As a result of the investigations and proposals mentioned above the genus *Serratia* was placed in the tribe Klebsielleae and redefined by Davis et al. (1957) and Ewing et al. (1959, 1962). These definitions were emended further and stated in conventional form by Ewing (1967), as follows:

> The genus *Serratia* is composed of motile bacteria that conform to the definitions of the family ENTEROBACTERIACEAE and the tribe KLEBSIELLEAE. A positive Voges-Proskauer reactions is given by *Serratia marcescens* subsp. *marcescens,* but *S. marcescens* subsp. *kiliensis* gives negative results in this test. Lipase is produced, gelatin is liquefied rapidly, and lysine and ornithine are decarboxylated. Sodium alginate is not utilized as a sole source of carbon. When gas is formed from fermentable substrates the volumes are small (10% or less). Acid is produced from sorbitol but rhamnose, arabinose, and raffinose are not fermented. The type species is *Serratia marcescens* Bizio.

This definition should be employed in conjunction with the definition of the tribe Klebsielleae (Chapter 1).

As mentioned in Chapter 13, it has been suggested by a few investigators (e.g., Le Minor, personal communication, 1970) that the bacteria included in the species *E. liquefaciens* should be transferred to the genus *Serratia.* The author and colleagues have no strong objections to this suggestion. On the contrary, such a transfer probably would help to solve a number of problems (v. inf.) and a transfer of some sort has been under consideration for some time. However, if this transfer is made in future the integrity of the species should be maintained, i.e., the bacteria should not be incorporated into the species *S. marcescens.* It should be recalled that these microorganisms were described and named *Aerobacter liquefaciens* by Grimes and Hennerty in 1931 and traditionally have been regarded as members of the genus *Aerobacter* (now *Enterobacter*) since that time. It has been known for many years that close biochemical affinities are extant between cultures of *E. liquefaciens* and *S. marcescens.* In fact, these bacterial types appear to resemble each other more closely than either resembles any other species in the family when their collective biochemical reactions are considered. Nevertheless with few exceptions members of these two species can be differentiated without difficulty (Ewing et al., 1959 et seq). Further, the growth temperature relationships and the fact that cultures of *E. liquefaciens* are more active biochemically when incubated at 22 C to 25 C than at 35 C to 37 C should be considered in connection with any proposed transfer of the microorganisms. Strains of *S. marcescens* are not known to exhibit such temperature relationships. It is for these reasons that the author and colleagues are of the opinion that if the suggested transfer is made in future, *E. liquefaciens* should be maintained as a separate species (i.e., *Serratia liquefaciens*). There are ancillary reasons for maintenance of the separate status of this species, e.g., apparent differences in the ecology of members of the two species. However, as stated above, the author and colleages are not averse to the suggested transfer provided that future studies confirm its desirability, and provided that some measure of agreement regarding its acceptance can be reached among concerned investigators. Although there is a considerable volume of data available that could be used to support a proposal for the above-mentioned transfer, it would be desirable if a representative number of characterized strains of each species could be investigated with respect to homology, or lack of it, of the DNA of each.[1]

For a number of years the author and co-workers

[1]The author has been informed that some work of this sort has been done but has no knowledge of its extent.

have collected and studied cultures that resemble *E. liquefaciens* as regards their biochemical characteristics but produce varying degrees of pink to red pigment. Some of these were received from various culture collections and were labeled *Serratia* of one species or another but were not *S. marcescens* (see Ewing et al., 1959). Other strains were submitted for identification and some were furnished by Dr. H. Lautrop of the State Serum Institute, Copenhagen. Although it has been felt that these cultures probably are pigmented biotypes of *E. liquefaciens,* a definite opinion has not been formed as to their taxonomic position. It is possible that these and the microorganisms described by Stapp (1940) as *Bacterium rubidaeum* might form a link between *E. liquefaciens* and *S. marcescens.* Among other things the nature of the pigment produced by such cultures should be determined.

Readers who are interested in the taxonomy and nomenclature of the genus *Serratia,* and in investigations of the biochemical reactions of serratiae as reported by others, are referred to the publications of Hefferan (1904), Sedlak et al. (1965), Colwell and Mandel (1965), Burger (1967), Grimont (1969), and Wilfert et al. (1970), as well as to the citations given in the preceding paragraphs.

BIOCHEMICAL REACTIONS

The biochemical reactions given by cultures of *Serratia marcescens, marcescens* are summarized in Table 140. The data given in the table are based upon the examination of 922 strains received for identification (Davis and Ewing, 1957; Ewing et al., 1959, 1962b; Fife et al., 1965; and unpublished data). The reactions given by *S. marcescens,* subspecies *kiliensis* are not tabulated since the only significant differences detected between cultures of this subspecies and *S. marcescens, marcescens* are the positive methyl red and negative Voges-Proskauer reactions yielded by strains of *S. marcescens, kiliensis.* Parenthetically, only about 30 strains of *S. marcescens, kiliensis* have been received during the last 15 years. The temperature at which cultures of *S. marcescens* of either subspecies are incubated (22 to 25 C or 35 to 37 C) apparently does not affect the results obtained in these two tests.

Hydrogen sulfide is not formed (Table 140) either rapidly or strongly in triple sugar iron agar, in media of similar composition, or in peptone iron agar. When gas is formed from fermentable carbohydrates the volumes are small and constitute 10 percent or less of the volume of the insert tubes. When it occurs production of urease is slow and weak (Table 140). Serratiae are lipolytic and produce extracellular deoxyribonuclease (DNase). With the exception of rare strains of *Pectobacterium* and *E. cloacae,* the only Entero-

TABLE 140

The Biochemical Reactions of *Serratia marcescens, marcescens*

Test or substrate	Sign	%+	(%+)*	Test or substrate	Sign	%+	(%+)*
Hydrogen sulfide	−	0		Rhamnose	−	0	
Urease	d	29.1[W]	(26.8)	Malonate	−	1.7	
Indol	−	0.2[W]		Mucate	−	0	
Methyl red	− or +	17.7		Christensen's citrate	+	99.2	
Voges-Proskauer	+	100		Jordan's tartrate	+	99.5	
Citrate (Simmons')	+	98.6	(0.5)	Sodium acetate	+	90	
KCN	+	99.1		Sodium alginate, synthetic	−	0	(3.7)
Motility	+	98.6		nutrient	−	0	
Gelatin (22 C)	+ or (+)	86.9	(13.1)	Lepases, corn oil	+	98.5	
Lysine decarboxylase	+	99.6		triacetin	d	3.2	(13.2)
Arginine dihydrolase	−	1.3[W]		tributyrin	d	89.7	(8.1)
Ornithine decarboxylase	+	99.5		Maltose	+	97.1	(1.3)
Phenylalanine deaminase	−	0		Xylose	d	8	(18.3)
Glucose acid	+	100		Trehalose	+	99.7	(0.3)
gas	− or +**	52.6		Cellobiose	d	20.8	(33.4)
Lactose	−	2.2	(6.2)	Glycerol	+	97	(2.6)
Sucrose	+	99.7		Alpha methyl glucoside	−	0.9	(0.6)
Mannitol	+	100		Erythritol	d	1.7	(22.8)
Dulcitol	−	0		Esculin	+	90.8	
Salicin	+,	95.1	(1.7)	Phenylpropionic acid agar	−	1.9[W]	
Adonitol	d	55.5	(16.4)	Cetrimide agar	d	14.3	(3.3)
Inositol	d	78.5	(8.2)	DNase	+	96.7	
Sorbitol	+	98.3		Oxidase	−	0	
Arabinose	−	0		Nitrate to nitrite	+	95.8	
Raffinose	−	1.7	(1.2)	Pegment	− or +	20.9	

*Numerals in parentheses indicate percentage of delayed reactions.
**Gas volumes 10% or less.

+　Positive (90% or more) within 1 or 2 days.
(+) Positive reaction after 3 or more days.
−　No reaction (90% or more).
+ or −　Majority of strains positive, some cultures negative.
− or +　Majority of cultures negative, some strains positive.
d　Different reactions: +, (+), −.
w　Weakly positive reaction.

bacteriaceae that produce detectable lipase (methods of Hugo and Beveridge, 1962) are *S. marcescens, E. liquefaciens, P. vulgaris,* and *P. mirabilis* (Davis and Ewing, 1964). Similarly the only microorganisms in the family that are known to produce detectable extracellular DNase (methods of Jeffries et al., 1957 and Smith et al., 1969) are *S. marcescens, E. liquefaciens,* and some members of the genera *Proteus* and

Providencia (Martin and Ewing, 1967 and unpublished data). However, large numbers of each of the genera of Enterobacteriaceae other than *Serratia* have not been examined for production of extracellular DNase.

Rapid utilization of lactose by serratiae is rare (Table 140) and strains that ferment it generally produce only slightly acid reactions after 5 to 21 days' incubation. Since most cultures of serratiae yield positive reactions in the ONPG test (see Chapter 18 for references), it appears that a lactase is formed but that permease is produced slowly under ordinary conditions.

Differentiation of cultures of *Serratia marcescens* from those Enterobacteriaceae to which they are more closely allied may be accomplished by the use of selected tests. The reactions given by serratiae in commonly employed tests are recorded in Tables 8 and 22 (Chapter 3) and tests of value in the differentiation of *S. marcescens*, *marcescens* and *E. liquefaciens* are listed in Table 21 (Chapter 3).

Cultures of *Serratia* may be differentiated from klebsiellae on the basis of their positive reactions in ornithine medium, failure to produce gas from inositol, glycerol, or cellobiose, slowness and reluctance to ferment lactose, failure to utilize malonate, rapid gelatin liquefaction, their motility, and by the small volumes of gas produced from fermentable substrates by aerogenic strains.

Tests and substrates that are helpful for differentiation of *S. marcescens*, *E. cloacae*, and *E. aerogenes* are listed in Table 141. The fact that cultures of *E. cloacae* do not decarboxylate lysine (Table 135, Chapter 13) also is of assistance in effecting this differentiation. Further, gas production from certain fermentable carbohydrates (e.g., cellobiose) by most cultures of *Enterobacter* serves to differentiate these from serratiae (Table 135). Means by which cultures of *S. marcescens* and *E. hafniae* may be differentiated are given in Table 142 (see also Tables 135 and 140).

In the differentiation of *S. marcescens* and *E. liquefaciens* (Table 21, Chapter 3), the production of larger volumes of gas from fermentable substrates by strains of the latter species should be borne in mind. Further, cultures of *E. liquefaciens* are more active biochemically when incubated at 22 C to 25 C than they are at 35 C to 37 C. Cultures of *S. marcescens* do not exhibit this preference for the lower temperature of incubation, for example:

Serratiae also must be differentiated from *Aeromonas* and from Erwineae. Aeromonads (and vibrios) yield positive reactions in indophenol oxidase tests, which serves to differentiate them from serratiae as well as from other Enterobacteriaceae (Ewing and Johnson, 1960). In addition, certain other reactions are helpful in the differentiation of *S. marcescens* and *Aeromonas hydrophila*. These differential reactions are recorded in Table 143. Strains of *Aeromonas shigelloides* are anaerogenic and ferment inositol but fail to produce acid from mannitol (Ewing and Johnson, 1960; Ewing et al., 1961). Lastly the flagellation of *Aeromonas* is polar. The bacteria included in the tribe Erwineae *(Bergey's Manual)* are mentioned in Chapter 15 (The Genus *Pectobacterium*).

SEROLOGICAL REACTIONS

Six somatic antigen groups of *Serratia* were established by Davis and Woodward (1957) and Ewing et al. (1959) extended the number of somatic groups to 9 and characterized 13 flagellar (H) antigens. The number of O antigen groups since has been extended to 15 and with these O and H antisera it has been possible to determine the serotype (i.e., both the O and the H antigens) of about 95 per cent of cultures from a variety of sources and from many localities (e.g., see Ewing et al., 1962b). It was possible to delineate 97 serotypes among 652 cultures (Davis, unpublished data, 1967). Sedlak et al. (1965) added a tenth O antigen group to the original 9 mentioned above and subdivided several existing groups. Hefferan (1906) reported antigenic relationships among cultures of serratiae, and though this investigator apparently did not distinguish between O and H agglutination reactions, as would be done now, the work is of interest because it is the earliest work dealing with serological reactions of *Serratia* known to the author.

The methods employed in the serological examination of cultures of *Serratia* are based upon those established in the study of other genera of Enterobacteriaceae such as *Escherichia* (Chapter 5). Antisera for the O antigens may be produced with 4- to 5-hr broth cultures, inoculated with smooth colonies selected from platings, and heated at 100 C for 2½ hr. Antisera for the H antigens of *Serratia* are produced by the injection of phenolized broth cultures. Strains used for H antiserum production should be passaged through semisolid agar medium to enhance motility

		S. marcescens, marcescens	E. liquefaciens
Lysine decarboxylase	37 C	+ (99.6%)	+or- (82.4%)
	22 C	+ (99.6%)	+ (100%)
Voges-Proskauer	37 C	+ (100%)	-or+ (30.9%)
	22 C	+ (100%)	+or- (79.4%)

TABLE 141

Differentiation of *S. marcescens*, *E. cloacae*, and *E. aerogenes*

Test or substrate	Serratia marcescens			Enterobacter					
				cloacae			aerogenes		
	Sign	%+	(%+)*	Sign	%+	(%+)*	Sign	%+	(%+)*
Arabinose	–	0		+	99.5		+	100	
Raffinose	–	1.7	(1.2)	+	97		+	96	
Rhamnose	–	0		+	92	(1.5)	+	98.7	
Alpha methyl glucoside	–	0.9	(0.6)	+ or (+)	84.2	(12.8)	+	96	(2)
Malonate	–	1.7		+ or –	80.6		+ or –	74.7	
Mucate	–	0		+ or –	75.6		+	94.7	
Lipase (corn oil)	+	98.5		–	0.8	(1.5)	–	0	

*Numerals in parentheses indicate percentage of delayed reactions.

+ Positive (90% or more) within 1 or 2 days.
(+) Positive reaction after 3 or more days.
– Negative (90% or more).
+ or – Majority of strains positive, some cultures negative.
– or + Majority of cultures negative, some strains positive.

TABLE 142

Differentiation of *S. marcescens* and *E. hafniae*

Test or substrate	S. marcescens			E. hafniae		
	Sign	%+	(%+)*	Sign	%+	(%+)*
Sorbitol	+	93.3		–	0	
Arabinose	–	0		+	96	
Rhamnose	–	0		+	93	(7)
Malonate	–	1.7		+ or –	74	
Lipase (Corn oil)	+	98.5		–	0	
Adonitol	d	55.5	(16.4)	–	0	
Inositol	d	78.5	(8.2)	–	0	

*Numerals in parentheses indicate percentage of delayed reactions.

+ Positive (90% or more) within 1 or 2 days.
(+) Positive reaction after 3 or more days.
– Negative (90% or more).
+ or – Majority of strains positive, some cultures negative.
– or + Majority of cultures negative, some strains positive.
d Different reactions: +, (+), –.

TABLE 143

Differentiation of *S. marcescens* and *A. hydrophila*

Test or substrate	S. marcescens			A. hydrophila		
	Sign	%+	(%+)*	Sign	%+	(%+)*
Indophenol oxidase	–	0		+	100	
Indol	–	0.2^w		+ or –	86.7	
Methyl red 37 C	– or +	17.7		+	94.7	
22 C	–	8.6		– or +	54.9	
Voges-Proskauer 37 C	+	100**		– or +	32.7	
22 C	+	100**		+ or –	66.1	
Simmons' citrate	+	98.6	(0.5)	d	52.2	(25.7)
Arabinose	–	0		d	52.2	(1.8)
Adonitol	d	55.5	(16.4)	–	0	
Inositol	d	78.2	(8.2)	–	0	
Sorbitol	+	98.3		d	13.3	(0.9)
Xylose	d	8	(18.3)	–	0	
Phenylalanine deaminase	–	0		– or +^{w}	24.8	
Lysine decarboxylase	+	99.6		–	0	
Arginine dehydrolase	–	1.3^w		d	75.2	(9.7)
Ornithine decarboxylase	+	99.5		–	0	

*Numerals in parentheses indicate percentage of delayed reactions.
**Cultures of *S. marcescens, kiliensis* are V-P negative.

+ Positive (90% or more) within 1 or 2 days.
(+) Positive reactions after 3 or more days.
 – Negative (90% or more).
+ or – Majority of strains positive, some cultures negative.
– or + Majority of cultures negative, some strains positive.
d Different reactions: +, (+), –.

and H antigen development prior to their use. The first such passages may be made in medium containing 0.2% agar and subsequent passages in medium containing 0.4% agar. (See Chapter 5 for details)

The agglutination reactions obtained with O and H antisera produced by the author and colleagues are given in Tables 144 and 145, respectively. The strains selected for production of antisera also are listed in the tables. As might be expected all strains of O group 2 do not cross react in 03 antiserum, and all O group 3 cultures are not agglutinated by 02 antiserum (Table 144).

Although slide agglutination may be used for deter-

mination of the O antigen group of many cultures of *Serratia* they cannot be relied upon in all instances. For this reason, Ewing (1959) suggested that the more accurate way to determine the O antigen group of strains of *Serratia* is by means of single tube agglutination tests followed by titrations and tests in absorbed antisera when the latter tests were received. For this purpose O antiserum pools are used for preliminary examinations. These pools are employed in single tube tests using 1.0 ml of broth culture, heated at 100 C for 1 hr, as antigen. The final dilution of each component antiserum in these tests is 1:100. Subsequently, antigen suspensions are tested in the individual antisera con-

TABLE 144

The Relationship of the O Antigens of *Serratia*

Antigen Suspensions 100 C, 1 Hour	O Antisera														
	1	2	3	4	5	6	7	8	9	10	11	12	13	14	15
O1 (866-57)	1280	0	0	0	0	0	0	0	0	0	0	0	0	0	0
O2 (868-57)	0	2560	40	0	0	0	0	0	80	0	0	0	0	0	0
O3 (863-57)	0	160	640	0	0	0	0	0	160	0	0	0	0	0	0
O4 (864-57)	0	0	0	1280	0	0	0	0	0	0	0	0	0	0	0
O5 (867-57)	0	0	0	0	2560	0	0	0	0	0	0	0	0	0	0
O6 (862-57)	0	0	0	0	0	640	80	0	0	0	0	0	0	160	0
O7 (843-57)	0	0	0	0	0	40	1280	0	0	0	0	0	0	0	0
O8 (1604-57)	0	0	0	0	0	0	0	5120	0	640	0	0	320	0	0
O9 (4534-60)	0	0	0	0	0	0	0	0	640	320	640	0	0	0	0
O10 (1289-59)	0	0	0	0	0	0	0	0	0	10240	2560	0	0	0	0
O11 (1914-63)	0	0	0	0	0	0	0	0	0	0	2560	0	0	0	0
O12 (6320-58)	0	0	0	0	0	0	0	0	0	0	0	640	0	320	0
O13 (3607-60)	0	0	0	0	0	0	0	0	0	0	0	0	1280	0	0
O14 (4444-60)	0	0	0	0	0	0	0	0	0	0	0	40	0	640	0
O15 (4523-60)	0	0	0	0	0	0	0	0	0	160	0	0	0	0	1280

Homologous titrations with strains selected as standard.

TABLE 145

Reactions of the H Antigens of *Serratia*

H Antigen Suspensions	H Antisera												
	1	2	3	4	5	6	7	8	9	10	11	12	13
	863	836	862	864	866	680	841	877	1783	2420	827	874	2436
1 863-57	6400	0	0	0	0	0	0	0	0	0	0	0	0
2 836-57	0	6400	0	0	0	0	0	0	0	0	200	0	0
3 862-57	0	0	6400	0	0	0	0	0	0	200	0	0	0
4 864-57	0	0	0	6400	0	0	0	0	0	0	0	0	0
5 866-57	0	0	0	0	6400	0	0	0	0	0	0	0	0
6 680-57	0	0	0	0	0	6400	0	200	0	0	0	0	800
7 841-57	0	0	0	0	0	0	6400	0	0	0	0	0	0
8 877-57	0	0	0	0	0	0	0	6400	200	200	0	0	0
9 1783-57	0	0	0	0	0	0	0	1600	6400	0	0	0	0
10 2420-57	0	0	0	0	0	0	200	200	0	12,800	0	0	0
11 827-57	0	0	0	0	0	0	0	0	0	0	12,800	0	0
12 874-57	0	0	0	0	0	0	0	0	0	0	0	6400	0
13 2436-57	0	100	0	0	0	0	0	0	0	0	0	0	6400

tained in the pool in which a reaction occurred. The latter tests also are made at a final antiserum dilution of 1:100. Tube tests for O agglutination are incubated in a waterbath at 48 to 50 C for 16 to 18 hr. The results obtained in the examination of numerous cultures in this manner showed that in many instances there were no cross reactions in these tests and confirmation of the O group of a culture could be obtained by simple titration in unabsorbed antiserum for the indicated O group. However, cross reactions often occurred between strains that belonged to O groups 2 and 3 and to 6 and 7. In these instances slide or single tube tests in appropriately absorbed O antisera often were required, in addition to titration in unabsorbed antisera, to complete the O group determinations.

In some instances it is possible to determine the H antigens of strains without recourse to passage through semisolid agar medium. However, many cultures do not yield typical, rapid, H agglutination until after they have been passaged through semisolid agar medium two or three times. This enhancement of motility and H antigen development may be carried out in the manner recommended in Chapter 5.

During the above-mentioned investigations, it was found that broth cultures (5 hours' incubation) that were preserved with an equal volume of 0.5% phenolized physiological saline solution reacted to higher titers (usually two dilutions) in H antisera than did portions of the same broth cultures preserved with formalin (0.3% final concentration). For this reason, broth cultures preserved and diluted with phenolized physiological saline solution are recommended for use as H antigen suspensions. A possible explanation for the difference in behavior of the two types of H antigen preparations may be found in the work of Leifson (1961), who studied the effect of formalin on the flagella of strains of *Serratia*. Leifson reported that formalin caused the flagella of *Serratia* to become tightly coiled. Conceivably, this alteration of the flagella could affect the character and extent of flagellar agglutination.

Pools of H antisera may be used for preliminary examination of the H antigens of *Serratia*. Following this, H antigen suspensions may be tested in single tube tests at dilutions of 1:500 or 1:1000 of the individual H antisera that comprise the pool in which a reaction occurred. The results of such tests should be confirmed by titrations, or by tests in absorbed H antisera when indicated, as in the differentiation of H6 and H13 or H8 and H9. Tests for agglutination of H antigens should be incubated in a waterbath at 48 C to 50 C and final readings should be made after 1 hr.

The occurrence of K antigens in cultures of *Serratia* has been noted (Ewing et al., 1959). Some strains are capsulated.

Serotyping of *Serratia* is of considerable value in the study of hospital-acquired infections in which members of this genus are involved (Ewing et al., 1962b; Wilfert et al., 1970). The occurrence of serratiae in a variety of specimens from man and lower animals has been reported (loc. cit., Fields et al., 1967)

REFERENCES

Bergey's Manual of determinative bacteriology, 5th ed., 1939. Edited by D. H. Bergey, R. S. Breed, E. G. D. Murray, and A. P. Hitchens. Baltimore: Williams & Wilkins.

————. Ibid., 6th ed., 1948. Edited by R. S. Breed, E. G. D. Murray, and A. P. Hitchens. Baltimore: Williams & Wilkins.

————. Ibid., 7th ed., 1957. Edited by R. S. Breed, E. G. D. Murray, and N. F. Smith. Baltimore: Williams and Wilkins.

Burger, G. 1967. Zentralbl. f. Bakt., 1. Ref., **207**, 365.

Colwell, R. R., and M. Mandel. 1965. J. Bacteriol., **89**, 454.

Davis, B. R., and J. M. Woodward. 1957. Can. J. Microbiol., **3**, 591.

Davis, B. R., and W. H. Ewing. 1957. Int. Bull. Bacteriol. Nomen. Tax., **7**, 151.

————. 1964. J. Bacteriol., **88**, 16.

Ewing, W. H. 1963. Int. Bull. Bacteriol. Nomen. Tax., **13**, 95.

Ewing, W. H. 1967. Revised definitions for the family Enterobacteriaceae, its tribes, and genera. CDC Publ.*

Ewing, W. H., B. R. Davis, and R. W. Reavis. 1959. Studies on the *Serratia* group. CDC Publ.*

Ewing, W. H., and J. G. Johnson. 1960. Inter. Bull. Bact. Nomen. Tax., **10**, 223.

Ewing, W. H., R. Hugh, and J. G. Johnson. 1961. Studies on the *Aeromonas* group. CDC Publ.*

Ewing, W. H., B. R. Davis, and J. G. Johnson. 1962a. Int. Bull. Bacteriol., Nomen. Tax., **12**, 47.

Ewing, W. H., J. G. Johnson, and B. R. Davis. 1962b. The occurrence of *Serratia marcascens* in noscomial infections. CDC Publ.*

Fields, B. N., M. M. Uwaydah, L. J. Kunz, and M. N. Swartz. 1967. Amer. J. Med., **42**, 89.

Fife, M. A., W. H. Ewing, and B. R. Davis. 1965. The biochemical reactions of the tribe Klebsielleae. CDC Publ.*

Grimont, P. A. D. 1969. Les *Serratia* etude taxométrique. Bergeret. Bordeaux.

Hefferan, M. 1904. Zentralbl. f. Bakt., 2. Abt., **11**, 311, 397, 456, 520.

*CDC Publ., Publication from the Center for Disease Control (formerly the Communicable Disease Center), Atlanta, Ga. 30333.

————— . 1906. Zentralbl. f. Bakt., 1. Orig., **41**, 553.

Hugo, W. B., and Beveridge, E. G. 1962. J. Appl. Microbiol., **25**, 72.

Jeffries, C. D., D. F. Holtman, and D. G. Guse. 1957. J. Bacteriol., **73**, 590.

Leifson, E. 1961. J. Gen. Microbiol., **25**, 131.

Martin, W. J., and W. H. Ewing. 1967. Can. J. Microbiol., **13**, 616.

Martinec, T., and M. Kocur. 1960. Int. Bull. Bacteriol. Nomen. Tax., **10**, 247.

————— . 1961a. Ibid., **11**, 7.

————— . 1961b. Ibid., **11**, 73.

————— . 1961c. Ibid., **11**, 87.

Sedlak, J., V. Dlabac, and M. Motlikova. 1965. J. Hyg. Epidemiol. Microbiol. Immunol., **9**, 45.

Stapp, C. 1940. Zentralbl. f. Bakt., 2 Abt., **102**, 251.

Stevenson, J. P. 1959. J. Gen. Microbiol., **21**, 366.

Smith, P. B., G. A. Hancock, and D. L. Rhoden. 1969. Appl. Microbiol., **18**, 991.

Wilfert, J. N., F. F. Barrett, W. H. Ewing, M. Finland, and E. H. Kass. 1970. Ibid., **19**, 345.

Chapter 15
The Genus *Pectobacterium*

This genus is composed of gram negative bacteria that possess all of the general characteristics of Enterobacteriaceae. They are peritrichously flagellated, fermentative microorganisms, which fail to produce indophenol oxidase but reduce nitrate to nitrite.

Following consultation with Dr. D. C. Graham of Scotland,[1] the author placed the pectobacteria in the tribe Klebsielleae and defined them as follows (Ewing, 1967):

The genus *Pectobacterium* is composed of motile or nonmotile bacteria that conform to the definitions of the family ENTEROBACTERIACEAE and the tribe KLEBSIELLEAE. Sodium pectate medium is liquefied. A minority of cultures produce indol, but the majority yield positive reactions in the methyl red test. Gelatin is liquefied although the reactions of a minority of strains may be somewhat delayed. Lysine, arginine, and ornithine are not decarboxylated. Sodium alginate is not utilized as a sole source of carbon. Gas is not formed from inositol or glycerol, and adonitol is not fermented. Sorbitol is fermented only very rarely but the majority of cultures produce acid from rhamnose, arabinose, and raffinose. The optimum growth temperature is about 25 C and cultures fail to grow or grow poorly at 37 C. The type species is *Pectobacterium carotovorum* (Jones) Waldee.

This definition should be used in conjunction with that given for the tribe Klebsielleae (Chapter 1).

The pectobacteria frequently have been referred to as the soft rot coliform bacteria because of the nature of the lesions they produce in plant tissues. On the basis of comparative studies Waldee (1945) concluded that the pectobacteria differed from other bacteria commonly classified in the genus *Erwinia* (e.g., various editions of *Bergey's Manual*) in many respects and stated that they resembled coliform bacteria with respect to their biochemical reactions. Waldee proposed the genus *Pectobacterium* for these particular pectinolytic bacteria and further proposed inclusion of

this genus in the family Enterobacteriaceae. As a result of his investigations Graham (1964) came to similar conclusions. It was his opinion that these pectolytic bacteria should be removed from the genus *Erwinia*, placed in the genus *Pectobacterium* Waldee, and included in the family Enterobacteriaceae. Waldee (1945) also proposed the formation of a new family Erwiniaceae to include the emended genus *Erwinia* represented by *E. amylovora, E. tracheiphila,* and *E. salcis* (i.e. the *Erwinia sensu stricto*). Waldee stated that yellow peritrichously flagellated plant bacteria occupied a doubtful position taxonomically but seemed to be close to *S. marcescens* (v. inf.).

The author and colleagues have been studying cultures classified as *Erwinia* since about 1955. These cultures were received from Dr. M. P. Starr of the University of California at Davis, from various culture collections, and more recently from Dr. D. C. Graham. As a result of these continuing investigations it has been concluded that the genus *Erwinia* as presented in the last several editions of Bergey's Manual is composed of at least four groups of different bacteria. First, the true *Erwinia,* which have been placed in a separate family Erwiniaceae by Waldee (1945). These bacteria are not Enterobacteriaceae and should not be included in that family. The true *Erwinia* as exemplified by *E. amylovora* differ from Enterobacteriaceae in their growth characteristics, their strong tendency toward oxidative utilization of glucose, failure to reduce nitrate to nitrite, and failure to grow at 37 C (Billings et al., 1961; unpublished data). According to Starr and Mandel (1950) and Billings et al. (1961) nicotinic acid is required by *E. amylovora.* Sutton and Starr (personal communications, 1956) reported that certain growth factors are required by the true *Erwinia* but not by pectobacteria or by the other Enterobacteriaceae examined.

The second of the above-mentioned groups is composed of the pectolytic bacteria formerly classified as *E. carotovorum* (v. inf.). The third group includes peritrichously flagellated microorganisms referred to by the author and co-workers as the Herbicola-Lathyri bacteria. These organisms have been given a number of different specific names (e.g. *E. herbicola, E. lathyri, E.*

[1]Department of Agriculture and Fisheries for Scotland, Agricultural Scientific Services, East Craigs, Edinburgh, Scotland.

318

milletiae, B. typhiflavum, etc.), but cultures received with these and other labels all appear to be related (Graham and Hodgkiss, 1967; Ewing and Fife, 1971[2]). In the author's experience about 65 per cent of cultures of these bacteria produce a yellow pigment which is said to be carotinoid-like. These bacteria possess many of the characteristics of Enterobacteriaceae (unpublished data), but opinion as to their taxonomic position is reserved for the present. In passing it is notable that although the Herbicola-Lathyri bacteria are anaerogenic, the author and colleagues have received a number of cultures that closely resemble the Herbicola-Lathyri bacteria but are aerogenic.

The fourth of the aforementioned groups of bacteria is composed of a heterogeneous collection of microorganisms which have been identified as *Enterobacter* of one species or another, *K. pneumoniae* etc. Two cultures labeled *E. carnegieana* when received are typical strains of *K. pneumoniae.* Clearly, these bacteria should be classified in the respective genera of Enterobacteriaceae in which they belong.

BIOCHEMICAL REACTIONS

The biochemical reactions given by 90 cultures of *Pectobacterium* received from Dr. D. C. Graham are recorded in Table 146. The reactions obtained when cultures were incubated at 22 C to 25 C and 37 C are included in Table 146 for comparison. Since some cultures did not grow on all substrates when incubated at 37 C, the signs listed in the column headed 37 C in Table 146 refer only to percentages of strains that grew at that temperature. The outstanding characteristic of pectobacteria, of course, is their ability to liquefy pectate gel, a characteristic that is not shared by other Enterobacteriaceae (e.g., see Davis and Ewing, 1964).

The reactions given by members of the genus *Pectobacterium* in commonly employed biochemical tests are recorded in Tables 8 and 22 (Chapter 3). The data given in these tables, together with that contained in Tables 135 (Chapter 13) and 140 (Chapter 14) should be helpful in the differentiation of pectobacteria. In the past cultures of *Klebsiella* and *Citrobacter* have been mistakenly identified as *Pectobacterium.* For this reason means by which these bacteria may be differentiated are listed in Tables 147 and 148.

For further information regarding the biochemical reactions given by members of the genus *Pectobacterium,* as well as the taxonomy and nomenclature of the microorganisms and many references, readers are referred to the publications of Martinec and Kocur (1963, 1968) in addition to the references cited in the preceding paragraphs.

The author is unaware of any systematic investigation of the serological characteristics of the pectobacteria. However, Elrod (1941a, 1941b) reported that serologically these bacteria were heterogeneous, in contrast to *E. amylovora* which was quite homogeneous.

During recent years at least six or eight cultures of *Pectobacterium,* which were isolated from extraintestinal sources in humans, have been received for identification. Considering the plant sources of these bacteria, it would seem likely that they would occur in stool specimens with some degree of frequency. However, there is no data available at present which indicates the frequency of occurrence of *Pectobacterium* in stool or extraintestinal specimens from man or lower animals.

REFERENCES

Billings, E., L. A. E. Baker, J. E. Crosse, and C. M. E. Garrett. 1961. J. Appl. Bacteriol., **24**, 195.

Davis, B. R., and W. H. Ewing. 1964. J. Bacteriol., **88**, 16.

Elrod, R. P. 1941a. Bot. Gaz., **103**, 123.

———. 1941b. Ibid., **103**, 266.

Ewing, W. H. 1967. Revised definitions for the family Enterobacteriaceae, its tribes, and genera. CDC Publ.*

Ewing, W. H., and M. A. Fife. 1971. *Enterobacter agglomerans,* the Herbicola-Lathyri bacteria. CDC Publ.*

Graham, D. C. 1964. Ann. Rev. Phytopathol., **2**, 13.

Graham, D. C. and W. Hodgkiss. 1967. J. Appl. Bacteriol., **30**, 175.

Martinec, T., and M. Kocur. 1963. Folia Fac. Sci. Natur. Univ. Purkyniana Brunensis, **4**, 3.

———. 1968. Ibid., **9**, 63.

Starr, M. P., and M. Mandel. 1950. J. Bacteriol, **60**, 669.

Waldee, E. L. 1945. Iowa State Coll. J. Sci., **19**, 435.

---

[2] The name *Enterobacter agglomerans* has been proposed for the Herbicola-Lathyri bacteria.

*CDC Publ., Publication from the Center for Disease Control (formerly the Communicable Disease Center), Atlanta, Ga. 30333.

TABLE 146

Summary of the Biochemical Reactions of *Pectobacterium*

Test or substrate	22 to 25 C			37 C			
	Sign	%+	(%+)*	Sign	%+	(%+)*	%NG
Hydrogen sulfide	–	0		–			28.6
Urease	d	30	(20)	d	21.4	(7.1)	28.6
Indol	– or +	20	(1.4)	– or +	20		24.3
Methyl red	+ or –	75.7		+ or –	45.7		20
Voges-Proskauer	– or +	47.1		– or +	22.9		15.7
Citrate (Simmons')	+ or (+)	71.5	(20)	d	45.8	(17.1)	27.1
KCN	+ or –	68.5	(12.9)	+ or –	56.1		
Motility	+ or –	82.9	(4.3)	+ or –	55.7		17.2
Gelatin (22 C)	+ or (+)	67.1	(32.9)				
Lysine decarboxylase	–	0		–	0		7.1
Arginine dihydrolase	– or (+)	0	(12.9)	–	0	(5.7)	7.1
Ornithine decarboxylase	–	0		–	0		
Phenylalanine deaminase	–	0		–	0		
Glucose acid	+	98.6	(1.4)	d	78.6	(2.8)	10
gas	d	18.6	(25.7)	– or +	15.7	(1.4)	10
Lactose	+ or (+)	75.7	(18.6)	d	52.9	(18.5)	12.9
Sucrose	+	97.1	(2.9)	+ or –	74.3		8.6
Mannitol	+ or –	97.1	(2.9)	+ or –	75.7	(11.4)	12.9
Dulcitol	–	0		–	0		12.9
Salicin	+	100		d	70	(1.4)	12.9
Adonitol	–	0		–	0		12.9
Inositol	–	0	(5.7)	–	0		12.9
Sorbitol	–	0	(1.4)	–	2.8		12.9
Arabinose	+	95.7	(4.3)	+ or –	71.4		12.9
Raffinose	+ or (+)	84.3	(12.8)	d	67.1	(1.4)	12.9
Rhamnose	d	71.4	(14.3)	d	64.3	(5.7)	12.9
Sodium acetate	+ or (+)	52.9	(44.3)		NT		
Malonate	– or +	22.9		– or +	21.4		17.1
Mucate	+ or (+)	78.6	(12.9)	d	45.7	(5.7)	27.1

TABLE 146 (Continued)

Summary of the Biochemical Reactions of *Pectobacterium*

Test or substrate	22 to 25 C			37 C			
	Sign	%+	(%+)*	Sign	%+	(%+)*	%NG
Pectate	+ or (+)	75.7	(24.3)	d	41.5	(17.1)	15.7
Sodium alginate							
synthetic	–	2.9	(1.4)	–	0		
nutrient	–	0		–	0		
Lipase (corn oil)	–	1.4		–	0		15.7
Maltose	d	1.4	(27.2)	–	0	(2.8)	12.9
Xylose	+ or (+)	92.9	(4.3)	d	70	(1.4)	11.5
Cellobiose acid	+ or (+)	81.5		d	68.5	(4.3)	12.9
gas	d	1.4	(45.7)	d	5.7	(14.3)	12.9
Glycerol acid	–	0	(8.6)	d	18.6	(22.8)	12.9
gas	–	0		–	0		
Alpha methyl glucoside	d	21.5	(11.4)	d	4.3	(1.4)	14.3
Erythritol	–	0		–	0		14.3
Esculin	+ or (+)	95.7	(4.3)	+ or –	71.4		14.3
Nitrate to nitrite	+	100		+	90		8.6
Oxidation – Fermentation	F	100		F	95.7		4.3
Oxidase	–	0		–	0		30
Pigment (transient blue)	–	5					
Cetrimide	d	74.3	(14.3)	d	37.1	(14.3)	15.7
Organic acids**							
citrate	+	91.4			NT		
D-tartrate	–	4.3			NT		

* Figures in parentheses indicate percentages of delayed reactions.
** Method of Kauffmann and Petersen.

+ 90% or more positive within one or two days' incubation.
(+) Positive reaction after 3 or more days.
 – No reaction (90% or more).
+ or –   Majority of strains positive, some cultures negative.
– or +   Majority of cultures negative, some strains positive.
d Different Reactions: +, (+), –.
NG   No growth.
NT   Not tested.

TABLE 147

Differentiation of *Pectobacterium* and *Klebsiella*

| Test or substrate | Pectobacterium | | | | | | Klebsiella | | | | | |
| | 25C | | | 37C | | | pneumonia | | | ozaenae | | |
	Sign	%+	(%+)*	Sign	%+	(%+)*	Sign	%+	(%+)*	Sign	%+	(%+)*
Lysine decarboxylase	-	0		-	0		+	97.2	(2.8)	- or +	48	
Urease	d	30	(20)	d	21.4	(7.1)	+	94.5		d	9.5	(10.3)
Voges-Proskauer	- or +	47.1		- or +	22.9		+	91.1		-	0	
Gelatin	+ or (+)	67.1	(32.9)	d	41.5	(17.1)	-	3.3		-	0	
Pectate	+ or (+)	75.7	(24.3)	+ or -	55.7		-	0		-	0	
Motility	+ or -	82.9		-			-	0		-	0	
Sodium alginate (synthetic)	-	2.9	(1.4)	-	0		+ or -	88.5	(9.2)	- or (+)	0	(37.5)
Sucrose	+	97.1	(2.9)	+ or -	74.3		+	98.9		d	16.3	(17.3)
Dulcitol	-	0		-	0		- or +	31.5		-	0	
Sorbitol	-	0	(1.4)	-	2.8		+	99.4	(0.3)	d	78	(10)
Adonitol	-	0		-	0		+ or -	87.7		-	98.3	(1.7)
Inositol	-	0	(5.7)	-	0		+	97.9	(0.8)	d	58.6	(21.6)
Gas from glycerol	-	0		-	0		+	92.5	(3.7)	d	30	(24)

*Figures in parentheses indicate percentages of delayed reactions (3 or more days).

+ 90% or more positive within one or two days' incubation.

(+) Positive reaction after 3 or more days.

- No reaction (90% or more).

+ or - Majority of strains positive, some cultures negative.

- or + Majority of cultures negative, some strains positive.

(+) or + Majority of reactions delayed, some occur within 1 or 2 days.

d Different reactions: +, (+), -.

TABLE 148

Differentiation of *Pectobacterium* and *C. freundii*

Test or Substrate	*Pectobacterium* 25C			*Pectobacterium* 37C			*C. freundii* 37C		
	Sign	%+	(%+)*	Sign	%+	(%+)*	Sign	%+	(%+)*
Arginine dihydrolase	– or (+)	0	(12.9)	–	0	(5.7)	d	43.6	(44.4)
Ornithine decarboxylase	–	0		–	0		d	17.2	(0.2)
Voges-Proskauer	– or +	47.1		– or +	22.9		–	0	
Gelatin	+ or (+)	67.1	(32.9)				–	0	(0.9)
Sodium acetate	–	2.9	(1.4)	–	0		+ or (+)	77.5	(13.5)
Pectate	+ or (+)	75.7	(24.3)	d	41.5	(17.1)	–	0	
Sucrose	+	97.1	(2.9)	+ or –	74.3		d	15.3	(9.4)
Dulcitol	–	0		–	0		d	59.8	(0.7)
Salicin	+	100		d	70	(1.4)	d	4.1	(23)
Sorbitol	–	0	(1.4)	–	2.8		+	98	(1)
Raffinose	+ or (+)	84.3	(12.8)	d	67.1	(1.4)	d	14.2	(0.8)
Maltose	d	1.4	(27.2)	–	0	(2.8)	+	98.5	(1.5)
Glycerol acid	–	0	(8.6)	d	18.6	(22.8)	+	97.9	(0.9)
gas	–	0		–	0		+ or –	83.5	
Esculin	+ or (+)	95.7	(4.3)	+ or –	71.4		–	0.9	(3)

*Figures in parentheses indicate percentages of delayed reactions (3 or more days).

+   90% or more positive within one or two days' incubation.

(+) Positive reaction after 3 or more days.

–   No reaction (90% or more).

+ or –   Majority of strains positive, some cultures negative.

– or +   Majority of cultures negative, some strains positive.

(+) or +   Majority of reactions delayed, some occur within 1 or 2 days.

d Different reactions: +, (+), –.

Chapter 16
The Genus *Proteus*

Rustigian and Stuart (1941, 1943a, 1943b, 1945) reviewed the literature on the subject of the genus *Proteus* and presented a classification which found wide acceptance. This classification divided the genus into four species: *Proteus vulgaris, P. mirabilis, P. morganii,* and *P. rettgeri.* Other systems of classification have been suggested and other nomenclature has been proposed for parts of the genus (see Ewing, 1958, 1962, and Rauss, 1962, for reviews of proposals) but the author has retained the classification and nomenclature of Rustigian and Stuart (1945) for use herein because it is practical and since it is familiar to the majority of workers.

The definition given in the 1957 edition of *Bergey's Manual* is unacceptable to the author because it incorporates the genus *Providencia* directly into the genus *Proteus,* in accordance with the proposal of Shaw and Clarke (1955). The author is of the opinion that this should not be done. The genus *Proteus* was defined by Ewing (1962) and in 1967 this definition was emended as follows:

The genus *Proteus* is composed of motile bacteria that conform to the definitions of the family ENTEROBACTERIACEAE and the tribe PROTEEAE. Urea is hydrolyzed rapidly. Two species, *Proteus vulgaris* and *Proteus mirabilis,* produce hydrogen sulfide rapidly and abundantly, liquefy gelatin, and swarm on moist agar media. The majority of cultures of these two species are lipolytic. Two other species, *Proteus morganii* and *Proteus rettgeri* do not possess these particular characteristics. Ornithine is decarboxylated by two species, *P. mirabilis* and *P. morganii.* Mannitol is fermented by the majority of strains of *P. rettgeri* but the remaining species fail to produce acid from this substrate. The type species is *Proteus vulgaris* Hauser.

This definition should be used in conjunction with the definition of the tribe Proteeae (Chapter 1).

BIOCHEMICAL REACTIONS

The more comprehensive publications on the biochemical reactions given by members of the genus *Proteus* that have appeared since 1945 were reviewed and summarized by Ewing et al. (1960a). The studies reviewed included those of Rustigian and Stuart (1945), Cook (1948), Proom and Woiwood (1951), Buttiaux et al. (1954), Singer and Bar-Chay (1954), Phillips (1955), Shaw and Clarke (1955), Colichon and Arana (1956), Keating (1956), Kauffmann (1956a,1956b), Namioka and Sakazaki (1958), and Rauss and Vörös (1959), in all of which 1677 cultures belonging to the four species of *Proteus* were studied. With few exceptions there was agreement in the results reported. The differences that were noted in results reported by some of the authors were for the most part due to the use of overly sensitive test methods (e.g., the determination of hydrogen sulfide production by means of lead acetate paper strips). Further, Ewing et al. (1960a) reported the results of studies with 255 cultures of *Proteus* in tests that have been introduced relatively recently as well as those in use for many years.

Summaries of the biochemical reactions given by 512 cultures that belonged to the four species of *Proteus* are given in Tables 149 and 150. These data are adapted from the publication of Ewing et al. (1960a) and from unpublished data. Further, the reactions given by strains of *Proteus* in commonly used tests are listed in Table 9 (Chapter 3) and tests that are of particular value for differentiation of the four species of *Proteus* are recorded in Tables 23, 24, and 25 (Chapter 3).

Rustigian and Stuart (1945) reported that a number of cultures of *P. mirabilis* yielded positive reactions in the Voges-Proskauer tests, and Suassuna et al. (1961) reported that a majority of strains of this species gave positive V-P tests when the cultures were incubated at 25 C for 4 days. However, not all cultures of *P. mirabilis* yielded positive tests even when tested by the Barritt method. The results obtained by Suassuna et al. (1961) may be summarized as follows:

	Voges-Proskauer	
	%+, 37 C (2 da)	%+, 25 C (4 da)
P. vulgaris	0	11.3
P. mirabilis	15.6	51.6
P. morganii	0	0
P. rettgeri	0	0

Since some strains of *P. mirabilis* may give positive V-P reactions when the cultures are incubated at 37 C, and since some isolants of *P. vulgaris* may yield positive results in this test when incubated at the lower temperature, keys for the identification of the species of *Proteus,* in which the V-P test is an integral part, are of little value. This is not meant to imply that the author believes the V-P test to be valueless in taxonomy. On the contrary, he is convinced that when the V-P test is performed under controlled conditions, it is of considerable value for differentiation within the family Enterobacteriaceae.

The decarboxylase reactions given by cultures of *Proteus* proved to be uniform and were of considerable value in the differentiation of the four species within the genus (Tables 149, 150). They were particularly useful as aids in the proper classification of aberrant cultures. In general, the results obtained confirmed and

TABLE 149

Summary of the Biochemical Reactions of *P. vulgaris* and *P. mirabilis*

Test or substrate	P. vulgaris			P. mirabilis		
	Sign	%+	(%+)*	Sign	%+	(%+)*
Hydrogen sulfide	+	94.7		+	94.2	(2.7)
Urease	+	94.7		+ or (+)	88.4	(1.9)
Indol	+	98.2		–	1.9	
Methyl red (37 C)	+	93		+	98.8	
Voges-Proskauer (37 C)	–	0		– or +	15.6	
Citrate (Simmons')	d	10.5	(14.1)	+ or (+)	58.7	(37.1)
KCN	+	100		+	98.6	
Motility	+	94.7		+	95.1	(0.8)
Gelatin (22 C)	+	90.6	(9.4)	+	91.8	
Lysine decarboxylase	–	0		–	0	
Arginine dihydrolase	–	0		–	0	
Ornithine decarboxylase	–	0		+	99.2	
Phenylalanine deaminase	+	100		+	99.6	
Glucose acid	+	100		+	100	
gas	+ or –	86		+	95.4	(0.4)
Lactose	–	0		–	1.5	
Sucrose	+	94.7		d	18.9	(63.2)
Mannitol	–	0		–	0	
Dulcitol	–	0		–	0	
Salicin	d	58.2	(10.9)	d	0.8	(29.8)
Adonitol	–	0		–	0	
Inositol	–	0		–	0	
Sorbitol	–	0		–	0	
Arabinose	–	0		–	0	
Raffinose	–	0		–	1	
Rhamnose	–	9.4		–	1.5	
Malonate	–	0		–	1.5	
Mucate	–	0		–	0	
Christensen's citrate	– or +	41.5		+	96.8	(1.6)

TABLE 149 (Continued)

Summary of the Biochemical Reactions of *P. vulgaris* and *P mirabilis*

Test or substrate	P. vulgaris			P. mirabilis		
	Sign	%+	(%+)*	Sign	%+	(%+)*
Jordan's tartrate	+	93.3		+ or –	88.1	
Pectate	–	0		–	0	
Sodium acetate	d	23.1	(61.5)	d	13.3	(60)
Ammonium salts glucose agar	d	2.6	(7.7)	d	9.5	(18.1)
Sodium alginate	–	0		–	0	
Lipases						
Corn oil	d	86.9	(2.6)	+	92.3	(7.7)
Triacetin	– or +	32		–	4	
Tributyrin	+ or (+)	84	(8)	d	80	(8)
Maltose	+	96.2	(1.9)	–	0.9	(0.4)
Xylose	+ or (+)	88.7	(1.9)	+	96.2	(3.8)
Trehalose	d	30.2	(60.4)	+	98.4	
Cellobiose	(+) or –	0	(57.7)	d	1.6	(46.6)
Glycerol	+ or (+)	69.8	(30.2)	+	90.3	(7.5)
Alpha methyl glucoside	d	79.5	(5.5)	–	0	
Erythritol	–	2.6		–	0	
Esculin	d	59	(2.6)	–	0	(0.9)
Nitrate to nitrite	+	100		+	93.5	
Oxidation – fermentation	F	100		F	100	
Oxidase	–	0		–	0	
Cetrimide	–	0		–	2.5	

Note: Volumes of gas produced by *Proteus* are small (bubble to 10 or 15%).

* Figures in parentheses indicate percentages of delayed reactions (3 or more days).

  + 90% or more positive within one or two days' incubation.

(+) Positive reaction after 3 or more days.

  – No reaction (90% or more).

+ or –   Majority of strains positive, occasional cultures negative.

– or +   Majority of cultures negative, occasional strains positive.

(+) or +   Majority of reactions delayed, some occur within 1 or 2 days.

 d Different reactions: +, (+), –.

 F Fermentative

extended the work of Moeller (1954, 1955). A few of the strains of *P. morganii* produced doubtful or very weak reactions in lysine medium after 3 or more days of incubation but the number of such reactions did not approach the 91 per cent of weak, delayed reactions reported by Moeller in this species. In the opinion of the author the decarboxylase reactions of *Proteus* cultures, and of other Enterobacteriaceae as well, were more clear-cut and of greater differential value when final readings were made after 3 or 4 days than they were when the tests were incubated for longer periods (Ewing et al., 1960b).

It has been known for many years that an occasional strain of *Proteus* may produce acid from lactose (see Rustigian and Stuart, 1945, for references). In the author's experience such cultures occur with about the same frequency as *Salmonella* that ferment lactose. However, such aberrant strains may become epidemic or endemic in institutions, in which instances the number of isolations from different patients during

a given period of time may lead to a false impression regarding the frequency of occurrence of aberrant cultures under natural conditions (see Chapter 4).

It has been known for a number of years that cultures of *Proteus* and *Providencia* oxidatively deaminate certain amino acids and that the keto acids formed react with ferric compounds to produce a color (see Ewing et al., 1957 for references). Oxidative

TABLE 150

The Biochemical Reactions of *P. morganii* and *P. rettgeri* Summarized

Test or substrate	P. morganii			P. rettgeri		
	Sign	%+	(%+)*	Sign	%+	(%+)*
Hydrogen sulfide (TSI agar)	−	0		−	0	
Urease	+	98.2	0.9	+	100	
Indol	+	100		+	100	
Methyl red (37 C)	+	97.1		+	93.3	
Voges-Proskauer (37 C)	−	0		−	0	
Citrate (Simmons')	−	0		+	95.6	(8.3)
KCN	+	99		+	96.7	
Motility	+ or −	84	(3.7)	+	94.4	
Gelatin (22 C)	−	0		−	0	(2.3)
Lysine decarboxylase	−	0	(1)	−	0	
Arginine dihydrolase	−	0		−	0	
Ornithine decarboxylase	+	97.1		−	0	
Phenylalanine deaminase	+	95.3		+	97.8	
Glucose acid	+	97.2	(2.8)	+	100	
gas	d	84.9	(0.9)	− or +	12.2	
Lactose	−	0		−	8.9	(1.1)
Sucrose	−	1	(2.9)	d	13.3	(56.7)
Mannitol	−	0		+ or −	88.5	
Dulcitol	−	0		−	0	
Salicin	−	0		d	30	(6.6)
Adonitol	−	0		d	80.9	(5.6)
Inositol	−	0		+	93.3	(4.5)
Sorbitol	−	0		d	1.2	(9.6)
Arabinose	−	0		−	0	
Raffinose	−	0		−	9.5	
Rhamnose	−	0		+ or −	67.9	
Malonate	−	4.8		−	1.2	
Mucate	−	0		−	0	
Christensen's citrate	+ or (+)	69.2	(26)	+	100	

TABLE 150 (Continued)

The Biochemical Reactions of *P. morganii* and *P. rettgeri* Summarized

Test or substrate	*P. morganii*			*P. rettgeri*		
	Sign	%+	(%+)*	Sign	%+	(%+)*
Jordan's tartrate	+	92.5		+	95.5	
Pectate	–	0		–	0	
Sodium acetate	–	0		d	59.1	(22.7)
Ammonium salts glucose agar	–	0		+	93.1	(5.2)
Sodium alginate	–	0		–	0	
Lipases						
Corn oil	–	0		–	0	
Triacetin	–	0		–	4	
Tributyrin	–	0		–	0	
Maltose	–	0		–	2.3	(2.4)
Xylose	–	0		– or +	15.1	
Trehalose	– or +	14.4		d	15.9	(1.2)
Cellobiose	–	0	(1.9)	d	3.7	(30.4)
Glycerol	d	4.9	(81.5)	+ or (+)	65.9	(32.9)
Alpha methyl glucoside	–	0		–	2.2	
Erythritol	–	0		d	78.3	(6.5)
Esculin	–	0		d	30.4	(8.7)
Nitrate to nitrite	+ or –	88.5		+	98.8	
Oxidation – fermentation	+ (F)	100		+ (F)	100	
Oxidase	–	0		–	0	
Cetrimide	–	0		– or +w	25	

*Figures in parentheses indicate percentages of delayed reactions (3 or more days).

+ 90% or more positive within one or two days' incubation.

(+) Positive reaction after 3 or more days.

– No reaction (90% or more).

+ or – Majority of strains positive, some cultures negative.

– or + Majority of cultures negative, some strains positive.

(+) or + Majority of reactions delayed, some occur within 1 or 2 days.

d Different reactions: +, (+), –.

w Weakly positive reaction.

deamination of lysine by cultures of these two genera takes place in the area of the slant of lysine iron agar (LIA). When lysine is deaminated the resulting keto acid yields an orange color when tested with ferric compounds such as ferric chloride. In LIA this orange color apparently blends with the bromcresol purple indicator to produce a red color in the area of the slant when the oxidative deamination process has occurred.

The majority of cultures of *Proteus* (except *P. morganii*) and *Providencia* give this red reaction on the slant of LIA and the reaction is very helpful in preliminary examination of colonies picked from plating media as outlined in Chapter 2. Strains of *P. morganii* usually do not produce a discernible red color within 24 hr. While the reaction given by most cultures of *Proteus* and *Providencia* in LIA is useful, as mentioned above, LIA should not be substituted for phenylalanine agar in definitive work (Tables 149, 150).

Susassuna (1963) examined the biochemical characteristics of 697 cultures of *Proteus* and compared the results obtained with those reported by Ewing et al. (1960a). Except for minor differences in percentages, the results of the two investigations were similar. None of the 697 strains reported upon by Suassuna (1963) fermented lactose. This investigator reported that mannose was useful in the differentiation of the species of *Proteus* when used in conjunction with other differential tests and substrates. All cultures of *P. morganii* and *P. rettgeri* fermented mannose, whereas all strains of *P. vulgaris* and *P. mirabilis* failed to utilize this substrate.

Most cultures of *P. morganii* produce a certain amount of brown discoloration in TSI and peptone iron agars after 48 or more hours of incubation. This discoloration appears as a spot at the point at which the stab inoculation was made. Such reactions may be the result of slight production of hydrogen sulfide from organic sources of sulfur, although some investigators (e.g., Tittsler and Sandholzer, 1937) have expressed doubt as to whether it is hydrogen sulfide production. In any event, these reactions do not occur within 24 hours and should not interfere with identification of cultures of *P. morganii*. This subject was reviewed by Rauss et al. (1964).

The occasional culture of *Proteus* that ferments lactose also yields positive results in the ONPG test, as might be expected. Otherwise, members of the genus *Proteus* are not known to produce beta galactosidase when the standard test is used (for references see ONPG test, Chapter 18).

Occasionally specific epithets other than the four mentioned above are seen in the literature. However, the bacteria described under these epithets are strains of *P. vulgaris* or *P. mirabilis*, or they are members of the genus *Aeromonas*. Thus, *P. ammoniae* and *P. americanus* actually are cultures of *P. mirabilis* and *P. paraamericanus* is *P. vulgaris* (see Rustigian and Stuart, 1941, 1945; Ewing et al., 1960a; and Suassuna and Suassuna, 1963). Rustigian and Stuart (1941, 1945) recognized that cultures labeled *P. hydrophila* and *P. icthyosmius* were not members of the genus *Proteus*, and Miles and Miles (1951) identified these bacteria as members of the genus *Aeromonas* (see also Ewing et al., 1961).

SEROLOGICAL REACTIONS

Extensive serological investigations have been made with cultures that belonged to the various species of *Proteus*, but there has been an almost total lack of correlative studies on antigenic interrelationships between the species. The notable exception to this was the work of Perch (1948, 1950) in which both *P. vulgaris* and *P. mirabilis* were included. This investigator established an antigenic schema for these two species and characterized 49 O antigen groups and 19 flagellar antigens. She also demonstrated (1950) that all of the serotypes described by Winkle (1944/1945) were either identical with or closely related to serotypes within the schema. Similarly, Rauss (1936) and Rauss and Vörös (1959) established an antigenic schema for *P. morganii*, within which 29 O groups, 19 H antigens, and 57 serotypes were delineated. In 1967 Rauss and Vörös (1967a) added five new O and six new H antigens to their schema. A schema for *P. rettgeri* also has been established by Namioka and Sakazaki (1958). This was composed of 34 O antigen groups, 26 flagellar antigens, and 45 serotypes. Other serological studies that have been reported included those of Castellani (1932) with *P. morganii*, Varela et al. (1944) with *P. vulgaris* and *P. morganii*, Rustigian and Stuart (1945), who described frequent interrelationships among four species, Belyavin (1951) with *P. vulgaris*, and Keating (1956). Readers are referred to the publications cited for further details and references to early work.

The short review given in the preceding paragraph indicated that there is a need for correlative studies which would include a thorough investigation of antigenic relationships among serotypes within the four species of *Proteus* and members of the genus *Providencia* as well. Such studies might lead to the eventual establishment of a single antigenic schema. Until such time as correlative work of this sort has been done it does not seem necessary or desirable to include the separate schemata in a book of this kind.

Perch (1950) and Kauffmann and Perch (1950) described numerous intergeneric antigenic relationships between the O antigens of *Proteus* and those of members of the genera *Salmonella*, *Arizona*, and *Escherichia*. Rauss (1957) and Rauss and Vörös (1967b) reported upon antigenic relationships between *P. morganii*, *E. coli* 0112a, 112c, and *S. dysenteriae* 2, and Sompolinsky (1957) described a relationship of certain cultures of *P. morganii* and *S. flexneri* 1. Further, many relationships are extant between strains of *Proteus*, particularly *P. morganii* and *P. rettgeri*, and *Providencia*.

REFERENCES

Belyavin, G. 1951. J. Gen. Microbiol., 5, 197.
Bergey's Manual of determinative bacteriology, 7th ed.

1957. Edited by R. S. Breed, E. G. D. Murray, and H. P. Hitchens. Baltimore: Williams & Wilkins.

Buttiaux, R., R. Osteux, R. Fresnoy, and J. Moriamez. 1954. Ann. Inst. Pasteur, **87**, 375.

Castellani, A. 1932. J. Trop. Med. Hyg., **35**, 160.

Colichon Arbulu, H., and J. Arana Sialer. 1956. Revisita Pathologica Clinica, **1**, 102.

Cook, G. T. 1948. J. Pathol. Bacteriol., **60**, 171.

Ewing, W. H. 1958. Int. Bull. Bacteriol. Nomen. Tax., **8**, 17.

———. 1962. Ibid., **12**, 93.

———. 1967. Revised definitions for the family Enterobacteriaceae, its tribes, and genera. CDC Publ.*

Ewing, W. H., B. R. Davis, and R. W. Reavis. 1957. Pub. Health Lab.,** **15**, 153.

Ewing, W. H., I. Suassuna, and I. R. Suassuna. 1960a. The biochemical reactions of members of the genus *Proteus*. CDC Publ.*

Ewing, W. H., B. R. Davis, and P. R. Edwards. 1960b. Pub. Health Lab.,***18**, 77.

Ewing, W. H., R. Hugh, and J. G. Johnson. 1961. Studies on the *Aeromonas* group. CDC Publ.*

Kauffmann, F. 1956a. Acta Pathol. Microbiol. Scand., **39**, 85.

———. 1956b. Ibid., **39**, 103.

Kauffmann, F., and B. Perch. 1950. Ibid., **27**, 182.

Keating, S. V. 1956. Med. J. Australia, (Aug), 1968.

Miles, E. M., and A. A. Miles. 1951. J. Gen. Microbiol., **5**, 298.

Moeller, V. 1954. Acta Pathol. Microbiol. Scand., **35**, 259.

———. 1955. Ibid., **36**, 158.

Namioka, S., and R. Sakazaki. 1958. Ann. Inst. Pasteur, **94**, 485.

Perch, B. 1948. Acta Pathol. Microbiol. Scand., **25**, 703.

———. 1950. Om Proteusgruppens Serologi. (Summary in English.) Copenhagen: Arnold Busck.

Phillips, J. 1955. J. Hyg., **53**, 26.

Proom, H., and A. J. Woiwood. 1951. J. Gen. Microbiol., **5**, 930.

Rauss, K. 1936. J. Pathol. Bacteriol., **42**, 183.

———. 1957. Zentralbl. f. Bakt. I. Orig., **170**, 428.

———. 1962. Int. Bull. Bacteriol. Nomen. Tax., **12**, 53.

Rauss, K., and S. Vörös. 1959. Acta Microbiol. Acad. Sci. Hungaricae, **6**, 233.

———. 1967a. Ibid., **14**, 195.

———. 1967b. Ibid., **14**, 199.

Rauss, K., S. Vörös, and T. Kontrohr. 1964. Ibid., **11**, 35.

Rustigian, R., and C. A. Stuart. 1941. Proc. Soc. Exper. Biol. Med., **47**, 108.

———. 1943a. Ibid., **53**, 231.

———. 1943b. J. Bacteriol., **45**, 198.

———. 1945. Ibid., **49**, 419.

Shaw, C., and P. H. Clarke. 1955. J. Gen. Microbiol., **13**, 155.

Singer, J., and J. Bar-Chay. 1954. J. Hyg., **52**, 1.

Sompolinsky, D. 1957. Ann. Inst. Pasteur, **92**, 343.

Suassuna, I., and I. R. Suassuna. 1960. An. Microbiol., **8**, 161.

Suassuna, I., I. R. Suassuna, and W. H. Ewing. 1961. Pub. Health Lab.,** **19**, No. 3, 38-46; No. 4, 67-74.

Suassuna, I. 1963. Estudos sobre o genero *Proteus*. Instituto de Microbiologia, Univ. Brasil, Rio de Janeiro.

Tittsler, R. P., and L. A. Sandholzer. 1937. Amer. J. Pub. Health, **27**, 1240.

Varela, G., J. Zozaya, and F. Imaz. 1944. Rev. Inst. Salub. Enferm. Trop., **5**, 15.

Winkle, S. 1944/45. Zentralbl. f. Bakt. I. Orig., **151**, 494.

*CDC Publ., Publication from the Center for Disease Control (formerly the Communicable Disease Center, Atlanta, Ga. 30333.
**Public Health Laboratory: Journal of the Conference of Public Health Laboratory Directors.

Chapter 17
The Genus *Providencia*

The name Providence group was suggested by Kauffmann (1951) for the microorganisms described by Stuart and co-workers (1943, 1946) under the designation 29911 bacteria. This name later was changed to *Providencia* by Kauffmann and Edwards, 1952 (v. inf.). Members of the genus *Providencia* occur in specimens from extraintestinal sources, particularly urinary tract infections, and have been isolated from small outbreaks and sporadic cases of diarrheal disease (e.g., Stuart et al., 1943, 1946; Brown, 1952; Singer and Bar-Chay, 1954; Ridge and Thomas, 1955; van Ros, 1961; Sen, 1962; and Fields et al., 1967).

The genus *Providencia* was defined by Ewing et al. (1954) and by the author in 1962, and those definitions were emended (1967) as follows:

The genus *Providencia* is composed of motile bacteria that conform to the definitions of the family ENTEROBACTERIACEAE and the tribe PROTEEAE. Urea is not hydrolyzed and hydrogen sulfide is not produced. Indol is produced and growth occurs on Simmons' citrate medium. Gelatin is not liquefied, lysine, arginine, and ornithine are not decarboxylated, and lipase is not produced. With rare exceptions mannitol is not fermented and acid is not produced from alpha methyl glucoside, erithritol, or esculin. The type species is *Providencia alcalifaciens* (De Salles Gomes) Ewing.

When used in conjunction with the definition for the tribe Proteeae (Chapter 1) this definition provides a convenient point of reference for a discussion of the bacteria of the genus *Providencia*.

The nomenclature and taxonomy of the bacteria of this genus were discussed at length by Ewing (1958, 1962) and details need not be repeated here. However, there is one point that requires comment. In 1969 the author was informed by the Chairman of the Judicial Commission of the International Committee on Nomenclature of Bacteria that the commission had ruled that the generic name *Arizona* (Kauffmann and Edwards, 1952) was not validly published because these authors did not propose a classification but merely suggested one that might be used. According to the above-mentioned ruling Kauffmann and Edwards (1952) did not *propose* any new names. In replying to the communication the author accepted the ruling, but pointed out the fact that the generic name *Arizona* was validly published and the genus and species characterized in several publications by the author and colleagues (see Chapters 1 and 10). Further, it was pointed out that if the name *Arizona* was not validly published by Kauffmann and Edwards (1952), then the generic name *Providencia* Kauffmann and Edwards 1952 also was illegitimate. However, it was mentioned that the genus *Providencia* was validly published, characterized, and defined by Ewing (1962).

CHARACTERISTICS AND BIOCHEMICAL REACTIONS

Members of the genus *Providencia* grow well on EMB, SS, and MacConkey agars and the colonies present appearances similar to those of salmonellae or shigellae. In contrast to shigellae, *Providencia* grow well on bismuth sulfite (Wilson-Blair) agar. On this medium, colonies may be colorless, light green, or olive green, and colonies of some cultures may be slightly blackened with occasional production of a metallic sheen. Of 100 strains tested on brilliant green agar, only one grew well and produced alkalinization of the medium. In Kligler's iron agar or triple sugar iron agar (TSI), the reactions simulate those of shigellae, except that evidence of a small amount of gas production sometimes is apparent in the butt of the tube. Although sucrose is utilized by most cultures of *Providencia*, fermentation usually is delayed and therefore is not apparent in TSI agar. The red color reactions that occur in the area of the slant in lysine iron agar medium inoculated with *Providencia* are discussed in Chapter 16.

Two species are recognized within the genus *Providencia*: *P. alcalifaciens* and *P. stuartii*. Differentiation of the species is based primarily upon production of gas from glucose and utilization of adonitol and inositol. Cultures of *P. alcalifaciens* usually produce a bubble to about 15 per cent of gas from glucose, usually ferment adonitol, and utilize inositol only rarely. In contrast, strains of *P. stuartii* are anaerogenic, utilize inositol, and usually do not produce acid from adonitol. Trehalose also is helpful for differentiation of members of the two species (Table 151). The biochemical reactions given by 633 cultures of *P. alca-*

331

TABLE 151

Summary of the Biochemical Reactions of *P. alcalifaciens* and *P. stuartii*

Test or substrate	P. alcalifaciens			P. stuartii		
	Sign	%+	(%+)*	Sign	%+	(%+)*
Hydrogen sulfide	–	0		–	0	
Urease	–	0		–	0	
Indol	+	99.5		+	98.7	
Methyl red (37 or 22 C)	+	99.8		+	100	
Voges-Proskauer (37 or 22 C)	–	0		–	0	
Citrate (Simmons')	+	97.9	(1.3)	+	95.6	(3.1)
KCN	+	98.8		+	98.6	
Motility	+	96.2	(0.3)	d	85.5	(2.5)
Gelatin (22 C)	–	0	(1.4)	–	0	(6.8)
Lysine decarboxylase	–	0		–	0	
Arginine dihydrolase	–	0		–	0	
Ornithine decarboxylase	–	0		–	0	
Phenlalanine deaminase	+	97.2		+	93.7	
Glucose acid	+	100		+	100	
gas	d	85.8	(0.6)	–	0	
Lactose	–	0.3		–	3.8	
Sucrose	d	13	(74.2)	+ or (+)	26	(65.8)
Mannitol	–	2	(0.2)	d	13.3	(1.3)
Dulcitol	–	0		–	0	
Salicin	–	0.6	(0.3)	–	1.9	
Adonitol	+	94.5	(0.2)	–	3.8	
Inositol	–	0.6		+	97.5	(2.5)
Sorbitol	–	0	(1.2)	d	4.1	(40.2)
Arabinose	–	0.7	(0.7)	–	5	(3.3)
Raffinose	–	1	(1)	–	6.2	(1.6)
Rhamnose	–	0	(1.3)	–	0	
Malonate	d	0.7		d	1.2	
Mucate	–	0.7		–	0	
Christensen's citrate	+	94.2	(1.5)	+	95.8	
Jordan's tartrate	+	100		+	96.2	
Pectate	–	0		–	0	
Sodium acetate	d	26.9	(38.5)	+	84	(8)
Sodium alginate	–	0		–	0	
Lipases						
Corn oil	–	0		–	0	
Triacetin	–	0		–	0	
Tributyrin	–	0		–	0	

TABLE 151 (Continued)

Summary of the Biochemical Reactions of *P. alcalifaciens* and *P. stuartii*

Test or substrate	P. alcalifaciens			P. stuartii		
	Sign	%+	(%+)*	Sign	%+	(%+)*
Maltose	−	0.7	(0.7)	−	3.2	
Xylose	−	0.8		d	10	(1.7)
Trehalose	−	4.3	(0.9)	+ or (+)	87.2	(6.4)
Cellobiose	−	1.5	(3)	d	12.5	(68.7)
Glycerol	d	10.9	(54.7)	+ or (+)	34.4	(65.6)
Alpha methyl glucoside	−	0		−	0	(2.6)
Erythritol	−	0		−	0	
Esculin	−	0		−	0	
Nitrate to nitrite	+	100		+	100	
Oxidation − fermentation	F	100		F	100	
Oxidase	−	0		−	0	
Cetrimide	d	5	(18.7)	d	5.3	(10.5)

*Figures in parentheses indicate percentages of delayed reactions (3 or more days).

+ 90% or more positive within one or two days' incubation.
(+) Positive reaction after 3 or more days.
− No reaction (90% or more).
+ or − Majority of strains positive, some cultures negative.
− or + Majority of cultures negative, some strains positive.
(+) or + Majority of reactions delayed, some occur within 1 or 2 days.
d Different reactions: +, (+), −.
w Weakly positive reactions.

lifaciens and 158 isolants of *P. stuartii* are summarized in Table 151. The data presented in this table are based upon past investigations of the author and colleagues and upon unpublished data. Attention is directed to the fact that of the 791 cultures examined five failed to produce indol and that seven others did not grow on Simmons' citrate agar. Three strains produced very weak, delayed (3 days or more) alkalinity in lysine medium, and eight fermented lactose (see comments made in Chapter 16 regarding reactions of this sort). Dulcitol, erythritol, and esculin are not utilized and several other carbohydrates (Table 151) are fermented rarely or seldomly.

Although it is possible to describe a large number of biotypes within the species of *Providencia*, only six principal biogroups are delineated (Table 152), primarily for purposes of discussion. The data in Table 152 indicated that:

(1) 80 per cent of the cultures reported upon belonged to the species *P. alcalifaciens* and 20 per cent were strains of *P. stuartii;*
(2) 86 per cent of the isolants of *P. alcalifaciens* were aerogenic, while 14 per cent were not;
(3) only 2.5 per cent of cultures of *P. alcalifaciens* failed to ferment either inositol or adonitol and these were aerogenic;
(4) only 2.7 per cent of the isolants of *P. alcalifaciens* were anaerogenic and failed to ferment either adonitol or inositol; and that,
(5) only 3.8 per cent of cultures of *P. stuartii* utilized both adonitol and inositol.

Therefore, with the exception of the 17 strains that composed biogroup 4 (Table 152), the division between the two species was clear. Conceivably the biogroup 4 cultures might be placed in a separate subspecies or species, but since there were so few, they were placed arbitrarily in *P. alcalifaciens*.

The reactions given by cultures of *Providencia* in

TABLE 152

Biogroups Within the Species of *Providencia*

Species and biogroup	No. Cultures	Gas +	Gas −	Adonitol +	Adonitol −	Inositol +	Inositol −	Percent of Species
P. alcalifaciens	633 (80%)							
Biogroup 1	531	531	0	531	0	4	527	83.9
2	69	0	69	69	0	0	69	10.9
3	16	16	0	0	16	0	16	2.5
4	17	0	17	0	17	0	17	2.7
Total	633	547	86	600	33	4	629	(100)
P. stuartii	158 (20%)							
Biogroup 5	152	0	152	0	152	152	0	96.2
6	6	0	6	6	0	6	0	3.8
Total	158	0	158	6	152	158	0	(100)

commonly employed tests are listed in Table 9 (Chapter 3) and tests and substrates of value for differentiation of the two species are recorded in Table 26 (Chapter 3). The data presented in Table 152 also should be useful for this purpose.

For additional information regarding the biochemical reactions given by strains of *Providencia* the reader is referred to the publications of Brooke (1951) and Buttiaux et al. (1953/54) in addition to the references cited above.

SEROLOGICAL REACTIONS

In 1954 Ewing et al. introduced an antigenic schema for the genus *Providencia* which was based upon 56 O antigen groups and 28 H antigens, as well as several K antigens, and which contained 125 serotypes. Since 1954 the number of O antigen groups has been extended to 62, the H antigens to 30, and the number of known serotypes to about 175.

The methods described in Chapter 5 *(Escherichia)* may be used for production of O, K, and H antisera for members of the genus *Providencia*. Further, the methods given in Chapter 5 for serological examination of cultures, absorption of antisera etc. may be adapted for use with cultures that belong to this genus.

Intergeneric Antigenic Relationships. A number of reciprocal relationships between the O antigens of members of the genus *Providencia* and those of *E. coli* were described by Ewing et al. (1956). These are recorded in Table 153 (see Chapter 5 for an explana-

tion of the meaning of a, b-a,c etc. relationships). Further, the O antigenic relationships of cultures of *Providencia* 06, *E. coli* 055, and *Arizona* 09 are related. These relationships may be expressed by the following arbitrary formulas:

Escherichia 055	a,b,e (b and e, minor factors)
Arizona 09	a,c,e (e, minor factor)
Providencia 06	a,d.

It should be noted that the relationship between the O antigens of *E. coli* 055 and those of *Arizona* 09 first was reported by Frantzen (1950). Strains of *Salmonella* O group 50 are known to be related to O antigen 9 of *Arizona* and to the O antigens of certain cultures of *Citrobacter* (see Chapter 10 and 11). Schmid and Velaudapillai (1953) reported that the O antigens of a culture of *Providencia* (later determined to be O group 57) were identical with those of 043 of salmonellae.

Some shigellae are known to be antigenically related to *Providencia* (Ewing et al., 1956). These O antigen relationships are as follows:

Shigella	*Providencia*	Nature of Relationship
S. boydii 5	047	a-a,b
S. boydii 6	041	a,b-a,c
S. dysenteriae 3	037	reciprocal, minor

Finally, it should be noted that the O and H antigens of some cultures of *P. morganii* and *P. rettgeri* are reciprocally related to those of *Providencia,* and the antigens of some cultures of other species of *Proteus*

TABLE 153

Relationship of the O Antigens of *Providencia* and *E. coli*

E. coli O Antigen group	*Providencia* O Antigen group	Variety of Relationship (Only Reciprocal Relationships are Listed)
5	5	a,b — a,c*
13	29	a,b — a,c
21	11	a,b — a,c
37	8	a,b — a,c
46	23	a,b — a,c
48	36a,36c	a, — a,b
55	6	a,b — a,c
58	16	a,b — a,c
64	4	a,b — a,c
64	10	a,b — a,c
65	18	a,b — a,c
70	5	a,b — a,c
70	18	a,b — a,c
70	40	a,b — a,c
71	53	a,b — a,c
76	41	a,b — a,c
78	50	a,b — a,c
79	47	a — a,b
88	36	a,b — a,c
91	17	a,b — a,c
92	37	a,b — a,c
108	49	a,b — a,c
120	2	a,b — a,c

* These arbitrary formulas are used merely as an aid to make the relationship clearer and they should not be interpreted as permanent designations for the antigens involved.

are related to the O and H antigens of *Providencia.* Clearly, correlative serological studies with members of these two genera are needed.

REFERENCES

Brooke, M. S. 1951. Acta Pathol. Microbiol. Scand., **29**, 1.

Brown, G. W. 1952. Med. J. Australia, (Nov.), 658.

Buttiaux, R., R. Fresnoy, and J. Moriamez. 1953/54. Ann. Inst. Pasteur, Lille, **6**, 62.

Ewing, W. H. 1958. Int. Bull. Bacteriol. Nomen. Tax., **8**, 17.

————. 1962. Ibid., **12**, 93.

————. 1967. Revised definitions for the family Enterobacteriaceae, its tribes and genera. CDC Publ.*

Ewing, W. H., K. E. Tanner, and D. A. Dennard. 1954. J. Infect. Dis., **94**, 134.

Ewing, W. H., H. W. Tatum, B. R. Davis, and R. W. Reavis. 1956. Studies on the serology of the *Escherichia coli* group. CDC Publ.*

*CDC Publ., Publication from the Center for Disease Control (formerly the Communicable Disease Center), Atlanta, Ga. 30333.

Fields, B. N., M. M. Uwaydah, L. J. Kunz, and M. N. Swartz. 1967. Amer. J. Med., **42**, 89.

Frantzen, A. 1950. Acta Pathol. Microbiol. Scand., **27**, 647.

Kauffmann, F. 1951. Enterobacteriaceae. 1st Ed. Copenhagen:Munksgaard.

Kauffmann, F., and P. R. Edwards. 1952. Int. Bull. Bacteriol. Nomen. Tax., **2**, 2.

Ridge, L. E., and M. E. M. Thomas. 1955. J. Pathol. Bacteriol., **69**, 335.

Schmid, E. E., and T. Velaudapillai. 1953. Jap. J. Med. Sci. Biol., **7**, 169.

Sen, R. 1962. Ind. J. Med. Res., **50**, 622.

Singer, J., and J. Bar-Chay. 1954. J. Hyg., **52**, 1.

Stuart, C. A., E. van Stratum, and R. Rustigian. 1945. J. Bacteriol., **49**, 437.

Stuart, C. A., K. M. Wheeler, and V. McGann. 1946. Ibid., **52**, 431.

Van Ros, G. 1961. Memoirs, Acad. Roy. Sci. d'Outre—Mer, N. S., **12**, 1.

Chapter 18
Media and Reagents

PRESERVATIVE AND TRANSPORT MEDIA FOR STOOL SPECIMENS

ENRICHMENT MEDIA

PLATING MEDIA

DIFFERENTIAL MEDIA AND REAGENTS

OTHER MEDIA AND SOLUTIONS

(Note that listings are alphabetical under each of the above-mentioned headings.)

Many of the solutions and media mentioned herein may be obtained from commercial sources and, as far as is known, these products are satisfactory if the directions that accompany them are followed. The ingredients of such media and details of their use also are given in manuals and other publications of the respective firms (e.g., BBL, Difco, etc.). Reference also is made to the compendium of Levine and Schoenlein (1930).

It will be noted that in most of the formulas in which peptone is listed no particular brand is named. Where this is the case any good peptone may be used. This also applies to extracts of meat and of yeast. In the few where certain brands of peptone are specified, the brand specified is critical and should be employed.

When dehydrated media are employed, they should be prepared in *exact* accordance with the directions that accompany them. When containers of such media first are opened they should be dated, and their contents should not be used after long periods of storage. Most dehydrated media are very hygroscopic and should be stored in a cool dry place. If these products have absorbed moisture, dyes and other constituents may be altered, in which instances aberrant results may be expected.

Control of all media and reagents always should be exercised. Most items that are in daily use may be controlled simply by constant observation of their behavior with diagnostic cultures of various kinds. For instance, if a culture yields typical reactions in all tests except one or two, other tubes of the media that supposedly gave atypical results should be inoculated with strains that are known to yield positive and negative reactions. Representative tubes of certain media (e.g., KCN and malonate media) should be tested with known cultures each time they are prepared.

PRESERVATIVE SOLUTIONS FOR STOOL SPECIMENS

In recent years several transport or preservative media have been introduced. Some of these have been evaluated quite thoroughly while others have not. These more recent additions are listed below along with preservatives that have been known and used for many years. Some are available from commercial sources. A review of transport methods was made by Ewing (1968).

AMIES' TRANSPORT MEDIUM (Amies and Douglas, 1965, Amies, 1967)

Distilled water	1000 ml
Agar	4 g

Heat until agar is dissolved, then add while hot:

Sodium chloride	3 g
Potassium chloride	0.2 g
Disodium phosphate (anhydrous)	1.15 g
(or Disodium phosphate; $12H_2O$)	(2.9 g)
Monopotassium phosphate	0.2 g
Sodium thioglycolate	1 g
(=Mercapto-acetic acid, sodium salt)	
Calcium chloride, 1% solution freshly prepared	10 ml
Magnesium chloride ($6H_2O$), 1% solution	10 ml

Stir until ingredients are dissolved. Add:

Charcoal, pharmaceutical neutral	10 g

Dispense in 6 ml screw-capped bottles or vials (stir continuously to keep charcoal evenly suspended). Tighten caps firmly. Sterilize at 121 C for 15 min. Invert bottles just before gelation occurs to keep charcoal distribution uniform. Retighten caps if necessary. Store in a cool place. The pH is 7.2-7.4. Buffer treated cotton swabs (without charcoal) are used. See Stuart's Transport Medium for preparation.

BUFFERED GLYCEROL-SALINE SOLUTION (Teague and Clurman, 1916; modified by Sachs, 1939)

Glycerol (neutral)	1000 ml
0.85% Sodium chloride solution	2000 ml
Saturated aqueous solution of phenol red in 5% solution of disodium phosphate	
Disodium phosphate solution (conc.)	

Method of preparation:

a) mix glycerol and saline solution
b) add enough phenol red solution to permit determination of pH with phenol red standards
c) adjust to pH 8.0 with disodium phosphate solution
d) bottle in 10 ml amounts in 1 oz. screw-capped bottles
e) sterilize fractionally or for *10 min at 116 C*

After sterilization the pH should be 7.4.

CARY AND BLAIR TRANSPORT MEDIUM (Cary and Blair, 1964)

Sodium thioglycolate	1.5 g
Disodium phosphate	1.1 g
Sodium chloride	5 g

These ingredients are added to 991 ml of demineralized or distilled water in the order given. Add:

Agar (Difco specified)	5 g

Heat medium only until clear. After mixture has cooled to 50 C, add

Calcium chloride, 1% solution	9 ml

Adjust pH to 8.4. Dispense in 7 ml quantities in 9 ml screw-capped vials (previously rinsed and sterilized). *Steam vials for 15 min,* cool, and tighten caps. (To prevent evaporation at room temperature celons may be placed over the caps.)

Cotton swabs are dipped in Sorenson's buffer (pH 8.1) previously heated to the boiling point, drained, placed in glassine envelopes, and sterilized at 121 C for 15 min. Swabs are stored without drying at room temperature.

EDTA SOLUTION (Shipe et al., 1960)

Disodium ethylenediamine tetraacetate (EDTA)	2.76 g
Glycerol	270 ml
Disodium phosphate	0.19 g
Monopotassium phosphate	4.35 g
Distilled water	700 ml

Adjust to pH 5.5 with 1 N sodium hydroxide (usually 2 to 3 ml) and add distilled water to make 1000 ml.

Dispense in 20 ml amounts in 1 oz. wide-mouthed screw-capped bottles and sterilize at *116 C for 15 min.*

STUART'S TRANSPORT METHOD (Stuart, 1959)

The method as outlined by Stuart follows:

"Transport Medium

Dissolve 6 gm Bacto agar in 1,000 ml chlorine-free distilled water. At the same time to another 900 ml distilled water, add 2 ml thioglycolic acid (Difco), 12-15 ml $\frac{N}{1}$ NaOH (to bring to approximately pH 7.2; the amount is selected according to local experience to avoid as far as possible any final pH adjustment), 100 ml solution of sodium glycerophosphate (commercial, 20 per cent w/v in water), and 20 ml solution of $CaCl_2$ (1 per cent w/v in water.) Add this mixture to the original melted agar. Check pH and if necessary adjust to pH 7.3-7.4. Add 4 ml methylene blue (0.1 per cent in water). Mix well and dispense in ¼ oz screwcapped bottles, filling to capacity. Screw caps on securely, but not tightly. Sterilize in flowing steam, avoiding overcrowding, for 1 hour. Tighten caps immediately following sterilization. After cooling, the transport medium should be colorless.

Note: Distillate from chlorinated water occasionally contains significant amounts of free chlorine. This must be checked rigorously. We pass all distilled water through an ion-exchange resin column before use. This water is used for all the above preparation work.

"Swabs

Prepare neat swabs with good quality absorbent cotton and applicator sticks. Prepared swabs are boiled in Sorensen's phosphate buffer solution $\frac{M}{15}$ pH 7.4 (approximately 500 swabs standing in 100 ml buffer in 1,000 ml beaker, boiled 5 minutes). Remove, shake off excess moisture, then dip in a 1 per cent suspension of finely powdered charcoal in water. We use activated charcoal (B.D.H.), but animal charcoal (Cenco), blood charcoal (B.D.H.) and Norit have been used successfully when powdered sufficiently fine. Swirl swabs around to coat thoroughly with charcoal (swabs should be quite black when wet). Shake off excess moisture, place in cotton-plugged tubes, dry in oven, and sterilize.

"Instructions Sent With Outfits

Take the specimen and insert the swab or swabs into the upper third of the medium in the small bottle. Cut off the protruding portion of the swab stick with scissors and screw the lid on the bottle tightly. This usually forces the swab down slightly and centers it in the transport medium. Label the bottle and return it with the swabs enclosed to the laboratory as soon as possible. Keep specimens in refrigerator until ready for shipment.

The swabs are sterile. They have been treated with charcoal to improve the conditions for culture.

"Laboratory Handling

Grasp the end of the short swab stick firmly with a long-shanked artery forceps and apply swab to culture media in the usual way. It is important that material from the swab itself is applied to the medium surface because adhering transport medium may be deposited on a culture plate and give a false impression of a satisfactory inoculum."

SPECIMEN PRESERVATIVE MEDIUM (Hajna, 1955)

Diammonium phosphate, $(NH_4)_2HPO_4$	4 g
Monopotassium phosphate, KH_2PO_4	2 g
Sodium chloride, NaCl	5 g
Magnesiumsulfate, $MgSO_4 \cdot 7H_2O$	0.4 g
Sodium citrate, $Na_3C_6H_5O_7 \cdot 2H_2O$	5 g
Yeast extract	1 g
Sodium desoxycholate	0.5 g
Distilled water	700 ml

Heat to dissolve ingredients, then add 300 ml glycerol. Mix well, dispense into suitable containers, and sterilize at *116 C for 10 min* (pH 7.0, without adjustment).

ENRICHMENT MEDIA

GN BROTH (Hajna, 1955)

Glucose	1 g
d-mannitol	2 g
Sodium citrate	5 g
Sodium desoxycholate	0.5 g
K_2HPO_4	4 g
KH_2PO_4	1.5 g
NaCl	5 g
Bacto tryptose	20 g
Distilled water	1000 ml

Dissolve ingredients by heat. Dispense in tubes in convenient amounts and sterilize at *116 C for 15 min*. Final pH is 7.

SELENITE BROTH (Leifson, 1936)

Tryptone or Polypeptone	5 g
Lactose	4 g
Disodium phosphate	10 g
Sodium acid selenite	4 g

Distilled water	1000 ml

(Final pH about 7)

Dispense in sterile tubes to a depth of at least 2 inches. Sterilization is unnecessary if medium is to be used immediately. To sterilize expose tubes to flowing steam for 30 min. *Do not autoclave.*

TETRATHIONATE BROTH (Muller, 1923; modified by Kauffmann, 1935)

Basal Medium

Polypeptone or Proteose peptone	5 g
Bile salts	1 g
Calcium carbonate	10 g
Sodium thiosulfate	30 g
Distilled water	1000 ml

Iodine Solution

Iodine	6 g
Potassium iodide	5 g
Distilled water	20 ml

The ingredients of the basal medium are heated to boiling temperature and cooled to less than 45 C, after which 2 ml of iodine solution is added to each 100 ml of base. Brilliant green then may be added to the medium (1 ml of 1:1000 solution per 100 ml of base medium) as recommended in Kauffmann's Combined Enrichment Medium (1935, 1966). The basal medium, with or without added brilliant green, may be tubed, sterilized at 121 C, and stored. In this case iodine solution is added (0.2 ml per 10 ml of medium) prior to use.

Sulfathiazol (0.125 mg per 100 ml of medium) may be added to prevent excessive growth of *Proteus* (Galton et al, 1952).

PLATING MEDIA

BISMUTH SULFITE AGAR (see Wilson and Blair, 1931) modified. Hajna (1951) and Hajna and Damon (1956) gave the following method for the preparation of bismuth sulfite agar medium with brilliant green.

A. Basal Medium

Beef extract	5 g
Thiotone or thiopeptone	10 g
Glucose	5 g
Sodium chloride	5 g
Agar	25 g
Distilled water	1000 ml

Heat, with agitation until ingredients are melted.

B. Bismuth Sulfite Mixture (stock solution)
   1. Prepare a 20% solution by dissolving 200 g anhydrous sodium sulfite in 1000 ml distilled water.

2. Prepare a 10% solution by dissolving 50 g bismuth ammonium citrate in 500 ml distilled water. Add 1 ml concentrated ammonium hydroxide and allow solution to stand until it is clear. The addition of several ml of ammonium hydroxide may be required. Add solution 2 to solution 1 and mix.

3. Add 100 g of anhydrous dibasic sodium phosphate and mix.

4. Prepare a 10% solution by adding 10 g of ferric ammonium sulfate to 100 ml distilled water. Add entire 100 ml to mixture (1, 2, and 3, above).

Heat the combined solution at 100 C for 2 or 3 min. Stopper flask or bottle with a rubber stopper and store in the dark at room temperature. Do not refrigerate the stock solution.

C. Completed Medium

To 1000 ml of melted agar base medium add 70 ml of bismuth sulfite stock solution. Shake thoroughly and add 4 ml of a 1% aqueous solution of brilliant green. Cool medium to 40-50 C and pour into plates.

Dehydrated bismuth sulfite agar medium is available from various commercial sources. These are known to be satisfactory if the directions for preparation are followed. These products should not be autoclaved.

BRILLIANT GREEN AGAR (Kristensen, Lester, and Juergens, 1925, modified by Kauffmann, 1935, 1966, and others)

Yeast extract	3 g
Peptone (Bacto, Proteose no. 3, or Polypeptone)	10 g
Sodium chloride	5 g
Lactose	10 g
Sucrose	10 g
Phenol red	0.08 g
Brilliant green	0.0125 g
Agar	20 g
Distilled water	1000 ml
(Final pH about 6.9-7)	

Sterilize at 121 C for exactly 15 min, cool and pour into plates. Some formulas for this medium incorporate meat extract (0.5%) instead of yeast extract. Dye solutions for use in the medium may be prepared as follows:

Brilliant green 0.5% in distilled water, use 2 ml per liter of medium; Phenol red 1 g, dissolved in 40 ml 0.1 N sodium hydroxide, add 460 ml distilled water, use 4 ml per liter of medium.

Galton et al., 1954, added sodium sulfadiazine (8-16 mg per 100 ml) to brilliant green agar used in

examination of foods such as sausage for salmonellae. This concentration markedly inhibited the growth of pseudomonads.

DESOXYCHOLATE AGAR (Leifson, 1935)

Distilled water	1000 ml
Agar	17 g
Peptone (Bacto or Polypeptone)	10 g
Sodium chloride	5 g
Lactose	10 g
Ferric ammonium citrate	2 g
Dipotassium phosphate	2 g
Sodium desoxycholate	1 g
Neutral red	0.033 g
(pH 7.3)	

Add peptone to water and after it is dissolved add sufficient sodium hydroxide to adjust pH to 7.3. Boil and filter through paper. Add agar to peptone water and allow agar to soak for about 15 min. Melt agar by boiling. To each liter add about 6 ml of 1.0 N sodium hydroxide, and then add other ingredients in the order given above, except for the neutral red. Adjust pH of medium to 7.3 using phenol red as indicator. After adjustment, add the neutral red to the medium (3 ml of 1% aqueous solution per liter). The medium may be heated in flowing steam long enough to kill vegetative cells (20 to 30 min), *but should not be overheated.* The ferric ammonium citrate may be omitted when the medium is to be used only as streak plates in the search for intestinal pathogens. This medium is available from commercial sources in slightly modified forms.

DESOXYCHOLATE CITRATE AGAR (Leifson, 1935, modified)

Meat infusion (pork, beef, or beef heart)	330 g
Peptone (Proteose no. 3, Thiotone)	10 g
Lactose	10 g
Sodium citrate	20 g
Sodium desoxycholate	5 g
Ferric ammonium citrate	2 g
Agar	20 g
Neutral red	0.02 g
Distilled water	1000 ml
(pH 7.3-7.5)	

This medium may be boiled for a minute or two to dissolve the ingredients, but it should not be subjected to excessive heat. *It should not be autoclaved.*

The original formulation given by Leifson (1935) employed 1 liter of infusion made from fresh pork adjusted to pH 7.5. To this the other ingredients were added in the order given (and as described under desoxycholate agar, above). Twenty-five g of sodium citrate (11 H$_2$O) were used and the neutral red was added after the final adjustment of the pH (2 ml of 1% solution per liter).

EOSIN METHYLENE BLUE AGAR (Holt-Harris and Teague, 1916, modified).

Peptone (Gelysate or Bacto)	10 g
Lactose	5 g
Sucrose	5 g
Dipotassium phosphate	2 g
Agar	13.5 g
Eosin Y	0.4 g
Methylene blue	0.065 g
Distilled water	1000 ml

This medium may be sterilized at 121 C for 15 min. The medium should be mixed frequently during the process of pouring plates.

HEKTOEN ENTERIC MEDIUM (King and Metzger, 1968)

Lactose	12 g
Sucrose	12 g
Salicin	2 g
Bile salts	20 g
Proteose peptone	12 g
Beef extract	3 g
Sodium chloride	5 g
Agar	14 g
Distilled water	1000 ml

Add 16 ml of 0.4% bromthymol blue and 20 ml of Andrade's indicator and adjust to pH 7.4 with NaOH. Boil until ingredients are completely dissolved and add 20 ml each of solution A and solution B:

Solution A	
Sodium thiosulfate	34 g
Ferric ammonium citrate	4 g
Distilled water	100 ml
Solution B	
Sodium deoxycholate	10 g
Distilled water	100 ml

Readjust pH to 7.5 *Do not autoclave this medium.*

MacCONKEY LACTOSE BILE SALT AGAR (MacConkey, 1905, 1908, modified).

Peptone (Bacto or Gelysate)	17 g
Peptone (Proteose or Polypeptone)	3 g
Lactose	10 g
Bile salts	1.5 g
Sodium chloride	5 g
Neutral red	0.03 g
Crystal violet	0.001 g

Distilled water	1000 ml
(pH 7.1)	

Sterilize at 121 C for 15 min. Sucrose in 1% concentration may be added, if desired (Holt-Harris and Teague, 1916).

The earlier formula of MacConkey (nutrient agar with sodium taurocholate, 0.5%; peptone, 2%; lactose, 1%; and neutral red) has been modified for special purposes by a number of investigators. For example, Harvey (1956) added brilliant green in final concentration of 1:30,000 to MacConkey agar base for use as a selective medium in the isolation of salmonellae. Also, Jameson and Emberley (1956) reported that substitution of the anionic detergent Teepol (0.1%) for the bile salts and bromthymol blue for neutral red in the MacConkey formula resulted in a very useful medium for the isolation of shigellae, salmonellae, and most other Enterobacteriaceae. The principal advantage was in the fact that bile salts are variable in their properties and are expensive, whereas Teepol (Shell) was said to be more reliable and cheaper.

SS AGAR (SHIGELLA-SALMONELLA AGAR). This medium was devised by investigators at Difco, Inc. The published formula (Difco Manual, 9th ed.) is as follows:

Beef extract	5 g
Proteose peptone	5 g
Lactose	10 g
Bile salts no. 3	8.5 g
Sodium citrate	8.5 g
Sodium thiosulfate	8.5 g
Ferric citrate	1 g
Agar	13.5 g
Brilliant green	0.00033 g
Neutral red	0.025 g
Distilled water	1000 ml

The medium is heated to a boiling temperature to dissolve the ingredients, cooled to 42-45 C and poured into petri dishes. Do not autoclave SS agar.

Shipolini et al. (1959) reported that substitution of 0.85% apocholic acid for the bile salts in a medium otherwise similar to SS agar resulted in a plating medium that was superior to Leifson's agar especially as regards isolation of shigellae. This medium and the Teepol agar medium mentioned above appeared to be worthy of further study and evaluation.

XYLOSE LYSINE DESOXYCHOLATE AGAR (Taylor, 1965)

Although there are several modifications of this medium in use, XLD agar is the one recommended by Taylor for general laboratory use, and especially for direct isolation of *Shigella* and *Providencia* from stool specimens.

XL Agar Base

Yeast extract	3 g
L-lysine	5 g
Xylose	3.75 g
Lactose	7.5 g
Sucrose	7.5 g
Sodium chloride	5 g
Phenol red	0.08 g
Agar	15 g
Distilled water	1000 ml

Heat mixture to boiling temperature to dissolve the ingredients. Sterilize at 121 C for 15 min, and then cool to 55 to 60 C. Aseptically add 20 ml of sterile solution containing:

Sodium thiosulfate	34 g
Ferric ammonium citrate	4 g
Distilled water	100 ml

Mix well to obtain a uniform suspension.

The completed XLD medium is prepared by adding 25 ml of 10% sterile solution of sodium desoxycholate per liter of the above-mentioned (cooled) medium. Mix well and adjust to pH 6.9.

Taylor and Schelhart (1968) do not recommend prepackaged complete XLD agar in which *all* of the ingredients are included for convenience. They stressed the use of the above-mentioned formulation, which adds the heat-labile solutions after sterilization and cooling of the basal medium for maximum performance.

DIFFERENTIAL MEDIA AND REAGENTS

ACETATE MEDIUM (Trabulsi and Ewing, 1962; Costin, 1965; Ewing, 1968)

This medium is the same as Simmons' citrate agar except that 0.25% sodium acetate is used instead of citrate.[1] It is inoculated, incubated, and interpreted in the same manner as Simmons' agar.

ALGINATE MEDIUM (Davis and Ewing, 1964)

Synthetic

This medium is prepared by substituting 0.25% sodium alginate for the citrate in Simmons' agar.* It is inoculated, incubated, and interpreted in the same way as is Simmons' agar.

[1]*The basal medium for Simmons' citrate agar, *without* sodium citrate, is available from commercial sources. To this sodium acetate, sodium alginate, or other source of carbon may be added.

Nutrient

Sodium alginate	10 g
Peptone	2 g
Yeast extract	1 g
Sodium nitrate	2 g
Potassium chloride	0.5 g
Magnesium sulfate	0.01 g
Potassium phosphate (monobasic)	1 g
Agar	15 g
Distilled water	1000 ml

Adjust to pH 7.0, if necessary. Dispense in tubes and sterilize at 121 C for 15 min, and allow medium to solidify in a slanted position (long slants).

Inoculation: Make a line inoculation from the bottom to the top of the slant. Use young (18-20 hr) agar slant culture.[2]

Incubation: 37 C. Examine daily for 7 days. Positive results are indicated by a zone of clearing along the line of inoculation. Enterobacteriaceae do not give this reaction.

CARBOHYDRATE FERMENTATION MEDIA

Fermentation Broth Base

Peptone	10 g
Meat extract	3 g
Sodium chloride	5 g
Andrade's indicator	10 ml
Distilled water	1000 ml

Adjust reaction to pH 7.1-7.2. Dispense in tubes with inverted insert tubes and sterilize at 121 C for 15 min. (See exceptions noted below.)

Glucose, lactose, sucrose, and mannitol are employed in a final concentration of 1%. Other carbohydrates such as dulcitol, salicin, etc., may be used in a final concentration of 0.5%. Glucose, mannitol, dulcitol, salicin, adonitol, and inositol may be added to the basal medium prior to sterilization. Medium containing neutral glycerol should be sterilized at *121 C for 10 min.* Disaccharides such as lactose, sucrose, and cellobiose (10% solution in distilled water, netural pH) should be sterilized by filtration or at *121 C for 10 min* and added to previously sterilized basal medium. Arabinose, xylose, and rhamnose also should be sterilized separately. If basal medium is tubed in 3.0 ml amounts, add 0.3 ml of sterile aqueous carbohydrate solution, i.e., one-tenth the volume.

Inoculation: Inoculate lightly from a young agar slant culture.

[2]Throughout this section a "young" agar slant or broth culture means one that has been incubated overnight, or about 18-20 hr.

Incubation: 37 C. Examine daily for 4 or 5 days. Note acid production and the volume of gas produced by aerogenic cultures. Negative tests should be observed at regular intervals thereafter for a total of 30 days. (See also Chapter 3).

ANDRADE'S INDICATOR

Distilled water	100 ml
Acid fuchsin	0.5 g
Sodium hydroxide (1.0 N)	16 ml

The fuchsin is dissolved in the distilled water and the sodium hydroxide is added. If, after several hours, the fuchsin is not sufficiently decolorized, add an additional 1 or 2 ml of alkali. The dye content of different samples of acid fuchsin varies quite widely, and the amount of alkali which should be used with any particular sample usually is specified on the label. The reagent improves somewhat on aging and should be prepared in sufficiently large amount to last for several years. The indicator is used in amount of 10 ml per liter of medium.

Other indicators such as bromthymol blue or bromcresol purple may be used instead of Andrade's. However the author prefers Andrade's indicator for several reasons: (a) It is colorless and early fermentation within insert tubes is observed easily. (b) It is not reduced easily. (c) Reversion from an acid to an alkaline condition is more clear-cut than is usually the case with other indicators. (d) It is cheap.

CETRIMIDE MEDIUM (King, personal communication, 1960)

Distilled water	1000 ml
Heart infusion agar	40 g

Heat until dissolved, then add 0.5% "Cetrimide" (cetyl-trimethylammonium bromide). Distribute in 5 ml amounts in 15 X 125 mm tubes, shaking the flask to ensure even distribution of the sediment.

Sterilize at 121 C for 15 min and then slant the tubes. The pH is 7.2 and should not need adjustment.

Inoculation: Inoculate the slant of the cetrimide medium *lightly* with a straight wire from a young agar slant culture.

Incubation: 37 C for 7 days. Growth of the organism on this medium is considered a positive result.

CITRATE AGAR, SIMMONS' (Simmons', 1926 modified)

Sodium chloride	5 g
Magnesium sulfate	0.2 g

Monoammonium phosphate	1 g
Dipotassium phosphate	1 g
Sodium citrate	2 g
Agar (washed vigorously for 3 days)	20 g
Distilled water	1000 ml

Add 40 ml of 1:500 bromthymol blue indicator solution. Sterilize at 121 C, 15 minutes, and slant the tubes so as to obtain a 1-inch butt and a 1.5-inch slant.

Inoculation: Prepare a saline suspension from a young agar slant culture and inoculate the slant of the citrate medium with a straight wire from the saline suspension. If desired, the butt of the medium may be stabbed.

Incubation: 37 C for 4 to 7 days. If equivocal results are obtained as sometimes happens with members of the genus, *Providencia* for example, medium should be reinoculated and incubated at room temperature for 7 days.

CITRATE AGAR (Christensen, 1949)

Test for citrate utilization in the presence of *organic* nitrogen

Sodium citrate	3 g
Glucose	0.2 g
Yeast extract	0.5 g
Cysteine monohydrochloride	0.1 g
Ferric ammonium citrate	0.4 g
Monopotassium phosphate	1 g
Sodium chloride	5 g
Sodium thiosulfate	0.08 g
Phenol red	0.012 g
Agar	15 g
Distilled water	1000 ml

Tube and sterilize at 121 C for 15 minutes and slant (1-inch butt, 1.5-inch slant).

The ferric ammonium citrate and sodium thiosulfate may be omitted from the formula, if desired, since they do not affect the value of the medium as an indicator of citrate utilization.

Inoculation: The medium is inoculated over the entire surface of the slant.

Incubation: 37 C for 7 days. Positive reactions are indicated by alkalinization of the medium and development of a red color, particularly on the slant of the agar.

CYSTEINE-GELATIN MEDIUM (Hinshaw, 1941; modified by Trabulsi and Edwards, 1962)

Basal medium	
Peptone (Bacto)	5 g
Beef extract	3 g

Gelatin	125 g
Bromothymol blue (0.2% solution)	40 ml
Distilled water	1000 ml

To the basal medium 1.5 g of cysteine hydrochloride is added slowly. The amino acid first may be dissolved in 25 ml of hot distilled water. The pH is adjusted to 7 by addition of NaOH. The medium is tubed in amounts of 4 ml in 13 X 100 mm screw-capped tubes and sterilized at 121 C for 20 min.

Inoculation: The medium is inoculated with a loopful of growth from an 18-24 hr agar slant culture (stab inoculation).

Incubation: The medium is incubated at 37 C for 7 days and read daily. A positive reaction is evidenced by production of acid and the appearance of a marked precipitate that appears first in the upper portion of the medium.

DECARBOXYLASE TEST MEDIA

Moeller Method (1954, 1955)

Basal medium

Peptone (Orthana special)[3]	5 g
Beef extract	5 g
Bromcresol purple (1.6%)	0.625 ml
Cresol red (0.2%)	2.5 ml
Glucose	0.5 g
Pyridoxal	5 mg
Distilled water	1000 ml
Adjust pH to 6 or 6.5	

The basal medium is divided into four equal portions, one of which is tubed without the addition of any amino acid. These tubes of basal medium are used for control purposes. To one of the remaining portions of basal medium is added 1% of L-lysine dihydrochloride; to the second, 1% of L-arginine monohydrochloride; and to the third portion, 1% of L-ornithine dihydrochloride. If DL amino acids are used they should be incorporated into the media in 2% concentration, since the microorganisms apparently are only active against the L forms. *The pH of the fraction to which ornithine is added should be readjusted after the addition and prior to sterilization.* The amino acid media may be tubed in 3 or 4 ml amounts in small (13 X 100 mm) screw-capped tubes and sterilized at *121 C for 10 minutes.* A small amount of floccular precipitate

may be seen in the ornithine medium. This does not interfere with its use.

Inoculation: Inoculate lightly from a young agar slant culture. After inoculation add a layer (about 10 mm in thickness) of sterile mineral (paraffin) oil[4] to each tube including the control. A control tube should be inoculated with each culture under investigation.

Incubation: 37 C. Examine daily for 4 days. Positive reactions are indicated by alkalinization of the media and a consequent change in the color of the indicator system from yellow to violet or reddish-violet. Weakly positive reactions may be bluish grey in color. The majority of positive reactions with Enterobacteriaceae occur within the first day or two of incubation but sufficient delayed reactions occur to warrant a 4-day incubation and observation period.

Falkow Method (1958). This method, which may be regarded as a modification of the Moeller method, was originally devised as a lysine medium only. However, Falkow (personal communication, 1957) stated that arginine could be substituted successfully for lysine, and the author has employed Falkow's basal medium with lysine, arginine, and ornithine. Comparative studies (Ewing, Davis and Edwards, 1960) have shown that the Falkow modification can be used successfully in most areas of the family. Results obtained with cultures that belong to the genera *Klebsiella* and *Enterobacter* may be equivocal, however, and the method cannot be recommended for use with members of these genera.

Basal medium

Bacto peptone	5 g
Yeast extract	3 g
Glucose	1 g
Bromcresol purple (1.6% solution)	1 ml
Distilled water	1000 ml

Adjust pH to 6.7 - 6.8, if necessary (medium with ornithine requires adjustment).

The basal medium is divided into four parts and treated in the same manner as that given above for the Moeller medium except that only 0.5% of L-amino acid is added. After the amino acids are added to three of the four portions of basal medium, the media are tubed in small (13 X 100

[3]Peptone Special "Orthana" Meat USP XV from A/S Orthana Kemish Fabrik, Copenhagen, Denmark. Dr. Hariette Vera informed the author (personal communication, 1960) that this peptone apparently is a copy of USP *Peptic Digest of Animal Tissues,* developed by BBL, and marketed under the name Thiotone.

[4]The mineral oil (heavy) should be sterilized in an autoclave at 121 C for 30 minutes to one hr, depending on the size of the containers. Large screw-capped test tubes, about half full of oil, are convenient. The cloudiness seen after sterilization disappears in a short time.

mm screwcapped) tubes and sterilized at *121 C for 10 min*. The remaining portion of basal medium, without amino acid, serves as a control.

Inoculation: Inoculate lightly from a young agar slant culture. Oil seals should be used with this method. A control tube should be inoculated with each culture under investigation.

Incubation: 37 C. Examine daily for 4 days. These media first become yellow because of acid production from glucose; later, if decarboxylation occurs, the medium becomes alkaline (purple). The control tubes should remain acid (yellow).

DNase TEST MEDIUM (Jeffries et al., 1957; Martin and Ewing, 1967)

Tryptose (Bacto)	20 g
Deoxyribonucleic acid	2 g
Sodium chloride	5 g
Agar	15 g
Distilled water	1000 ml

Heat mixture to boiling temperature with frequent agitation (swirling). Sterilize at 121 C for 15 min, cool, and pour into petri dishes.

Inoculation: Inoculate from a young agar slant culture by spot inoculation. Several cultures may be tested simultaneously on one plate of medium.

Incubation: 37 C for 18-24 hr. Flood plate with N hydrochloric acid. A zone of clearing around a spot of growth indicates DNase activity.

DNase TEST MEDIUM (Alternate method: Smith, Hancock, and Rhoden, 1969)

1. Dissolve 500 mg of methyl green (Fischer M295) in 100 ml of distilled water. Extract repeatedly with chloroform employing approximately 100 ml each time until chloroform layer becomes colorless. This procedure removes methyl violet, which is an impurity in the methyl green. About six extractions should suffice.
2. Prepare DNA test agar in the manner outlined above. Add one ml of methyl green solution per 100 ml of melted medium and sterilize at 121 C for 15 min. Plates may be kept for several days (about a week), if refrigerated. If desired, the dye solution can be filter-sterilized and added aseptically.
3. Plates should be inoculated (see above), incubated overnight, and examined by transmitted light against a white background. A zone of clearing around the colony indicates a positive reaction.

ESCULIN BROTH (Vaughn and Levine, 1942)

Peptone	5 g
Dibasic potassium phosphate	1 g
Andrade's indicator	10 ml
Distilled water	1000 ml
Add:	
Esculin	3 g
Ferric citrate	0.5 g

Heat to dissolve ferric citrate. Tube and sterilize at *121 C for 10 min*.

Inoculate and Incubate: As with any carbohydrate medium. Positive results are indicated by production of a black precipitate.

GELATIN LIQUEFACTION

Nutrient gelatin

Beef extract	3 g
Peptone	5 g
Gelatin	120 g
Distilled water	1000 ml

Sterilize at *121 C, 12 min*.

Inoculation: Inoculate by stabbing the medium with a wire, using inoculum from an agar slant culture.

Incubation: 20 to 22 C for 30 days.

It is believed that the nutrient gelatin method should be employed as the standard method in taxonomic work, since the rate of gelatin liquefaction is important in the characterization of members of certain genera and species within the family Enterobacteriaceae. Some of the rapid methods (see below) are excellent for diagnostic work in which one is interested only in whether a culture liquefies gelatin or not, but they are of little or no value as differential tests in those areas of the family where the rate of gelatin liquefaction is of differential importance (e.g., within the tribe Klebsielleae). In those areas where the rate of gelatin liquefaction is of differential value, positive tests obtained by means of rapid methods should be repeated using the conventional method. If the above-mentioned limitations are borne in mind certain rapid methods can be recommended.

GELATIN LIQUEFACTION DEMONSTRATION BY RAPID METHODS (Kohn 1953)

Preparation of denatured charcoal gelatin:

1. Dehydrated nutrient gelatin medium is mixed with water in the proportion of 15 g to 100 ml of water and the mixture is heated.
2. Finely powdered charcoal is added to the

melted gelatin medium in the proportion of 3-5 g to 100 ml of medium.

3. The charcoal gelatin medium is thoroughly mixed and poured into petri dishes, or other suitable flat containers, to form a layer about 3 mm thick. Parenthetically, it is advisable to apply a thin film of vaseline to the containers before pouring the mixture to prevent it from sticking. The mixture should be cooled somewhat before it is poured, and then chilled quickly so that the charcoal does not sediment. It is helpful to chill the containers prior to pouring the mixture and to place them in a refrigerator until gelation occurs.

4. After the medium has set thoroughly, the entire sheet is covered with 10% formalin for about one hr. Then it is removed from the container and placed in 10% formalin for 24 hr.

5. The formalin-gelatin sheet then is cut into pieces about 10 mm by 5-8 mm. These pieces are wrapped in gauze and washed in running tap water for 24 hr.

6. The pieces then are placed in sterile jars or wide-mouthed screw-capped bottles and covered with distilled water. Toluene should be added to each container.

Inoculation:

1. The growth from an 18-24 hr agar plate culture is suspended in 3 ml of sterile 0.85% sodium chloride solution containing 0.01 M calcium chloride in a small sterile test tube (e.g., 13 X 100 mm). The suspension should be very dense, since the rapidity of the reactions in this test appear to be functions of density and of temperature of incubation. The agar plates should be thick, i.e., they should contain 35-40 ml of infusion agar medium. Three or 4 drops of a broth culture of the strain to be tested should be spread over the entire surface of the agar medium so as to obtain a maximum, confluent growth.

2. With aseptic precautions add a piece of denatured charcoal-gelatin to each dense suspension prepared as outlined above.

3. Add 0.1 ml of toluene to each suspension and shake the tubes.

Incubation: 37 C. Examine after 5 or 6 hr, and daily for 14 days. Positive reactions are indicated by the release of charcoal particles which collect in the bottom of the tube.

Toluene is added because it appears to have an activating effect upon the reactions of strains that

ordinarily are slow liquefiers of gelatin. However, advantage may also be taken of its bacteriocidal effect: the suspending fluid (physiological saline solution with 0.01 M calcium chloride) can be distributed in 3 ml amounts in small test tubes and sterilized at 121 C for 15 min. A piece of charcoal-gelatin and 0.1 ml of toluene are placed in each tube, after which the cotton stoppers are replaced with corks that have been soaked in hot paraffin. The tubes are then stored until needed.

GLYCEROL FUCHSIN BROTH (Stern, 1916)

Solution 1

Meat extract	10 g
Peptone	20 g
Tap water	1000 ml
Adjust pH to 8	

Solution 2
Saturated alcoholic solution of basic fuchsin (10%).

Solution 3
Fresh 10% aqueous solution of anhydrous sodium sulfite.

Mixture

Solution 1	100 ml
Solution 2	0.2 ml
Solution 3	1.66 ml
Glycerol	1 ml

Tube and sterilize at *121 C for 12 min.*

Inoculation: Inoculate from a young agar slant culture.

Incubation: 37 C. Observe daily for 8 days, noting the development of a deep red color as compared to an uninoculated control tube incubated with cultures under examination. Final readings of commercially available medium should be made at 24 hr.

HYDROGEN SULFIDE TESTS

Triple Sugar Iron (TSI) agar (Sulkin and Willett, 1940). This medium may be considered a modification of Kligler's iron agar, and other similar media. It also is employed as a differential medium for the examination of colonies picked from plating media. pH changes in the butt and on the slant of the medium are recorded after 18-24 hr only.

Beef extract	3 g
Yeast extract	3 g
Peptone (Bacto or polypeptone)	15 g
Proteose peptone (Difco)	5 g
Lactose	10 g

Sucrose	10 g
Glucose	1 g
Ferrous sulfate	0.2 g
Sodium chloride	5 g
Sodium thiosulfate	0.3 g
Agar	15 g
Phenol red	0.024 g
Distilled water	1000 ml

After sterilization at 121 C, 15 min, the medium is slanted with a deep butt (1-inch butt, 1.5-inch slant).

Inoculation: The butt of the medium is stabbed and the slant is streaked.

Incubation: 37 C. Observe after 18-24 hr and 40-48 hr for blackening caused by hydrogen sulfide production. Cultures may be observed daily for 7 days, but care should be exercized in interpreting weak and delayed and weak reactions.

KLIGLER'S IRON AGAR (KI agar, Kligler, 1917, modified).

The formulation of this medium, as generally given, is essentially the same as for TSI agar (above) except that KI agar does not contain sucrose.

Inoculation and Incubation: as with TSI agar.

PEPTONE IRON AGAR (Levine et al., 1934, modified: Tittsler and Sandholser, 1937)

Peptone (Bacto)	15 g
Proteose peptone	5 g
Ferric ammonium citrate	0.5 g
Dipotassium phosphate	1 g
Sodium thiosulfate	0.08 g
Agar	15 g
Distilled water	1000 ml

Inoculation and Incubation: as with TSI agar.

INDOL TEST

Peptone water

Peptone (Bacto)	20 g
Sodium chloride	5 g
Distilled water	1000 ml

Leave reaction unadjusted and sterilize at 121 C, 15 min. One per cent Bacto tryptone may be substituted for the 2% Bacto peptone, if desired. Or, 1.5% casitone or trypticase may be used.

Inoculation: Inoculate lightly from a young agar slant culture.

Incubation: 37 C, 40-48 hr. Test with Kovacs' reagent. Add about 0.5 ml of reagent, shake tube gently. A deep red color develops in the presence of indol.

Kovacs' Reagent

Pure amyl or isoamyl alcohol	150 ml
Paradimethylaminobenzaldehyde	10 g
Concentrated pure hydrochloric acid	50 ml

Dissolve aldehyde in alcohol, and then slowly add acid. The dry aldehyde should be light in color. Kovacs' reagent should be prepared in small quantities and stored in a refrigerator when not in use.

Tests for indol production may be made after 24 hr of incubation, but if this is to be done, 1 or 2 ml of culture should be removed aseptically from the tube and the test made on this sample. If the test is negative, the remaining portion of the peptone water culture should be reincubated for an additional 24 hr.

Paper Strip Tests for Indol

Strips of filter paper (about 5 X 60 or 70 mm) may be soaked in freshly prepared Kovacs' reagent, dried, and stored in brown bottles. These strips are suspended over peptone water or lysine iron agar cultures. When positive (pink or red color), further tests are unnecessary. If no change occurs, tests should be made by the conventional method (above).

Gillies (1956) described a method by which filter paper strips are saturated with the following reagent:

p-dimethylaminobenzaldehyde	5 g
Methanol	50 ml
o-phosphoric acid	10 ml

The strips are dried at 70 C for a minimum period, and stored. They are used in the same way as papers prepared with Kovacs' reagent. The best grade of filter paper available should be used.

INDOPHENOL OXIDASE TEST (Gaby and Hadley, 1957; modified by Ewing and Johnson, 1960)

Nutrient agar slant cultures incubated at 37 C, or at a lower temperature if required, are recommended. After incubation for 18 to 24 hr (older cultures should not be used), 2 or 3 drops of each reagent are introduced and the tube tilted so that the reagents are mixed and flow over the growth on the slant. Positive reactions are indicated by rapid development of a blue color in the growth within 2 min. The majority of positive cultures produce strong reactions within 30 sec. Any very weak or doubtful reaction that occurs after 2 min should be ignored. Plate cultures (nutrient or infusion agar) may be tested by allowing an equal parts mixture of the reagents to flow over isolated colonies.

Test Reagents

A. Ethyl alcohol, 95-96%	100 ml
Alpha naphthol	1 g
B. Distilled water	100 ml
Para-aminodimethylaniline HCl	
(or oxylate)	1 g

(Reagent B should be prepared frequently and stored in a refrigerator when not in use.)

LIPASE MEDIUM. CORN OIL (Hugo and Beveridge, 1962; Davis and Ewing, 1964)

Peptone	10 g
Yeast extract	3 g
Sodium chloride	5 g
Agar	20 g
Distilled water	900 ml

Adjust to pH 7.8 and add 100 ml of Victoria blue (aqueous solution, 1:1, 500). Corn oil is added in 5% (v/v) concentration and mixed thoroughly into the medium. This may be accomplished by means of a magnetic stir-plate, or a blender, or by *brief* treatment with a sonifier. The medium is distributed in tubes, sterilized at *115 C for 30 min,* and allowed to solidify in a slanted position (long slants). The completed medium is light in color.

Inoculation: Using a young agar slant culture as source of inoculum, a line inoculation is made from the bottom to the top of the slant.

Incubation: 37 C. Read daily for 7 days.

Positive reactions are indicated by the development of a dark blue color in the medium, in the growth, or in both. The acid salt of Victoria blue is water-soluble and blue, whereas the free base is red and soluble in fats. When Victoria blue and a fat are mixed in a melted medium at about pH 8, the dye is extracted by the fat as the red base. With gelation of the agar, a light pinkish gray medium is obtained. When the fat is hydrolyzed, fatty acid is formed which converts the dye to the acid (blue) salt and liberates it. The success of the method depends upon the use of substrates that are insoluble in water.

LYSINE IRON AGAR (Edwards and Fife, 1961)

Peptone (Bacto)	5 g
Yeast extract	3 g
Glucose	1 g
L-lysine	10 g
Ferric ammonium citrate	0.5 g
Sodium thiosulfate	0.04 g
Bromcresol purple	0.02 g
Agar	15 g

Distilled water	1000 ml

(Adjust to pH 6.7)

Dispense in 4 ml amounts in 100 X 13 mm tubes and sterilize at *121 C for 12 min.* Slant tubes so as to obtain a deep butt and a short slant.

Inoculation: Inoculate with a straight wire by stabbing twice to the base of the butt and by streaking the slant.

Incubation: 37 C 18-24 hr (48 hr, if necessary).

This medium was designed originally for the examination of blackened colonies on bismuth sulfite agar plates, but it has since been found to have wider applications (Johnson et al., 1966). This medium is *not* a substitute for the standard (Moeller) method.

MALONATE BROTH (Leifson, 1933, modified; modified, Ewing et al., 1957)

Test for utilization of malonate

Yeast extract	1 g
Ammonium sulfate	2 g
Dipotassium phosphate	0.6 g
Monopotassium phosphate	0.4 g
Sodium chloride	2 g
Sodium malonate	3 g
Glucose	0.25 g
Bromthymol blue	0.025 g
Distilled water	1000 ml

Sterilize at 121 C for 15 min.

Inoculation: Inoculate from a young agar slant or broth culture. (A 3-mm loopful of broth culture is preferred.)

Incubation: 37 C for 48 hr. Observe after 24 and 48 hr.

Positive results are indicated by a change in the color of the indicator from green to Prussian blue.

METHYL-RED TEST (MR)

Buffered peptone glucose broth. Several modifications of the Clark and Lubs formula are available. The two that follow are known to be satisfactory. The medium selected for MR tests also may be used for Voges-Proskauer (VP) reactions.

MR – VP medium (BBL)

Dipotassium phosphate	5 g
Polypeptone	7 g
Glucose	5 g
Distilled water	1000 ml

Suspend ingredients in water and heat slightly to dissolve them. Tube and sterilize at *116 C for 15 min* or *121 C for 10 min.* Medium should be water clear; if yellow, it is unsatisfactory.

MR – VP medium (Difco)

Buffered peptone	7 g
Glucose	5 g
Dipotassium phosphate	5 g
Distilled water	1000 ml

Tube and sterilize at *116 C for 15 min,* or at *121 C for 10 min.*

Inoculation: Inoculate lightly from a young agar slant culture.

Incubation: Standard methods generally recommend 5 days' incubation at 30 C. However, incubation at 37 C for 48 hr is sufficient for the determination of the MR reactions of the majority of cultures. *Tests should not be made with cultures incubated less than 48 hr.* If the results of tests incubated at 37 C for 48 hr are equivocal, the tests should be repeated with cultures that have been incubated for 4 or 5 days. In such instances duplicate tests should be incubated at 22-25 C.

Test Reagent

Methyl red	0.1 g
Ethyl alcohol (95 - 96%)	300 ml

Dissolve dye in the alcohol and then add sufficient distilled water to make 500 ml. Use 5 or 6 drops of reagent per 5 ml of culture. Read reactions immediately. Positive tests are bright red, weakly positive tests are red-orange, and negative tests are yellow or orange.

MOTILITY TEST MEDIUM

Beef extract	3 g
Peptone	10 g
Sodium chloride	5 g
Agar	4 g
Distilled water	1000 ml

Adjust reaction to pH 7.4 and tube, about 8 ml per tube, and sterilize at 121 C, 15 min.

Inoculate: Inoculate by stabbing into the top of the column of medium to a depth of about 5 mm.

Incubation: 37 C for 1 or 2 days. If negative, follow with further incubation at 22-25 C for 5 days.

For special purposes such as enhancement of motility and flagellar development in poorly motile cultures, it is advisable to passage cultures first through a semisolid medium containing 0.2% agar in Craigie tubes or in U-tubes. Subsequent passages may be made in the medium listed above (0.4% agar).

NITRATE REDUCTION TEST MEDIA

Although the nitrate reduction test is of little or no value for differentiation within the Enterobacteriaceae, it is included because a negative test is often a value in the exclusion of otherwise doubtful strains. The results of negative nitrate tests always should be confirmed. This may be done by adding a minute amount of zinc dust to the tube. The development of a red color indicates the presence of unreduced nitrate.

Tryptone	5 g
Neopeptone	5 g
Agar	2.5 g
Distilled water	1000 ml
Boil and adjust pH to 7.3-7.4 then add	
potassium nitrate (nitrite free)	1 g
Glucose	0.1 g

Sterilize at 121 C for 15 min.

Inoculation: The medium may be inoculated by stabbing into the column of semisolid agar medium.

Incubation: 37 C, 24 hr. The occasional culture that gives apparently equivocal results should be reinoculated and tested after 1, 2, 3, and 4 days' incubation. Fluid medium tubed with inverted insert vials may be used in lieu of the semisolid medium given above (e.g., meat extract, 3 g; peptone, 5 g; potassium nitrate 1 g; and distilled water 1000 ml).

Test Reagents:

A. Dissolve 8 g of sulfanilic acid in 1000 ml of 5 N acetic acid.
B. Dissolve 5 g of alpha naphthylamine in 1000 ml of 5 N acetic acid.

Immediately before use, equal parts of the solutions A and B are mixed and 0.1 ml of the mixture is added to each culture.

Positive tests for reduction of nitrate to nitrite are indicated by the development of a red color within a few minutes.

ONPG TEST. Test for Beta-D-galactosidase activity (LeMinor and Ben Hamida, 1962; Lubin and Ewing, 1964; Buelow, 1964; Costin, 1966)

Reagent

1. Monosodium phosphate solution 1.0 M, pH 7. Dissolve 6.9 g $NaH_2PO_4 \cdot H_2O$ in approximately 45 ml distilled water. Add approximately 3 ml of 30% (w/v) NaOH

solution and adjust to pH 7. Bring volume to 50 ml with distilled water and store in a refrigerator (about 4 C).

2. 0.75 M o-nitrophenyl-β-D-galactopyranoside (ONPG) in 0.25 M monosodium phosphate solution, pH 7. Dissolve 80 mg ONPG in 15 ml distilled water at 37 C. Add 5 ml of 1.0 M NaH_2PO_4 solution (1, above). Solution should be colorless and should be stored in a refrigerator. Prior to its use an appropriate portion (sufficient for the number of tests to be done) of the buffered 0.75 M ONPG solution should be warmed to 37 C.

Procedure

Cultures to be tested are inoculated onto triple sugar iron agar (TSI) slants and incubated for 18 hr at 37 C (or at another appropriate temperature if required). Nutrient agar slants containing 1.0% lactose also may be used. A large loopful of growth from each culture is emulsified in 0.25 ml of physiological saline solution (heavy suspension).

One drop of toluene is added to each tube and the tubes shaken well (to aid liberation of the enzyme). Following this the tubes are allowed to stand for about five min at 37 C.

0.25 ml of buffered 0.75 M ONPG solution is added to each suspension to be tested, and the tests are incubated in a waterbath at 37 C.

Readings should be made after 30 min, 1 hr, and 24 hr of incubation. Positive results are indicated by the development of a yellow color.

ORGANIC ACID MEDIA (Kauffmann and Petersen, 1956, as modified by Ellis et al., 1957)

Basal medium	
Peptone (Bacto)	10 g
Bromthymol blue (0.2% solution)	12 ml
Distilled water	1000 ml

One percent of the following test substances is added to equal portions of basal medium:

D-tartrate	— Sodium potassium tartrate, C.P.
Citrate	— Sodium citrate, C.P.
Mucate	— Mucic acid, practical (Eastman)

After addition of the tartrate and citrate the pH is adjusted to 7.4 with 5 N or 10 N sodium hydroxide and the two media are tubed in 3 ml amounts in 100 X 13 mm test tubes and sterilized at 121 C for 15 min. The basal medium to which mucic acid is to be added is sterilized in bulk. The mucic acid is weighed with aseptic precautions and added to the base while it is hot. Sufficient sodium hydroxide (5 N or 10 N) then is added slowly to bring the mucic

acid into solution as sodium mucate and to adjust the final pH of the medium to 7.4. The medium is tubed in 3 or 4 ml amounts and incubated for sterility. If sufficient care is taken to make certain that all of the mucic acid is in solution, the mucic acid may be added to the basal medium prior to sterilization at *121 C for 10 min.*

Inoculation: Each of the three above-mentioned organic acid media should be inoculated with a 3 mm loopful of inoculum from an overnight nutrient broth culture. An uninoculated tube of each medium should be incubated along with the sets of tests done in any one day.

Incubation: 37 C. After 20-24 hr incubation, 0.5 ml of a 50% neutral lead acetate solution is added to the tubes of D-tartrate and citrate media, as well as to uninoculated control tubes. Utilization of these substances is indicated by a diminution in the volume of precipitate formed in comparison with that in the control tubes. Utilization of mucate is indicated by an acid reaction (yellow color). Multiple tubes of the above-mentioned substrates, as well as *i*-tartrate and *ℓ*-tartrate, may be inoculated and tested at different intervals of incubation, as recommended by Kauffmann and Petersen (1956). The malonate medium mentioned elsewhere may be used to advantage in conjunction with these organic acid media.

OXIDATION-FERMENTATION TEST MEDIUM (Hugh and Leifson, 1953)

This method aids in the differentiation of microorganisms that utilize carbohydrates oxidatively rather than fermentatively and therefore is helpful in the identification of pseudomonads and members of the tribe Mimeae. It also aids in the identification of microorganisms that do not utilize glucose in either way (e.g., *Pseudomonas alcaligenes*). Only a glucose medium is included here but other carbohydrates may be substituted for special purposes (Hugh and Leifson, 1953).

Peptone	2 g
Sodium chloride	5 g
Dibasic potassium phosphate	0.3 g
Agar	3 g
Bromthymol blue (1% *aqueous* solution)	3 ml
Distilled water	1000 ml
Adjusted pH to 7.1	

Distribute the basal medium in test tubes (e.g., 13 X 100 mm), 3 or 4 ml per tube, and sterilize at 121 C for 15 min. After sterilization 1% of glucose is added to each tube. A 10% solution of glucose in distilled water sterilized by filtration is convenient for this purpose.

Inoculation: Two tubes of medium are inoculated (stab) with each culture under investigation. The medium should be inoculated lightly with inoculum from a young agar slant culture. After inoculation a layer (about 10 mm) of sterile melted petrolatum or of sterile paraffin oil (see Decarboxylase media) is added to *one* of the tubes. A stiff petrolatum seal is recommended in the study of pseudomonads and allied bacteria (Hugh, personal communication). Therefore, if equivocal results are obtained using oil seals, tests should be repeated following Hugh's recommendation.

Incubation: 37 C. Observe daily for 3 or 4 days. Acid formation in the open tube only indicates oxidative utilization of glucose. Acid formation in both the open and the sealed tube is indicative of a fermentation reaction. Lack of acid production in either tube indicates that the microorganism being tested does not utilize glucose by either method.

PECTATE MEDIUM (Starr, 1947)

Certain bacteria formerly referred to as *Erwinia* and now classified as *Pectobacterium* occasionally may be seen, and their differentiation from other Enterobacteriaceae may be somewhat difficult. However, these bacteria produce pectinases and other Enterobacteriaceae are not known to possess them. Hence, a medium containing pectin is of value in the differentiation of these bacteria. A suitable pectate medium (Starr, 1947) may be prepared as follows:

Place 100 ml distilled water in a beaker and add:

0.5 g yeast extract

0.9 ml of 1.0 N sodium hydroxide

0.6 ml 10% calcium chloride solution ($CaCl_2 \cdot 2H_2O$)

1.25 ml Bromthymol blue (0.2% solution)

Add 3 g sodium polypectate, no. 24 (Sunkist Growers, Inc., Products Division, 616 East Grove St., Ontario, Calif.)

Stir to wet pectate.

Heat in a boiling-water bath to dissolve pectate (constant stirring). The color should now be blue-green (about pH 7.3). Further adjustment of pH cannot be made. Distribute in 3 to 4 ml amounts in small test tubes and sterilize at 121 C for 15 min. The color should now be yellowish-green (about pH 6.4). Allow tubes to cool in an upright position. *This medium cannot be reheated.*

Inoculation: Inoculate from a young agar slant culture by stabbing deep into the column of medium.

Incubation: 37 C for 7 days. Observe daily for evidence of liquefaction as with nutrient gelatin medium. Changes in the color of the medium may occur during growth of Enterobacteriaceae, e.g., to yellow. This is of no importance so far as Enterobacteriaceae are concerned. Such changes probably are caused by small amounts of glucose in the polypectate.

PECTATE MEDIUM, MODIFIED (Martin and Ewing, unpublished data)

Place 100 ml distilled water in a beaker and add:

0.5 g yeast extract

0.9 ml N sodium hydroxide

0.5 ml 10% calcium chloride solution ($CaCl_2 \cdot 2H_2O$)

1.25 ml Bromthymol blue (0.2% solution)

Add 1 g sodium polypectate no. 24 (source given above) *and* 2 g of agar. Stir thoroughly to wet, then heat mixture in a boiling water bath to dissolve ingredients as much as possible. While mixture is still warm, place it in an autoclave and sterilize at *121 C for 5 min.* Pour thin plates (20-25 ml per plate). The advantage of this method is the relative ease in which the sodium polypectate goes into solution.

Inoculation: Inoculate from a young agar slant culture by spot inoculation onto a plate. Several cultures can be tested on a single plate, i.e., 10 cultures per plate.

Incubation: 37 C for 1-3 days. Evidence of liquefaction of sodium polypectate is shown by a depression in the medium surrounding growth. Cultures that do not utilize sodium polypectate will not show this reaction. Adequate control cultures should be included on each plate.

PHENYLALANINE AGAR (Ewing et al., 1957)

Test for deamination of phenylalanine to phenylpyruvic acid.

Yeast extract	3 g
DL-phenylalanine	2 g
(or L-phenylalanine)	(1 g)
Disodium phosphate	1 g
Sodium chloride	5 g
Agar	12 g
Distilled water	1000 ml

Tube and sterilize at *121 C for 10 min* and allow to solidify in a slanted position (long slant).

Test reagent: 0.5 M (about 13% w/v) solution of ferric chloride

Inoculation: Inoculate the slant of the PA agar with a fairly heavy inoculum from an agar slant culture.

Incubation: 4 hr or, if desired, 18-24 hr at 37 C. Following incubation, 4 or 5 drops of ferric chloride reagent are allowed to run down over the growth on the slant. If phenylpyruvic acid has been formed, a green color develops in the syneresis fluid and in the medium.

PHENYLPROPRIONIC ACID MEDIUM (see D'Alessandro and Comes, 1956), modified.

Peptone (Bacto)	10 g
Yeast extract	3 g
Disodium phosphate	1 g
Sodium chloride	5 g
Phenylproprionic acid	0.2 g
Agar	15 g
Distilled water	1000 ml
Adjust to pH 7.2	

Distribute in tubes and sterilize at 121 C for 15 min. Allow agar to gel while tubes are in a slanted position. The medium may also be used as a semisolid agar medium, in which case the agar content may be reduced to 5 or 6 g per liter. The color reactions may be intensified by the addition of 0.2% proline.

Inoculation: The medium may be inoculated over the entire surface of the slant using a young culture as the source of inoculum. Stab inoculations may be made into the depths of semisolid agar medium if it is employed.

Incubation: 37 C. Examine daily for 4 or 5 days. Positive reactions are indicated by development of a rose to violet color in the medium.

NB: The author usually refers to this medium as HCA (hydrocinnamic acid) agar to avoid confusion with phenylalanine agar and phenylpyruvic acid production.

PIGMENT PRODUCTION MEDIA For enhancement of pigment formation by *Pseudomonas* (King et al., 1954)

These media are of value in the differentiation of pseudomonads that do not produce detectable amounts of pigment on ordinary nutrient agar or broth media. Medium A is particularly useful for the detection of pyocyanin by *Pseudomonas aeruginosa* strains, whereas medium B enhances production of fluorescin by *P. aeruginosa* and other varieties of *pseudomonads* (E. O. King, personal communication, 1959). The determination of fluorescence may be aided by the use of an ultraviolet light.

Medium A

Peptone (Bacto)	20 g
Glycerol C.P.	10 ml
Magnesium chloride	1.4 g
Potassium sulfate	10 g
Agar	15 g
Distilled water	1000 ml

Medium B

Proteose peptone No. 3	20 g
Glycerol C.P.	10 ml
Dibasic potassium phosphate	1.5 g
Magnesium sulfate	1.5 g
Agar	15 g
Distilled water	1000 ml

The peptones listed for these media are critical. Adjust pH to 7.2 if necessary, tube and sterilize at 121 C for 15 min. Slant tubes so as to obtain a deep butt in the medium.

Inoculation: Each medium may be inoculated with a loopful of a young broth culture or with an inoculum from a young agar slant culture.

Incubation: Medium A, 37 C, 48 hr or longer.

Medium B, 37 C, 24 hr followed by 2 to 3 days at room temperature. (Some strains form pigment at the latter temperature and not at the former.)

POTASSIUM CYANIDE (KCN) MEDIUM

Test for growth in the presence of KCN (Moeller, 1954; Edwards and Fife, 1956; Edwards et al., 1956).

Peptone, Orthana special[5]	10 g
Sodium chloride	5 g
Monobasic potassium phosphate	0.225 g
Dibasic sodium phosphate	5.64 g
Distilled water	1000 ml
Adjust to pH 7.6	

The basal medium is sterilized at 121 C for 15 min, then refrigerated until thoroughly chilled. To the cold medium is added 15 ml of 0.5% potassium cyanide solution (0.5 g KCN dissolved in 100 ml *cold* sterile distilled water). The medium is then tubed in approximately 1 ml amounts in sterile tubes (12 X 150 mm or 13 X 100 mm) and

[5]Peptone Special Orthana Meat USP XV from A/S Orthana Kemish Fabrik, Copenhagen, Denmark. (See Decarboxylase test media.)

stoppered quickly with corks that have been soaked in hot paraffin. The medium in such tubes can be stored safely for 2 weeks at 4 C. The final concentration of KCN in the medium is 1:13,300. It has been found that 0.3% Bacto Proteose Peptone no. 3 may be substituted for Orthana Peptone.

Inoculation: The tubes are inoculated with 1 loopful (3 mm loop) of a 20-24 hr broth culture grown at 37 C. Quickly reinsert cork stoppers.

Incubation: Tests are incubated at 37 C and observed daily for 2 days. Positive results are indicated by growth in the presence of KCN.

TARTRATE AGAR (Jordan and Harmon, 1928) This medium also is known as phenol red tartrate agar.

Peptone (Trypticase or Bacto)	10 g
Sodium potassium tartrate	10 g
Sodium chloride	5 g
Agar	15 g
Phenol red	0.024 g
Distilled water	1000 ml

Adjust pH to 7.8.

Dispense in tubes, about 10 ml per tube and sterilize at *116 C for 20 min.* Cool tubes in an upright position.

Inoculation: Inoculate by stabbing deep into the column of medium with a straight wire, using a young agar slant culture as source of inoculum.

Incubation: 37 C. Make readings after 24 and 48 hr incubation. Positive tests are indicated by an acid reaction (definite yellow color).

UREASE TEST MEDIA

Christensen's urea agar (1944, personal communication; 1946)

Basal Ingredients

Peptone	1 g
Sodium chloride	5 g
Glucose	1 g
Monobasic potassium phosphate	2 g
Phenol red	0.012 g (6 ml of 1:500 solution)

Urea concentrate

Urea	20 g
Distilled water	100 ml

Adjust to pH 6.8-6.9. Filter sterilize.

Dissolve 15 g agar in 900 ml distilled water, add basal ingredients, and sterilize at 121 C for 15 min. Cool to 50-55 C, then add 100 ml urea concentrate (above). Mix and distribute in sterile tubes. The medium is slanted with a deep butt. This medium may be employed in fluid form if desired.

Inoculation: The medium is inoculated heavily over the entire surface of the slant.

Incubation: 37 C. Examine at 2 hr, 4 hr, 6 hr, and after overnight incubation. Negative tubes should be observed daily for 4 to 7 days in order to detect delayed reactions given by members of certain genera other than *Proteus.* Urease positive cultures produce an alkaline reaction in the medium evidenced by a red color.

Alternate Method. Stuart et al., 1945, modification of the highly buffered medium of Rustigian and Stuart, 1941.

Yeast extract	0.1 g
Monobasic potassium phosphate	0.091 g
Dibasic sodium phosphate	0.095 g
Urea	20 g
Phenol red	0.01 g
Distilled water	1000 ml

This medium is filter-sterilized and tubed in sterile tubes in 3 ml amounts. The basal medium (without urea) may be prepared in 900 ml of distilled water and sterilized at 121 C, 15 min. After cooling, 100 ml of 20% sterile urea solution are added and the medium dispensed in sterile tubes in 3 ml amounts.

Inoculation: Three loopsful (2-mm loop) from an agar slant culture are inoculated into a tube of medium and the tube is shaken to suspend the bacteria.

Incubation: Tests are incubated in a water bath at 37 C, and the results are read after 10 min, 60 min, and 2 hr.

VOGES-PROSKAUER TEST (V-P) Test for production of acetymethylcarbinol (acetoin).

Buffered peptone glucose broth. The same medium used for the MR reaction may be used for the V-P test.

Inoculation: Inoculate lightly from a young agar slant culture.

Incubation: 37 C for 48 hr. If equivocal results are obtained under these circumstances, the tests should be repeated with cultures incubated at 25 C. (See remarks under MR test.)

Test Reagent O'Meara, modified).

Potassium hydroxide	40 g
Creatine	0.3 g
Distilled water	100 ml

Dissolve the alkali in distilled water and add creatine. Use the reagent in the proportion of 1 ml to 1 ml of culture. Tests may be placed at 37 C or left at room temperature. In either case final readings are made after 4 hr. Tests should be aerated by shaking the tubes. Positive reactions are indicated by development of an eosin pink to red color.

Note: The O'Meara reagent for V-P tests should be prepared frequently and should be refrigerated when not in use. If refrigerated the reagent may be used for 2-3 weeks but deteriorates rapidly thereafter (Levine et al., 1934; Suassuna et al., 1961).

Alternate Test Reagent (Barritt)

Add 0.6 ml of 5% alpha naphthol to absolute alcohol and 0.2 ml of 40% KOH to 1 ml of culture. Shake well after the addition of each reagent. Positive reactions occur at once or within 5 min (final reading) and are indicated by the production of a red color. The development of a copper color in some tubes should be disregarded.

OTHER MEDIA AND SOLUTIONS

BROTH MEDIUM FOR H (FLAGELLAR) ANTIGENS OF SALMONELLAE

Although any good infusion medium may be used, the author and co-workers has found that an equal mixture of trypticase soy and tryptose broths (available commercially) results in a good medium for growth of salmonellae and development of their H antigens. This mixture is tubed in 10 ml amounts in large tubes (18 X 140 to 150 mm) and sterilized (121 C, 15 min). After incubation an equal volume of formalinized saline solution (0.5 or 0.85% sodium chloride plus 0.6% formalin) is added to each culture. After thorough mixing of culture and formalinized saline solution, the suspensions are allowed to remain at room temperature for at least *one hour,* during which time the bacteria are killed.

Ordinarily such broth cultures are incubated for 18-20 hr at 35-37 C in an incubator. However, if necessary or desirable, the broth medium may be inoculated heavily and incubated in a waterbath at 35-37 C for 4-6 hr, treated with formalinized saline solution (1 hr), and tested in H antisera.

MERCURIC IODIDE SOLUTION (Bridges, 1951) For preparation of suspensions for slide agglutination tests.

Stock solution

Mercuric iodide	1 g
Potassium iodide	4 g
Distilled water	100 ml

Working solution (1:1000)

Stock solution	10 ml
0.5 or 0.85% sodium chloride solution	90 ml
Formalin	0.05 ml

SEMISOLID MEDIUM FOR ISOLATION OF PHASES. (Phase reversal medium)

Peptone	10 g
Gelatin	80 g
Meat extract	3 g
Agar	4 g
Sodium chloride	5 g
Distilled water	1000 ml

Dissolve the gelatin in 600 ml of water and the remaining ingredients in 400 ml of water. Mix, leave reaction unadjusted, tube, and sterilize at 121 C for 15 min. Cool quickly. Medium should be clear.

STOCK CULTURE MEDIA

Any good plain infusion agar medium *without* added carbohydrate may be used for this purpose. Blood agar base medium, which is available from commercial sources, is an excellent choice. The medium is tubed in 3-4 ml amounts in 13 X 100 mm tubes and sterilized at 121 C for 15 min., after which the tubes are placed in a slanted position (relatively short slant and deep butt) until gelatin of the agar occurs. Inoculation should be made by stabbing the butt of the medium once or twice, after which the slant is streaked. The tubes should be sealed with cork stoppers (no. 3) that have been soaked in hot paraffin. *Stock cultures should be kept at room temperature in the dark.*

Alternatively, a medium prepared by mixing equal parts of meat extract agar and nutrient broth may be used. This medium is not slanted. Stab inoculations are made; otherwise sealing, storage, etc., are the same as noted above.

WORFEL-FERGUSON MEDIUM (Modified; personal communication, 1953)

For enhancement of capsule production in cultures of *Klebsiella.*

Sodium chloride	2 g
Potassium sulfate	1 g
Magnesium sulfate	0.25 g
Sucrose	20 g
Yeast extract	2 g
Agar	15 g
Distilled water	1000 ml

Reaction is left unadjusted. Sterilize at 121 C for 15 min. The agar may be omitted as in the original formula of Worfel and Ferguson.

REFERENCES

Amies, C. R. 1967. Can. J. Pub. Health, **58**, 296.

Amies, C. R., and J. I. Douglas. 1965. Ibid., **56**, 27.

BBL Manual of products and laboratory procedures, 5th Ed. 1968. BBL, Division of BioQuest, Cockeysville, Maryland.

Bridges, R. F. 1951. Brit. Med. Bull., 7, 200.

Buelow, P. 1964. Acta Pathol. Microbiol. Scand., 60, 387.

Cary, S. G., and E. B. Blair. 1964. J. Bacteriol., 88, 96.

Christensen, W. B. 1946. Ibid., 52, 461.

――――. 1949. Res. Bull., Weld County Health Dept. (Greeley, Colo.), 1, 3.

Costin, I. D. 1965. J. Gen. Microbiol., 41, 23.

Davis, B. R., and W. H. Ewing. 1964. J. Bacteriol., 88, 16.

D'Alessandro, G., and Comes, R. 1956. Boll. Inst. Sieroter., Milan, 35, 202.

Difco Supplementary Literature. 1962. Difco Laboratories, Detroit, Michigan. (See also Difco Manual, 9th Ed., 1953).

Edwards, P. R., and M. A. Fife. 1956. Appl. Microbiol., 4, 46.

――――. 1961. Appl. Microbiol., 9, 478.

Edwards, P. R., M. A. Fife, and W. H. Ewing. 1956. Amer. J. Med. Technol., 22, 28.

――――. 1961. Pub. Health Lab.,**19, 85.

Ellis, R. J., P. R. Edwards, and M. A. Fife. 1957. Pub. Health Lab.,** 15, 89.

Ewing, W. H. 1968. Transport methods for Enterobacteriaceae. CDC Publ.*

Ewing, W. H., B. R. Davis, and R. W. Reavis. 1957. Pub. Health Lab.,**15, 153.

Ewing, W. H., and J. G. Johnson. 1960. Int. Bull. Bacteriol. Nomen. Tax., 10, 223.

Ewing, W. H., B. R. Davis, and P. R. Edwards. 1960. Pub. Health Lab.,**18, 77.

Falkow, S. 1958. Amer. J. Clin. Pathol., 29, 598.

Gaby, W. L. and C. Hadley. 1957. J. Bacteriol., 74, 356.

Galton, M. M., J. E. Scatterday, and A. V. Hardy. 1952. J. Infect. Dis., 91, 1.

Galton, M. M., W. D. Lowery, and A. V. Hardy. 1954. Ibid., 95, 232.

Gillies, R. R. 1956. J. Clin. Pathol., 9, 368.

Hajna, A. A. 1951. Pub. Health. Lab.,**9, 48.

Hajna, A. A. 1955. Ibid., 13, 59; 13, 83.

Hajna, A. A., and S. R. Damon. 1956. Appl. Microbiol., 4, 341.

Harvey, R. W. S. 1956. Mo. Bull. Min. Hlth., Gt. Brit., 15, 118.

Hinshaw, W. R. 1941. Hilgardia, 13, 583.

Holt-Harris, J. E., and O. Teague. 1916. J. Infect. Dis., 18, 596.

Hugh, R., and E. Leifson. 1953. J. Bacteriol., 66, 24.

Hugo, W. B., and E. G. Beveridge. 1962. J. Appl. Bacteriol., 25, 72.

Jameson, J. E., and N. W. Emberley. 1956. J. Gen. Microbiol., 15, 198.

Jeffries, C. D., D. F. Holtman, and D. G. Guse. 1957. J. Bacteriol., 73, 590.

Jordan, E. O., and P. H. Harmon. 1928. J. Infect. Dis., 42, 238.

Kauffmann, F. 1935. Z. f. Hyg., 117, 26.

Kauffmann, F. 1966. The bacteriology of Enterobacteriaceae. Copenhagen: Munksgaard.

Kauffmann, F., and A. Petersen. 1956. Acta Pathol. Microbiol. Scand., 38, 481.

King, E. O., M. K. Ward, and D. E. Raney. 1954. J. Lab. Clin. Med., 44, 301.

King, S., and W. I. Metzger. 1968. Appl. Microbiol., 16, 577; 16, 579.

Kligler, I. J. 1917. Amer. J. Pub. Health. 7, 805; 7, 1042.

Kohn, J. 1953. J. Clin. Pathol., 6, 249.

Kristensen, M., V. Lester, and A. Juergens. 1925. Brit. Exptl. Pathol., 6, 291.

Leifson, E. 1933. J. Bacteriol., 26, 329.

Leifson, E. 1935. J. Pathol. Bacteriol., 40, 581.

Leifson, E. 1936. Amer. J. Hyg., 24, 423.

LeMinor, L., and F. Ben Hamida. 1962. Ann. Inst. Pasteur, 102, 267.

Levine, M., and H. W. Schoenlein. 1930. A compilation of culture media for the cultivation of microorganisms. Society of American Bacteriologists, Monographs on Systematic Bacteriology, Vol. II. Baltimore: Williams & Wilkins.

Levine, M., S. S. Epstein, and R. H. Vaughn. 1934. Amer. J. Pub. Health, 24, 505.

Lubin, A. H., and W. H. Ewing. 1964. Pub. Hlth. Lab.** 22, 83.

MacConkey, A. T. 1905. J. Hyg., 5, 333.

MacConkey, A. T. 1908. Ibid., 8, 322.

Martin, W. J., and W. H. Ewing. 1967. Can. J. Microbiol., 13, 616.

Moeller, V. 1954. Acta Pathol. Microbiol. Scand., 34, 115.

Moeller, V. 1955. Ibid., 36, 158.

Muller, L. 1923. Comp. rend. soc. biol., 89, 434.

Sachs, A. 1939. J. Roy. Army Med. Corps. 73, 235.

Shipe, E. L., A. Fields, and J. R. Shea. 1960a. Pub. Health Lab.,**18, 95.

――――. 1960b. Ibid., 18, 104.

Shipolini, R., G. Konstantinow, A. Triponowa, and S. Atanassowa. 1959. Zentralbl. f. Bakt. I. Orig., 174, 75.

*CDC Publ., Publication from the Center for Disease Control (formerly the Communicable Disease Center), Atlanta, Ga. 30333.

**Public Health Laboratory: Journal of the Conference of Public Health Laboratory Directors.

Simmons, J. S. 1926. J. Infect. Dis., **39**, 209.

Smith, P. B., G. A. Hancock, and D. L. Rhoden. 1969. Appl. Microbiol., **18**, 991.

Starr, M. P. 1947. Phytopathology, **37**, 291.

Stern, W. 1916. Zentralbl. f. Bakt., I. Orig., **78**, 481.

Stuart, C. A., E. van Stratum, and R. Rustigian. 1945. J. Bacteriol., **49**, 437.

Sulkin, S. E., and J. C. Willett. 1940. J. Lab. Clin. Med., **25**, 649.

Taylor, W. I. 1965. Amer. J. Clin. Path. **44**, 471.

Taylor, W. I., and D. Schelhart. 1968. Appl. Microbiol. **16**, 1383.

Teague, O. and A. W. Clurman. 1916. J. Infect. Dis., **18**, 653.

Tittsler, R. P., and L. A. Sandholzer. 1937. Amer. J. Pub. Hlth., **27**, 1240.

Trabulsi, L. R., and P. R. Edwards. 1962. Cornell Vet. **52**, 563.

Trabulsi, L. R., and W. H. Ewing. 1962. Pub. Health. Lab.**20**, 137.

Vaughn, R. H., and M. Levine. 1942. J. Bact. **44**, 487.

Wilson, W. J., and E. M. Blair. McV. 1931. J. Hyg., **31**, 139.

Index

Antisera
  absorption
    capsular, 297
    H, 94,95,181-191
    K, 90,91
    O, 82-87,123-136,178-184
    polyvalent, 122
  production
    Arizona, H, 267,273
      O, 263,273
      polyvalent, 273,274
    Citrobacter, 287,288
    Escherichia, H, 105
      K, 104,105
      O, 100,103,104
      polyvalent, 73,80-82,90,94
    Klebsiella, 299
    Providencia, 334
    Salmonella, H, 202,203
      O, 198,202,203
      polyvalent, 203-205
      Vi, 202
    Serratia, 311,316
    Shigella, 120,121
Acetate medium, 342
Aerobacter, groups A,B,C, 301
Aerobacter aerogenes, 301
A. cloacae, 301
A. liquefaciens, 309
Aeromonads, 10,21
Aeromonas, 21,329
Aeromonas hydrophila, 311,313
A. margarita, 308
A. shigelloides, 136,138,139,311
Alcaligenes, 308
Alcaligenes fecalis, 21
Alginate medium, 342
Alkalescens-Despar (A-D) bacteria, 18,67,72,98-103
Alpha antigen, 52,287,302,307,334
Alphabetical list, salmonellae, 209,213-224
Alternate biochemical methods, 23,342-354
Amies' transport medium, 337
Andrade indicator, 343
Animal infections, *Escherichia*, 101-103
Antigens (*see also* variational phenomena)
  A, 48,87-91
  alpha, 52,287,307
  Arizona, H, 265-268
    M, 265
    O, 264-266
  B, 48,87-91
  Beta, 52,53
  common hapten, 54,55
  Citrobacter, H, 286,287
    O, 282-286
    Vi, 282,283
  Edwardsiella, 145
  Enterobacter, 302,307

Escherichia, general, 67,70,71
  H, 91-95
  K, 87-91
  O, 73-87,98-103
fimbrial, 53,54,131,168
general characters, 48
Klebsiella, capsular, 294-299
  O, 295,296
M, 52,167,265,295
Proteus, 329
Providencia, 334
Salmonella, H, 184-195
  M, 52,167
  O, 174-184
  Vi, 184
Serratia, H, 311,313,315,316
  K, 316
  O, 311,313-316
Shigella, 123,136-139
  O, 123-136
X, 54
Antigenic schemata
  Arizona, 265,269-272
  Citrobacter, 282,283
  Edwardsiella, 145
  Enterobacter, 302,307
  Escherichia, 67,70,71,99
  Proteus, 329
  Providencia, 334
  Salmonella, 208-212,225-257
  Serratia, 311,313,316
  Shigella, 111
Apocholic acid, 341
Arizona, 2,22,259-275
  antigenic schema, 265,269-272
  antiserum production
    cultures, H, 267,273
    cultures, O, 263,267
  A. arizonae, 259
  biochemical reactions, 26,32,34,161,259-262,281
  decarboxylase reactions, 148,149
  definition, 4,259
  determination
    H antigens, 265
    O antigens, 262,264
  H antigen relationships, 268
  A. hinshawii, 2,259
    biochemical reactions, 260
  immunochemistry, 56
  intergeneric relationships, q.v.
  KCN reactions, 148
  M antigens, 265
  malonate, 149,150,151
  ONPG, 150,152,261
  polyvalent antisera, 273,274
  O antigen relationships, 264,266
  serological characterization, 262
  variational phenomena, 262

Atypical Enterobacteriaceae, 62

Bacteria, type 14011, 282
    type 29911,331
    type 32011,301
Bacterium alkalescens, 99,100
B. asiaticus, 301
B. cadaveris, 301
B. ceylonensis A, 118
B. rabaulensis, 117
B. rubidaeum, 309
B. typhiflavum, 319
B. shigae, 111
Bacteroides, 8
Bacteriophage mediated variations, 59,60,62,168
Ballerup bacteria, 276,282
Barritt reagent, 354
Basal medium carbohydrates, 342
Beta antigen, 52
Beta galactosidase, 349
Bethesda bacteria, 276,282
Bile, 13
Biochemical reactions
    Aeromonas hydrophila, 311,313
    Arizona hinshawii, 26,32,34,161,259-262,281
    Citrobacter freundii, 26,32,34,161,261,276-281,323
    Edwardsiella tarda, 23,25,31,143,144
    Enterobacter, 27,28,302-306
        E. aerogenes, 27,39,40,42,303-306,312
        E. cloacae, 27,38,39,42,303-306,312
        E. hafniae, 27,40,42,302-306,312
        E. liquefaciens, 27,28,41,42,303-306,311
    Escherichia coli, 23,25,30,31,67-72,119
        A-D bioserotypes, 72,119
    Klebsiella, 23,27,28,42,290-294,322
        K. ozaenae, 37,291-293
        K. pneumoniae, 27,37,38,42,291-293
        K. rhinoschleromatis, 37,291-293
    Pectobacterium caratovorum, 28,42,319-323
    Proteus, 23,324-329
        P. mirabilis, 29,43,44,325,326
        P. morganii, 29,43,45,46,327,328
        P. rettgeri, 29,43,45,46,327,328
        P. vulgaris, 29,43,44,325,326
    Providencia, 331-334
        P. alcalifaciens, 29,46,332-334
        P. stuartii, 29,46,332-334
    Salmonella, 23,26,32,34,146-165,261,281
        S. cholerae-suis, 33,154,157,164
        S. enteritidis, 33,35,156,157,159
            bioser Decatur, 164
            bioser Gallinarum, 36,160
            bioser Java, 163
            bioser Miami, 165
            bioser Paratyphi-A, 35,159,163
            bioser Paratyphi-B, 163
            bioser Paratyphi-C, 164
            bioser Pullorum, 36,160
            bioser Sendai, 165
            bioser Typhisuis, 164
        S. typhi, 33,155,157,165
    Serratia, 28,41,42,309-313
        S. marcescens, 310-313
    Shigella, 25,30,113-119
        S. boydii, 115,119
        S. dysenteriae, 115,119
        S. flexneri, 115-117,119
        S. sonnei, 115,117,119
        sub judice types, 116
Bismuth sulfite agar, 10
Blood agar, 10
Blood cultures, 12

Body fluids, 12
Brilliant green agar, 9,10
Brom thymol blue lactose agar, 10
Broth medium, H antigens, 354
Buffered glycerol saline, 7,338
Buffer treated swabs, 8,338

Capsule reactions, *Klebsiella,* 298
Carbohydrate
    sterilization, 342
    fermentation media, 342
Cary and Blair medium, 8,338
Cetrimide, 343
Character of agglutination H, 49
    K, 48
    O, 49
chemotypes, 56,57
Chopped meat media, 14
Christensen's medium, citrate, 119,343
    urea 15,353
Chromobacterium, 308
Citrate media
    Christensen, 119,343
    Simmons, 343
Citrobacter, 2,4,56,276-289
    biochemical reactions, 276,279,280
    decarboxylase reactions, 148,149
    definition, 4,276
    C. diversus, 282
    C. freundii, 2,4,56,276-289
    immunochemical studies, 56
    intergeneric relations, 283-286
    KCN reactions, 148
    ONPG reactions, 150,152
    Production of antisera, 287
    serological reactions, 282,286,287
    Vi antigen, 18,282
Classification, 1-6
Cloaca, 301
C. cloacae, 301
Colicines, 60,61
Colobactrum freundii, 276
Common antigen (common hapten), 54,55
Commonly used tests, 25-29
Conjugation, 60
Corn oil medium, 348
Cysteine-gelatin medium, 343
Cystine-selenite medium, 9

Decarboxylase media, 22,344
Definitions
    family, 3
    genera, 3-6
    tribes, 3-6
Dehydrated media storage, 337
Decarboxylase reactions; *see* biochemical reactions
Detection of Vi antigen, 185
Determination
    hydrogen sulfide, 21
    motility, 21,22
Desoxycholate agar, 10,340
Desoxycholate citrate agar, 10,340
Differentiation
    biochemical, 15,17,23-46
    of tribes, 23-31
Differential counts, 13
DNase media, 345

Edwardsiella, 2,143-145
    biochemical reactions, 143-145
    definition, 4,143
    E. tarda, 2,143-145

EDTA solution, 338
Endotoxin, 55,56
Enrichment media, 339
    limitations, 8-10
    use, 9,10
Enterobacter, 2,5,301-307
    biochemical reactions, 302-304,306
    definition, 5,301
    nomenclature, 301
    serological reactions, 302,307
Enterobacter aerogenes, 2,39,40,42,302-306
E. agglomerans, 319
E. alvei, 301
E. cloacae, 2,38,39,42,302-306
    aberrant, 281,282
E. hafniae, 2,40,42,302-306
E. liquefaciens, 2,41,42,302-306,309
Eosin methylene blue agar, 10,341
Episomes, 62,67
Erwinia, 318
E. amylovora, 318
E. caratovorum, 318
E. carnegiana, 319
E. herbicola, 318
E. lathyri, 319
E. milletiae, 319
E. salcis, 318
Erwiniaceae, 318
Escherichia, 1,19,67-105
    absorption of antisera, q.v.
    antigens, 67,70,71,73,76-79,91-95
    antiserum production, 103-105
    biochemical reactions, 25,67-72,98-103
    definition, 3,67
    extrageneric relations, 138,139,273,286,329,334,335
    E. freundii, 276
    H antigens, 91-95
    H antiserum pools, 94
    immunochemistry, 56
    infections in animals, 101-103
    K antigens, 87-91
    K antiserum pools, 90
    K determination, 90,91-94
    O antigens, 73-87
    O antiserum pools, 73,80-82
    O determination, 73-87
    serogroups, diarrheal disease, 83-98
        further examination, 97,98
        general considerations, 95
        isolation, 95
        preliminary identification, 95-97
Escherichieae, definition, 3
Esculin broth, 345
Essential results, biochemical, 22
Exudates, 12

Falkow method, 344
Fecal specimens, 7
Fermentation media, 342
Ferric chloride reagent, 352
Filter papers, 8
Fimbriae, 53,54,131,168
Flagellar antigens; *see* H antigens
    broth medium for, 354
    phase reversal medium, 354
Flagellin, 59
Fluorescent antibody techniques, 63
Form variation, 51

Gall stones, 13
Gelatin liquefaction, *Klebsiella,* 290
Gelatin media, 345

Genetic recombinations, 59-62,168
Glycerol fuchsin medium, 346
GN medium, 10,339
Gram stains, 15

H (flagellar) antigens; *see* antigens
    immunochemistry, 59
Hauch, 49
Heat labile antigens, 48,49
Heat stable antigens, 48
Hemophilus pertussis, 169
Herbicola–Lathyri bacteria, 319
Herellea, 21
Hektoen enteric agar, 10,341
HO-O variation, 48
Hospital-aquired infections, 22,316
Hydrocinnamic acid medium, 352
Hydrogen sulfide determination, 18,21

Identifying information, 16,195
Immunochemical studies, 55-59
Incidence, serotypes *Escherichia* (diarrheal disease), 98
Incubation periods, 22,342-354
Indicators, 343
Indol test medium, 347
Indophenol oxidase test, 347
Infusion agar, 10
Inhibition
    of *Proteus,* 339
    of pseudomonads, 340
Inositol, 22
Intergeneric (antigenic) relationships, 56
    Arizona, 265,273,274,284,285,329,334
    Citrobacter, 273,283-287
    Enterobacter, 299
    Escherichia, 138,139,273,286,329,334,335
    Klebsiella, 52,167,299
    Proteus, 329
    Providencia, 273,334,335
    Salmonella, 273,274,283-285,287,329
    Shigella, 138,139,273,329,334
Intermediate antigens, 57
Isolation of N forms from MN, 167,168

Jordan's tartrate medium, 353

K antigens, 48,51-54
    immunochemistry, 57-59
    of *Escherichia,* 87-89
    of *Shigella,* 136-139
    see also antigens
KO-O variation, 51,52,87-91
    see also variational phenomena
Kapsule, 48
Katwijk type, 113
KCN medium, 148,352
KDO, 56
Klebsiella, 2,5,290-300
    antigens, 294-299
    antiserum
        absorption, 297
        pools, 296,297
        production, 299
    biochemical reactions, 290-294
        K. pneumoniae, 290-293
        K. ozaenae, 290-293
        K. rhinoschleromatis, 291-294
    capsule antigens, 295-297
    definition, 5,290
    immunochemistry, 57,58
    intergeneric relationships, 296
    M antigens, 295

O antigens, 295,296
    production of antisera, 299
    quellung reactions, 296
    serological reactions, 294
Klebsielleae, 2
    definition, 5
Kligler's medium, 14,347

L antigen, 48,87-91
Large-Sachs group, 109,111
Lipase medium, 348
Lipopolysaccharides, 55-58
    basal components, 56
    biosynthesis, 57
    chemotypes, 56,57
    K antigens, 57,58
    O (S) antigens, 56,57
    R antigens, 56,57
    T antigens, 57
Lysine-iron-agar medium, 11,16
    reactions, 17
Lysogenic conversion 59-61
    counterparts, 61

MacConkey agar medium, 10,341
Manchester bacillus, 18, 116
Malonate medium, 149,150,348
Media
    control, 337
    preparation and use, 337-354
    see individual listings
M forms, 18,51
    determination
        Arizona, 265
        Salmonella, 52,167,51,52
M-N (MO-O) variation, 51,52
    see also Klebsiella
Millorganite, 9
Mima, 21
Mineral oil, sterilization, 344
Moeller's decarboxylase media, 16,344
Motility determination, 21
    medium, 11
Mercuric iodide solution, 354
Methyl red medium, reagent, test, 348
Mucic acid, 350
Multiphasic *Arizona,* 265
Multiphasic *Salmonella,* 61,172-174

Newcastle bacillus, 18,116
Nomenclature, 1-6
Nonfimbriate-fimbriate variation, 53,54,131,168
Nosocomial infections, 22,316
Nitrate reduction, 349
Nutrient gelatin, 345

O antigens; *see* antigens
Occurrence of multiple phases
    Arizona, 265
    Salmonella, 61,172-174
Occurrence of Vi antigen, 52
ohne Hauch, 49
O'Meara reagent, 353
ONPG test, 23,117,150,152,261,349
Organic acid media, 34,149-151,350
Oxidase test, 21,347
Oxidation-fermentation medium, 21,350
Oxytoca biotype, 290
Outline of classification, 1,2

Papaine, 13
Padewskia, 282

Paper strip test, indol, 347
Paracolobactrum, 276
Pectate medium, 351
Pectobacterium, 2,5,319-323
    biochemical reactions, 319-323
    P. carotovorum, 318
    definition 5,318
Peptone-iron-agar medium, 347
Phase reversal medium, 354
    variation of Andrewes, 169-174
        usage of terminology, 170
Phenol red lactose agar, 10
    tartrate agar, 353
Phenylalanine medium, 351
Phenylproprionic acid medium, 352
Pigment production, media, 352
Pili, pilus, 54
Plasmids, 63
Plating media, 10,12-14,339-342
Polagglutination, 49
Polyvalent antisera
    see individual generic listings and antiserum production
Potassium cyanide medium, 352
Preliminary serologic examinations, 16-19
Preservative (transport) media, 337-339
Proteeae, 2
    definition, 5
Proteus, 11,16,324-330
    aberrant strains, 327
    P. americanus, 329
    P. ammoniae, 329
    biochemical reactions, 43-46,324-329
    classification, 2,324
    decarboxylase reactions, 325,326
    definition, 5,324
    P. hydrophila, 329
    P. icthyosmius, 329
    inhibition, 339
    lysine iron agar, 327-329
    P. mirabilis, 2
        biochemical reactions, 43,44,325,326
    P. morganii, 2
        biochemical reactions, 43,45,46,327,328
    P. paraamericanus, 329
    P. rettgeri, 2
        biochemical reactions, 43,45,46,327,328
    serological reactions, 329
    P. vulgaris, 2
        biochemical reactions, 43,44,325,326
    X19,49
Primary differentiation, 14
Providencia, 2,19,113,331-336
    P. alcalifaciens, 2,331
        biochemical reactions, 46,332-334
    biogroups, 333,334
    cultural characters, 331
    definition, 6,331
    intergeneric relations, 334,335
    lysine iron agar, 331
    production of antisera, 334
    serological reactions, 334,335
    P. stuartii, 2,331
        biochemical reactions, 46,332-334
Pseudomonads, 10,350,352
    inhibition of, 340
Pseudomonas, 21
    P. aeruginosa, 352
    P. alcaligenes, 21,350
Purulent exudates, 14
Pyrolysis-gas-liquid chromatography, 56

Salmonella, 4,10,19,21,22,146-205,208-257

abbreviated antigenic schema, 199,200
absorption of H antisera, 189-191
     sterilization, 194
(absorption of) O antisera, 178-183
     single factor, 179-182
alphabetical list, 209,213-224
antigenic analysis, 174-195
(antigenic) change induced by bacteriophage, 59,62,168
(antigenic) schema, 225-257
antisera
     for common serotypes, 201
     for H antigens, 185
S. arizona, 259
atypical strains, 147
S. ballerup, 166,276,282
biochemical reactions, 26,35,36,146-165
     see also Biochemical reactions
biochemical reactions of antigenically related, 160-185
S. cholerae-suis, 2,8,9,13,22
     biochemical reactions, 154
S. coli, 166,174
decarboxylase reactions, 148,149
definition, 4,146
determination of Vi antigen, 165-167
development, antigenic
     schema, 208-212
differentiation
     of species, 157,158
     within tribe, 159,160
S. enteritidis, 2,22,156
     bioserotypes, 8-10,158-160
form variation, 168,169
H antigens, 185-195
     g . . . complex, 193,194
H antisera, 184
HO-O variation, 163
hydrogen sulfide production, 147
immunochemistry, 55-57
irreversible variation, 172-174
isolation of phases, 194-195,202
KCN reactions, 148
malonate reactions, 149,150
M-N variation, 167-169
multiple phases, 172-174
nomenclature, 146,147
nonfimbriate-fimbriate variation, 168
ONPG reactions, 150,152
O antisera
     for identification, 175
     for serogroups A-E, 197
phase reversal, 194,195
phase variation, 169-174
production of antisera, 198,201-205
reactions, H antisera, 187-188
selection of H antisera, 189,191-193
serological identification, 162-204
smooth-rough variation, 50,51,163,164,169
somatic (o) antigens, 174-184
     subdivisions, 174,175
species, 2,3
stages in serological classification, 195-201
sterilization of absorbed antisera, 194
subgenera of Kauffmann, 146,147
T forms, T antigens, 164,165,169
S. typhi, 2,7-10,13,18,155
     see also biochemical reactions
unnamed serotypes, 223
use of H antisera, 186-189
use of O antisera, 176-178
use of single factor sera, 183,189-195
variational phenomena, 162-174
V–W (Vi-O) variation, 52,165-167,169

Salmonelleae, 2
     definition, 4
Selection of colonies, 14
     plating media, 10-12
Selenite medium, 8,9,339
Serotypes of Escherichia, diarrheal disease, 95-98
Serratia, 2,5,308-316
     S. anolium, 308
     antiserum production, 311
     biochemical reactions, 309-311
          see also biochemical reactions
     definition, 5,308
     differentiation, 309-311
          from E. liquefaciens, 311
          from Aeromonas, 313
     effect of formalin on H antigens, 316
     S. essyana, 308
     S. fuchsina, 308
     H antigen determination, 311,313,315,316
          antiserum pools, 316
     hospital-aquired infections, 316
     immunochemical studies, 56
     S. indica, 308
     S. marcescens, 308
          kiliensis, 2,308
     nomenclature, 2,308,309
     O antigens, 311,314
     S. piscatorum, 308
     S. pyoseptica, 308
     relationship to E. liquefaciens, 309
     S. saponaria, 308
     serological reactions, 311,313-316
Shigella, 2,10,15,21,108-139
     S. alkalesens, 99,100,108
     S. ambigua, 109,111
     S. arabinotarda, 111
     biochemical reactions, 113-120
     S. boydii, 2,18,111-113,115,119
          biochemical reactions, 115,119
          serological reactions, 131-135
     S. ceylonensis A, 100,111,118
     S. ceylonensis B, 99,100
     classification, 2,108-113
          used in USSR, 112
     definition, 4,108
     S. dispar, 99,100,108
     S. dysenteriae, 2,18,109,111,108
          biochemical reactions, 115,119
          serological reactions, 123,124
     S. etousa, 111
     extrageneric relations, 138,139
     S. flexneri, 2,18,110-113,115
          absorption of antisera, 128
          biochemical reactions, 115-117,119
          serological reactions, 126,127,129,130
          variations, 128,129, 131
          serotypes, subserotypes, 110
     fimbrial antigens, 131
     genetic change, 61
     immunochemistry, 36
     K antigens, 136-139
     lysogenic conversion, 61
     S. madampensis, 99,100
     S. newcastle, 111
     nomenclature, 2,111
     S. paradysenteriae, 113
     prelimenary serology, 18
     production of antisera, 120,121
     S. rabaulensis, 117
     S. rio, 111,117
     S. saigonensis, 111,117
     S. schmitzii, 111

serological reactions, 120-139
    S. boydii, 131-135
    S. dysenteriae, 123-124
    S. flexneri, 126,127-131
    S. sonnei, 135,136
sodium acetate medium, 119,342
  S. sonnei, 2,18,111-113,115,135,136
    biochemical reactions, 135,136
    serological reactions
  S. tieté, 99,100
  use of antisera, 123
Shigella-Salmonella agar (medium), 10,341
Shipment of cultures, 16,19,195,354
Signs used in tables, explanation, 22
Simmons' citrate medium, 343
Slime-wall, 167
Smooth-rough variation, 48,50,51,134-136
  see also antigens and variational phenomena
Sodium mucate, 119,350
Sodium potassium tartrate, 350
Sorensen buffer, 338
Specimen preservative medium, 8,339
Specimens, 7,11-14
  blood, 12,13
  bile, 13
  collection, processing, 7,11
  exudates, tissues, 12,13,14
  feces, 7
  urine, 13
Sterilization of antisera (absorbed), 194
Stern's glycerol fuchsin medium, 346
Stock culture media, 354
Storage
  of cultures, 354
  of dehydrated media, 337
Streptokinase, 13
Stuart's medium, 8,338
Sulfadiazine, 340
Sulfathiazole, 339

Tartrate agar medium, 353
Taxonomy, 1-6
Teepol, 341
Temperature of incubation, 14,342-354
Tetrathionate media, 8,9,339
T forms; *see* antigens, 50
Tissues, 12
Transduction, 59,60
Transmission of atypical characters, 62

Transport media, 7,8,337-339
Tribes, 1,2
  definitions, 3-6
  differentiation, q.v.
Tribe Mimeae, 350
Triple sugar iron agar, 11,14,15,346
  reactions, 15
Type 14011, 282
Type 29911, 331
Type 32011, 301
Trypsin, 13

Urea media, 11,15,16,353
Urine cultures, 13

Variational phenomena, 48-55
  flagella
    Citrobacter, 50
    HO—O, 48,49,163
    irreversible, 172-174
    multiple phases, 172-174,265
    phase (Andrewes), 49,50,169-172,265
    phage mediated, 168
  alpha-O, 52
  beta-O, 52,53
  capsule antigens, 295,296
  form variation, 51,168
  KO-O, 51
  MO-O (M-N), 51,167,168
  nonfimbriate-fimbriate, 53,54,131,168
  smooth to rough, 50,51,134-136,163-165
  variation in *S. flexneri*, 128-131
  Vi antigen, 48,58,165-167
    detection, 184
    antiserum, 17,18,48,184
    production, 202
  Vi-O (V-W), 52,165-167
Vibrio, 21
Vibrio cholerae, 7
Voges-Proskauer medium, 353
  reagents, 353,354

Wilson-Blair medium, 10,339
Worfel-Ferguson medium, 296,354

X antigen, 54
Xylose-lysine-desoxycholate medium, 10,341

Zinc dust, 349